HANDBOOK OF NANOFABRICATION

HANDBOOK OF NANOFABRICATION

Editor
Gary P. Wiederrecht

ELSEVIER

AMSTERDAM • BOSTON • HEIDELBERG • LONDON • NEW YORK • OXFORD
PARIS • SAN DIEGO • SAN FRANCISCO • SINGAPORE • SYDNEY • TOKYO

Elsevier B.V.
Radarweg 29, 1043 NX Amsterdam, the Netherlands

First edition 2010

Copyright © 2010 Elsevier B.V. All rights reserved

Permissions may be sought directly from Elsevier's Science & Technology Rights Department in Oxford, UK: phone (+44) (0) 1865 843830; fax (+44) (0) 1865 853333; email: permissions@elsevier.com. Alternatively visit the Science & Technology website at www.elsevierdirect.com/rights for further information

Notice
No responsibility is assumed by the publisher for any injury and/or damage to persons or property as a matter of products liability, negligence or otherwise, or from any use or operation of any methods, products, instructions or ideas contained in the material herein. Because of rapid advances in the medical sciences, in particular, independent verification of diagnoses and drug dosages should be made

British Library Cataloguing in Publication Data
A catalogue record for this book is available from the British Library

Library of Congress Control Number: 2009937755

ISBN: 978-0-12-375176-8

For information on all Elsevier publications
visit our website at elsevierdirect.com

Printed and bound in Spain

09 10 11 12   10 9 8 7 6 5 4 3 2 1

**Working together to grow
libraries in developing countries**

www.elsevier.com | www.bookaid.org | www.sabre.org

ELSEVIER   BOOK AID International   Sabre Foundation

CONTENTS

Contents	v
Editorial Advisory Board	vi
Preface	vii
Contributors	ix

1 Directed Assembly of Nanostructures
  J M MACLEOD, *Università degli studi di Trieste, Trieste, Italy*
  F ROSEI, *Université du Québec, Varennes, QC, Canada* ... 1

2 Bio-Mediated Assembly of Ordered Nanoparticle Superstructures
  W L CHENG, S J TAN, M J CAMPOLONGO, M R HARTMAN, J S KAHN and
  D LUO, *Cornell University, Ithaca, NY, USA* ... 57

3 Chiral Molecules on Surfaces
  C J BADDELEY, *University of St. Andrews, St. Andrews, UK*
  G HELD, *University of Reading, Reading, UK* ... 91

4 Electron Beam Lithography of Nanostructures
  D M TENNANT and A R BLEIER, *Cornell University, Ithaca, NY, USA* ... 121

5 Status of UV Imprint Lithography for Nanoscale Manufacturing
  J CHOI, P SCHUMAKER and F XU, *Molecular Imprints, Inc., Austin, TX, USA*
  S V SREENIVASAN, *Molecular Imprints, Inc., Austin, TX, USA,*
  University of Texas at Austin, Austin, TX, USA ... 149

6 Picoliter Printing
  E GILI, M CAIRONI and H SIRRINGHAUS, *University of Cambridge, Cambridge, UK* ... 183

7 Molecular Printboards: From Supramolecular Chemistry to Nanofabrication
  R SALVIO, J HUSKENS and D N REINHOUDT, *University of Twente, Enschede,*
  The Netherlands ... 209

8 Molecular Machines and Motors
  A CREDI, *Università di Bologna, Bologna, Italy* ... 247

Index ... 291

EDITORIAL ADVISORY BOARD

ASSOCIATE EDITORS

Yugang Sun
Center for Nanoscale Materials, Argonne National Laboratory

Takao Someya
Professor, Department of Electric and Electronic Engineering School of Engineering,
The University of Tokyo, 7-3-1 Hongo, Bunkyo-ku,
Tokyo 113-8656 JAPAN

John Rogers
University of Illinois at Urbana/Champaign, 1304 W. Green St, Urbana, IL 61801

Preface

All of the grand goals for nanoscience are dependent upon reliable ways to fabricate nanostructures. The challenges to nanofabrication are many, beginning with the incredibly broad range of applications, materials, and geometries that have been proposed for nanoscale structures. Applications include nanoelectronics, nanophotonics, nanomechanics, nanocatalysis, nanoantennae, and nanosensors, to name only a few. Materials are needed that posess almost every conceivable range of properties: metallic to insulating, hard to soft, inert to reactive, luminescent to quenched, crystalline to glassy - the list goes on. As a result, an immeasurable number of elements, compounds, and alloys have been subject to nanostructuring and nanofabrication tools. Add to this the range of geometrical nanostructures required: disks, rods, holes, pyramids, etc., and a range of tunability in the degree of interaction between nanoparticles to be isolated or closely coupled. The degree of long range ordering, either random or periodic, can also be a critical consideration, as well as whether that ordering extends in one, two, or three dimensions. It is clear that nanofabrication is a daunting task.

The range of nanofabrication routes towards these structures is almost as diverse as the materials, applications, and geometries needed for next-generation applications. The approaches can begin to be compartmentalized by separation into either a "top-down" or "bottom-up" approach. Top-down fabrication refers to methods where one begins with a macroscopically dimensioned material, such as a thin film, into which is placed nanostructured defects. Top-down thus refers to approaches such as electron beam lithography (EBL) or focused ion beam lithography (FIB). In these cases, either focused electrons or ions are used to carve nanostructures out of larger structures. Alternatively, in the bottom-up approach, one begins to assemble nanostructures from smaller units. Examples include colloidal synthesis, where frequently the colloids are literally assembled from single ions in solution that are chemically neutralized so as to produce an aggregation process resulting from a sudden lack of solubility of the now neutral atom. Such processes are frequently used for the creation of plasmonic metal nanoparticles or semiconductor quantum dots. Bottom-up assembly also refers to the assembly of larger hierarchical structures, where, for example, colloids self-assemble into larger structures for a particular purpose. These can be for diverse applications such as the creation of three dimensionally periodic photonic bandgap structures or periodic structures for the study of charge transport processes at the nanoscale.

Equally important to the patterning steps described above are the deposition methods. In many cases it is not enough to produce a pattern from top-down or bottom-up methods. In these cases, the patterned material simply serves as a template for the deposition of the true nanostructured material of interest. This is frequently the case for photonic crystals where, for example, the self-assembled colloids serve only as a template for the high refractive index material that is deposited in the interstitial regions of the template. Another example is plasmonic nanostructures, where metal is deposited into the EBL patterned holes in polymer films.

In many cases, these nanofabrication advances have produced stunning physical and chemical discoveries and phenomena that have no analog in larger scale structures. One example of new phenomena is the confinement of electronic states in semiconductor materials to produce quantized electronic transitions rather than band structure behavior that typifies bulk semiconductors. Entirely new optical and electronic behavior is produced in such materials compared to their bulk counterparts, such as narrow linewidths for emission, multiexciton generation for solar energy conversion, nanoscale light sources, etc. Other examples of new phenomena include the use of nanostructuring to produce materials with a negative index of refraction at visible wavelengths (although, at this date, for the longest wavelengths in the visible spectrum and for materials with significant absorption loss). These metamaterials have no analogs in nature as negative index materials do not exist otherwise.

In addition to basic science advances, nanofabrication is leading the way towards solving technological challenges of great importance. For example, entirely new architectures in electronic integrated circuitry will likely be needed soon, simply because silicon transistors are approaching sizes that will rely on the transport of a single electron. Furthermore, enormous engineering challenges are present at this length scale, as the high density of transistors create tremendous heat loads and the lithography requires ever more complex and costly fabrication tools. As transistors scale down to the 10 nm length scale, nanofabrication advances have enabled totally different materials to be explored as next generation transistors. These include carbon nanotube or graphene-based transistors, or even single molecule transistors. Nanofabrication is also proving critical for next generation energy conversion and storage devices, such as nanostructured lithium in batteries for improved recharging and power delivery properties. The high surface-to-volume ratio of nanoparticles relative to bulk materials is also proving important for new, efficient catalytic processes.

This book describes the recent work of true leaders in the field of nanofabrication. The work described herein covers both the bottom-up and top-down lithography approaches. In the top-down category, Tennant and Bleier describe the state-of-the-art in electron beam lithography, perhaps the most used means to create nanostructures from the top-down. Novel approaches to patterning nanostructures reliably and over large areas is described in the chapter on nanoimprint lithography by Sreenivasan et al. Routes that have great potential in manufacturing, such as inkjet printing, are described in the chapter by Sirringhaus et al. In the bottom-up category, methods that produce nanostructures with elegant long range order are described. These include directed assembly (Rosei and Macleod), bio-mediated assembly (Luo et al.) and patterned molecular binding (Reinhoudt et al.) methods. Important nanofabrication methods towards next generation nanoscale applications and devices are also described. These include routes towards optically active structures with applications in catalysis (Baddeley and Held) and nanoscale molecular motors (Credi).

It is assured that the importance and number of nanofabrication approaches will grow dramatically in the coming years with the rise of nanotechnology. It is clear that scientists and engineers working from many different directions and finding their inspiration from biological, chemical, and physical sources, are all contributing greatly to the field of nanofabrication. Thus, it is our belief that readers from all fields will find material of interest in this multidisciplinary topic, and perhaps even find additional inspiration to invent the next-generation nanofabrication methods and tools.

CONTRIBUTORS

C J Baddeley
University of St. Andrews, St. Andrews, UK

A R Bleier
Cornell University, Ithaca, NY, USA

M Caironi
University of Cambridge, Cambridge, UK

M J Campolongo
Cornell University, Ithaca, NY, USA

W L Cheng
Cornell University, Ithaca, NY, USA

J Choi
Molecular Imprints, Inc., Austin, TX, USA

A Credi
Università di Bologna, Bologna, Italy

E Gili
University of Cambridge, Cambridge, UK

M R Hartman
Cornell University, Ithaca, NY, USA

G Held
University of Reading, Reading, UK

J Huskens
University of Twente, Enschede, The Netherlands

J S Kahn
Cornell University, Ithaca, NY, USA

D Luo
Cornell University, Ithaca, NY, USA

J M MacLeod
Università degli studi di Trieste, Trieste, Italy

D N Reinhoudt
University of Twente, Enschede, The Netherlands

F Rosei
Université du Québec, Varennes, QC, Canada

R Salvio
University of Twente, Enschede, The Netherlands

P Schumaker
Molecular Imprints, Inc., Austin, TX, USA

H Sirringhaus
University of Cambridge, Cambridge, UK

S V Sreenivasan
Molecular Imprints, Inc., Austin, TX, USA and University of Texas at Austin, Austin, TX, USA

S J Tan
Cornell University, Ithaca, NY, USA

D M Tennant
Cornell University, Ithaca, NY, USA

F Xu
Molecular Imprints, Inc., Austin, TX, USA

1 Directed Assembly of Nanostructures

J M MacLeod, Università degli studi di Trieste, Trieste, Italy

F Rosei, Université du Québec, Varennes, QC, Canada

© 2010 Elsevier B.V. All rights reserved.

1.1 Introduction

The realization that nanoscale matter often behaves differently with respect to the same materials in the bulk form has prompted a wealth of research aimed at understanding, characterizing, describing, and predicting 'nano' [1,2]. However, while 'nanotechnology' has been a buzzword for almost two decades, it has delivered fairly little so far in terms of new technologies, that is, new products that are commercialized and used by the general public.

One of the great promises of nanotechnology is the ability to do more in the same space: to advance our current technologies through miniaturization, so that each crop of electronics is smaller, faster, and more powerful than the one before. This is the manifestation of Moore's law [3], the now-famous 1965 empirical prediction by Gordon Moore (who later went on to co-found Intel) that the number of transistors accommodated in a chip of given size doubles roughly every two years. The semiconductor industry has used this prediction as a roadmap over the last three decades. As the limits of this down-scaling approach the dimensions of single molecules and atoms, the discrepancy between nanoscale and bulk behavior has become evident. While this is detrimental in some situations, for example, in scaled-down versions of larger transistors that can exhibit problematic behaviors, such as unexpected leakiness, at nanoscale dimensions [4], it opens the door to opportunities for custom-designing new circuit architectures to exploit behaviors unique to the nanoscale. For example, quantum size effects [5,6], confinement of excitons [7,8], and high surface-to-volume ratios [9] can all impart new, unexpected, and potentially useful behavior to nanoscale systems.

To capitalize on the full potential of nanostructured materials and their properties, it is necessary to develop the ability to purpose-build nanoscale systems, a task which hinges on the precise placement of appropriate nanoscale building blocks in two and three dimensions (2D and 3D). This approach is generally referred to as 'bottom-up', implying the spontaneous formation of a desired architecture. This approach provides a diametric counterpoint to the 'top-down' techniques (typically lithographic techniques, which are very precise but must adhere to Rayleigh's equation, and therefore cannot resolve fine nanoscale features [10]) used in the contemporary fabrication of semiconductor devices [11], and provides an intuitive mechanism for building architectures from countable numbers of atoms or molecules.

The use of molecules as the basic building blocks of nanoscale structures capitalizes on a wealth of knowledge that can be obtained from the study of biological systems [12–14]. Supramolecular chemistry [15], applied to nanoscale design [16–18], additionally benefits from the capabilities of synthetic chemists, since molecules can essentially be custom-designed for form and functionality salient to specific systems and devices [19,20].

The aim of this article is to provide an overview of the tools and techniques available for building nanoscale architectures from molecular building blocks, limiting ourselves primarily to a discussion of the geometry of molecular assemblies at surfaces, that is, structures confined to 2D. Outside of our focus will be atomic structures [21–24], clusters [25,26], and quantum dots [27–29], all of which provide their own unique set of challenges and rewards. Our focus will be on the major experimental advances made via surface physics and chemistry over the past 25 years. The majority of the investigations that we describe have been performed with scanning probe microscopies (SPM), specifically, scanning tunneling microscopy (STM) [30–34]. The STM is a remarkably versatile instrument capable of imaging conducting and semiconducting surfaces [35], probing their electronic characteristics [36], investigating the vibrational characteristics of adsorbed molecules [37,38], interacting with the surface or adsorbates to produce new geometric and electronic configurations [39–41], or even to initiate chemical bond formation [42,43]. Many excellent books and reviews are available, describing various facets and uses of SPM [44–64].

After briefly discussing the fundamentals of directing nanoscale assembly of surfaces, as well as the most salient experimental techniques for probing these systems, we will provide an overview of significant experiments grouped by the type of interaction used to pattern the molecules: strong bonding between the molecules and the underlying surface, molecular self-assembly driven by hydrogen bonding, and metal–organic coordination, using inclusion networks to position molecules, and, finally, surface-confined polymerization for producing robust, covalently bonded structures. An emerging area that we will unfortunately neglect due to space limitations is the formation of ordered multicomponent assemblies driven by curved surfaces. We refer interested readers to the relevant literature [65–77].

1.2 Fundamentals of Directing Nanoscale Assembly at Surfaces

There are two competing types of interactions that control the formation of patterns at surfaces: (1) molecule–molecule and (2) molecule–substrate [78]. In most cases, bottom-up assemblies depend on the balance between (1) and (2); however, depending on the choice of surface (2) can be either the dominant interaction or can be almost suppressed, with various intermediate regimes. For example, graphite surfaces are essentially inert and therefore their participation in pattern formation is usually minimal besides providing regular and planar array of adsorption sites. On the other hand, reconstructed silicon surfaces are characterized by a high density of reactive unsaturated dangling bonds (DBs) that interact strongly with molecules upon adsorption, often causing the molecules to fragment as in the case of cyclo-addition reactions [79]. With respect to (1), most intermolecular interactions used so far are noncovalent in nature, that is, they may induce the formation of long-range ordered patterns, yet, are easily disrupted because of their weak bonding. This aspect has several advantages, including the 'self-repair' mechanisms that are well known in supramolecular chemistry: defects tend to disappear as the interactions locally break up the pattern forming a new ordered one devoid of defects. Notably, hydrogen bonding and metal–organic coordination are noncovalent interactions frequently employed to form ordered patterns in both 2D and 3D. van der Waals forces alone can usually lead to the formation of local patterns, yet their lack of directionality is usually a barrier to producing long-range patterns.

Stronger molecule–molecule interactions can lead to the formation of covalent bonds. While these are often desirable to obtain more robust structures with interesting mechanical and electronic properties, they are significantly more difficult to direct and their use for nanostructure formation at solid surfaces has been explored only in the last decade. Some elegant examples of covalent architectures, together with a discussion of their challenges and limitations, will be provided in Section 1.7.

1.2.1 Noncovalent Interactions between Molecules

1.2.1.1 Hydrogen bonding

Hydrogen bonds are formed between an electronegative atom and a hydrogen atom bonded to a second electronegative atom [80]. The strength of the hydrogen bond depends on the electronegativity of the atoms; Table 1 classifies hydrogen bonds as very strong (e.g., [F...H...F]$^-$), strong (e.g., O—H...O=C), or weak (e.g., C—H...O) depending on the bond energy, which ranges from 40 to <4 kcal mol^{-1}, respectively. The directionality of the bond increases with strength. For crystal engineering, the 'strong' hydrogen bond is perhaps the most useful type [81,82]. For example, in the systems we describe in this article, hydrogen bonds between carboxylic groups (O—H...O=C) are often used to drive self-assembly.

1.2.1.2 Metal–organic coordination

The attraction between an organic ligand and a metal center provides an alternative route to creating directional associations at surfaces. Metal–organic coordination provides a stronger association than the commonly employed modes of hydrogen bonding (typically in the order of 50–200 kJ mol^{-1} in 3D compounds) [84], and can confer various geometrical motifs due to the flexibility of the coordination modes available in transitional metals [85].

1.2.2 Molecule–Surface Interactions

The attraction between a molecule and an underlying surface is typically characterized as either physisorption or chemisorption, depending on the strength of the interaction. Physisorption generally refers to van der Waals interaction, and has a maximum interaction energy of about 60 kJ mol^{-1} for a small molecule [86]. Chemisorption implies higher interaction energy due to a significant charge rearrangement in the adsorbed molecule to facilitate the formation of a covalent or ionic bond with the surface. This is obviously the dominant case for molecules on semiconductor

Table 1 Properties of very strong, strong, and weak hydrogen bonds (X—H...A)

	Very strong	Strong	Weak
Bond energy (kJ mol^{-1})	63–167	17–63	<17
Examples	[F...H...F]$^-$	O—H...O=C	C—H...O
	[N...H...N]$^+$	N—H...O=C	N—H...F—C
	P—OH...O=P	O—H...O—H	O—H...π
Bond lengths	H—A ≈ X—H	H...A > X—H	H...A >> X—H
Lengthening of X—H (Å)	0.05–0.2	0.01–0.05	≤0.01
X...A range (Å)	2.2–2.5	2.5–3.2	3.0–4.0
H...A range (Å)	1.2–1.5	1.5–2.2	2.0–3.0
Bonds shorter than vdW	100%	Almost 100%	30–80%
X—H...A bond angle range (°)	175–180	130–180	90–180
kT at room temperature	>25	7–25	<7
Effect on crystal packing	Strong	Distinctive	Variable
Utility in 3D crystal engineering	Unknown	Useful	Partly useful
Covalency	Pronounced	Weak	Vanishing
Electrostatics	Significant	Dominant	Moderate

Adapted with permission from Desiraju GR and Steiner T (2001) *The Weak Hydrogen Bond in Structural Chemistry and Biology*. Oxford: Oxford University Press.

surfaces; on metal surfaces chemisorption strength depends on whether the metal has a d-band (the absence of which leads to relatively weak chemisorption), on the filling of the adsorbate–metal antibonding d-states, and on the orbital overlap between the adsorbate and the surface [87,88]. In general, for transition metal substrates, the reactivity of the surface decreases from left to right across the periodic table, and from top-to bottom [89]. We avoid using the total adsorption energy to distinguish between physisorption and chemisorption since complex molecules can interact with the surface over a large area, leading to a total physisorption energy that can be comparable to chemisorption energies for smaller molecules even in the absence of charge rearrangement.

1.2.2.1 Common surfaces for studies of molecular assembly

1. *Highly oriented pyrolytic graphite*. While its constituent graphene sheets have been the source of extremely intense investigation recently [90–98], highly oriented pyrolytic graphite (HOPG) holds its own place of importance in ambient and solution studies of molecules on surfaces. One reason for this is the ease with which a clean, flat surface can be prepared and used: an HOPG crystal can be cleaved with adhesive tape, and the exposed clean surface remains stable in air for hours. These properties are a direct consequence of the bonding between the carbon atoms. Each atom is sp^2 hybridized and bonded in-plane to three nearest-neighbors to form graphene sheets, with the remaining electron contributing to a delocalized π-bond between the sheets.

The π system at the graphite surface is advantageous for studies of aromatic molecules, since π–π interactions stabilize the molecules on the surface. Long-chain alkanes can also adsorb stably, with their molecular axis oriented parallel to the HOPG surface [99,100]. In this configuration, the periodicity of the alternate methylene groups along the alkane molecule (2.51 Å) is very nearly commensurate with the spacing of the hexagons in the graphite surface (2.46 Å), providing strong molecule–surface interaction [101]. The subsequent formation of an alkane monolayer is stabilized by van der Waals interactions.

It is important to note that care must be taken in the interpretation of STM images obtained from HOPG surfaces, since defects in the graphite (e.g., those introduced by the rotation of subsurface graphene, etc.) [102–105] can be easily misinterpreted as molecular features [106,107].

2. *Cu(110)/Cu(111)*. The face-centered cubic (fcc) structure of copper leads to different atomic geometries on its low-energy faces: Cu(100) is fourfold symmetric, Cu(110) is twofold symmetric, comprising atomic rows, and Cu(111) is threefold symmetric, comprising close-packed atoms. The (110) and (111) surfaces (**Figure 1**) are extensively used as substrates in molecular assembly experiments. The surfaces present distinct characteristics to molecular adsorbates: besides the obvious difference in geometry, the open structure of (110) is associated with a higher adsorption reactivity than the closed (111) structure

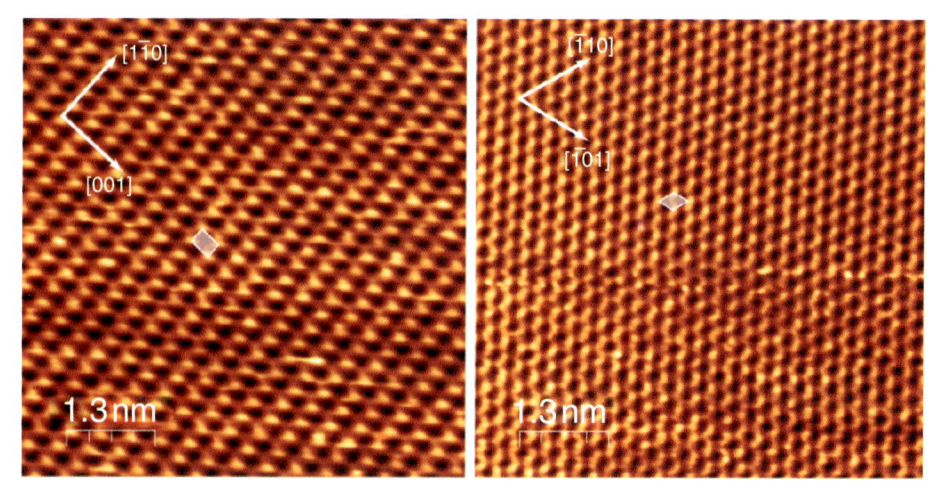

Figure 1 STM images of Cu(110) (left) and (111) (right). Unit cells are outlined in white, and surface directions are specified. STM image parameters: $V_b = -0.4$ V, $I_t = 1.5$ nA (left), $V_b = -0.2$ V, $I_t = 0.8$ nA (right). Courtesy of J. A. Lipton-Duffin.

[108]. The two surfaces are accordingly used in different contexts, with (111) being used to favor molecule–molecule interactions and (110) being used to impose a twofold symmetry on the molecular building blocks (see Section 1.5 for examples). The anisotropy of (110) has additional implications for the diffusion of adsorbates, which is enhanced parallel to the [1$\bar{1}$0] direction [109,110].

3. *Au(111)*. The inclusion of an extra Au atom once every 22 atoms along [1$\bar{1}$0] leads to a massive, strain-induced (22 × $\sqrt{3}$) reconstruction of Au(111) [111,112]. The unit cell comprises both hexagonal close-packed (hcp) and fcc sites, with a narrow band of bridge sites between the two. These bridge sites form the characteristic herringbone bands evident in STM images (**Figure 2**). The spatial and electronic [113,114] inhomogeneities introduced by this reconstruction can lead to a templating effect for adsorbates introduced to the surface. Both atomic [115,116] and molecular [117] species have been demonstrated to preferentially adsorb at the 'elbows' of the herringbones.

Although the Au(111)-(22 × $\sqrt{3}$) surface can be prepared on a single-crystal metal sample using the usual ultrahigh vacuum (UHV) techniques [118], it can also be prepared on thin films of gold deposited on mica [119–121]. The films can be easily prepared by vacuum deposition of gold onto mica, with subsequent flame-annealing to improve the quality of the substrate, or can be purchased commercially. Gold on mica substrates is amenable to study in ambient, or in solution.

Figure 2 STM image of the Au(111)-(22 × $\sqrt{3}$) herringbone reconstruction. Obtained from http://ipn2.epfl.ch/LSEN/jvb/collection/coll_au111.htm.

4. *Si(001)*. The electronic properties of Group IV semiconductors have made them the cornerstone of modern devices [122]. Growing useful nanostructures on a semiconductor surface is therefore of considerable interest, since this approach offers the opportunity to integrate novel technologies with established ones. We focus here the Si(100) surface, which is nearly ubiquitous in the microelectronics market [123].

The Si(001) surface has been described in many insightful reviews [124–126,123], so we will provide only a brief sketch of the surface structure. The

Si(001) surface reconstruction is c(4 × 2) at low temperature, and (2 × 1) at room temperature [127, 128]. This surface periodicity results from formation of asymmetric silicon dimers [129], with the 2× periodicity aligned along [$\bar{1}$10] (traditionally, a surface that is 2× along the equivalent [110] is referred to as (1 × 2).) The dimerization results in a reduction of the number of surface DBs from one per atom to one per dimer. These remaining DBs mean that even the dimerized surface is quite reactive; exposure to hydrogen passivates the surface through the formation of one of several hydrogen terminations, depending on conditions of preparation [130]. Reactive DB sites can persist as defects after passivation [131], or can be introduced in a controlled manner by desorbing a hydrogen atom with a voltage pulse from the STM tip [132].

1.3 Patterned Bonding between Molecules and Surfaces

In this section, we will provide an overview of two rather different methods of spatial control over molecular bonding at surfaces. In Section 1.4.1, we will discuss chemical chain reactions that lead to the formation of molecular lines covalently bonded to a silicon surface. In Section 1.4.2, we will discuss a more general technique for molecular positioning through the selective patterning of substrates to delineate reactive regions on a surface. This approach can be used to guide covalent molecule–substrate bonding or to confine molecular self-assembly to a predefined area.

1.3.1 Chemical Chain Reactions

Self-propagating directed growth via chemical chain reactions represents an extremely appealing mechanism for fabricating molecular architectures. As a general schema, the growth of a molecular architecture via a chemical chain reaction requires a nucleation site, at which a molecule will attach in such a way so as to create a second nucleation site. Attachment of a molecule at this site will create a third nucleation site, and so on. The power of this technique lies in the simplicity of its execution: once the initial nucleation sites are created, the subsequent chemical chain reaction can be carried out quickly, and in parallel, that is, the growth of multiple structures can be carried out simultaneously.

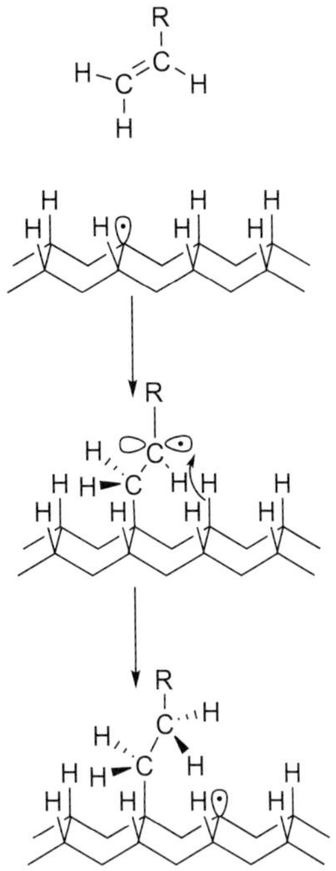

Figure 3 Schematic illustration of an alkene reacting at a silicon dangling bond in the first step of a chemical chain reaction. From Lopinski GP, Wayner DDM, and Wolkow RA (2000) Self-directed growth of molecular nanostructures on silicon. *Nature* 406: 48–51.

The seminal work on molecular architecture formation through a chemical chain reaction is described by Lopinski *et al.* [133]. The nucleation site for the chemical chain reaction was a single silicon DB on an otherwise H-terminated Si(001). **Figure 3** shows a schematic of the reaction of an alkene at the DB site: the alkene moiety (CH$_2$=CH—R) interacts with the surface DB to form a C-centered radical at the carbon–carbon double bond, and the radical abstracts a hydrogen from a neighboring silicon dimer (this mechanism was confirmed with density functional theory modeling [134]). After exposing a slightly defective H:Si(001) surface to 3 l of styrene (**1**), Lopinski *et al.* observed molecular lines up to 130 Å in length. High-resolution imaging confirmed the spacing of the features within the line to be 3.8 Å, corresponding to the dimer spacing on Si(100). The longest observed lines therefore indicate a 34× propagation of the chemical chain reaction.

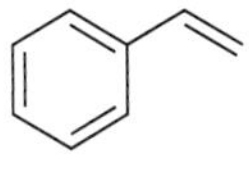

Lopinski *et al.* essentially provided a recipe for growing chain-reaction architectures. By changing the substrate geometry or molecular substituents, architectures with different geometries or functionalities have been grown via the same principle. We will discuss some of these below; for a discussion of chemical chain reactions to produce molecular monolayers, we direct the reader to a recent review by Lopinski [135].

The growth of compact styrene islands was demonstrated by exploiting the geometry of the H-terminated Si(111) surface [136]. In contrast to the unidirectional propagation fostered on the dimer rows on H:Si(100), the H:Si(111) surface yields a (1 × 1) hexagonal array. DB defects were created by selectively using the STM tip to desorb hydrogen [137] from a high-quality H:Si(111) surface created via wet-chemical methods [138]. Subsequent exposure to 12 l of styrene resulted in the growth of a two-dimensional (2D) island at each of the DB sites. The islands appeared to be self-limiting in size, terminating after ~20 reactions, likely turning in on themselves due to an attractive interaction between the phenyl rings in the growing island.

Although island growth on Si(111) provides a neat extension of the chemical chain reaction principle, most efforts were directed toward growing linear chains, which may play an important role in molecular electronics, on Si(100), the 'device' face of silicon.

The styrene chemical chain reaction proceeds along a silicon dimer row, but growth across the dimer rows has also been demonstrated [139]. The key to tailoring the growth direction lies in matching the physical parameters of the reactant site in the molecule with the periodicity of the silicon surface (3.84 Å along the dimer row, 7.68 Å across the rows) [139]. With styrene, for example, the C-centered radical abstracts a hydrogen from a neighboring silicon dimer. However, for molecules where R is CH_2—SH, Hossain *et al.* [139] propose that the C-centered radical can be transferred to the sulfur, resulting in the creation of a thiyl radical. The thiyl radical subsequently reacts with the surface to abstract a hydrogen and create a new silicon DB. Since it is located further from the anchored site of the molecule, it is favorable for the thiyl to abstract a hydrogen from the neighboring dimer row ~5 Å away, rather than from the closer neighboring site along the same dimer row. **Figure 4** shows a schematic for this reaction, together with an STM image of an allyl mercaptan (**2**) line traversing the H:Si(100) dimer rows. For comparison, the schematic and an STM image of a styrene line are also shown.

SH
2

In a follow-up work, Hossain *et al.* described a beautiful experiment in which they interconnected styrene and allyl mercaptan lines on the H:Si(100) surface [140]. **Figure 5** displays a series of STM images showing the steps involved. After creating a H:Si(100) surface with a dilute concentration of DB defects, lines were grown by exposing the surface to 5 l of allyl mercaptan. Using the STM tip, Hossain *et al.* created a DB defect immediately adjacent to one of the allyl mercaptan lines. This defect served as the nucleation position for a styrene line, which, upon exposure to 5 l of styrene, grew perpendicular to the initial allyl mercaptan line, terminating at a parallel adjacent line.

More recent work in this field has focused on creating contiguous molecular lines incorporating 90° angles, a geometry potentially useful for molecular wiring and electronics applications. It was demonstrated that on the (3 × 1)-H:Si(100) surface the DB terminating a styrene line could be used to seed the growth of a perpendicular line of trimethylene sulfide, creating a contiguous, L-shaped structure [141]. Lines comprising sections directed both parallel and perpendicular to the (2 × 1)-H:Si(100) surface have also been formed from a single molecule, acetophenone [142]. The capability for a single molecule to propagate via chain reaction in perpendicular directions on this silicon surface is unique, and the authors suggest that it may be a manifestation of chirality effects in the adsorbed molecule, together with the buckling of the underlying silicon dimer [142].

1.3.2 Selectively Patterned Surfaces

An alternative route to guiding the bonding between a molecule and substrate relies on the prepatterning of the substrate through a preliminary process, such as the growth of a specific surface phase. This can result in the formation of spatially limited reactive zones on the surface, and molecules introduced in a secondary growth step will selectively bond to the reactive regions.

For this purpose, the Cu(110)/CuO surface provides an ideal template. The reaction of even small amounts of oxygen with Cu(110) produces a (2 × 1) periodicity [143]. The reconstruction, comprising Cu—O strings aligned along [001] and spaced by two lattice constants (<0.512 nm) along [1$\bar{1}$0], is formed from the chemisorption of oxygen with

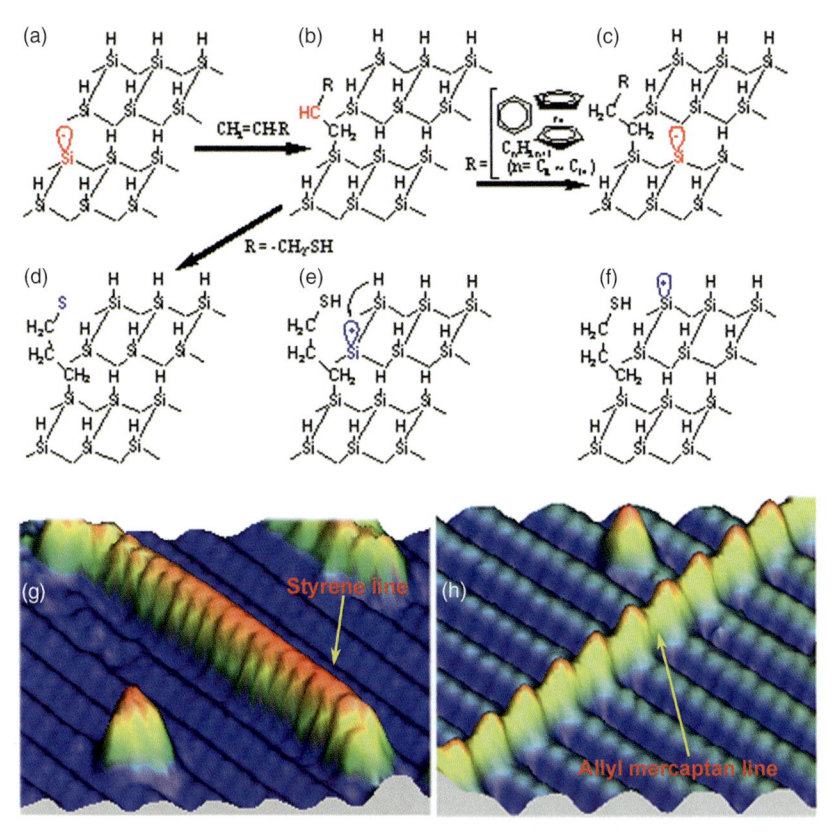

Figure 4 Schematic mechanisms for the formation of molecular lines via chemical chain reaction propagating parallel and perpendicular to the underlying Si dimer rows on H:Si(100)–(2 × 1) (a–f), and STM images showing (g) styrene (parallel) and (h) allyl mercaptan (perpendicular) lines. STM image parameters: $V_b = -1.1$ V, $I_t = 0.2$ nA, image area = 9.2×6.3 nm^2 (g), $V_b = -1.8$ V, $I_t = 0.2$ nA, image area = 7.0×5.5 nm^2 (h). From Hossain MZ, Kato HS, and Kawai M (2005) Fabrication of interconnected 1D molecular lines along and across the dimer rows on the Si(100)-(2×1)–H surface through the radical chain reaction. *Journal of Physical Chemistry B* 109: 23129–23133.

copper atoms liberated from step edges [144–146]. Kern *et al.* [147] showed that under appropriate preparation conditions the CuO regions define a periodic grating of regularly spaced stripes. The surface can therefore be thought of as a periodic array of structurally and chemically distinct regions, and can be used to investigate and guide molecular adsorption. Notable molecular species investigated on this surface include CF_4 [148], lander molecules [149], alpha-quinquethiophene [150], rubrene [151] and para-sexiphenyl [152]. We direct the interested reader to a more thorough overview of molecular absorption on Cu(110)/CuO, recently published by Cicoira *et al.* [153].

Recently, Lu *et al.* [154] published the first account of molecular adsorption on a differently patterned substrate, the nitrogen-induced reconstruction of Cu(100) [155]. Exposing Cu(100) to activated nitrogen produces a c(2 × 2) reconstruction [156]. Although historically the subject of some debate [157–159], the reconstruction is now believed to form from islands of Cu_2N, which has a lattice constant ~3% larger than that of Cu(100) [160]. To relieve strain, the Cu_2N forms into roughly square islands. Lu *et al.* worked with a 0.35-ML (monolayer) coverage of nitrogen, forming an array of ~5 nm squares of Cu_2N, separated by narrow strips of clean copper. They found that C_{60} fullerenes deposited on the surface preferentially adsorbed onto the copper grid, as shown in **Figure 6**. The fullerenes continue to adsorb exclusively on the grid until it saturates, at ~0.28 ML C_{60}, at which point molecules begin to adsorb on the c(2 × 2) regions.

A novel type of patterning has been used to control the placement of self-assembled monolayers (SAMs) [161–163], 2D ordered films of molecules that rely on a strong covalent association between the substrate and the molecular headgroup (e.g., Au—S bonds, as in thiol layers on gold) [164].

8 Directed Assembly of Nanostructures

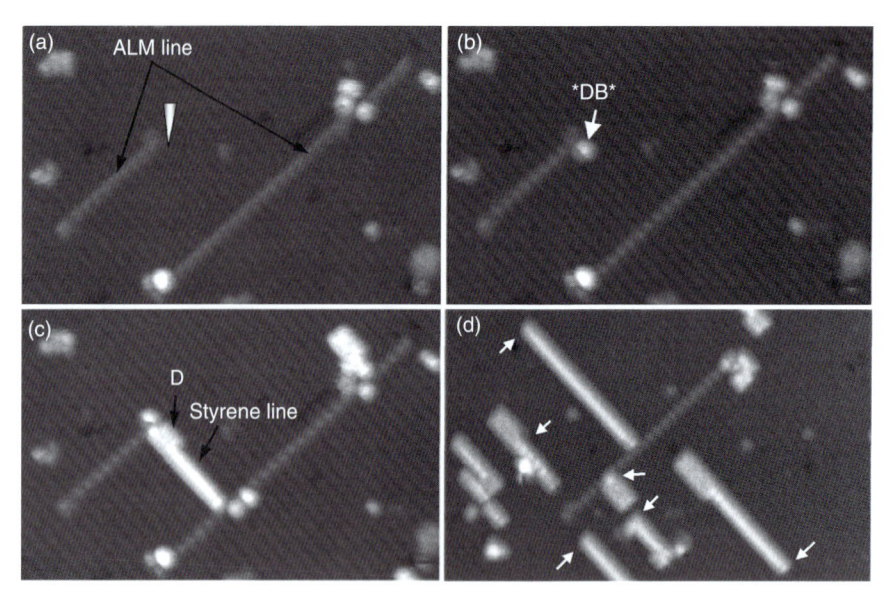

Figure 5 A dangling bond (DB) defect formed with the STM tip is used to nucleate the growth of a styrene line spanning two allyl mercaptan (ALM) lines. STM image parameters: $V_b = -2.6\,V$, $I_t = 0.2\,nA$, image area $= 26 \times 16\,nm^2$ (a–c), $30 \times 20\,nm^2$ (d). From Hossain MZ, Kato HS, and Kawai M (2005) Fabrication of interconnected 1D molecular lines along and across the dimer rows on the Si(100)-(2×1)-H surface through the radical chain reaction. *Journal of Physical Chemistry B* 109: 23129–23133.

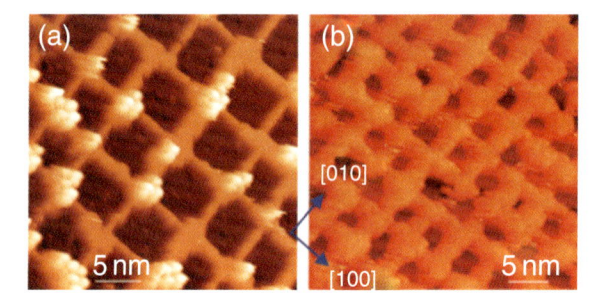

Figure 6 C_{60} fullerenes selectively adsorb onto the clean Cu grid defined by nitrogen-adsorbed regions of a Cu(001)-c(2 × 2) surface. (a) It shows 0.08 ML fullerene coverage; and (b) shows ~0.28 ML coverage. STM image parameters: $V_b = 0.5\,V$, $I_t = 0.15\,nA$, image area $= 25 \times 25\,nm^2$ (a), $V_b = 0.5\,V$, $I_t = 0.15\,nA$, image area $= 35 \times 35\,nm^2$ (b). From Lu B, Iimori T, Sakamoto K, Nakatsuji K, Rosei F, and Komori F (2008) Fullerene on nitrogen-adsorbed Cu(001) nanopatterned surfaces: From preferential nucleation to layer-by-layer growth. *Journal of Physical Chemistry C* 112: 10187–10192.

Traditionally, patterning of SAMs is achieved through a top-down technique like soft lithography [165] or irradiation (see Ref. [166] for a review). The spatial limitation in using a top-down approach is obviously dictated by the resolution of the patterning technique, and can often exceed the nanoscale.

However, Madueno *et al.* [167] presented a demonstration of controlled nanoscale templating of SAMs by using a porous hexagonal supramolecular network of 1,3,5-triazine-2,4,6-triamine (melamine), and perylene-3,4,9,10-tetracarboxylic di-imide (PTCDI) to selectively mask a Au(111) substrate. Immersion of the supramolecularly patterned substrate in a thiol solution created a discrete pattern of SAM, defined by the pores of the supramolecular structure where the gold substrate was exposed. **Figure 7** shows a schematic of this patterning approach, and STM images of the SAMs formed from three different thiol molecules.

1.4 Guiding Supramolecular Assembly

In this section, we will provide an overview of the approaches available for custom-designing the geometry of self-assembled molecular structures at

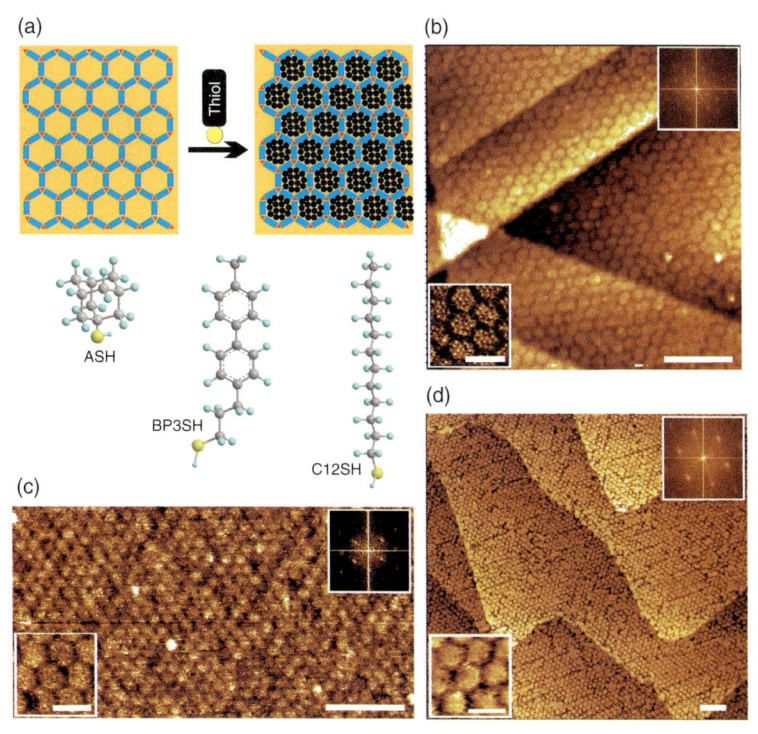

Figure 7 Patterning thiol SAMs with a supramolecular network. (a) shows a schematic of the approach used to create a hexagonally patterned SAM of the thiols shown in (b). The STM images (c–e) show typical SAMs obtained with ASH, BP3SH, and C12SH, respectively. The inset FFTs confirm the hexagonal pattern. Scale bars for the large images are 20 nm; for the insets, the scale bars are 5 nm. From Madueno R, Raisanen MT, Silien C, and Buck M (2008) Functionalizing hydrogen-bonded surface networks with self-assembled monolayers. *Nature* 454: 618–621.

surfaces. We emphasize that a number of excellent reviews, covering various facets of self-assembly at surfaces, are available to the interested reader [168, 169, 78, 2, 170–173].

Although by Whitesides' definition [18] molecular self-assembly can be governed by any noncovalent interaction, here, we focus on the directional bonding provided by hydrogen bonding and by metal–organic coordination. Use of these directional associations, when combined with a careful choice of the molecular substituents, can provide a high degree of control over the geometry of the resulting self-assembled structure.

1.4.1 Hydrogen-Bonded Architectures

1.4.1.1 Overview

Descriptions of the controlled self-assembly of molecules on surfaces began to emerge at the turn of the twenty-first century. Soon after initial demonstrations of the formation of clusters and chains [117,174], the geometrical control of self-assembled hydrogen-bonded structures through the design of the molecular building blocks was documented by Yokoyama and coworkers [175]. By systematically varying the placement of the functional groups on their porphyrin molecular framework, Yokoyama and coworkers were able to predictably fashion molecular assemblies into trimers, tetramers, and chains.

The interactions between the molecules were tailored via the arrangement of cyanophenyl groups around a porphyrin core; the cyanophenyl groups have an asymmetric charge distribution that introduces dipole–dipole interactions between the molecules in addition to hydrogen bonding. Starting from 5,10,15,20-tetrakis-(3,5-di-tertiarybutylphenyl) propyrin (H_2-TBPP) (**3a**), the authors replace one of the four di-tertiarybutylphenyl (tBP) substituents with a cyanophenyl group to create (cyanophenyl)-tris(di-tertiarybutyl-phenyl) porphyrin (CTBPP, **3b**) and replace two of the tBP substituents, in either *cis*- or *trans*-geometries, to form the bis(cyanophenyl)-bis(di-tertiarybutylphenyl) porphyrin (BCTBPP, **3c** and **3d**, respectively.)

The supramolecular architectures produced by these molecules on Ag(111) are dictated by the location

Structure of the porphyrins used by Yokoyama and co-workers: (a) H$_2$-TBPP, (b) CTBPP, (c) cis-BCTBPP; and (d) trans-BCTBPP. From Yokoyama T, Yokoyama S, Kamikado T, Okuno Y, and Mashiko S (2001) Selective assembly on a surface of supramolecular aggregates with controlled size and shape. Nature 413: 619–621.

and number of the cyano substituents of the porphyrin molecules. In the absence of the cyano groups, the molecules assemble into a close-packed overlayer stabilized by the van der Waals interactions between the tBP substituents (**Figures 8(e)** and **8(i)**). The single cyano group in the CTBPP molecules precipitates an entirely new bonding architecture: the molecules form into trimers arranged around the cyano groups and adsorbed at the elbows of the herringbone reconstruction (**Figures 8(b)**, **8(f)**, and **8(j)**). The tetramers shown in **Figure 8(c)**, **8(g)**, and **8(k)** are produced by introducing two cyano groups at adjacent positions on the porphyrin (*cis*-BCTBPP), leading to cyclical bonding between the BCTBPP molecules within the tetramer. Finally, the two diametrically opposed cyano groups of the *trans*-BCTBPP favor the formation of the molecular chain shown in **Figures 8(d)**, **8(h)**, and **8(i)**.

This case study very elegantly lays out a framework for the rational control of self-assembled architectures. In the rest of this section, we will provide examples of systems assembled according to molecular functionality and geometry, as well as discussing an approach to breaking the symmetry imposed by the substituent molecules.

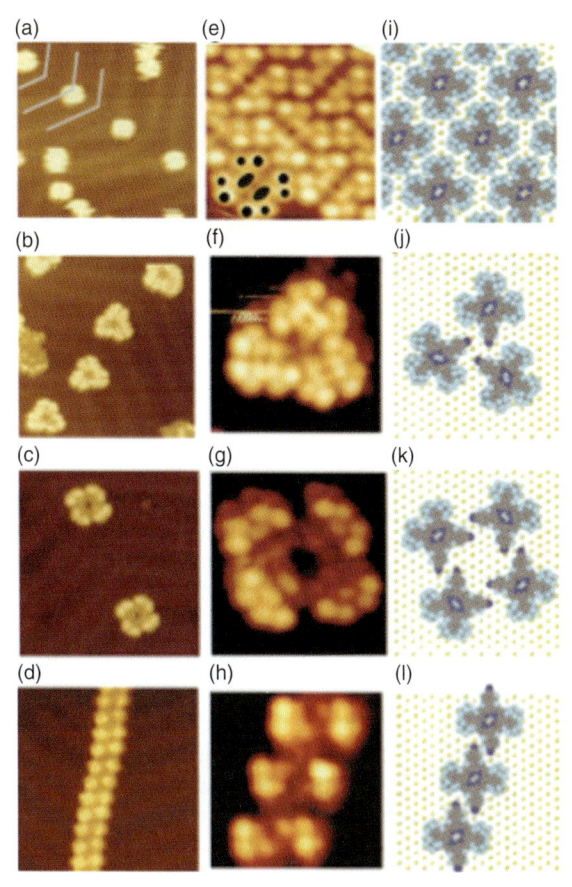

Figure 8 STM images and structural models for the hydrogen-bonded architectures formed by H_2-TBPP (a, e, i), CTBPP (b, f, j), cis-BCTBPP (c, g, k) and trans-BCTBPP (d, h, l). STM image parameters: $T = 63$ K, image area $= 20 \times 20$ nm^2 (a–d), 5.3×5.3 nm^2 (e–h). From Yokoyama T, Yokoyama S, Kamikado T, Okuno Y, and Mashiko S (2001) Selective assembly on a surface of supramolecular aggregates with controlled size and shape. Nature 413: 619–621.

1.4.1.2 Basic geometries in hydrogen-bonded structures

1. *Chains.* Detailed accounts of the formation of hydrogen-bonded chains at surfaces had already been published prior to the work of Yokoyma et al. First, Böhringer et al. showed that relatively high coverages of 1-nitronaphthalene (NN, **4**) form into domains of parallel lines on the Au(111) surface [117]. Barth et al. subsequently reported on the self-assembly of 4-[trans-2-(pyrid-4-yl-vinyl)]benzoic acid (PVBA, **5**) into chains on the same surface [174]. We will briefly describe both of these systems here, since they present two conceptually different routes to forming linear hydrogen-bonded structures.

Böhringer et al. demonstrated that two different types of hydrogen-bonded architectures, shown in **Figure 9**, could be formed from NN on Au(111), depending on the coverage range [117]. At coverages from 0.3 to 0.75 ML, the double chains shown in **Figure 9(a)** are formed. Lower coverages (0.05–0.2 ML) result in isolated clusters (**Figure 9(b)** and **9(c)**).

The authors' model for the chains consists of two rows of antiparallel NN molecules, stabilized by H bonds between the O atoms of one molecule and the CH groups of its neighbor. The doubling of the chains is caused by the inhomogenous electrostatic field of the molecules (inset of **Figure 9(a)**). Once doubled, the chains have a uniformly positive field at their outer edges, resulting in a repulsive interaction between neighboring chains.

Based on density functional theory calculations, the authors found that the doubled chains represent the lowest-energy structure of NN on the

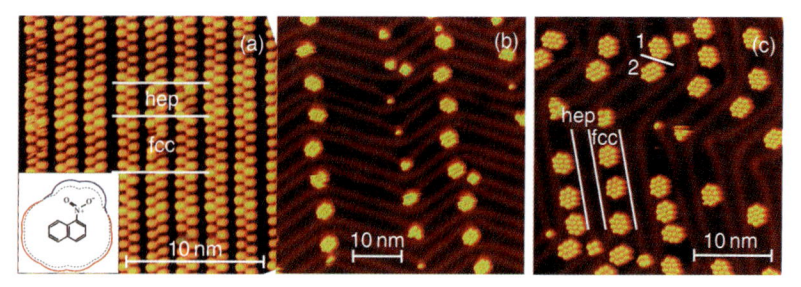

Figure 9 Low-temperature STM images of NN on Au(111). (a) Chain structure (0.7 ML coverage, imaged at 50 K). (b) Clusters arranged at the elbows of the herringbone reconstruction (0.1 ML coverage, imaged at 65 K). (c) Higher coverage of clusters (0.2 ML coverage, imaged at 50 K). From Bohringer M, Morgenstern K, Schneider WD, et al. (1999) Two-dimensional self-assembly of supramolecular clusters and chains. *Physical Review Letters* 83: 324–327.

surface, with a cohesive energy of 0.1394 eV per molecule. However, short chains, with unsaturated bonds at their ends, are energetically unfavorable with respect to the formation of the clusters shown in **Figures 9(b)** and **9(c)**, which, for the hexameric structure, have a cohesive energy of 0.1277 eV per molecule. Hence, below some critical coverage, the formation of clusters over chains is favored.

This work demonstrates the flexibility that can be inherent in hydrogen-bonded supramolecular structures at surfaces: depending on the preparation conditions, different geometries can be facilitated. However, the coverage-dependent metastability of the NN lines with respect to cluster formation may not be a desirable attribute in a purpose-designed molecular system. The weak hydrogen bonds formed between NN molecules, and the geometry of the molecules themselves, are key factors in this behavior. Conversely, the molecular chains described by Barth et al. [174] are based on PVBA, a molecule designed to form strong head-to-tail hydrogen bonds [176,177], and, at least on Ag(111), the formation of PVBA chains is accordingly predictable.

Barth and coworkers did a meticulous job of demonstrating the importance of minimizing substrate–molecule interactions in surface supramolecular assembly. In UHV experiments, they evaporated PVBA molecules onto different substrates held at low temperature. On Pd(110), the flat-lying molecules form a strong π-bond to the surface, and align in two orientations according to the high-symmetry directions of the surface even after annealing to 450 K. The Cu(111) surface presents similar problems: at 160 K, although some dendritic networks are formed from flay-lying, head-to-tail-bonded molecules, some molecules are adsorbed in a standing-up orientation. After annealing, only flat-lying molecules remain, but the formation of an ordered chain structure is hindered by interactions with Cu atoms liberated from step edges.

It is on the Ag(111) surface, where the influence of the substrate is negligible, that Barth and coworkers observe the formation of ordered chain structures. At 125 K, the aggregation of the PVBA molecules is diffusion limited, resulting in the formation of tangled, curved PVBA chains. By providing sufficient thermal energy to the molecules, either by annealing or depositing at 300 K, arrays of parallel PVBA chains, as shown in **Figure 10**, are formed.

The one-dimensionality of the chains is dictated by the head-to-tail hydrogen bonding expected from the location of the carboxylic acid and the pyridil group. Interestingly, the length of the OH...N bond is longer than expected for similar hydrogen bonds in other molecules (2.5 vs 1.5–2.1 Å), indicating that the bonding is weak. The doubling of the chains stems from an additional weak CH...OC hydrogen bond between molecules in adjacent chains (**Figure 11(b)**). The authors ascribe the nearly regular spacing of the twinned lines (e.g., **Figure 10(a)**) to a weak, long-range dipole–dipole interaction.

Later work by the same group unraveled some interesting properties of the PVBA lines. As illustrated in **Figure 11**, the PVBA molecule is prochiral (i.e., it becomes chiral when confined to the 2D plane of the surface, according to the orientation of the carboxylic acid moiety). Based on high-resolution STM images (**Figure 11**) Weckesser et al. determined that each PVBA line is homochiral [178]. The homochirality within the lines can be explained by stereochemistry, since the intrachain hydrogen bonds are geometrically favored when the chirality of the molecules on both sides of the chain is

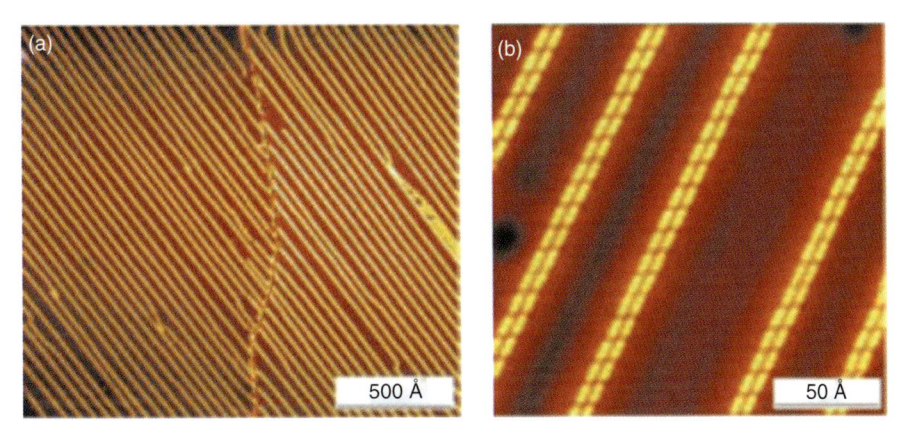

Figure 10 Hydrogen-bonded PVBA lines on the Ag(111) surface. STM images obtained at 77 K show: (a) large-scale and (b) molecule-by-molecule ordering. From Barth JV, Weckesser J, Cai CZ, *et al.* (2000) Building supramolecular nanostructures at surfaces by hydrogen bonding. *Angewandte Chemie International Edition* 39: 1230–1234.

identical. However, stereochemical effects cannot account for the fact that the homochirality extends over entire domains of PVBA chains, since the chains themselves are not in close contact with one another (the chains are spaced by ~120 Å on Au(111), but can be spaced more closely on the Ag(111) surface.) The authors attribute the formation of homochiral domains to the growth mechanism of the chains, which initially involves the formation of quadrupled homochiral chains that eventually split in to the twinned chains routinely observed in STM images. The spacing between the twinned chains is driven by different effects on the different substrates: the PVBA chains on the Ag(111) are spaced according to the elbows of the Au(111) herringbone reconstruction, whereas on Ag(111), which lacks a surface reconstruction, the spacing can be tuned between 25 and 50 Å by varying the PVBA coverage [178].

Further investigations with the similar molecule PEBA (**6**) demonstrated just how sensitive supramolecular assembly is to perturbations to the stereochemistry of the molecule [179]. Although the low-temperature (125 K) growth of PEBA on Ag(111) is qualitatively similar to that of PVBA, outside of this diffusion-limited regime the structures formed are much different. The achiral, linear PEBA molecule forms into compact homochiral sheets, rather than the twinned chain structure of PVBA. The authors suggest that PEBA chain formation is inhibited because the linear PEBA molecule does not support the smooth steric match of the interchain hydrogen bonds seen in the PVBA chains, even though the same head-to-tail hydrogen bonding exists. Clearly, the design of molecular nanostructures is complicated by rather subtle interactions, even when major interactions (e.g., head-to-tail bonding) have been designed to favor a specific geometry.

6

2. *Networks.* Molecules favoring head-to-tail bonding can similarly be used in the formation of molecular chains, higher symmetry molecules can be designed to form into 2D networks. In particular, C_3-symmetric molecules can be tailored to form into hexagonal meshes, which we will discuss here both as an example of a 2D structure, and because of the importance of porous 2D networks as inclusion architectures (discussed extensively in the next section.)

One of the best examples of a C_3-symmetric hydrogen bonding unit is 1,3,5-benzenetricarboxylic acid (trimesic acid, TMA, **7**). Hydrogen bonding between the carboxylic groups leads to the formation of TMA dimers and trimers, both of which can be incorporated into hexagonal meshes (see below). The aromaticity of the molecule means that it is well suited to surface studies, since aromaticity is often associated with adsorption parallel to the surface, leaving the carboxyl groups free to interact with one another. Crystallographic studies dating back to 1969 show that, even in 3D crystals, TMA forms into a planar, hydrogen-bonded hexagonal mesh, referred to as the chicken-wire structure [180,181].

14 Directed Assembly of Nanostructures

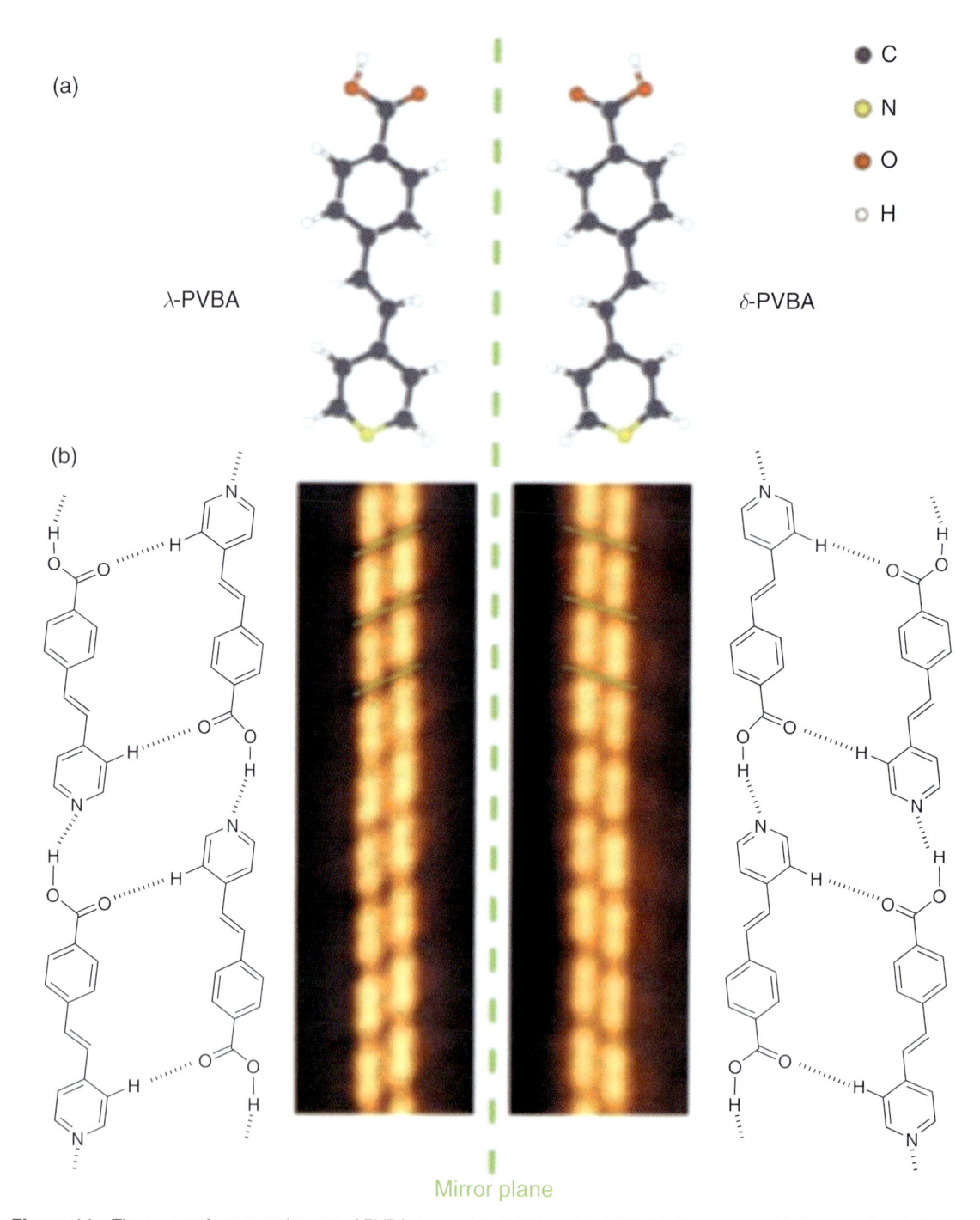

Figure 11 The two surface enantiomers of PVBA, termed λ-PVBA and δ-PVBA (a), form homochiral molecular chains, shown schematically and in STM images in (b). STM image parameters: $T = 77$ K, image area $= 4.0 \times 13.5$ nm^2. From Weckesser J, De Vita A, Barth JV, Cai C, and Kern K (2001) Mesoscopic correlation of supramolecular chirality in one-dimensional hydrogen-bonded assemblies. *Physical Review Letters* 87: 096101.

Griessl et al. investigated the surface self-assembly of TMA by evaporating the molecule onto a clean HOPG surface in UHV, and studying the result with STM at 25 K [182]. They found that the TMA assembled into two different porous meshes: the chicken-wire structure, previously identified in 3D crystals, and the flower structure, both of which are shown schematically in **Figure 12**. Both meshes contain pores bounded by six TMA molecules. In the chicken-wire structure, each TMA molecule is dimerically associated with three other molecules. In the flower structure, each TMA molecule forms two dimeric associations, defining the pores, and one trimeric association, which creates a densely packed region between the pores. Dmitriev et al. found that nearly the same chicken-wire structure was formed when TMA was deposited onto the Cu(100) surface at low tempeature (240 K) [183]. The structure was slightly distorted due to the underlying surface symmetry, and two distinct domain orientations could be identified due to this distortion. The chicken-wire structure did not persist to room temperature, where the authors instead found that the TMA formed a stripe phase, consisting of molecules in an upright geometry. Dmitriev et al. suggest that the chemisorption of the upright TMA, specifically the interaction between the copper surface and the oxygens in the carboxyl groups, makes up for the loss of the π-interaction in the flat-lying geometry.

A later study showed that the TMA chicken-wire and flower structures could be selectively formed at the solution/HOPG interface [184]. By preparing solutions of TMA in a series of alkanoic acids of increasing length ($CH_3(CH_2)_nCOOH$, $2 \leq n \leq 7$), and observing the corresponding TMA network with STM, Lackinger et al. determined that either the chicken-wire or the flower structure would be favored depending on the solvent. **Figure 13** summarizes the relationship between the TMA solubility and the polymorph formed and the number of carbons in the alkanoic acid: TMA is more soluble in shorter acids, which favor the formation of flower structure, and less soluble in longer acids, which favor the formation of chicken-wire patterns.

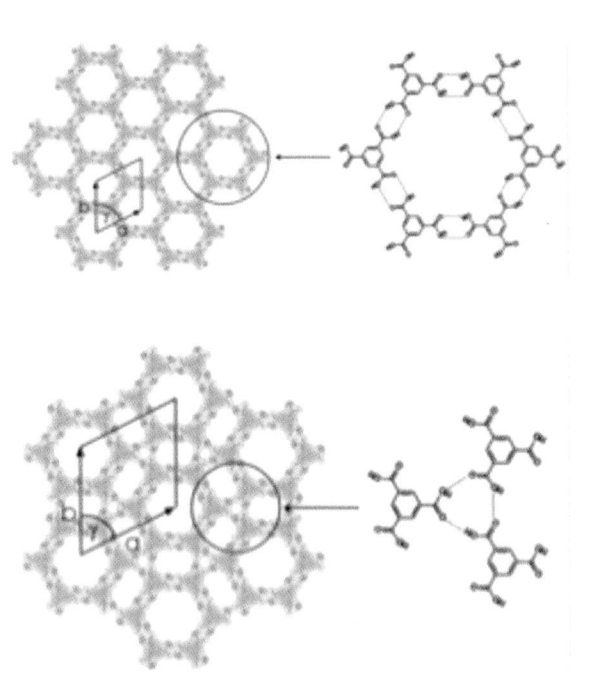

Figure 12 Structure of the TMA chicken-wire (top) and flower (bottom) meshes. Dimeric and trimeric association of TMA molecules is shown. In both unit cells, $\gamma = 60°$. For chicken-wire, $a = b = 1.7 \pm 0.1$ nm, and for flower structure, $a = b = 2.7 \pm 0.1$ nm. From Griessl S, Lackinger M, Edelwirth M, Hietschold M, and Heckl WM (2002) Self-assembled two-dimensional molecular host–guest architectures from trimesic acid. *Single Molecules* 3: 25–31.

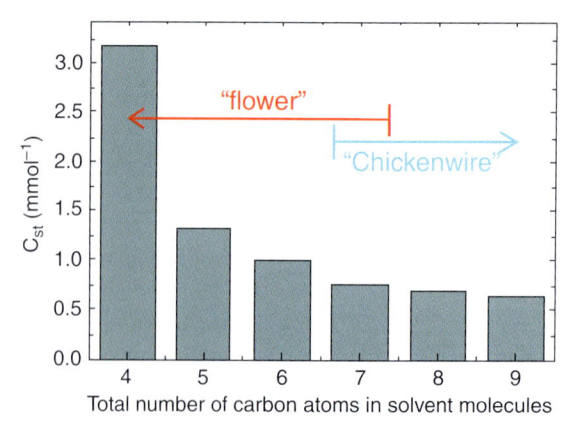

Figure 13 TMA solubility in alkanoic acids. The TMA polymorph formed at the solution/HOPG interface is indicated for each acid. From Lackinger M, Griessl S, Heckl WA, Hietschold M, and Flynn GW (2005) Self-assembly of trimesic acid at the liquid-solid interface – a study of solvent-induced polymorphism. *Langmuir* 21: 4984–4988.

Although the authors speculate about the possible formation of 'seed' trimers in the shorter-chain acids, and suggest that solvent co-adsorption (not explicitly observed) may play a role, the solvent-induced polymorphism of TMA may be more simply explained via the different solubility, and therefore different concentration, of TMA in the different acids. The concentration of a molecule has recently been shown to drive the formation of different polymorphs at the solution/solid interface [185]. Lei *et al.* demonstrated that either a porous or a densely packed mesh of alkoxylated dehydrobenzo[12]annulenes (DBAs) would be formed at the 1,2,4-trichlorobenzene (TCB)/HOPG interface depending on the concentration of DBAs in the TCB solution. They elegantly showed that the polymorph formed depends on the molecular concentration and the stability and molecular density of the possible polymorphs, and suggested a general approach of thermodynamic control over 2D molecular structure at surfaces via concentration.

Ye *et al.* [186] took a similar approach to investigating the formation of the TMA polymorphs in a series of UHV experiments. By carefully controlling the amount of TMA deposited onto a Au(111) surface, they were able to use the TMA surface coverage to select which polymorph was formed. The chicken-wire and flower structures are only two members of the infinite number of mathematically possible TMA polymorphs comprising dimeric and trimeric associations [187], which range from the fully dimeric chicken-wire structure to the fully trimeric 'superflower' structure predicted by Lackinger *et al.* [184]. By varying the TMA coverage between 0.3 (chicken-wire structure) and 1.0 ML (superflower structure), Ye *et al.* could control the unit cell size of the TMA polymorph, as shown in **Figure 14**. Each observed polymorph, with the exception of the superflower structure, contains the same dimerically bounded pores contained in the chicken-wire structure. In higher polymorphs, the pores are separated from one another by rows of dimerized TMA molecules, and triangular regions of trimerically associated TMA. Through this approach, Ye *et al.* were able to vary the pore spacing from 1.6 to 8.2 nm; in theory, complete control of the pore spacing, in step sizes of ∼0.93 nm, should be possible.

The chicken-wire structure is a common motif in hydrogen-bonded networks self-assembled from C_3-symmetric molecules. For example, 1,3,5-benzenetribenzoic acid (BTA, **8**) can form a chicken-wire mesh at the solution/ HOPG interface and under UHV conditions on Au(111) [188,189]. Combinations of C_3-symmetric molecules can similarly form chicken-wire meshes, as demonstrated by STM images obtained from a binary solution of TMA and 1,3,5-tris(4-pyridyl)-2,4,6-triazine (TPT, **9**) at the solution/HOPG interface [190]. The co-adsorption of a C_3-symmetric molecule with a linear molecule can result in each C_3 molecule being linked to three linear spacers, creating a hexagonal porous mesh with large pores. The co-adsoprtion of perylene tetra-carboxylic di-imide (PTCDI, **10**) and 1,5,3-triazine-2,4,6-triamine (melamine, **11**) can produce just such a network at the Si(111)–Ag($\sqrt{3} \times \sqrt{3}$)R30 surface [191]. This particular network will be discussed in detail in Section 1.6.1.

image in **Figure 15** [192, 193]. By obtaining STM images of the self-assembled layer formed by a 1-undecyl monoester of TMA prepared *ex situ*, the authors establish that the pattern is not caused by esterification, and can instead be attributed to hydrogen bonding: dimerization of TMA and association between a carboxyl group of the TMA and the —OH group of the alcohol.

Since the formation of the self-assembled pattern depends only on the —OH group in the alcohols, Nath *et al.* were able to demonstrate control over the spacing of the TMA dimer rows by incorporating different alcohols of the type $C_nH_{2n+1}OH$. Surprisingly, the spacing of the TMA dimer rows did not show a simple dependence on n; it also depended on the parity of n, that is, whether the incorporated an even or an odd number of carbons into its methylene backbone. The reason for this can be traced to the adsorption geometry of the alcohols, which align with their backbones nearly perpendicular to the TMA dimer row for alcohols with an odd carbon count, but form an angle closer to 60° for alcohols with an even-carbon count, to allow for closer packing of the alcohols on the surface.

The TMA-alcohol tapes were found to coexist with domains of pure TMA, either in the chicken-wire or flower structure, depending on the solvent used. Interestingly, small regions of the superflower structure were also observed at domain boundaries of the TMA-alcohol, suggesting that it is possible to geometrically stabilize polymorphs that are not thermodynamically favored [187].

1.4.2 Metal–Organic Coordination

The coordination of organic ligands with metal atoms provides an alternative route to fabricating low-dimensional molecular architectures. The metal–organic interaction is directional, is generally stronger than the interaction in hydrogen-bonded structures, and the incorporation of metal atoms can confer useful characteristics (e.g., magnetism [194]) to the assemblies. Some recent reviews of metal–organic coordination networks (MOCNs) at surfaces can be consulted to provide a thorough discussion of the field [172, 195–197]; here, we will simply provide an overview of some notable structures.

Some of the same principles from designing molecules for hydrogen bonding can be extended to MOCNs: the placement of the functional groups (dentates) can guide the geometry of the resulting structure. For example, bidentate linear molecules

The formation of a large-pored chicken-wire network is not the only possible outcome of the co-adsorption of a linear molecule with a C_3-symmetric linker. Nath and coworkers showed that the co-adsorption of linear alcohols with TMA at the solution/HOPG interface breaks the symmetry imparted by the TMA molecules and results in the formation of a linear array of TMA dimer rows separated by alcohol lamellae, as shown in the STM

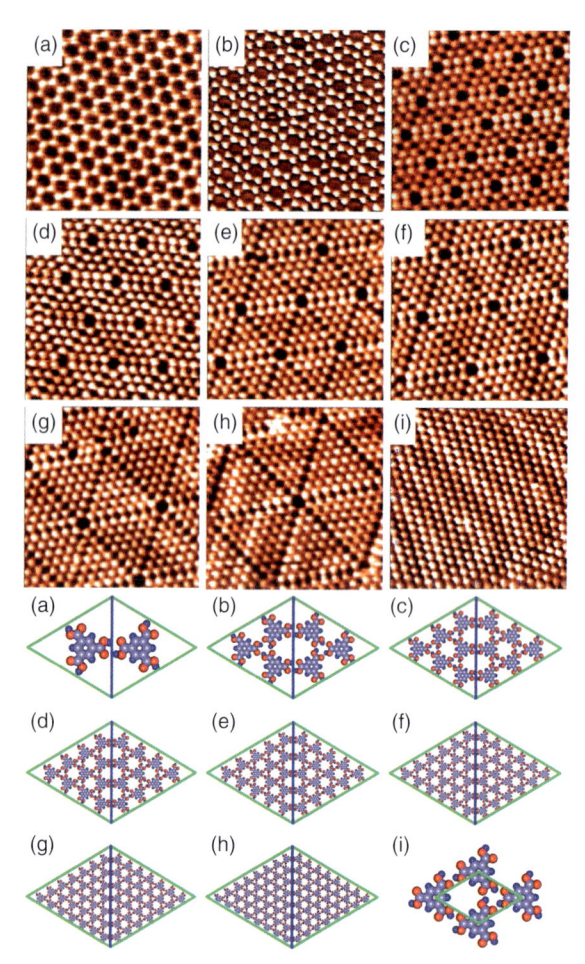

Figure 14 STM images (A–I) and corresponding unit cell models (a–i) for TMA polymorphs formed with increasing TMA coverage on the Au(111) surface in UHV. STM image area = 16.5 × 16.5 nm^2. From Ye YC, Sun W, Wang YF, et al. (2007) A unified model: Self-assembly of trimesic acid on gold. *Journal of Physical Chemistry C* 111: 10138–10141.

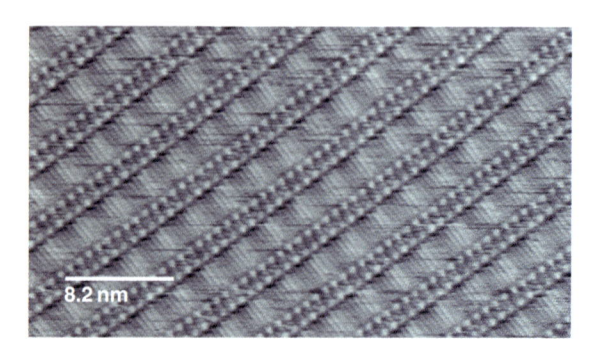

Figure 15 A typical linear pattern produced by the co-adsorption of TMA and alcohols at the solution/HOPG interface. This self-assembled network is formed from TMA and C$_{17}$H$_{35}$OH at the heptanoic acid/HOPG interface. STM image parameters: $V_b = -0.8$ V, $I_t = 20$ pA.

arranged with the functional groups at opposite ends of the molecule can coordinate into chains, as is the case for chains formed by the 1:1 coordination of either **12** or **13** with Cu on Cu(001) [198]. However, the symmetry of the molecule does not guarantee the symmetry of the MOCN, since the coordination sites of the metal center must also be considered. Nonlinear geometries have been demonstrated from bidentate linear molecules. Stepanow et al. [199] reported on the formation of honeycomb meshes from the coordination of linear molecules **14** and **15** with iron and cobalt metal centers, respectively. To form the meshes, each metal atom coordinates three linker molecules. Schlickum et al. demonstrated a similar result on Ag(111), where

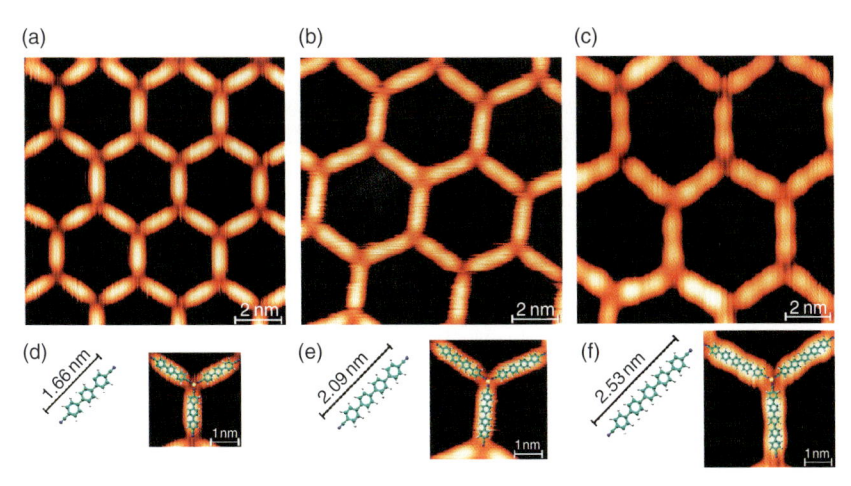

Figure 16 Honeycomb meshes formed by coordinating Co with ditopic dicarbonitrile-polyphenyl molecular linkers on Ag(111). STM images (a–c) and molecular models (d–f) are shown for three linkers of increasing length. STM image parameters: $V_b = 0.9\,V$, $I_t = 0.1\,nA$ (a), $V_b = 1.0\,V$, $I_t = 0.1\,nA$ (b), $V_b = 2.0\,V$, $I_t = 0.1\,nA$ (c). From Schlickum U, Decker R, Klappenberger F, et al. (2007) Metal–organic honeycomb nanomeshes with tunable cavity size. *Nano Letters* 7: 3813–3817.

co-depositing Co with each of the ditopic dicarbonitrile-polyphenyl molecular linkers shown in **Figure 16** results in the formation of a honeycomb mesh [200]. Bidentate linear linkers also commonly form into orthogonally oriented fourfold symmetric meshes, such as those formed by Fe–carboxylate networks with di-iron centers [201–203].

Substrate effects can also have important consequences for the geometry of MOCNs. For example, although molecules **12** and **13** differ only by one benzene, **12** assembles into long-chain segments coordinated with Cu on Cu(001), whereas **13** forms only short segments [198]. The difference can be explained by the molecules' commensurability with the substrate. For a Cu—N bond length of 2.0 Å, chains of **12** aligned along [110] have their constituent copper atoms identically placed in fourfold hollow sites. Conversely, chains of **13** aligned along the same direction can only achieve reasonable Cu—N bond lengths with every third Cu atom in a fourfold hollow site. The other Cu atoms sit just off twofold bridge sites, in an unfavorable absorption geometry. The observed disparity in the chain lengths for the two molecules demonstrates the importance of commensurability for MOCN stability. The substrate symmetry can also assert itself in the MOCN, breaking the symmetry of the organic molecules. This is the case for the C_3-symmetric TMA molecule, which coordinates into both [-Cu-TMA-Cu-]$_n$ and [-Fe-TMA-Fe-]$_n$ linear chains on Cu(110) [204]. Strong coupling to the substrate, rather than the symmetry of the TMA, defines the geometry of this system.

Finally, we highlight a recent work that presented a novel approach to stabilizing and positioning MOCNs by combining two modes of self-assembly. Making use of hierarchical self-organization, Langner *et al.* were able to control the spacing between metal–organic copper–pyridil chains with a hydrogen-bonded 1D array of linear carboxylic acids [205]. **Figure 17** shows an STM image of a region comprising copper-coordinated chains of **12** spaced by 1,4-benzoic-dicarboxlic acid (**16**) alongside a domain of pure **16** on Cu(001), along with

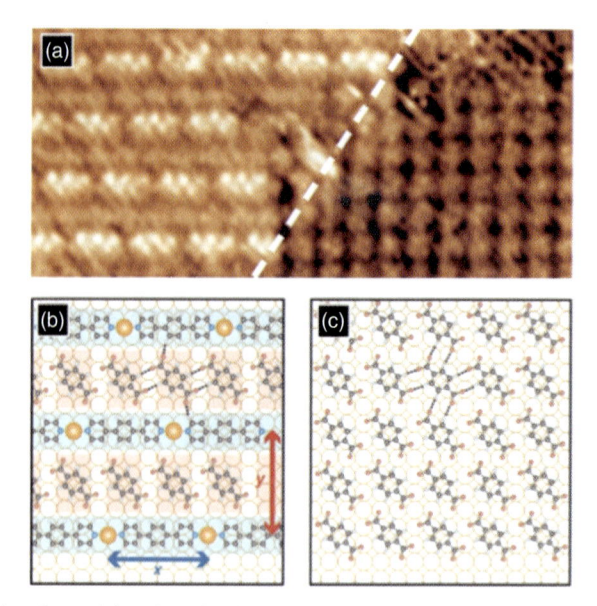

Figure 17 STM image (a) showing at left a domain of copper–pyridil (**12**) metal–organic chains stabilized by interstitial hydrogen-bonded carboxylic acids (**16**) bordering a pure domain of carboxylic acid (right), with structural models for both phases (b and c, respectively.) STM image area = 14 × 6.3 nm^2. From Langner A, Tait SL, Lin N, Chandrasekar R, Ruben M, and Kern K (2008) Ordering and stabilization of metal–organic coordination chains by hierarchical assembly through hydrogen bonding at a surface. *Angewandte Chemie-International Edition* 47: 8835–8838.

models for the metal–organic associations and the hydrogen bonding. In the absence of **16**, the spacing of the copper–pyridil chains can only be controlled via a substrate-mediated repulsion that occurs at very high coverages [198]. However, using this co-crystallization approach, Langner *et al.* showed that the spacing of the metal–organic lines could be controlled by changing the carboxylic acid (e.g., to **17**). The addition of the hydrogen-bonded regions had the additional benefit of stabilizing copper–pyridil chains of **13**, which are rendered unstable by their incommensurability with the substrate in the absence of the interstitial hydrogen bonding. In the case of both **12** and **13**, the additional hydrogen bonding does not appear to perturb the metal–organic bonding. The authors have combined metal–organic coordination and hydrogen bonding to control the spacing of MOCN chains, with the added benefit of stabilizing structures that may not be energetically favorable on their own.

16

17

1.5 Templated Physisorption: Molecular Organization via Self-Assembled Inclusion Networks

1.5.1 Designing Host–Guest Networks at Surfaces

The growth of well-defined, porous, self-assembled networks, such as those described in Section 1.5.1.2, has created a powerful tool for controlling the position of molecular adsorbates. The cavities within a porous network can effectively trap molecules, usually via physisorption, including those that are not amenable to self-assembly on their own. In analogy to 3D networks, these architectures are referred to as 'host–guest' or 'inclusion' networks.

The host–guest relationship, in fact, depends on the relatively weak interaction of the guests. The formation of the host architecture and its subsequent population with guest molecules relies on a hierarchical ordering of interaction energies: the host network must be bound by a higher interaction energy than the one stabilizing the guest molecules within its pores. This can be accomplished with self-assembled networks stabilized via hydrogen bonds, dipole–dipole interaction, metal–organic complexation, etc., or by assembling large, preformed pore-containing molecules into 2D networks. The choice of guest molecule is then dictated by the caveat that the guests interact weakly with the host, at least with respect to the intramolecular interactions stabilizing the host architecture.

Within these constraints, a remarkable variety of porous meshes have been made, exhibiting an equally remarkable variety of cavity periodicities, shapes and sizes, as well as functionalization of the cavities. Since we will focus our discussion on the templating capabilities of porous architectures, we refer the reader to Section 1.3 for a more general overview of the design of self-assembled networks.

We emphasize that this breadth of architectures reflects just over 5 years of progress. The work reported by Theobald *et al.* [191] in 2003 is largely considered the seminal paper in this field. Therein, they described the formation of a two-component hydrogen-bonded network from melamine (**11**) and PTCDI (**10**). The self-assembly of the network was accomplished via successive deposition of PTCDI and melamine onto the Si(111)–Ag($\sqrt{3} \times \sqrt{3}$)R30 surface in UHV. Within the network, each melamine molecule is hydrogen-bonded to three PTCDIs, forming an extended structure of hexagonal pores with sides defined by the PTCDI molecules. The cooperative assembly of the two molecular types results in the formation of pores with dimensions much larger than either of the two molecules, shown in **Figure 20(a)**. Subsequent deposition of fullerenes demonstrated the pores to be amenable hosts to heptamers of C_{60} molecules (**Figure 18(b)**), with each heptamer exhibiting ordering not observed in the absence of the melamine-PTCDI mesh.

Although the self-assembly of porous meshes from a single component had been previously demonstrated [206,207,183,182] (including the formation of TMA networks [183,182], which will be further discussed below), the formation of the large-pored network, and their successful demonstration of organization of C_{60} within that network, inspired widespread interest in the field [2,170,172,153,197].

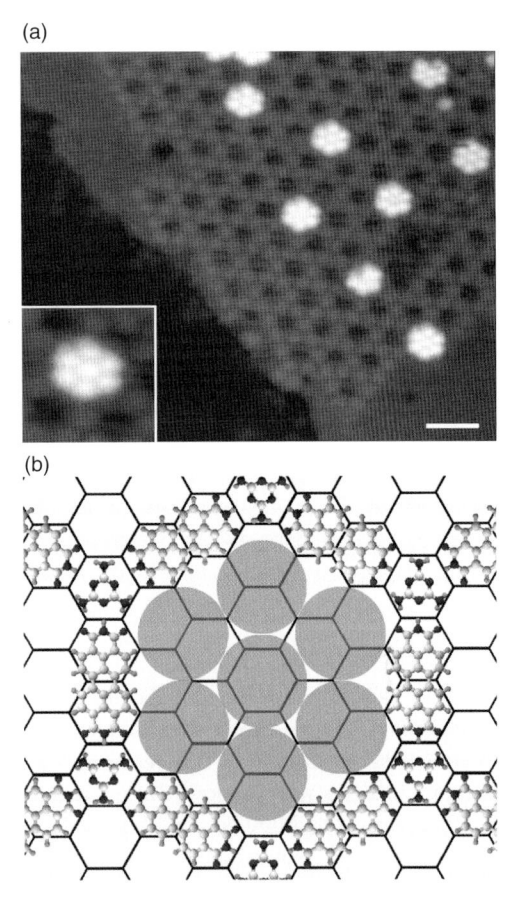

Figure 18 UHV–STM image C_{60} heptamers in a PTCDI-melamine network (a), with a schematic for the structure of a single heptamer (b). STM image parameters: $V_b = -2.0$ V, $I_t = 0.1$ nA. Scale = 5 nm. From Theobald JA, Oxtoby NS, Phillips MA, Champness NR, and Beton PH (2003) Controlling molecular deposition and layer structure with supramolecular surface assemblies. *Nature* 424: 1029–1031.

1.5.2 Patterning Arrays of Fullerenes

The fact that fullerenes [208] are probably the most often-employed guest molecules in 2D inclusion architectures is due to a number of factors. Fullerenes can possess a wealth of interesting chemical [209] and physical [210,211] properties. A fullerene can be modified through the insertion of a metal atom or molecule within its framework [212–214], or through functionalization [215,216], which can potentially lead to interesting biological applications [217]. As 3D electron acceptors, they present a variety of opportunities for fabricating novel donor–acceptor systems [218]. The potential applications of supramolecular fullerene architectures are vast and exciting [219]. On a pragmatic level, fullerenes can be easily evaporated in UHV experiments, or solvated/suspended for experiments at the solution/solid interface. Finally, the most common type of fullerene, C_{60}, has a spherical shape which allows for interaction with both the pore and the substrate, and a size ($d\sim6.80$ Å) that is of the order of the size of typical building blocks for self-assembled networks, and therefore generally matches quite well with the cavity size.

Theobald et al. first demonstrated the surface confinement of fullerenes using a self-assembled bicomponent network. The use of self-assembly to create host networks imparts an inherent versatility, since the geometry and functionality of the network can be modified in a straightforward manner by using slightly different molecules as building blocks. Such changes to the host network can introduce corresponding changes to the confinement of guest molecules. Stepanow et al. [220] who describe a set of lattices self-assembled via metal–organic coordination, provide an elegant illustration of this concept. By creating MOCNs from three molecules of different backbone length and functionality, they investigated the adsorption of fullerenes into pores of different size, as well as pores that incorporate —COOH groups. The MOCNs were formed by evaporating either 1,4-dicarboxylic benzoic acid (terephthalic acid, TPA, **18**), 1,2,4-tricarboxylic benzoic acid (trimellitic acid, TMLA, **19**) or 4,1′,4′,1″-terphenyl-1,4″-dicarboxylic acid (TDA, **20**) onto a Cu(100) surface in UHV, subsequently depositing Fe and annealing the surface for 5 min at 450 K. The shape of each MOCN is determined though interaction with the substrate and through the coordination of Fe atoms by the carboxylate groups at the ends of the molecule (the carboxylic side group of the TMLA remains uncoordinated, and extends into the cavity.) The TPA lattice is particularly sensitive to the number of Fe atoms available for coordination: at low relative Fe coverages, the ladder-shaped lattice contains regions of both rectangular (long side ~10 Å) and square (5.5 Å) pores, whereas at saturation Fe coordination, the lattice uniformly consists of larger square (7 Å) pores. **Figure 19** illustrates the adsorption preference shown by the fullerenes, which individually adsorb exclusively in the larger rectangular cavities of the ladder-shaped MOCN, as well as adsorbing within the larger square pores of the saturated MOCN. The authors postulate that the exclusion of the smaller square sites is driven by size considerations – the fullerenes are simply too large to adsorb stably. **Figure 20** shows the accommodation of fullerenes in the TMLA and TDA lattices. As was the case for the larger TPA pores, individual fullerenes adsorb in the pores of the TMLA MOCN. However, the much larger pores (long side ~20 Å) of the TDA lattice are found to host up to four fullerenes at once, although fullerene trimers, dimers, and individual molecules are more often observed.

18

19

20

By heating the fullerene-containing networks, and observing the temperature at which the fullerenes were ejected from the pores, the authors investigated the strength of the adsorption of the fullerenes in each of the MOCNs. They found that maximizing the interaction between the guest fullerenes and the Cu(100) substrate, as was the case in the larger-pored TDA network, leads to stable adsorption. C_{60} molecules are known to adsorb quite strongly on copper surfaces, with a desorption temperature of 730 K from Cu(110) [221]. The ejection temperature from the TDA network was found to be above 500 K, the temperature at which the network itself became unstable. Similarly, the ejection of fullerenes adsorbed in the TMLA network was not observed; the fullerenes adsorbed irreversibly in the network, and were not ejected even up to temperatures of ~450 K, where the host–guest MOCN was destroyed. Conversely, the fullerenes adsorbed less stably in the smaller-pored TPA networks, and were ejected at temperatures near 370 K (larger pores) or even below 300 K (small-square

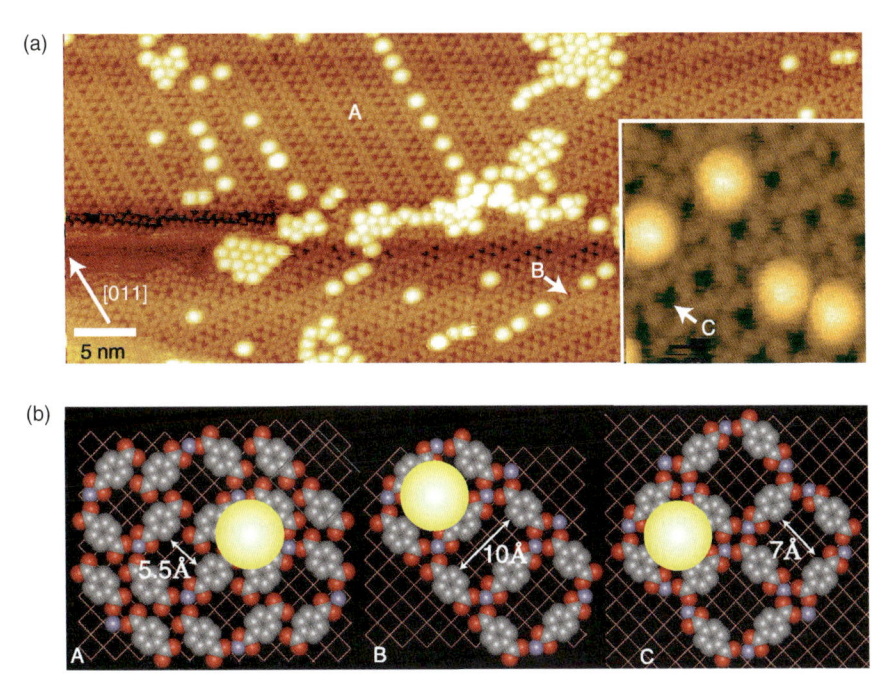

Figure 19 Single C$_{60}$ molecular guests in a MOCN. Both the type-B (ladder) and type-C Fe–TPA neworks can host single fullerene guests, as shown in (a) and (b). From Stepanow S, Lingenfelder M, Dmitriev A, et al. (2004) Steering molecular organization and host–guest interactions using two-dimensional nanoporous coordination systems. *Nature Materials* 3: 229–233.

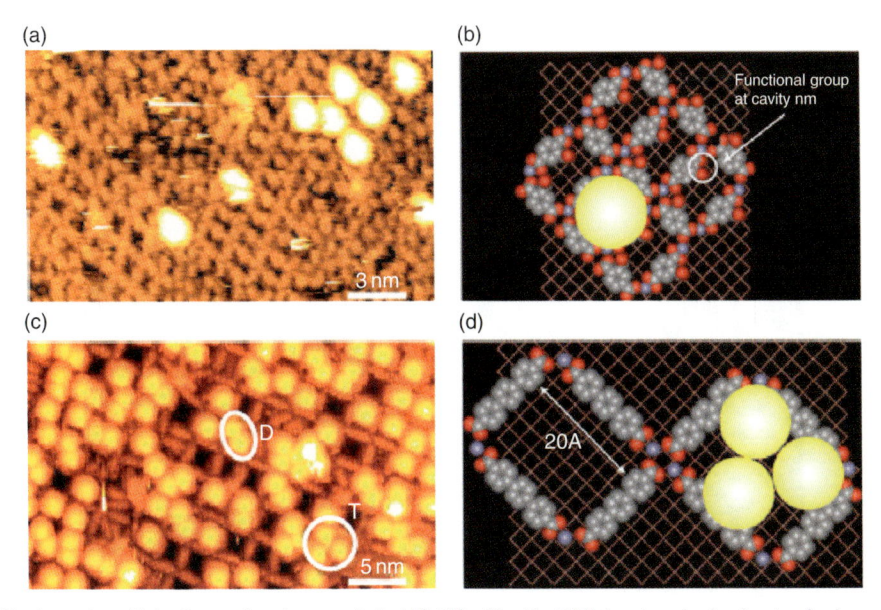

Figure 20 Single and multiple C$_{60}$ molecular guests in MOCNs. The Fe–TMLA network also hosts single guests (a–b), but the larger pores of the Fe–TDA network can host monomers, dimers (D), and trimers (T) of C$_{60}$ (cd). From Stepanow S, Lingenfelder M, Dmitriev A, et al. (2004) Steering molecular organization and host–guest interactions using two-dimensional nanoporous coordination systems. *Nature Materials* 3: 229–233.

pores.) The variability in these desorption temperatures highlights the possibility for tailoring the properties of the host–guest architecture; the stability of the network can be manipulated through the size of the cavity, the nature of the substrate, and the presence or absence of functional groups at the cavity rim.

Recently, the same type of site selectivity demonstrated for TPA–Fe–MOCNs has been demonstrated for fullerenes within a kagome network containing two distinct cavity types [222]. The cavities of the hydrogen-bonded network formed by the custom-designed tetra-acidic azobenzene molecule (NN4A, **21**) on HOPG are termed as A and B type, and have internal van der Waals diameters of 12.0 Å and 8.6 Å, respectively, as shown in **Figure 21**. The shapes of the cavities are also distinct, with A-type cavities having a hexagonal shape, and B-type cavities being roughly triangular. Introducing C_{60} fullerene molecules to the solution/solid interface, where the network forms reveals that the C_{60} will populate either the A- or B-type cavities within a domain, but not both simultaneously. The authors postulate that repulsive interaction between the fullerenes prevents the simultaneous population of A- and B-type sites.

21

The authors then show that the A-type sites can be exclusively populated by exploiting their larger size; by changing the guest molecule to the larger C_{80} fullerene ($d \sim 8.22$ Å), the theoretical C_{80}–NN4A–HOPG interaction energy becomes much higher for the A-type cavity than for the B type. Correspondingly, STM images reveal that the fullerenes reside only in the A-type cavities. Similar site selectivity is demonstrated for $Sc_3N@C_{80}$, and the authors postulate that the increased electronegativity of the metal-atom-containing fullerene leads to an increased interaction between the fullerenes and the host template.

Whereas creating a lattice of two different sizes of cavity sites can reveal a size-selected site preference [222,223], it may also be possible to induce a preference for certain sites within a lattice of identical cavities via conformational changes to the host network. In the case of a network based on the porphyrins shown in **22**, guest C_{60} fullerene molecules were observed to exhibit an attractive interaction with one another, leading to the local formation of chains and islands of fullerenes within the network, as seen in **Figure 22** [224]. Since the intercavity spacing of 3.3 ± 0.1 nm exceeds the range of van der Waals interactions [225], the authors instead surmise that conformational changes in the network, specifically rotation of the 3,5-di(tert-butyl)phenyl substituents, introduces intramolecular coupling through the network. As with any substrate-supported architecture, the influence of the

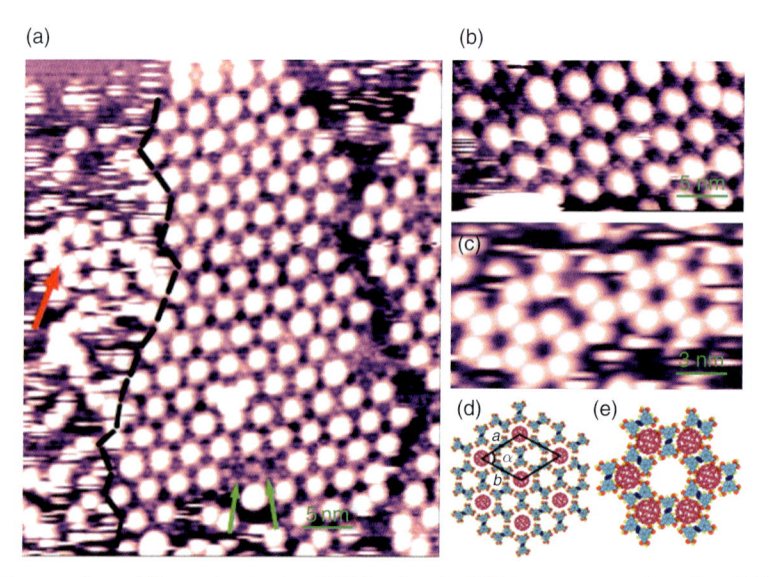

Figure 21 (a) STM image of guest C_{60} molecules in a NN4A network. STM image parameters: $V_b = 0.9$ V, $I_t = 0.12$ nA. (b, d) and (c, e) show C_{60} occupation in A- and B-type networks (see text), respectively. From Li M, Deng K, Lei SB, et al. (2008) Site-selective fabrication of two-dimensional fullerene arrays by using a supramolecular template at the liquid–solid interface. *Angewandte Chemie-International Edition* 47: 6717–6721.

underlying surface must be considered; charge transfer between the fullerenes and the underlying Ag(111) substrate may create electron standing waves in the metal, which are also known to introduce long-range interactions between adsorbates [226]. The authors postulate that both mechanisms are likely at work in this system [227].

22

Although the interaction between guest molecules can impart a degree of order to the filling subsaturation inclusion complexes, it is also possible to more directly influence the arrangement of fullerene guests through interaction with the STM tip. This phenomenon was first demonstrated by Griessl and coworkers, who used the STM tip to transfer C_{60} fullerenes to adjacent pores within the 1,3,5-benzenetricarboxylic acid (TMA) chicken wire mesh [183,182,228] formed at the solution/HOPG interface [229]. The fullerene manipulation was accomplished by temporarily increasing the STM tunneling current to ~150 pA, bringing the tip closer to the surface, and, the authors hypothesize, pushing the fullerene into an adjacent pore. The same type of manipulation has been demonstrated in UHV, where Stöhr and coworkers were able to displace C_{60} fullerene guests within a mesh formed from dehydrogenated 4,9-diaminoperylene-quinone-3,10-diimine (DPDI, **equation 1**) [230–232]:

DPDI   Dehydro-DPDI   (1)

From Stohr M, Wahl M, Galka CH, Riehm T, Jung TA, and Gade LH (2005) Controlling molecular assembly in two dimensions: The concentration dependence of thermally induced 2D aggregation of molecules on a metal surface. *Angewandte Chemie International Edition* 44: 7394–7398.

As shown in **Figure 23**, they were able to first move a fullerene molecule into a pore that already contained a guest zinc octaethylporphyrin (ZnOEP, **23**) molecule, creating a sort of molecular ball bearing, and subsequently into an empty pore. These short displacements were accomplished by moving the tip toward the surface, however, as opposed to the 'brushing' action cited by Griessl *et al.*, Stöhr reports inciting the fullerene displacement by approaching the tip 0.4 nm toward the pore into which the fullerene was to move. The fullerene would then move from a nearest-neighbor pore into the desired final location. Stöhr and coworkers describe an additional technique for fullerene manipulation, also shown in **Figure 23**, which facilitates the displacement of fullerenes across longer distances. By approaching the STM tip 0.4 nm closer to a pore occupied by a fullerene, they are able to pick up the molecule, and can then move it to an arbitrary pore simply by positioning the tip over the pore and approaching the STM tip 0.4 nm closer to the surface. In addition to showing the initial and final positions of the fullerene molecule, **Figure 23**

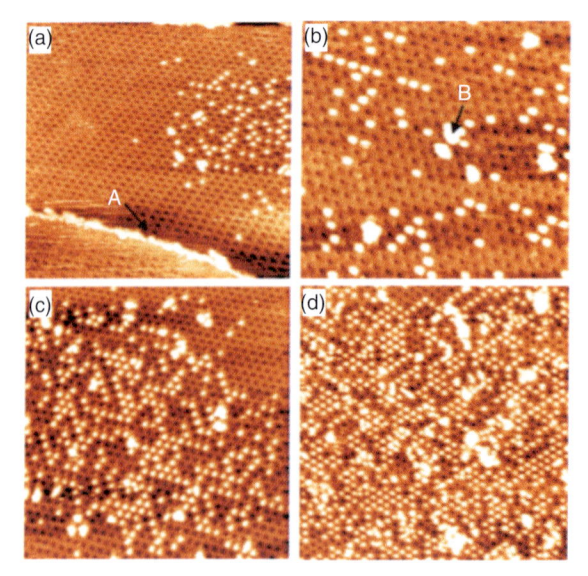

Figure 22 Distribution of guest C_{60} molecules in a porphyrin network at various coverages. Hallmarks of an attractive interaction between the C_{60} molecules can be observed, including chains and islands of C_{60}, as well as clusters (B) formed at defect sites. Coverage of 0.02 ML (a,b); (c) 0.07 ML; and (d) 0.1 ML. STM image parameters: $V_b = 3.0$ V, $I_t = 10$ pA, image area $= 93 \times 88$ nm^2 (a), $V_b = 2.9$ V, $I_t = 11$ pA, image area $= 68 \times 64$ nm^2 (b), $V_b = 3.0$ V, $I_t = 9$ pA, image area $= 100 \times 100$ nm^2 (c), $V_b = 3.0$ V, $I_t = 13$ pA, image area $= 111 \times 112$ nm^2 (d). From Spillmann H, Kiebele A, Stohr M, et al. (2006) A two-dimensional porphyrin-based porous network featuring communicating cavities for the templated complexation of fullerenes. *Advanced Materials* 18: 275–279.

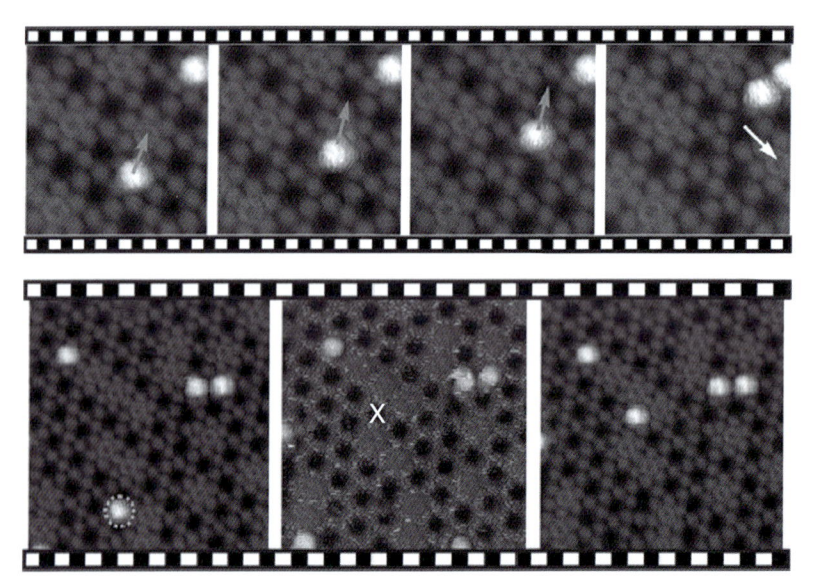

Figure 23 Using the STM tip to manipulate C_{60} guests. Top: a C_{60} is moved onto a ZnOEP guest molecule, and then into an empty pore. The gray arrows indicate the direction of movement of the C_{60}. Following the removal of the C_{60} the ZnOEP moves to an adjacent pore, indicated by the white arrow. Bottom: the C_{60} molecule indicated by the dashed circle is picked up, and set down in the pore marked by the X in the second frame. The second frame appears fuzzy because it was obtained with the C_{60} molecule adsorbed on the tip. STM image parameters: $V_b = 0.5$ V, $I_t = 90$ pA, image area $= 10 \times 10$ nm^2 (top), $V_b = 0.5$ V, $I_t = 90$ pA, image area $= 20 \times 20$ nm^2 (bottom). From Stohr M, Wahl M, Spillmann H, Gade LH, and Jung TA (2007) Lateral manipulation for the positioning of molecular guests within the confinements of a highly stable self-assembled organic surface network. *Small* 3: 1336–1340.

additionally shows an image collected with the fullerene adsorbed on the tip, which is sufficiently different from normal images so as to substantiate the mechanism for this type of 'vertical manipulation'.

23

Finally, we highlight here the possibilities for the formation of host–fullerene–donor–acceptor complexes within 2D inclusion architectures. In particular, fullerenes have been shown to form donor–acceptor complexes within networks formed from shape-persistent oligothiophene macrocycles [233,234]. The oligothiophene–fullerene system has generated technological interest for applications in solar conversion and molecular electronic devices [235].

When fullerenes are introduced to a network of self-assembled shape-persistent oligothiophene macrocycles, the donor–acceptor interaction is dominant in stabilizing the fullerenes [233,234]. For example, when macrocycle **24** is drop-cast onto HOPG, it forms a network that contains not only the inherent cavities of the macrocycles ($d\sim$1.4 nm, labeled as A in **Figure 24**), but additional elongated cavities due to the packing of the macrocycles on the surface ($\sim$2.5 nm $\times$ 1.8 nm, labeled as B in **Figure 24**). Yet, fullerene molecules added to the network are found to inhabit neither of these cavities; instead, they adsorb on top of the bithiophene units within the macrocycle backbone, maximizing the thiophene–fullerene interaction.

The localized charge transfer between the oligothiophene and the fullerene can have long-range implications for the ordering of the host–guest architecture, as is illustrated in the case of fullerene

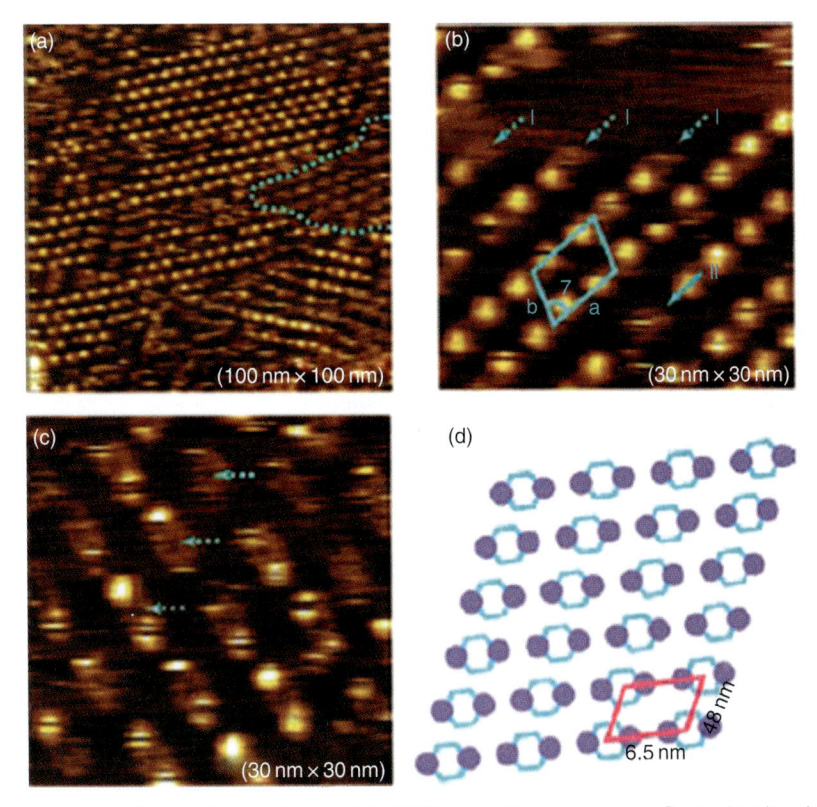

Figure 24 C_{60} guests in a self-assembled network of 21 STM images (a–c) reveal two C_{60} molecules sitting on each macrocycle; (d) shows a structural model. STM image parameters: $V_b = 0.8$ V, $I_t = 0.3$ nA (a), $V_b = 1.0$ V, $I_t = 0.3$ nA (b), $V_b = 1.0$ V, $I_t = 0.3$ nA (c). From Pan GB, Cheng XH, Hoger S, and Freyland W (2006) 2D supramolecular structures of a shape-persistent macrocycle and co-deposition with fullerene on HOPG. *Journal of the American Chemical Society* 128: 4218–4219.

inclusion within a cyclo [12]thiophene (c[12]T) network [233]. The circular c[12]T molecule self-assembles on HOPG from a solution of 1,2,4-trichlorobenzene [236], forming a van der Waals stabilized network of pores with an inner diameter of 1.2 nm, surrounded by a sixfold symmetric ring of thiophenes. Deposition of C_{60} fullerene molecules onto this network reveals two distinct adsorption locations: the fullerenes are either confined within a cavity (C type), or on the thiophene rim surrounding the pore (R type). In STM images, the C-type molecules appear as broad (d~1.6 nm) bright protrusions occupying the entire cavity, whereas the R-type molecules appear as distinct, bright spots with a smaller diameter (d~1 nm). The more localized appearance of the R-type molecules is due to the relatively strong π–π coupling between the fullerene and the thiophene of the rim, which also prevents the R-type molecules from being displaced during scanning by the STM tip (unlike the C-type molecules, which can be removed from the cavity during normal scanning.) At low fullerene coverages, both C-type sites and any of the six equivalent R-type sites are populated stochastically. The ordering implications of the fullerene–thiophene complexation in the R-type adsorption becomes evident at coverages approaching a 1:1 macrocycle:fullerene ratio, where the fullerenes adsorb exclusively in the same R-type position on each macrocycle. This behavior is described in terms of the dipole associated with the formation of the donor–acceptor complex, which creates an electron deficiency in the molecule away from the adsorbed fullerene, preventing the adsorption of more than one electron-accepting fullerene on each macrocycle. Further, the close proximity of each c[12]T to its nearest neighbors (~0.4 nm between macrocycle backbones) means that this dipole induces a quadrupole in the surrounding macrocycles, introducing an affinity for fullerene adsorption on the same site on each c[12]T molecule.

1.5.3 Patterning Other Molecules

We now briefly focus on interesting results from 2D inclusion architectures comprising nonfullerene guests. As in the case of fullerenes [229], the first architecture used to stabilize nonfullerene guests at the solution/HOPG interface was the TMA chicken wire mesh [237]. Griessl and coworkers who showed that coronene molecules could be inserted into the TMA network: coronene (d ~1 nm, **25**), is a planar molecule consisting of seven interconnected, hydrogen-saturated benzene rings. Although it has been shown to self-assemble into ordered overlayers on HOPG under UHV conditions [238,239], the authors report that coronene did not self-assemble at the heptanoic acid/HOPG interface in the absence of TMA. By adding a droplet of coronene/heptanoic acid solution to a preexisting TMA mesh, the authors were able to populate the mesh with coronene guests. STM images (e.g., **Figure 25**) reveal that the coronene–TMA lattice exhibits a periodic contrast modulation. The hexagonal superstructure has a periodicity of ∼5 nm, and manifests via a modulation in both the contrast and the appearance of the coronene molecules. Where the molecules appear bright, they also appear to have six discrete lobes at their periphery, consistent with the six benzene rings at the exterior of the molecule. Conversely, the darker molecules lack discrete lobes, and instead appear as two concentric rings, consistent with first-principles calculations of the molecular frontier orbitals [238]. The authors suggest that this effect arises from the incommensurability of the TMA and HOPG lattices, which creates a moiré-type modulation in the coronene–HOPG interaction. The brighter molecules (indicated as 1 in **Figure 25**) correspond to sites where the coronene–HOPG interaction is quite strong, stabilizing the molecule close to the HOPG surface. The darker molecules (indicated as 2 in **Figure 25**) occupy sites where the coronene–HOPG interaction is weaker, leading to a reduced apparent contrast and allowing for the rotation of the coronene within the cavity.

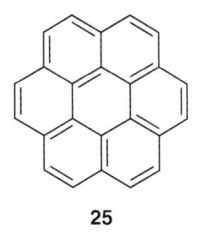

25

The authors liken the rotating coronene to a molecular bearing, and point out that the ability to remove or manipulate the guest species might be an important aspect of such a molecular widget. As was the case with C_{60}, the guest molecules (both type 1 and type 2), coronene can be removed from a cavity by interaction with the STM tip (**Figure 25**). However, the near-saturation of the TMA with coronene shown here means that the removed guest molecules are removed completely and not transitioned to a neighboring pore, since the availability of vacant pores is low. This demonstration raises the interesting possibility of using the STM tip to create a lattice of exclusively rotating (or, alternatively, fixed) coronene molecules, although the difficulty of removing a coronene molecule to solution will doubtless increase as the number of nearby vacant pores increases.

Stabilization within a host network can also provide insight into the intermolecular interactions between guest molecules. Recently, it was shown that the heterocyclic circulenes, octathio[8]circulene ('sulflower', **26**) [240] and its selenium analog tetraselenotetrathio[8]circulene ('selenosulflower', **27**), can be placed as guest molecules in the porous

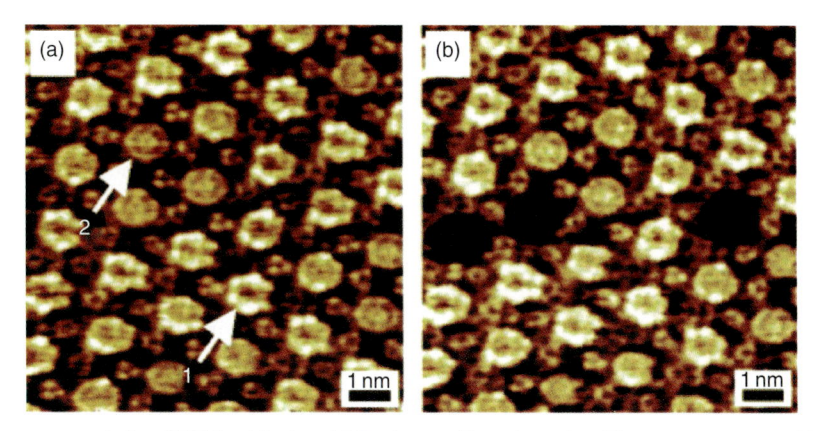

Figure 25 Coronene guests in a TMA host lattice. (a) Stationary (1) and rotating (2) coronene guests; (b) structure after several guest molecules are removed. STM image parameters: $V_b = 0.8$ V, $I_t = 147$ pA. From Griessl SJH, Lackinger M, Jamitzky F, Markert T, Hietschold M, and Heckl WA (2004) Incorporation and manipulation of coronene in an organic template structure. *Langmuir* 20: 9403–9407.

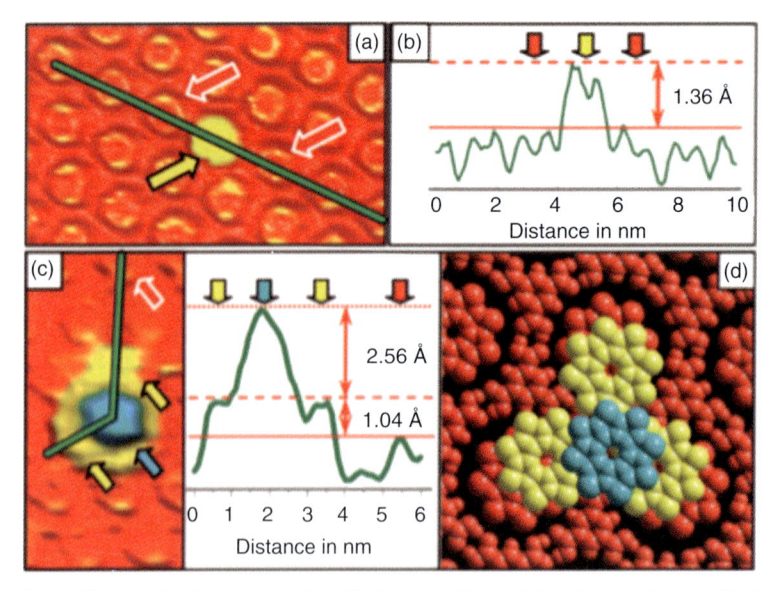

Figure 26 Aggregations of heterocirculene guests in a TMA host lattice. (a) A sulflower dimer, with the corresponding line profile in (b); (c) a pyramidal sulflower island and its line profile. A model for the pyramidal island is shown in (d). STM image parameters: $V_b = 1.2$ V, $I_t = 130$ pA, image area = 6×10 nm^2 (a), $V_b = 1.2$ V, $I_t = 130$ pA, image area = 4.1×8.0 nm^2 (c). From Ivasenko O, MacLeod JM, Chernichenko KY, et al. (2009) Supramolecular assembly of heterocirculenes in 2D and 3D. *Chemical Communications* 10: 1121–1280.

TMA network at the solution/solid interface [241]. Following the saturation of the TMA mesh with either 'sulflower' or 'selenosulflower', the formation of π–π-stacked cirulene dimers was observed, including pyramidal complexes with up to three molecular layers (see **Figure 26**). Interestingly, the circulenes also exhibit an affinity for the underlying HOPG surface despite not self-assembling in the absence of TMA: the TMA polymorph is driven toward the higher porosity chicken-wire structure, even in the presence of solvents that favor the flower phase. The authors interpret this as a maximization of circulene–substrate interaction, since the areal density of cavities in the chicken-wire structure (39%) is higher than that of the flower structure (28%). A similar phenomenon had also been reported for coronene guest molecules, which were shown to favor the formation of a porous kagome network, rather than the denser networks formed in their absence [242].

26

27

Finally, we note that nonplanar molecules are also amenable guests, as demonstrated in a recent work detailing the adsorption of biomolecules within a TMA–Fe–MOCN in UHV [243]. The TMA–Fe network formed on Cu(100) presents an array of x-shaped cavities that are ~1 nm wide [244]. Stepanow and co-workers populated these cavities with LL-cystine, an amino acid dimer formed from two cytosine molecules, and also with LL-diphenylalanine (Phe-Phe), an unnatural amino acid. Following deposition at room temperature, two cystine molecules are found to adsorb per cavity, each oriented with its long axis perpendicular to the surface (standing upright.) Although spectroscopic information is not available to unequivocally establish the interaction between the molecules and the substrate, the authors hypothesize that a carboxylate–substrate interaction stabilizes the molecules. Annealing the host–guest network to 430 K liberates a single cystine from each pore; this single-occupancy network is stable to 490 K, when the remaining cystine molecules are ejected. When a single cystine molecule occupies a pore, it occupies the center of the pore, implying a repulsive interaction between the cystine and the cavity. If Phe-Phe molecules are deposited onto the MOCN, STM images reveal bright, irregular protrusions within the cavities. The authors attribute this appearance to an interaction between the STM tip and the protruding end of the 1.3 nm-long Phe-Phe molecules, which is likely adsorbed into the cavity at one terminus. The Phe-Phe molecules are ejected from the cavities after annealing to 450 K, indicating that they are slightly less stably adsorbed than the single cystine molecules.

1.6 Covalently Bonded Structures: Surface-Confined Polymerization

As documented in Sections 1.5 and 1.6, molecular self-assembly and hierarchical ordering can be used to create surface assemblies with highly adaptable geometries and functionalities. The degree of perfection of these assemblies can theoretically be very high, since the intermolecular bonding within the self-assembled structures is reversible, and defects can be self-corrected. Conversely, the formation of nonreversible covalent bonds between molecules may have negative implications for the perfection of the architecture since defects are effectively locked in by the nonreversible bonding, but has the benefit of imparting mechanical and thermal and electronic robustness. In addition to these attractive physical attributes, covalent bonding can confer useful electronic characteristics through the formation of conjugated (π-electron) polymeric systems, particularly when conjugation extends in two dimensions [245]. Although the best route to producing extended 2D polymer sheets is not clear [246], significant work has already been undertaken in using surface science approaches to create 1D and 2D polymers [245,246,247].

One of the challenges of studying polymerization at a surface lies in demonstrating that polymerization has, in fact, been achieved. Often, this can be directly observed in STM or AFM images as a change of periodicity in the assembled monomers, or as a change in the contrast due to a new geometric or electronic configuration. One of more tangible forms of confirmation can be demonstrated by using the SPM to manipulate the supposed polymer, since covalently bonded structures are sufficiently robust to withstand almost all manner of pushing and prodding from a probe tip.

A nice example of this type of confirmation by manipulation is provided by Barner *et al.* [248]. After spin-coating a solution of polymers onto the HOPG surface, they used the probe tip of an AFM to manipulate two separate dendronized polymers into an X shape, shown in **Figures 27(a)** and **27(b)**. By subsequently irradiating the surface with UV light (**Figure 27(c)**), they were able to fuse the two polymers together into a single aggregate. The strength of the bonding between the polymers is demonstrated by the 'mechanical challenge' issued by aggressive AFM tip manipulation, shown in **Figures 27(d)–27(f)**), and allows the authors to interpret the bonding between the polymers as covalent.

In this section, we will provide an overview of experiments in which individual monomers have been assembled onto a surface, and subsequently polymerized by using various methods. We have divided the section according to the dimensionality of the polymers, first looking at 1D polymers (formed from monomers with two reactive sites), and then at some recent results in 2D polymerization (monomers with three or more reactive sites).

1.6.1 Polymer Lines

1.6.1.1 *UV light and STM tip-induced polymerization of diacetylenes*

In 1997, Grim *et al.* published an account of the UV light-induced polymerization within a molecular network of a diacetylene-containing isophthalic acid (ISA) derivative (**28**) at the solution/HOPG interface [249]. The polymerization of diacetylene compounds (i.e., the conversion of n(R-C≡C—C≡C-R′) to (R-C-C≡C-C-R′)$_n$, where R and R′ are substituent groups) had been previously reported in bulk structures [250,251], for Langmuir–Blodgett films [252] and for self-assembled monolayers on gold [253], however, this was the first instance where diacetylene molecules were self-assembled into a single-layer network on HOPG, and where their pre and postpolymerization structure was directly observed with an STM. The images obtained by Grim *et al.* clearly show the structure and periodicity of the hydrogen-bonded network formed by the ISA derivative, which forms interdigitated lamellae of the alkyl chains and diacetylene moities interspersed with lamellae of the coadsorbed

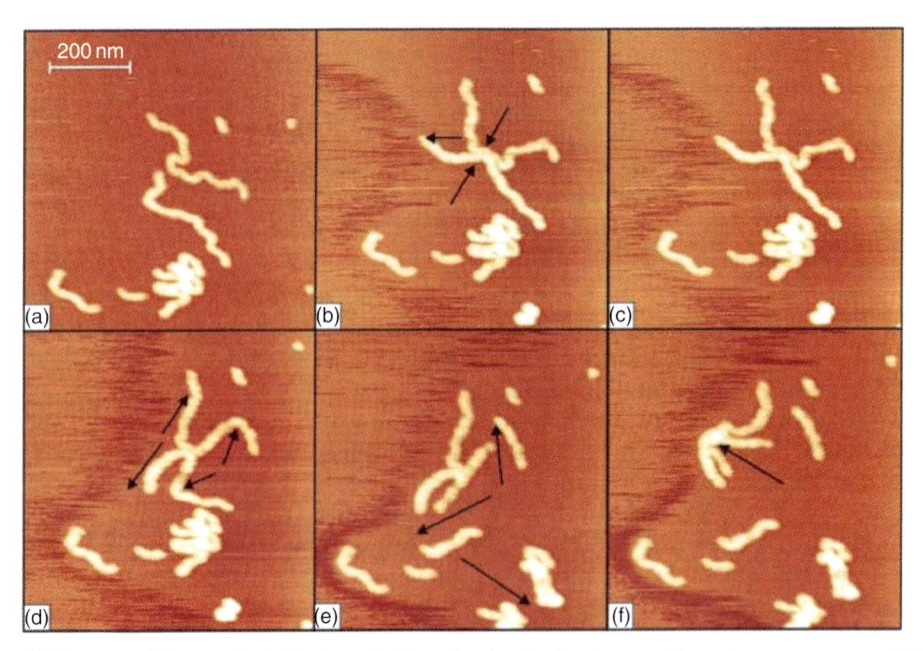

Figure 27 AFM images of the covalent attachment of two dendronized polymers. The polymers are moved into contact with one another (a, b) with the AFM tip, exposed to UV light (c), and subsequently tested (d–f) to demonstrate the strength of the bond between the polymers. The arrows show the direction of AFM tip manipulation. From Barner J, Mallwitz F, Shu LJ, Schluter AD, and Rabe JP (2003) Covalent connection of two individual polymer chains on a surface: an elementary step toward molecular nanoconstructions. *Angewandte Chemie-International Edition* 42: 1932–1935.

1-undecanol solvent molecules. Prior to polymerization, the diacetylene units are clearly discernible, spaced by 4.72 Å, and oriented at 50° to their stacking axis. These parameters are critical, as the diacetylene polymerization process is topochemical [250] (i.e., it depends on the orientation of the molecules), and known to occur for diacetylene spacings of ~5 Å and orientations of ~45° with respect to their stacking axis [254]. Following UV irradiation, the STM images reveal a different structure, where a bright, contiguous feature identified as the polydiacetylene (PDA) backbone has replaced the individual diacetylene units. The backbone periodicity of 4.91 Å further strengthens the polymerization interpretation, since this value matches crystal structure data for polymerized diacetylenes well [254].

28

The major advantage of using STM to study this type of system was revealed a few years later, when Okawa and Aono demonstrated that the nucleation of PDA within an self-assembled molecular network (SAMN) could be initiated with the STM tip [255, 256]. Okawa and Aono worked with networks of 10,12-nonacosadiynoic acid [255] (**29**) or 10,12-pentacosadiynoic acid [256] (**30**) on HOPG, which, like the ISA derivative studied by Grim *et al.*, organizes into interdigitated lamellae with the diacetylene units arranged parallel to one another, spaced by 0.47 nm [256], and angled near 45° with respect to their packing axis (shown schematically in **Figure 28(a)**). In analogy to the three-dimensional (3D) situation [257], they proposed that the photopolymerization of diacetylene compounds in films progressed according to the scheme shown in **Figure 28**, where a diacetylene moiety is excited into a diradical via photabsorption (**Figure 28(b)**). The unpaired electrons at either end of the diradical are highly reactive, and thermal excitations of a neighboring diacetylene moiety may bring it close enough to foster the formation of a radical dimer (**Figures 28(c) and 28(d)**). The persistence of unpaired electrons at the ends of the polymerized chain results in chain-reaction progression of the polymerization, and the PDA backbone thus grows out more or less symmetrically from the nucleation point. Okawa and Aono showed that a voltage pulse with a magnitude greater than ~3 V from the STM tip could be used to create the initial diradical, allowing for spatial control over the nucleation site of the PDA line.

Using the STM tip to disrupt the SAMN at a specific location, effectively predefining an interrupt for the polymerization chain reaction, can allow for even further control of the spatial localization of the PDA lines. **Figure 29** illustrates this scenario, where a 6 nm hole acts as the terminus for three successively created PDA lines. The 'control' offered by this technique is not complete however, since the pulse used to introduce the artificial defect (application of a voltage pulse of 3–8 V for 1–100 ms with the sample positive relative to the tip) [258] could also potentially induce the polymerization chain reaction.

According to the authors [259], the polymerization reaction occurs with a probability increasing with voltage for pulses over 3 V. Subsequent work by Sullivan *et al.* correlates an increase in polymerization probability with increased pulse length (in the range from milliseconds to microseconds) for a fixed pulse height of 6 V (sample negative) over networks of 10,12-tricosadiynoic acid (**31**) at the solution/HOPG interface [259], so it is conceivable that a regime can be identified where voltage pulses can reliably disrupt the molecular layer without routinely causing polymerization.

Sullivan *et al.* [259] have further shown that the wires can also be constrained spatially through the use of pits in the HOPG as 'molecular corrals'. Using either ion bombardment or heat treatment, the authors created pits in the HOPG substrate which defined spatially (height) isolated domains of the 10,12-tricosadiynoic acid network. Following tip-induced polymerization, PDA lines were observed to terminate at both the step-up and step-down boundaries between the pits and the substrate. Since the lamellae were not observed to traverse these steps, Sullivan *et al.* suggest that the topochemical condition for polymerization is interrupted at the pit edges. Takajo *et al.* make a similar observation of PDA chain polymerization spatially limited by the edges of the topmost molecular layer in multilayer films of 10,12-nonacosadiynoic acid and 10,12-pentacosadiynoic acid at the solution/HOPG interface, where they observed a single PDA line extending to a submicron length across a large, ripened molecular domain [260].

The polymerization reaction and the stability of the polymers may be somewhat less predictable at

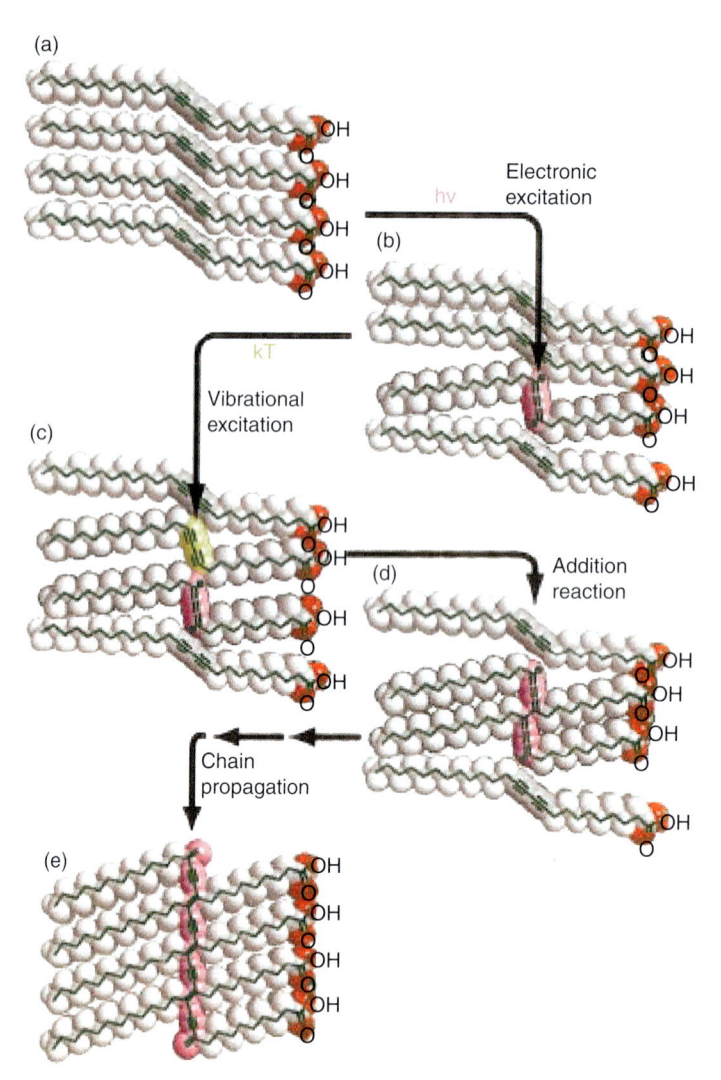

Figure 28 Schematic illustration of the photopolymerization of diacetylene moieties in a SAMN. From Okawa Y and Aono M (2001) Linear chain polymerization initiated by a scanning tunneling microscope tip at designated positions. *Journal of Chemical Physics* 115: 2317–2322.

the solution/solid interface than in ambient or vacuum. In particular, more than one polymerization event may be associated with a single voltage pulse [261,259], an effect that may be attributable to the excitation of solvent molecules during voltage pulsing and their subsequent interaction with the molecular layer [261]. Additionally, the dynamic exchange of molecules between solution and surface can also result in noninfinite residency times for polymers at the surface [259]. Neither of these traits is particularly copasetic with the use of PDA lines in controlled applications, such as for interconnects in nanoscale electronics devices.

The elegant control and potential utility of this one-dimensional (1D) system provoke questions about its possible extension to two dimensions. However, the topochemical nature of the polymerization, and the use of long alkyl chains around the diacetylene moiety to stabilize the interaction with the substrate HOPG make it difficult to conceive of a method of fabricating intersecting PDA lines. To date, the most promising approach to integrating PDA lines into a 2D network involves designing the substituent monomers such that the PDA lines are cross-coupled via a covalent link. Miura and coworkers fabricated this type of cross-coupled PDA lines from monomers containing two diacetylene moieties, shown in polymerized form in **Figure 30(c)**. Adjacent PDA lines are cross-coupled via the alkyl chain/terephthalic acid/alkyl chain

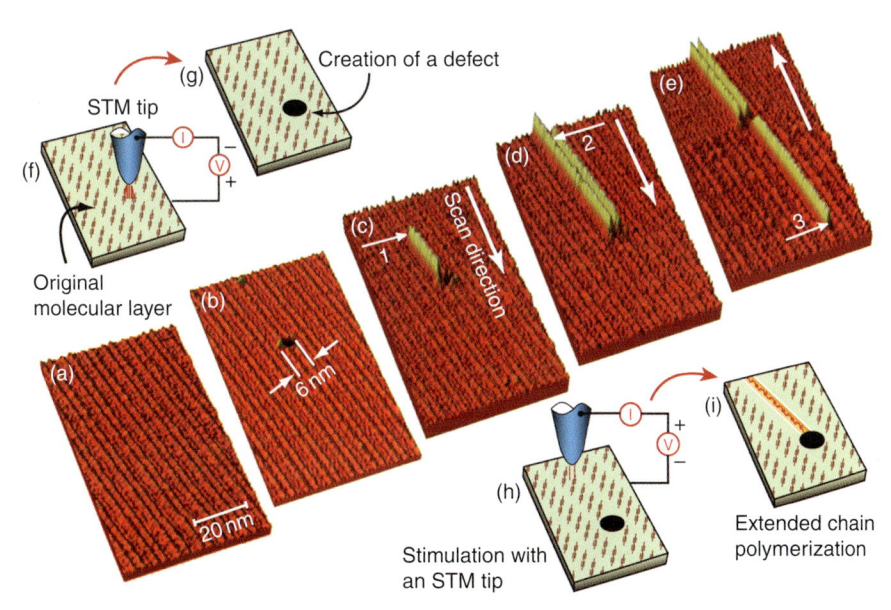

Figure 29 STM images (a–e) and schematic illustration (f–i) of STM tip-induced polymerization in a molecular layer of 10,12-pentacosadiynoic acid. Following the formation of an artificial defect (b), a voltage pulse is used to nucleate the polymerization chain reaction at points 1 (c), 2 (d), and 3 (d). From Okawa Y and Aono M (2001) Nanoscale control of chain polymerization. *Nature* 409: 683–684.

linkage at the center of the molecule. **Figures 30(a)** and **30(b)**) shows images taken before and after using the STM tip to initiate polymerization such that the each of the two diacetylene moieties from within a single lamellae are converted to PDA. This approach presents a neat solution to the problem of creating a 2D-linked sheet of PDA lines.

Takami *et al.* [263] who described the photopolymerization of the molecular network shown in **Figure 31**, had previously explored this approach. Following UV irradiation of the vapor-deposited self-assembled network formed by 1,15,17,31-dotriacontatetrayne (DTTY), the authors used Penning ionization electron spectroscopy (PIES) to confirm that the π molecular orbitals of the acetylene and diacetylene units had been replaced with the signature of the conjugated polyacetylene and PDA. Accompanying STM images indicate that the unit cell of the irradiated network is consistent with expectations for polymerization, although the complete polymerization of the acetylene and diacetylene units cannot be easily implied from the images.

1.6.1.2 Electrochemical formation of polythiophenes

Sakaguchi and coworkers have demonstrated the linear polymerization of thiophenes on Au(111) via a process that they call electrochemical epitaxial polymerization (ECEP) [264, 265]. As the name implies, ECEP creates polythiophenes that grow in registry with the substrate, iodine-terminated Au(111). **Figure 32(a)** illustrates the experimental procedure: 150 ms, 1.4 V pulses, spaced by 1 s, are applied to the Au(111) working electrode in an electrochemical cell containing a solution of iodine and the thipophene monomer (3-butoxy-4-methylthiophene, or BuOMT, **32**) [264]. Sakaguchi *et al.* proposed that trimers of BuOMT form in solution due to oxidation by iodine, and that these trimers act as nucleation centers for the polymer lines, which subsequently grow epitaxially via the addition of cation radical of the monomer on the iodine-covered gold surface [264, 265]. The STM images in **Figures 32(b)–32(d)** show the surface after 5, 13, and 15 V pulses. The epitaxial nature of the growth is implied by the growth directions of the polymer lines, which follow the threefold symmetry of the underlying Au(111)-I surface. Histograms of the polymer lengths for each of these images are shown in **Figures 32(f)–32(h)**, and indicate growth up to 70 nm. The length of the polymers appears to be limited by competition between the three growth directions; the longest lines can be seen to terminate at nonparallel lines in the high-coverage regime of **Figure 32(d)**.

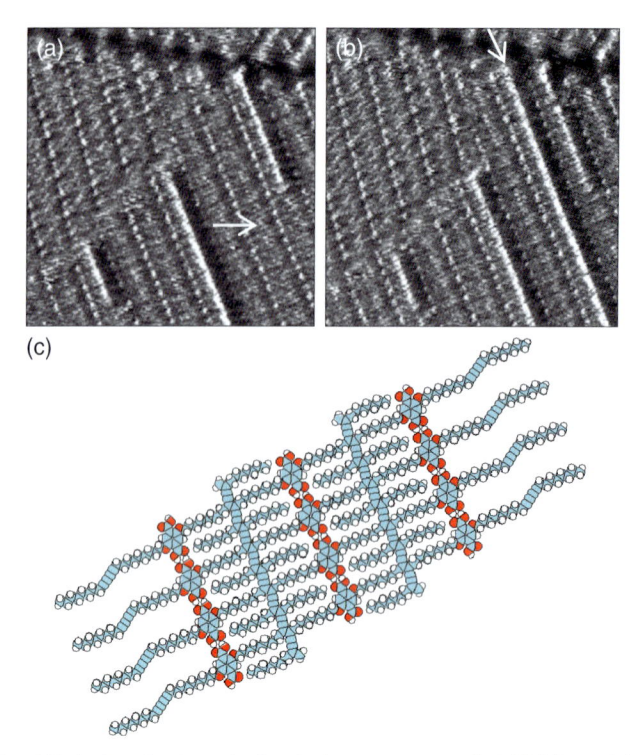

Figure 30 Combined UV and tip-induced polymerization to form cross-coupled PDA lines. (a) shows the surface after UV irradiation and (b) shows the line formed via a voltage pulse at the position indicated in (a). A schematic of the polymerized layer is shown in (c). STM image parameters: $V_b = -0.40$ V, $I_t = 0.4$ nA, image area $= 27.1 \times 27.1$ nm^2 (a, b). From Miura A, De Feyter S, Abdel-Mottaleb MMS, et al. (2003) Light- and STM-tip-induced formation of one-dimensional and two-dimensional organic nanostructures. *Langmuir* 19: 6474–6482.

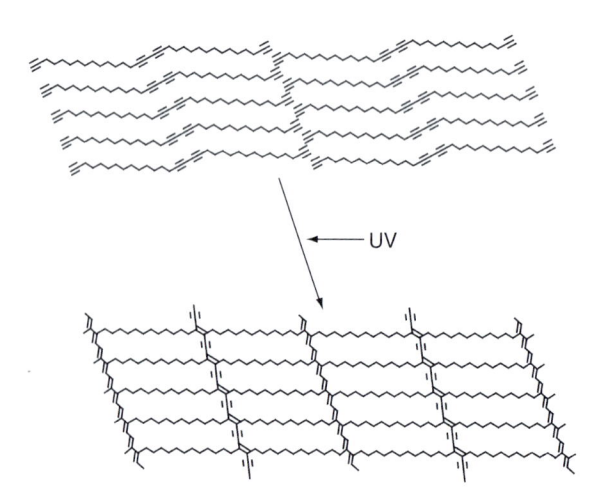

Figure 31 Schematic illustration of the UV-induced polymerization of 1,15,17,31-dotriacontatetrayne into alternating polydiacetylene and polyacetylene lines. From Takami T, Ozaki H, Kasuga M, et al. (1997) Periodic structure of a single sheet of a clothlike macromolecule (atomic cloth) studied by scanning tunneling microscopy. *Angewandte Chemie-International Edition* 36: 2755–2757.

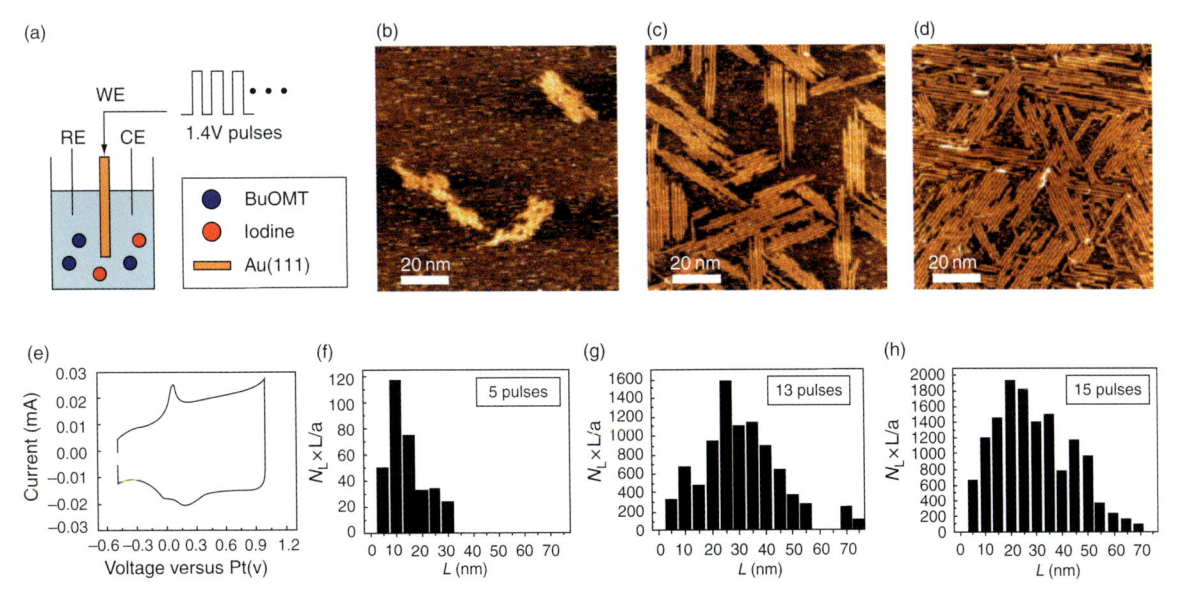

Figure 32 Polythiophene formation through electrochemical epitaxial polymerization on Au(111). (a) A schematic diagram of the experiment. (b–d) show STM images of the surface after 5, 13, and 15 voltage pulses, respectively, and (f–h) show the corresponding histograms of wire length. (e) A cyclic voltammogram obtained from a sample after 30 voltage pulses. From Sakaguchi H, Matsumura H, and Gong H (2004) Electrochemical epitaxial polymerization of single-molecular wires. *Nature Materials* 3: 551–557.

The authors have demonstrated the formation of polythiophenes by ECEP for a number of different monomers. They have also shown that the technique can be extended to the creation of di-, tri-, and multiblock polymers, which comprise discrete segments of two different polymers. The formation of these structures can be facilitated by two different techniques. In the first technique, the gold working electrode is moved between two different electrochemical cells, each containing a solution of a single species of thiophene monomer. In each cell, the appropriate voltage pulses for polymer growth are applied. The result of successive application of this technique can be seen in the STM images shown in **Figures 33(a)–33(c)**, where alternating sections of C8OMT (3-octyloxy-4-methylthiophene, shown in **equation 2**, which appears bright and solid in images) and C8MT (3-octyl-4-methylthiophene, also shown in **equation 3**, which appears as distinct round features due to periodic torsion of the thiophene units) can be distinguished:

In the second technique, illustrated schematically in **Figure 33(d)**, two monomer species are present in a single solution in the electrochemical cell. Applying voltage pulses in the usual manner results in the growth of polymers comprising alternating blocks of the two polymer types, as seen in the STM image in **Figure 33(e)**. Sakaguchi *et al.* attribute the formation of polymer blocks, rather than homogenously mixed polymers, to the affinity of each monomer type for bonding to itself.

1.6.1.3 Addition polymerization of carbenes

Polymerization at surfaces can also be facilitated through the selection of a monomer amenable to addition reaction, as demonstrated by Matena *et al.* [266]. Their density functional theory calculations show that the tautomerization and subsequent dimerization of two 1,3,8,10-tetraazaperopyrene (TAPP) molecules, shown in **equation 4**, is nearly thermoneutral:

$$+53.1 \text{ kcal mol}^{-1} \quad +53.4 \text{ kcal mol}^{-1}$$

$$2\times \quad -106.4 \text{ kcal mol}^{-1} \tag{4}$$

From Matena M, Riehm T, Stohr M, Jung TA, and Gade LH (2008) Transforming surface coordination polymers into covalent surface polymers: Linked polycondensed aromatics through oligomerization of *N*-heterocyclic carbene intermediates. *Angewandte Chemie International Edition* 47: 2414–2417.

Furthermore, the calculations predict that the formation of trimers should be exothermic by approximately 18.6 kcal mol^{-1}, and that larger oligomers should exhibit increasing stability with length. Experiments on Cu(111) revealed that TAPP forms a coordination complex with free Cu adatoms when deposited at 150 °C, but that heating this network to 250 °C fosters the intended polymerization reaction. **Figure 34** includes two STM images of the TAPP chains on the copper surface; measurements made from images like these confirm that the periodicity of these surface-confined TAPP chains (1.23 ± 0.12 nm) is in agreement with the gas-phase value predicted by density functional theory (1.27 nm). Further confirmation of the polymerization is given by the authors' ability to manipulate the surface chains with the STM tip, and by the difference in the N 1s XPS spectra between the coordination complex, which contains only equivalent N atoms, and therefore a single N 1s peak, and the polymer, which has an additional peak attributed to the N—H bonds:

It is also interesting to note that the authors have observed a hybrid structure formed by the Cu coordination of two parallel polymer chains, indicating that it may be possible to couple adjacent chains to form suprapolymeric architectures.

1.6.1.4 Polyphenylene lines via Ulmann dehalogenation

In the 1D polymer structures described thus far, the presence of the surface, although imperative to confining the molecules in proximity to one another in 2D, has been more or less incidental to the process of polymerization (the exception here is the ECEP polymerization of thiophene described by Sakaguchi *et al.*, which seems to require the presence of the iodine-covered gold substrate) [264, 265]. Lipton-Duffin and coworkers have recently described the formation of polyphenylenes via a dehalogenation reaction that makes use of the underlying Cu(110) surface to catalyze the polymerization.

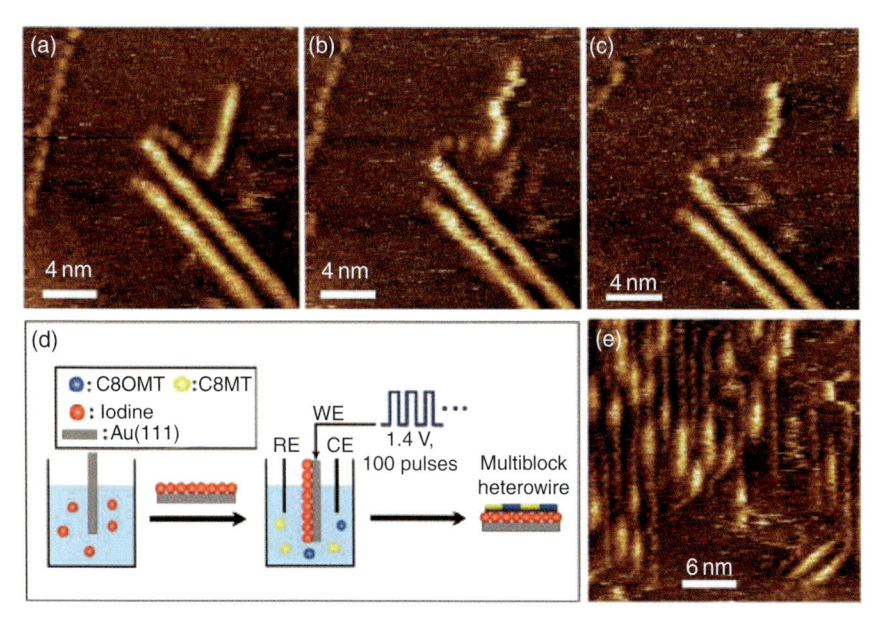

Figure 33 Multiblock polymers produced by electrochemical epitaxial polymerization on Au(111). (A–C) Images acquired from a sample alternated between solutions of C8OMT and C8MT. The images were acquired every 2 min. (D) A schematic of a mixed-solution polymer growth experiment, and (E) an STM image of multiblock heterowires produced from a mixed solution. From Sakaguchi H, Matsumura H, Gong H, and Abouelwafa AM (2005) Direct visualization of the formation of single-molecule conjugated copolymers. *Science* 310: 1002–1006.

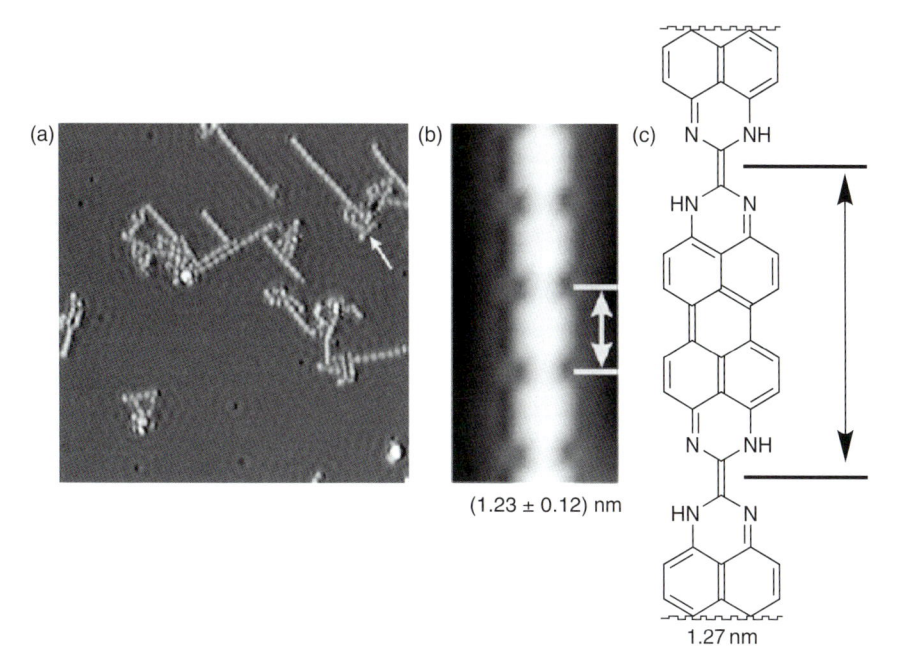

Figure 34 TAPP dimer chains on Cu(111): (a) an STM image of TAPP chains on the surface; and (b) a high-resolution image of a chain, with a corresponding structural model in (c). STM image parameters: $V_b = 0.1$ V, $I_t = 20$ pA, image area $= 50 \times 50$ nm^2 (a), $V_b = 0.1$ V, $I_t = 20$ pA, image area $= 2.4 \times 5$ nm^2 (b). From Matena M, Riehm T, Stohr M, Jung TA, and Gade LH (2008) Transforming surface coordination polymers into covalent surface polymers: Linked polycondensed aromatics through oligomerization of N-heterocyclic carbene intermediates. *Angewandte Chemie-International Edition* 47: 2414–2417.

Ulmann coupling, first identified more than 100 years ago [267], refers to the dehalogenation and subsequent C—C bond formation between alyl halides in the presence of copper. Despite a massive amount of work in the field [268], it is only in recent years that a surface-science approach has been undertaken to investigate the Ullmann reaction. For example, Xi and Bent have used AES, temperature-programmed reaction (TPR), and HREELS to study the formation of biphenyl from iodobenzene on Cu(111), finding that the dehalogenation can occur at temperatures as low as 175 K, and that the subsequent coupling onsets at a coverage-dependent temperature (300–400 K for sub-ML coverages, and as low as 210 K for 1–2 ML) [269, 270]. Using the same molecule on the same substrate, Hla et al. [43] demonstrated that each individual step of the Ullmann coupling (dehalogenation, lateral motion, and C–C bond formation) could be induced with the STM tip. The possibility of forming Ullmann-synthesized polymers requires a monomer that is at minimum bifunctional; McCarty and Weiss used STM to demonstrate the dehalogenation of 1,4-diiodobenzene at room temperature on Cu(111) [271]. However, they did not observe the subsequent coupling of the molecules, and instead found that the reacted monomer aligned into 'protopolymer' chains. Lipton-Duffin et al. observed the same type of 'protopolymer' chains following room temperature deposition of 1,4-diiodobenzene onto Cu(110) and suggest that it may be a metal–organic complex of the type [-Ph-Cu-]$_n$ [272].

The formation of poly(p-phenylene) (PPP), equation 5, was reported by Lipton-Duffin and coworkers following room temperature deposition of 1,4-diiodobenzene and subsequent annealing to ∼500 K for 5–10 min.

(5)

PPP   PMP

From Lipton-Duffin JA, Ivasenko O, Perepichka DF, and Rosei F (2009) Synthesis of polyphenylene molecular wires by surface-confined polymerization. *Small* 5: 592–597.

STM images show that the PPP lines grow epitaxially along [1$\bar{1}$0], occurring in discrete groups where each line is separated from its neighbors by a row of iodine atoms. The 4.1 Å periodicity measured from the STM images is in agreement with density functional theory calculations for the phenyl–phenyl spacing. The authors strengthen their argument for the formation of the polymer by demonstrating that 1,3-diiodobenzene forms into kinked chains of poly(m-phenylene) (PMP) under the same deposition conditions (**equation 5**). **Figure 35** shows an STM image of PMP on Cu(110) acquired at 102 K. Several PMP chains, primarily aligned along <21> and <2-1>, can be seen in the image. Further, the low substrate temperature reveals the presence of m-phenylene macrocycles which were not seen in room temperature-STM images. The authors hypothesize that the small size of the macrocycles (8.3 ± 0.8 Å diameter, comprising, on average, six m-phenyls) renders them mobile at room temperature, whereas the polymer chains are fixed on the surface due to their larger size and due to increased interaction at the chain ends.

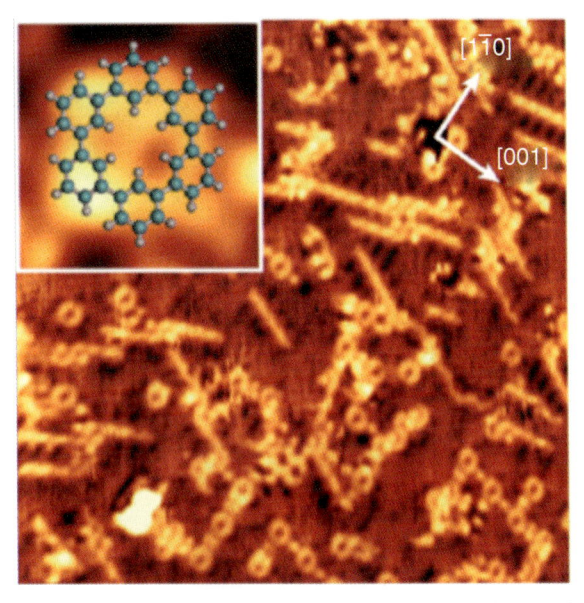

Figure 35 Poly(*m*-phenylene) lines and *m*-phenylene macrocycles produced by Ulmann dehalogenation on Cu(110). The inset shows a structural model of a macrocycle overlaying an STM image. STM image parameters: $T = 102$ K, $V_b = 1.61$ V, $I_t = 0.66$ nA, image area $= 30 \times 30$ nm^2. From Lipton-Duffin JA, Ivasenko O, Perepichka DF, and Rosei F (2009) Synthesis of polyphenylene molecular wires by surface-confined polymerization. *Small* 5: 592–597.

1.6.2 Two-Dimensional Polymers

1.6.2.1 Porphyrin networks

Two recent publications have demonstrated the creation of covalently bonded porphyrin networks on surfaces.

In the first study, Grill *et al.* [273] employed a porphyrin monomer, tetra(phenyl)porphyrin (TPP), with a customizable number of reactive bromine legs (shown in **Figures 36(a)–(c)**), under the assumption that the relatively weak carbon–halide bonds could be thermally cleaved without disrupting the porphyrin molecular framework. In a series of UHV experiments where they deposited the porphyrins onto an Au(111) surface, they demonstrated that the carbon–halide bonds could be preferentially cleaved by two different methods. In the first approach, the porphyrins were evaporated intact onto a room temperature surface at an evaporation temperature below 550 K, and subsequently the substrate temperature was increased sufficiently to liberate the bromines. In the second approach, evaporation temperatures above 590 K were used to liberate the bromines in the evaporation cell, and reactive dehalogenated monomers were deposited directly onto the room temperature surface. In either case, architectures consistent with the number of reactive sites on the monomer were formed: dimers from BrTPP (**Figures 36(g)** and **36(j)**), chains from *trans*-Br$_2$TPP (**Figures 36(h)** and **36(k)**), and square networks from Br$_4$TPP (**Figures 36(i)** and **36(l)**).

Grill *et al.* verified the covalent nature of the bonding through multiple methods. Manipulation of dimers, chains, and networks with the STM tip revealed them to be robust. The measured periodicity of the porphyrin-porphyrin spacing, $(17.2 \pm 0.3$ Å$)$, is consistent with the value of 17.1 Å determined by density functional theory. Finally, the authors very elegantly used spectroscopic STM to localize and measure the covalent bond between the porphyrins. Consistent with density functional theory predictions, the STM measurements suggest the presence of a localized orbital between the molecules, with an energy of about 3 eV.

The methodology presented by Grill *et al.* takes advantage of hierarchical bonding strengths to create active sites in a predetermined geometry on specific molecular framework. As Champness [274] pointed out, this approach is quite general, and should be translatable to other molecular systems. The first indication of this generality was provided by Veld *et al.* [275], who described the covalent

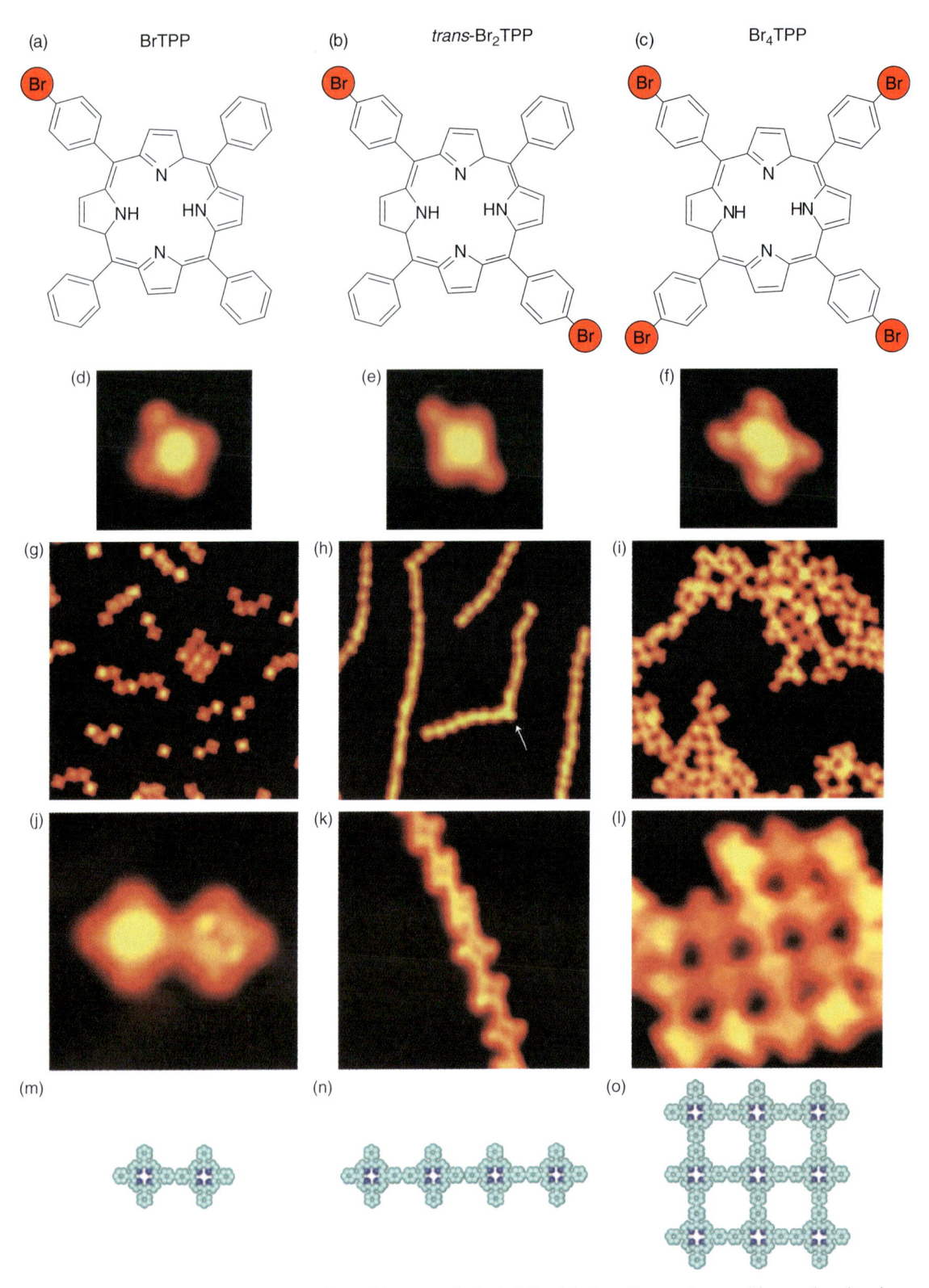

Figure 36 Covalently bonded structures formed from porphyrin-building blocks with one, two, and four active sites (a–c, respectively). STM images show the corresponding monomers (d–f), and large (g–i) and small (j–l) area scans of the covalent architectures. Models for the covalent structures are shown in (m–o). STM image sizes: $3.5 \times 3.5\,nm^2$ (d–f), $30 \times 30\,nm^2$ (g–i), $5 \times 5\,nm^2$ (j), $10 \times 10\,nm^2$ (k), $8.5 \times 8.5\,nm^2$ (l). From Grill L, Dyer M, Lafferentz L, Persson M, Peters MV, and Hecht S (2007) Nano-architectures by covalent assembly of molecular building blocks. *Nature Nanotechnology* 2: 687–691.

coupling of the tetra(mesityl)porphyrin (**33**) on the Cu(110) surface. In UHV experiments, the porphyrins were evaporated onto the room temperature surface from an evaporator held at 150 °C. STM images and reflection absorption infrared spectroscopy (RAIS) confirmed that the monomer remained unpolymerized immediately upon deposition, but STM images obtained after annealing the surface to 150–200 °C show that very few isolated porphyrins remain on the surface, the majority having aligned into chains or grids. By observing the surface diffusion of a porphyrin trimer at elevated temperature, the authors confirm the robust nature of the bonding. The mechanism for the polymerization, however, is less straightforward to confirm. Based on the observation of hydrogen evolution in thermal desorption spectroscopy, the authors speculate that reaction with the copper substrate may form CH_2 radicals at the para (relative to the bond to the porphyrin) position of the benzene rings that subsequently homocouple the molecules to one another.

1.6.2.2 Condensation polymerization via dehydration

The condensation reaction of an amine (—NH_2) and an aldehyde (—CHO) to form an imine bond (—CH=N—) presents a unique opportunity for growing polymers: the reaction is reversible, creating the possibility for self-repair of defects. The reversibility of the reaction depends, however, on the presence of water. The first examples of surface polymers grown via this reaction route have both been undertaken in UHV, where the H_2O evolved in the dehydration reaction is quickly desorbed and pumped off and is therefore unavailable to drive the reaction in reverse. However, the success of these early experiments will likely inspire others to find alternate environments in which to conduct the polymer growth, perhaps leading to higher-quality structures.

Weigelt and coworkers have published two works demonstrating amine–aldehyde covalent coupling in UHV. In their first study, they used STM and near edge x-ray absorption fine structure (NEXAFS) to show that the dehydration reaction shown in **equation 6** progresses to the same product in solution and on the Au(111) surface:

By successively depositing the dialdehyde and the amine on to the room temperature surface, they were able to produce a highly ordered overlayer of diimene molecules which was functionally identical to the overlayer produced by depositing *ex situ*-prepared diimene molecules onto the surface. NEXAFS provided a similar indication that the surface reaction produced the same product as the *ex situ* (solution) synthesis of the diimene.

Building on this premise, the authors subsequently used the condensation reaction shown in **equation 7**, between an aromatic three-spoked trisalicylaldehyde (trialdehyde) and 1,6-diaminohexane (diamine), to produce 2D covalently linked structures on the same surface [277]:

Weigelt *et al.* found that by dosing the trialdehyde onto a room temperature Au(111) surface, and subsequently heating the surface to 400 K before depositing the diamine, the covalent links between trialdehyde molecules could be fostered with a probability of 0.70 ± 0.03 per spoke. The authors found that they could kinetically control this linkage probability through their choice of deposition parameters: in a scheme where the diamine deposition was carried out at 120–160 K, forming multilayers on top of the previously deposited trialdehyde, and subsequently annealed to 450 K, the linkage probability per spoke was only 0.49 ± 0.03. This lower connectivity is attributed to the high probability of saturating the trialdehyde spokes with diamine under this regime, reducing the chance that a linkage to another molecule will be formed.

(a)

Aldehyde Amine   Hemiaminad   Imine
                   (tetrahedral intermediate)

(b)

(6)

From Weigelt S, Busse C, Bombis C, *et al.* (2007) Covalent interlinking of an aldehyde and an amine on a Au(111) surface in ultrahigh vacuum. *Angewandte Chemie International Edition* 46: 9227–9230.

The disordered covalently linked structures produced through the former experimental method contain domains of 1D chain-type linking intermixed with domains of 2D porous networks. **Figure 37** shows STM images typical of each of these regions. The authors report that the porous regions are more abundant. Individual pores can vary in diameter from ∼3 to 10 nm, corresponding to four to eight trialdehydes at their periphery.

Adapted from Weigelt S, Busse C, Bombis C, *et al.* (2008) Surface synthesis of 2D branched polymer nanostructures. *Angewandte Chemie International Edition* 47: 4406–4410.

Treier *et al.* used the same approach to demonstrate the thermally activated covalent linking of either 4,4′-diamino-p-terphenyl (DATP, **34**) or 2,4,6-tris(4-aminophenyl)-1,3,5-triazine (TAPT, **35**) by 3,4,9,10-perylenetetracarboxylic-dianhydride (PTCDA, **36**), to form 1D and 2D polyimides, respectively. Their experiments were conducted in UHV, where the molecules were deposited onto a room temperature surface. The authors found that PTCDA and DTAP co-assembled into a two-component hydrogen-bonded network that was stable up to 470 K. The formation of polymers could be initiated increasing the annealing temperature to 570 K for 15 min. Low-temperature (77 K) STM was used to investigate the reaction products. Deposition and annealing of PTCDA and DTAP produced chains of the expected periodicity for covalent linkage (~2.5 nm). The result of depositing and annealing PTCDA and TAPT is shown in **Figure 38**: the threefold symmetric TAPT molecules bond with the PTCDA to form a disordered, but highly interconnected porous network. The 3:2

46 Directed Assembly of Nanostructures

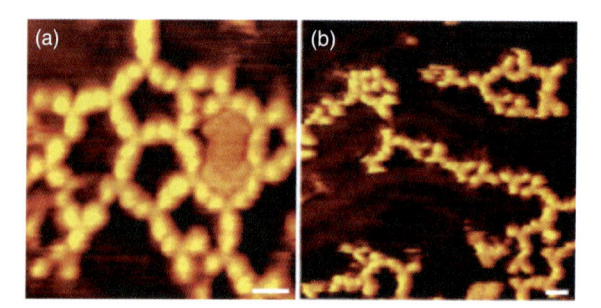

Figure 37 Branched polymers formed from the condendsation of a trialdehyde and a diamine. STM images show: (a) network and (b)chain structures. STM image parameters: $V_b = -2.0$ V, $I_t = 0.31$ nA (a); $V_b = -2.0$ V, $I_t = 0.33$ nA (b). In each image, the scale bar is 2 nm. From Weigelt S, Busse C, Bombis C, et al. (2008) Surface synthesis of 2D branched polymer nanostructures. *Angewandte Chemie-International Edition* 47: 4406–4410.

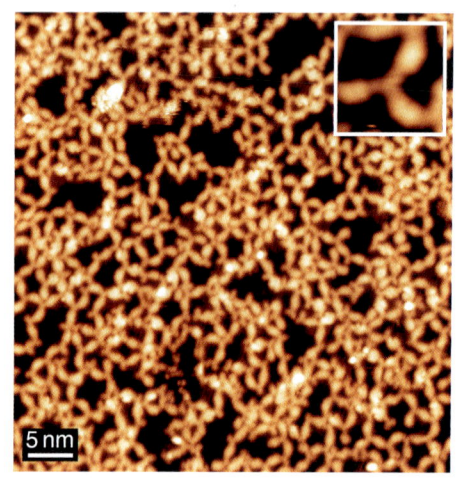

Figure 38 TAPT–PTCDA polyimide network. The inset shows a single TAPT molecule bonded to three PTCDA spokes. STM image parameters: $V_b = -0.8$ V, $I_t = 20$ pA. From Treier M, Richardson NV, and Fasel R (2008) Fabrication of surface-supported low-dimensional polyimide networks. *Journal of the American Chemical Society* 130: 14054–14055.

TAPT:PTCDA stoichiometry used should theoretically be conducive to the formation of an ordered network of hexagonal pores with a 3 nm periodicity. The disorder in the network may perhaps be resolved through perturbations to the preparation conditions, however, a subsequent publication by the same authors shows that the formation of such a network may also be hindered by a preference for forming iso-amide products, which lead to the formation of triangular, rather than hexagonal, pores [278].

Zwaneveld et al. [280] also investigated the dehydration reactions of molecules expected to form a covalent hexagonal porous network. Working on Ag(111), the authors demonstrated near-monolayer surface coverage by two different networks formed from the covalent linking of 1,4-benzenediboronic acid (BDBA). Following the dehydration of BDBA (**Figure 39(a)**) to form boroxine (**Figure 39(b)**), hexagonal covalent rings (**Figure 39(c)**, inset), can be formed via another dehydration reaction of the boroxine. The STM image in **Figure 39(c)** shows a network produced through this reaction. Similar networks were formed at substrate temperatures ranging from room temperature to 500 K, and following annealing up to 770 K. The covalent bonding between the molecules was confirmed by the STM-measured pore–pore spacing for six-membered rings, 15 ± 1 Å, which is in agreement with the value predicted by density

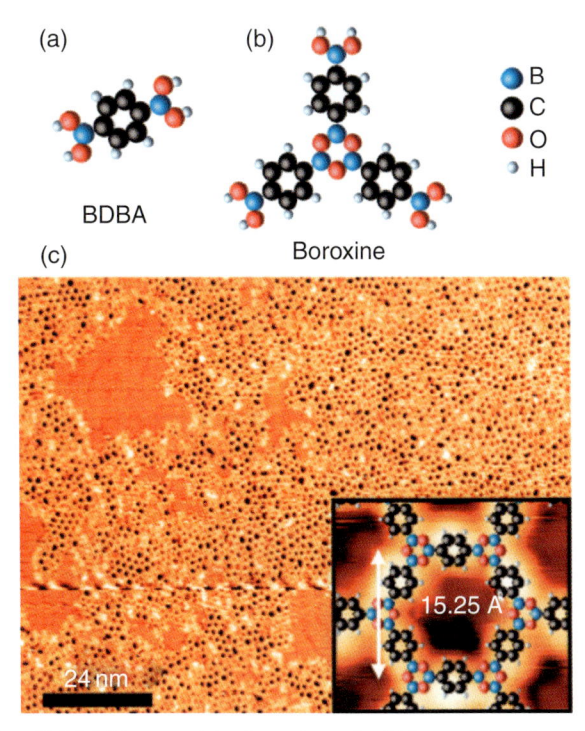

Figure 39 Covalent networks of BDBA (a), comprising boroxine units (b) formed from three BDBA monomers. STM images (c and inset) show a large-scale view of a network and the proposed chemical structure for a single pore. STM image parameters: $V_b = -2.1$ V, $I_t = 0.4$ nA, 120×90 nm^2 (c). From Zwaneveld NAA, Pawlak R, Abel M, et al. (2008) Organized formation of 2D extended covalent organic frameworks at surfaces. *Journal of the American Chemical Society* 130: 6678–6679.

functional theory for the structure shown in the inset of **Figure 39(c)**. The authors demonstrated that a network with larger pore sizes could be grown by co-depositing the threefold symmetric molecule 2,3,6,7,10,11-hexahydroxytriphenylene (HHTP), which reacts with the BDBA to form a dioxaborole heterocycle with a ~30 Å pore size. The network produced through this reaction was found to exhibit higher local order than the exclusively BDBA-based network. The authors ascribe this attribute to the favorable kinetic pathway available to the bimolecular reaction leading to the dioxaborole.

1.7 Conclusions and Outlook

In this article, we have given a broad overview of the most important examples of directed molecular assembly at solid surfaces over the last three decades. This is a fascinating area of modern science which has emerged only recently thanks to the advent of new characterization tools, namely the various forms of scanning probe microscopy, which allow surfaces to be probed with molecular and even atomic resolution. These tools have additionally created the possibility for atomic scale manipulation and fabrication, with particularly striking capabilities in creating nucleation sites for chemical chain reactions (Section 1.4.1), manipulating guest molecules in 2D inclusion architectures (Section 1.6) and for inducing polymerization in diacetylene films (Section 1.6.1.1).

Major advances in this field are undoubtedly also due to the wonders of organic synthesis. Modern synthetic chemistry allows chemists to create aromatic building blocks with a high degree of flexibility, leading to a variety of surface architectures. By balancing intermolecular forces and molecule–substrate interactions, the controlled formation of molecular architectures can be facilitated. Noncovalent interactions have been widely used to create supramolecular architectures ordered over long ranges, confining to 2D systems and structures that had been previously investigated in 3D. In this context, hydrogen bonding is the most extensively employed directional interaction. However, metal–organic coordination is also emerging as a powerful and flexible approach to engineering 2D architectures, and, when combined with hydrogen bonding (as in the recent work of Langner et al. [205]), may provide extremely precise control over molecular positioning at surfaces.

Another very interesting paradigm that has recently emerged is the use of nanotemplates, namely substrates that are patterned at the nanoscale, where specific characteristics provide suitable surface cues that can guide the formation of atomic or molecular patterns. We discussed surface reconstructions useful for this approach in Section 1.4.2. Taking a different approach, Theobald et al. opened new perspectives in this area by creating a multicomponent supramolecular pattern defined by a porous network [191]. In such systems, the pores can act as hosts for the inclusion of guest molecules such as fullerenes (the simplest case, see Section 1.6.2) and more complex ones like heterocyclic circulenes [241] and biomolecules [243].

The emerging frontier in molecular assembly is the directed formation of surface-confined polymers. The Holy Grail of this field is the controlled surface synthesis of 2D conjugated polymers, which would constitute a new class of materials of which graphene represents the zeroth order [245, 246, 247]. This is obviously an enormous challenge, which requires close interaction between synthetic chemists and surface scientists to optimize the choice of organic building blocks and the mechanism of reaction that will induce polymerization. The fact itself that graphene is the only known example of a 2D conjugated polymer implies that nature does not favor such structures. However, results reported in recent years are encouraging, and point to a promising and exciting field. While surface-confined oligomers have been demonstrated (see Section 1.7), one of the great difficulties toward creating extended structures relates to overcoming defects; since the bonding is so robust, the famous and flexible mechanisms of self-repair that are so useful in supramolecular chemistry generally do not apply here.

A constant backdrop to the continuing progress in molecular assembly is the question of how to integrate these architectures into useful devices. For example, although polydiacetylene lines described in Section 1.6.1.1 are often introduced in the context of molecular electronics [256,262,281], STS measurements of lines formed from Langmuir–Blodgett-deposited 10,12-nonacosadiynoic acid on HOPG reveal a seminconducting bandgap (~1.3 eV) which is obviously not ideal for most electronics applications [282]. Direct measurement of conductivity via two-tip STM reveals 1D conduction along the PDA backbone, with an estimated conductivity of $(3-5) \times 10^{-6}$ S cm^{-1}, comparable to intrinsic elemental semiconductors Si and Ge [281]. Some improvement can be made through iodine doping, which one study showed to result in a 10^3 increase to the conductivity [283].

The growth of perpendicular criss-crossed lines by chemical chain reaction [140] represents a step toward using molecular architectures to mimic current top-down device components. In a similar vein, a recent manuscript described the intersection of styrene lines with perpendicular bismuth nanolines [284]; however, since the bismuth nanolines are known to have a larger gap than the silicon surface on which they are grown [285], this experiment provides a proof-of-principle rather than an architecture immediately useful to devices.

As our knowledge of how to control molecular assembly at surfaces grows, we are confident that we will see a concomitant increase in the real-world applications of assembled molecular architectures. Our understanding of molecular assembly is constantly evolving, and novel approaches to molecular organization are still emerging. For example, Harikumar et al. have recently demonstrated the formation of lines of 1,5-dichloropentane on Si(100) via dipole-directed assembly, wherein the displacement of surface charge by the dipolar molecule creates a self-propagating growth mechanism [286]. Considering the wealth of approaches described in this article, and with new methodologies still emerging, the next years should see the emergence of exciting examples of molecular nanosystems tailored for specific form and function.

Acknowledgments

J.M.M. acknowledges salary support from NSERC of Canada in the form of a postdoctoral fellowship. F.R. is grateful to the Canada Research Chairs program for partial salary support. We thank J. Lipton-Duffin for providing STM images of copper.

References

1. Moriarty P (2001) Nanostructured materials. *Reports on Progress in Physics* 64: 297–381.
2. Rosei F (2004) Nanostructured surfaces: Challenges and frontiers in nanotechnology. *Journal of Physics-Condensed Matter* 16: S1373–S1436.
3. Moore GE (1965) Cramming more components onto integrated circuits. *Electronics* 38: 114–117.

4. Wong HSP, Frank DJ, Solomon PM, Wann CHJ, and Welser JJ (1999) Nanoscale CMOS. *Proceedings of the IEEE* 87: 537–570.
5. Halperin WP (1986) Quantum size effects in metal particles. *Reviews of Modern Physics* 58: 533–606.
6. Yoffe AD (1993) Low-dimensional systems – quantum-size effects and electronic-properties of semiconductor microcrystalites (zero-dimensional systems) and some quasi-2-dimensional systems. *Advances in Physics* 42: 173–266.
7. Jaskolski W (1996) Confined many-electron systems. *Physics Reports-Review Section of Physics Letters* 271: 1–66.
8. Scholes GD and Rumbles G (2006) Excitons in nanoscale systems. *Nature Materials* 5: 683–696.
9. Jortner J (1992) Cluster size effects. *Zeitschrift Fur Physik D-Atoms Molecules and Clusters* 24: 247–275.
10. Ito T and Okazaki S (2000) Pushing the limits of lithography. *Nature* 406: 1027–1031.
11. Kelly MJ (1995) *Low-Dimensional Semiconductors: Materials, Physics, Technology, Devices*. Oxford: Oxford University Press.
12. Sarikaya M, Tamerler C, Jen AKY, Schulten K, and Baneyx F (2003) Molecular biomimetics: Nanotechnology through biology. *Nature Materials* 2: 577–585.
13. Sarikaya M, Tamerler C, Jen AKY, Schulten K, and Baneyx F (2003) Molecular biomimetics: Nanotechnology through biology. *Nature Materials* 2: 577–585.
14. Sanchez C, Arribart H, and Guille MMG (2005) Biomimetism and bioinspiration as tools for the design of innovative materials and systems. *Nature Materials* 4: 277–288.
15. Lehn J-M (1995) *Supramolecular Chemistry: Concepts and Perspectives*. Weinheim: Wiley-VCH.
16. Lehn JM (1988) Supramolecular chemistry – scope and perspectives molecules, supermolecules and molecular devices. *Angewandte Chemie (International Edition in English)* 27: 89–112.
17. Lehn JM (1990) Perspectives in supramolecular chemistry – from molecular recognition toward molecular information processing and self-organization. *Angewandte Chemie (International Edition in English)* 29: 1304–1319.
18. Whitesides GM, Mathias JP, and Seto CT (1991) Molecular self-assembly and nanochemistry – a chemical strategy for the synthesis of nanostructures. *Science* 254: 1312–1319.
19. Nakamura T, Matsumoto T, Tada H, and Sugiura K-I (eds.) (2002) *Chemistry of Nanomolecular Systems: Towards the Realization of Molecular Devices*. New York: Springer.
20. Goddard W, III, Brenner DW, Lyshevski SE, and Iafrate GJ (eds.) (2007) *Handbook of Nanoscience, Engineering and Technology*. Boca Raton, FL: CRC Press.
21. Roder H, Hahn E, Brune H, Bucher JP, and Kern K (1993) Building one-dimensional and two-dimensional nanostructures by diffusion-controlled aggregation at surfaces. *Nature* 366: 141–143.
22. Teichert C (2002) Self-organization of nanostructures in semiconductor heteroepitaxy. *Physics Reports-Review Section of Physics Letters* 365: 335–432.
23. Wiley B, Sun YG, and Xia Y (2007) Synthesis of silver wanostructures with controlled shapes and properties. *Accounts of Chemical Research* 40: 1067–1076.
24. McLean AB, Hill IG, Lipton-Duffin JA, MacLeod JM, Miwa RH, and Srivastava GP (2008) Nanolines on silicon surfaces. *International Journal of Nanotechnology* 5: 1018–1057.
25. Aiken JD and Finke RG (1999) A review of modern transition-metal nanoclusters: Their synthesis, characterization, and applications in catalysis. *Journal of Molecular Catalysis A: Chemical* 145: 1–44.
26. Schmid G, Baumle M, Geerkens M, Helm I, Osemann C, and Sawitowski T (1999) Current and future applications of nanoclusters. *Chemical Society Reviews* 28: 179–185.
27. Ashoori RC (1996) Electrons in artificial atoms. *Nature* 379: 413–419.
28. McDonald SA, Konstantatos G, Zhang SG, et al. (2005) Solution-processed PbS quantum dot infrared photodetectors and photovoltaics. *Nature Materials* 4: 138–142.
29. Sargent EH (2005) Infrared quantum dots. *Advanced Materials* 17: 515–522.
30. Binnig G and Rohrer H (1982) Scanning tunneling microscopy. *Helvetica Physica Acta* 55: 726–735.
31. Binnig G, Rohrer H, Gerber C, and Weibel E (1982) Tunneling through a controllable vacuum gap. *Applied Physics Letters* 40: 178–180.
32. Binning G, Rohrer H, Gerber C, and Weibel E (1982) Surface studies by scanning tunneling microscopy. *Physical Review Letters* 49: 57–61.
33. Binnig G and Rohrer H (1986) Scanning tunneling microscopy. *IBM Journal of Research and Development* 30: 355–369.
34. Binnig G and Rohrer H (1987) Scanning tunneling microscopy – from birth to adolesence. *Reviews of Modern Physics* 59: 615–625.
35. Binnig G, Rohrer H, Gerber C, and Weibel E (1983) 7×7 reconstruction on Si(111) resolved in real space. *Physical Review Letters* 50: 120–123.
36. Crommie MF, Lutz CP, and Eigler DM (1993) Imaging standing waves in a 2-dimensional electron gas. *Nature* 363: 524–527.
37. Stipe BC, Rezaei MA, and Ho W (1998) Coupling of vibrational excitation to the rotational motion of a single adsorbed molecule. *Physical Review Letters* 81: 1263–1266.
38. Stipe BC, Rezaei MA, and Ho W (1998) Single-molecule vibrational spectroscopy and microscopy. *Science* 280: 1732–1735.
39. Eigler DM, Lutz CP, and Rudge WE (1991) An atomic switch realized with the scanning tunneling microscope. *Nature* 352: 600–603.
40. Crommie MF, Lutz CP, and Eigler DM (1993) Confinement of electrons to quantum corrals on a metal surface. *Science* 262: 218–220.
41. Moon CR, Mattos LS, Foster BK, Zeltzer G, and Manoharan HC (2009) Quantum holographic encoding in a two-dimensional electron gas. *Nature Nanotechnology* 4: 167–172.
42. Lee HJ and Ho W (1999) Single-bond formation and characterization with a scanning tunneling microscope. *Science* 286: 1719–1722.
43. Hla S-W, Bartels L, Meyer G, and Rieder K-H (2000) Inducing all steps of a chemical reaction with the scanning tunneling microscope tip: Towards single molecule engineering. *Physical Review Letters* 85: 2777.
44. Hansma PK and Tersoff J (1987) Scanning tunneling microscopy. *Journal of Applied Physics* 61: R1–R23.
45. Frommer J (1992) Scanning tunneling microscopy and atomic force microscopy in organic chemistry. *Angewandte Chemie (International Edition in English)* 31: 1298–1328.
46. Chen CJ (1993) *Introduction to Scanning Tunneling Microscopy*. New York: Oxford University Press.

47. Hansma HG and Hoh JH (1994) Biomolecular imaging with the atomic-force microscope. *Annual Review of Biophysics and Biomolecular Structure* 23: 115–139.
48. Besenbacher F (1996) Scanning tunnelling microscopy studies of metal surfaces. *Reports on Progress in Physics* 59: 1737–1802.
49. Carpick RW and Salmeron M (1997) Scratching the surface: Fundamental investigations of tribology with atomic force microscopy. *Chemical Reviews* 97: 1163–1194.
50. Poirier GE (1997) Characterization of organosulfur molecular monolayers on Au(111) using scanning tunneling microscopy. *Chemical Reviews* 97: 1117–1127.
51. Zhang ZY and Lagally MG (1997) Atomistic processes in the early stages of thin-film growth. *Science* 276: 377–383.
52. Brune H (1998) Microscopic view of epitaxial metal growth: Nucleation and aggregation. *Surface Science Reports* 31: 121–229.
53. Itaya K (1998) *In situ* scanning tunneling microscopy in electrolyte solutions. *Progress in Surface Science* 58: 121–247.
54. Wiesendanger R (1998) *Scanning Probe Microscopy: Analytical Methods*. Berlin: Springer.
55. Binnig G and Rohrer H (1999) In touch with atoms. *Reviews of Modern Physics* 71: S324–S330.
56. Liu GY, Xu S, and Qian YL (2000) Nanofabrication of self-assembled monolayers using scanning probe lithography. *Accounts of Chemical Research* 33: 457–466.
57. Voigtlander B (2001) Fundamental processes in Si/Si and Ge/Si epitaxy studied by scanning tunneling microscopy during growth. *Surface Science Reports* 43: 127–254.
58. Giessibl FJ (2003) Advances in atomic force microscopy. *Reviews of Modern Physics* 75: 949–983.
59. Rosei F (2003) Scanning probe microscopy studies of Ge/Si surfaces. In: *Science, Technology and Education of Microscopy: An Overview*, pp. 84–92. Badajoz: Formatex.
60. Rosei F and Rosei R (2003) Scanning tunnelling microscopy studies of elementary surface processes. In : Science, Technology and Education of Microscopy: An Overview, pp. 24–33. Badajoz: Formatex.
61. Schunack M, Rosei F, and Besenbacher F (2003) The scanning tunneling microscope as a unique tool to investigate the interaction between complex molecules and metal surfaces. In: *Science, Technology and Education of Microscopy: An Overview*, pp. 43–51. Badajoz: Formatex.
62. Samori P (2004) Scanning probe microscopies beyond imaging. *Journal of Materials Chemistry* 14: 1353–1366.
63. Miwa JA and Rosei F (2006) Molecular self-assembly: Fundamental concepts and applications. In: *The MEMS Handbook: MEMS, Design and Fabrication*. Boca Raton, FL: CRC Press.
64. Otero R, Rosei F, and Besenbacher F (2006) Scanning tunneling microscopy manipulation of complex organic molecules on solid surfaces. *Annual Review of Physical Chemistry* 57: 497–525.
65. Jackson AM, Myerson JW, and Stellacci F (2004) Spontaneous assembly of subnanometre-ordered domains in the ligand shell of monolayer-protected nanoparticles. *Nature Materials* 3: 330–336.
66. Jackson AM, Hu Y, Silva PJ, and Stellacci F (2006) From homoligand- to mixed-ligand-monolayer-protected metal nanoparticles: A scanning tunneling microscopy investigation. *Journal of the American Chemical Society* 128: 11135–11149.
67. Centrone A, Hu Y, Jackson AM, Zerbi G, and Stellacci F (2007) Phase separation on mixed-monolayer-protected metal nanoparticles: A study by infrared spectroscopy and scanning tunneling microscopy. *Small* 3: 814–817.
68. DeVries GA, Brunnbauer M, Hu Y, et al. (2007) Divalent metal nanoparticles. *Science* 315: 358–361.
69. Perepichka DF and Rosei F (2007) Metal nanoparticles: From "artificial atoms" to "artificial molecules". *Angewandte Chemie International Edition* 46: 6006–6008.
70. Singh C, Ghorai PK, Horsch MA, et al. (2007) Entropy-mediated patterning of surfactant-coated nanoparticles and surfaces. *Physical Review Letters* 99: 4.
71. Bauer CA, Stellacci F, and Perry JW (2008) Relationship between structure and solubility of thiol-protected silver nanoparticles and assemblies. *Topics in Catalysis* 47: 32–41.
72. Carney RP, DeVries GA, Dubois C, et al. (2008) Size limitations for the formation of ordered striped nanoparticles. *Journal of the American Chemical Society* 130: 798–799.
73. Centrone A, Penzo E, Sharma M, et al. (2008) The role of nanostructure in the wetting behavior of mixed-monolayer-protected metal nanoparticles. *Proceedings of the National Academy of Sciences of the United States of America* 105: 9886–9891.
74. DeVries GA, Talley FR, Carney RP, and Stellacci F (2008) Thermodynamic study of the reactivity of the two topological point defects present in mixed self-assembled monolayers on gold nanoparticles. *Advanced Materials* 20: 4243–4247.
75. Hu Y, Uzun O, Dubois C, and Stellacci F (2008) Effect of ligand shell structure on the interaction between monolayer-protected gold nanoparticles. *Journal of Physical Chemistry C* 112: 6279–6284.
76. Nakata K, Hu Y, Uzun O, Bakr O, and Stellacci F (2008) Chains of superparamagnetic nanoplarticles. *Advanced Materials* 20: 4294–4299.
77. Verma A, Uzun O, Hu YH, et al. (2008) Surface-structure-regulated cell-membrane penetration by monolayer-protected nanoparticles. *Nature Materials* 7: 588–595.
78. Rosei F, Schunack M, Naitoh Y, et al. (2003) Properties of large organic molecules on metal surfaces. *Progress in Surface Science* 71: 95–146.
79. Yoshinobu J (2004) Physical properties and chemical reactivity of the buckled dimer on Si(100). *Progress in Surface Science* 77: 37–70.
80. McNaught AD and Wilkinson A (1997) *IUPAC Compendium of Chemical Terminology*. Oxford: Blackwell Science.
81. Aakeroy CB and Seddon KR (1993) The hydrogen bond and crystal engineering *Chemical Society Reviews* 22: 397–407.
82. Prins LJ, Reinhoudt DN, and Timmerman P (2001) Noncovalent synthesis using hydrogen bonding. *Angewandte Chemie International Edition* 40: 2383–2426.
83. Desiraju GR and Steiner T (2001) *The Weak Hydrogen Bond in Structural Chemistry and Biology*. Oxford: Oxford University Press.
84. Faul CFJ and Antonietti M (2003) Ionic self-assembly: Facile synthesis of supramolecular materials. *Advanced Materials* 15: 673–683.
85. Blake AJ, Champness NR, Hubberstey P, Li WS, Withersby MA, and Schroder M (1999) Inorganic crystal engineering using self-assembly of tailored building-blocks. *Coordination Chemistry Reviews* 183: 117–138.
86. Petty MC (2007) *Molecular Electronics: From Principles to Practice*. Chichester: Wiley-Interscience.
87. Hammer B and Norskov JK (1995) Electronic factors determining the reactivity of metal surfaces. *Surface Science* 343: 211–220.
88. Hammer B and Norskov JK (1995) Why gold is the noblest of all metals. *Nature* 376: 238–240.

89. Hammer B and Norskov JK (2000) Theoretical surface science and catalysis – calculations and concepts. In: *Advances in Catalysis*, vol. 45, pp. 71–129. San Diego, CA: Academic Press.
90. Hashimoto A, Suenaga K, Gloter A, Urita K, and Iijima S (2004) Direct evidence for atomic defects in graphene layers. *Nature* 430: 870–873.
91. Novoselov KS, Geim AK, Morozov SV, et al. (2004) Electric field effect in atomically thin carbon films. *Science* 306: 666–669.
92. Berger C, Song ZM, Li XB, et al. (2006) Electronic confinement and coherence in patterned epitaxial graphene. *Science* 312: 1191–1196.
93. Ohta T, Bostwick A, Seyller T, Horn K, and Rotenberg E (2006) Controlling the electronic structure of bilayer graphene. *Science* 313: 951–954.
94. Peres NMR, Guinea F, and Neto AHC (2006) Electronic properties of disordered two-dimensional carbon. *Physical Review B* 73: 125411.
95. Son YW, Cohen ML, and Louie SG (2006) Energy gaps in graphene nanoribbons. *Physical Review Letters* 97: 216803.
96. Stankovich S, Dikin DA, Dommett GHB, et al. (2006) Graphene-based composite materials. *Nature* 442: 282–286.
97. Geim AK and Novoselov KS (2007) The rise of graphene. *Nature Materials* 6: 183–191.
98. Han MY, Ozyilmaz B, Zhang YB, and Kim P (2007) Energy band-gap engineering of graphene nanoribbons. *Physical Review Letters* 98: 4.
99. Couto MS, Liu XY, Meekes H, and Bennema P (1994) Scanning tunneling microscopy studies on *n*-alkane molecules adsorbed on graphite. *Journal of Applied Physics* 75: 627–629.
100. Faglioni F, Claypool CL, Lewis NS, and Goddard WA (1997) Theoretical description of the STM images of alkanes and substituted alkanes adsorbed on graphite. *Journal of Physical Chemistry B* 101: 5996–6020.
101. Cyr DM, Venkataraman B, and Flynn GW (1996) STM investigations of organic molecules physisorbed at the liquid–solid interface. *Chemistry of Materials* 8: 1600–1615.
102. Chang HP and Bard AJ (1991) Observation and characterization by scanning tunneling microscopy of structures generated by cleaving highly oriented pyrolytic graphite. *Langmuir* 7: 1143–1153.
103. Liu CY, Chang HP, and Bard AJ (1991) Large-scale hexagonal domainlike structures superimposed on the atomic corrugation of a graphite surface observed by scanning tunneling microscopy. *Langmuir* 7: 1138–1142.
104. Rong ZY and Kuiper P (1993) Electronic effects in scanning tunneling microscopy – moire pattern on a graphite surface. *Physical Review B* 48: 17427–17431.
105. Pong WT and Durkan C (2005) A review and outlook for an anomaly of scanning tunnelling microscopy (STM): Superlattices on graphite. *Journal of Physics D-Applied Physics* 38: R329–R355.
106. Clemmer CR and Beebe TP (1991) Graphite – a mimic for DNA and other biomolecules in scanning tunneling microscope studies. *Science* 251: 640–642.
107. Heckl WM and Binnig G (1992) Domain walls on graphite mimic DNA. *Ultramicroscopy* 42: 1073–1078.
108. Norskov JK, Stoltze P, and Nielsen U (1991) The reactivity of metal surfaces. *Catalysis Letters* 9: 173–182.
109. Schunack M, Linderoth TR, Rosei F, Laegsgaard E, Stensgaard I, and Besenbacher F (2002) Long jumps in the surface diffusion of large molecules. *Physical Review Letters* 88: 156102.
110. Miwa JA, Weigelt S, Gersen H, Besenbacher F, Rosei F, and Linderoth TR (2006) Azobenzene on Cu(110): Adsorption site-dependent diffusion. *Journal of the American Chemical Society* 128: 3164–3165.
111. Barth JV, Brune H, Ertl G, and Behm RJ (1990) Scanning tunneling microscopy observations on the reconstructed Au(111) surface – atomic structure, long-range superstructure, rotational domains and surface-defects. *Physical Review B* 42: 9307–9318.
112. Narasimhan S and Vanderbilt D (1992) Elastic stress domains and the herringbone reconstruction on Au(111). *Physical Review Letters* 69: 1564–1567.
113. Chen W, Madhavan V, Jamneala T, and Crommie MF (1998) Scanning tunneling microscopy observation of an electronic superlattice at the surface of clean gold. *Physical Review Letters* 80: 1469–1472.
114. Burgi L, Brune H, and Kern K (2002) Imaging of electron potential landscapes on Au(111). *Physical Review Letters* 89: 4.
115. Chambliss DD, Wilson RJ, and Chiang S (1991) Nucleation of ordered Ni island arrays on Au(111) by surface-lattice distortions. *Physical Review Letters* 66: 1721–1724.
116. Meyer JA, Baikie ID, Kopatzki E, and Behm RJ (1996) Preferential island nucleation at the elbows of the Au(111) herringbone reconstruction through place exchange. *Surface Science* 365: L647–L651.
117. Bohringer M, Morgenstern K, Schneider WD, et al. (1999) Two-dimensional self-assembly of supramolecular clusters and chains. *Physical Review Letters* 83: 324–327.
118. Czanderna AW, Powell CJ, and Madey TE (1999) *Specimen Handling, Preparation, and Treatments in Surface Characterization*. New York: Springer.
119. Reichelt K and Lutz HO (1971) Hetero-epitaxial growth of vacuum evaporated silver and gold. *Journal of Crystal Growth* 10: 103–107.
120. Chidsey CED, Loiacono DN, Sleator T, and Nakahara S (1988) STM study of the surface morphology of gold on mica. *Surface Science* 200: 45–66.
121. Derose JA, Thundat T, Nagahara LA, and Lindsay SM (1991) Gold grown epitaxially on mica – conditions for large area flat faces. *Surface Science* 256: 102–108.
122. Grove AS (1967) *Physics and Technology of Semiconductor Devices*. New York: Wiley.
123. Zandvliet HJW (2000) Energetics of Si(001). *Reviews of Modern Physics* 72: 593–602.
124. Haneman D (1987) Surfaces of silicon. *Reports on Progress in Physics* 50: 1045–1086.
125. Duke CB (1996) Semiconductor surface reconstruction: The structural chemistry of two-dimensional surface compounds. *Chemical Reviews* 96: 1237–1259.
126. Srivastava GP (1997) Theory of semiconductor surface reconstruction. *Reports on Progress in Physics* 60: 561–613.
127. Schlier RE and Farnsworth HE (1959) Structure and adsorption characteristics of cleran surfaces of germanium and silicon. *Journal of Chemical Physics* 30: 917–926.
128. Tabata T, Aruga T, and Murata Y (1987) Order–disorder transition on Si(001) – c(4×2) to (2×1) *Surface Science* 179: L63–L70.
129. Landemark E, Karlsson CJ, Chao YC, and Uhrberg RIG (1992) Core-level spectroscopy of the clean Si(001) surface – charge-transfer within asymmetric dimers of the 2×1 and c(4×2) reconstructions. *Physical Review Letters* 69: 1588–1591.
130. Boland JJ (1992) Role of bond-strain in the chemistry of hydrogen on the Si(100) surface. *Surface Science* 261: 17–28.

131. Boland JJ (1993) Scanning tunneling microscopy of the interaction of hydrogen with silicon surfaces. *Advances in Physics* 42: 129–171.
132. Lyding JW, Shen TC, Hubacek JS, Tucker JR, and Abeln GC (1994) Nanoscale patterning and oxidation of H-passivated Si(100)-2×1 surfaces with an ultrahigh-vacuum scanning tunneling microscope. *Applied Physics Letters* 64: 2010–2012.
133. Lopinski GP, Wayner DDM, and Wolkow RA (2000) Self-directed growth of molecular nanostructures on silicon. *Nature* 406: 48–51.
134. Cho JH, Oh DH, and Kleinman L (2002) One-dimensional molecular wire on hydrogenated Si(001). *Physical Review B* 65: 4.
135. Lopinski G (2008) Organics on silicon; single molecules, nanostructures and monolayers. *International Journal of Nanotechnology* 5: 1247–1267.
136. Cicero RL, Chidsey CED, Lopinski GP, Wayner DDM, and Wolkow RA (2002) Olefin additions on H-Si(111): Evidence for a surface chain reaction initiated at isolated dangling bonds. *Langmuir* 18: 305–307.
137. Shen TC, Wang C, Abeln GC, *et al.* (1995) Atomic-scale desorption through electronic and vibrational excitation mechanisms. *Science* 268: 1590–1592.
138. Higashi GS, Chabal YJ, Trucks GW, and Raghavachari K (1990) Ideal hydrogen termination of the Si(111) surface. *Applied Physics Letters* 56: 656–658.
139. Hossain MZ, Kato HS, and Kawai M (2005) Controlled fabrication of 1D molecular lines across the dimer rows on the Si(100)(2×1)-H surface through the radical chain reaction. *Journal of the American Chemical Society* 127: 15030–15031.
140. Hossain MZ, Kato HS, and Kawai M (2005) Fabrication of interconnected 1D molecular lines along and across the dimer rows on the Si(100)-(2×1)-H surface through the radical chain reaction. *Journal of Physical Chemistry B* 109: 23129–23133.
141. Zikovsky J, Dogel SA, Haider MB, DiLabio GA, and Wolkow RA (2007) Self-directed growth of contiguous perpendicular molecular lines on H-Si(100) Surfaces. *Journal of Physical Chemistry A* 111: 12257–12259.
142. Hossain MZ, Kato HS, and Kawai M (2008) Self-directed chain reaction by small ketones with the dangling bond site on the Si(100)-(2×1)-H surface: Acetophenone, a unique example. *Journal of the American Chemical Society* 130: 11518–11523.
143. Ertl G (1967) Untersuchung von oberflächenreaktionen mittels beugung langsamer elektronen (LEED): I. Wechselwirkung von O₂ und N₂O mit (110)-, (111)- und (100)-Kupfer-Oberflächen. *Surface Science* 6: 208–232.
144. Coulman DJ, Wintterlin J, Behm RJ, and Ertl G (1990) Novel mechanism for the formation of chemisorption phases – the (2×1)O-Cu(110) added-row reconstruction. *Physical Review Letters* 64: 1761–1764.
145. Jensen F, Besenbacher F, Laegsgaard E, and Stensgaard I (1990) Surface reconstruction of Cu(110) induced by oxygen-chemisorption. *Physical Review B* 41: 10233–10236.
146. Besenbacher F, Jensen F, Laegsgaard E, Mortensen K, and Stensgaard I (1991) Visualization of the dynamics in surface reconstructions. *Journal of Vacuum Science and Technology B* 9: 874–878.
147. Kern K, Niehus H, Schatz A, Zeppenfeld P, George J, and Comsa G (1991) Long-range spatial self-organization in the adsorbate-induced restructuring of surfaces – Cu(110)-(2×1)O. *Physical Review Letters* 67: 855–858.
148. Zeppenfeld P, Diercks V, Tolkes C, David R, and Krzyzowski MA (1998) Adsorption and growth on nanostructured surfaces. *Applied Surface Science* 130–132: 484–490.
149. Otero R, Naitoh Y, Rosei F, *et al.* (2004) One-dimensional assembly and selective orientation of lander molecules on an O-Cu template. *Angewandte Chemie International Edition* 43: 2092–2095.
150. Cicoira F, Miwa JA, Melucci M, Barbarella G, and Rosei F (2006) Ordered assembly of alpha-quinquethiophene on a copper oxide nanotemplate. *Small* 2: 1366–1371.
151. Cicoira F, Miwa JA, Perepichka DF, and Rosei F (2007) Molecular assembly of rubrene on a metal/metal oxide nanotemplate. *Journal of Physical Chemistry A* 111: 12674–12678.
152. Oehzelt M, Grill L, Berkebile S, Koller G, Netzer FP, and Ramsey MG (2007) The molecular orientation of para-sexiphenyl on Cu(110) and Cu(110) p(2×1)O. *Chemphyschem* 8: 1707–1712.
153. Cicoira F, Santato C, and Rosei F (2008) Two-dimensional nanotemplates as surface cues for the controlled assembly of organic molecules. In: *STM and AFM Studies on (Bio)Molecular Systems*, vol. 285, pp. 203–267. Berlin: Springer.
154. Lu B, Iimori T, Sakamoto K, Nakatsuji K, Rosei F, and Komori F (2008) Fullerene on nitrogen-adsorbed Cu(001) nanopatterned surfaces: From preferential nucleation to layer-by-layer growth. *Journal of Physical Chemistry C* 112: 10187–10192.
155. Ecija D, Trelka M, Urban C, *et al.* (2008). *Applied Physics Letters* 92: 223117.
156. Lee RN and Farnsworth HE (1965) LEED studies of adsorption on clean (100) copper surfaces. *Surface Science* 3: 461–479.
157. Leibsle FM, Flipse CFJ, and Robinson AW (1993) Structure of the Cu(100)-c(2×2)N surface – a scanning tunneling microscopy study. *Physical Review B* 47: 15865–15868.
158. Driver SM and Woodruff DP (2001) Nitrogen adsorption structures on Cu(100) and the role of a symmetry-lowering surface reconstruction in the c(2 x 2)-N phase. *Surface Science* 492: 11–26.
159. Hirjibehedin CF, Lutz CP, and Heinrich AJ (2006) Spin coupling in engineered atomic structures. *Science* 312: 1021–1024.
160. Choi T, Ruggiero CD, and Gupta JA (2008) Incommensurability and atomic structure of c(2×2)N/Cu(100): A scanning tunneling microscopy study. *Physical Review B* 78: 5.
161. Laibinis PE, Whitesides GM, Allara DL, Tao YT, Parikh AN, and Nuzzo RG (1991) Comparison of the structures and wetting properties of self-assembled monolayers of normal alkanethiols on the coinage metal surfaces, Cu, Ag, Au. *Journal of the American Chemical Society* 113: 7152–7167.
162. Ulman A (1996) Formation and structure of self-assembled monolayers. *Chemical Reviews* 96: 1533–1554.
163. Love JC, Estroff LA, Kriebel JK, Nuzzo RG, and Whitesides GM (2005) Self-assembled monolayers of thiolates on metals as a form of nanotechnology. *Chemical Reviews* 105: 1103–1169.
164. Tour JM, Jones L, Pearson DL, *et al.* (1995) Self-assembled monolayers and multilayers of conjugated thiols, alpha,omega-dithiols, and thioacetyl-containing adsorbates – understanding attachments between potential molecular wires and gold surfaces. *Journal of the American Chemical Society* 117: 9529–9534.

165. Xia YN and Whitesides GM (1998) Soft lithography. *Annual Review of Materials Science* 28: 153–184.
166. Diegoli S, Hamlett CAE, Leigh SJ, Mendes PM, and Preece JA (2007) Engineering nanostructures at surfaces using nanolithography. *Proceedings of the Institution of Mechanical Engineers Part G-Journal of Aerospace Engineering* 221: 589–629.
167. Madueno R, Raisanen MT, Silien C, and Buck M (2008) Functionalizing hydrogen-bonded surface networks with self-assembled monolayers. *Nature* 454: 618–621.
168. Barlow SM and Raval R (2003) Complex organic molecules at metal surfaces: Bonding, organisation and chirality. *Surface Science Reports* 50: 201–341.
169. De Feyter S and De Schryver FC (2003) Two-dimensional supramolecular self-assembly probed by scanning tunneling microscopy. *Chemical Society Reviews* 32: 139–150.
170. Barth JV, Costantini G, and Kern K (2005) Engineering atomic and molecular nanostructures at surfaces. *Nature* 437: 671–679.
171. Samori P (2005) Exploring supramolecular interactions and architectures by scanning force microscopies. *Chemical Society Reviews* 34: 551–561.
172. Barth JV (2007) Molecular architectonic on metal surfaces. *Annual Review of Physical Chemistry* 58: 375–407.
173. Auwarter W, Schiffrin A, Weber-Bargioni A, Pennec Y, Riemann A, and Barth JV (2008) Molecular nanoscience and engineering on surfaces. *International Journal of Nanotechnology* 5: 1171–1193.
174. Barth JV, Weckesser J, Cai CZ, et al. (2000) Building supramolecular nanostructures at surfaces by hydrogen bonding. *Angewandte Chemie International Edition* 39: 1230–1234.
175. Yokoyama T, Yokoyama S, Kamikado T, Okuno Y, and Mashiko S (2001) Selective assembly on a surface of supramolecular aggregates with controlled size and shape. *Nature* 413: 619–621.
176. Cai CZ, Bosch MM, Muller B, et al. (1999) Oblique incidence organic molecular beam deposition and nonlinear optical properties of organic thin films with a stable in-plane directional order. *Advanced Materials* 11: 745–749.
177. Cai CZ, Muller B, Weckesser J, et al. (1999) Model for in-plane directional ordering of organic thin films by oblique incidence organic molecular beam deposition. *Advanced Materials* 11: 750–754.
178. Weckesser J, De Vita A, Barth JV, Cai C, and Kern K (2001) Mesoscopic correlation of supramolecular chirality in one-dimensional hydrogen-bonded assemblies. *Physical Review Letters* 87: 096101.
179. Barth JV, Weckesser J, Trimarchi G, et al. (2002) Stereochemical effects in supramolecular self-assembly at surfaces: 1-D versus 2-D enantiomorphic ordering for PVBA and PEBA on Ag(111). *Journal of the American Chemical Society* 124: 7991–8000.
180. Duchamp DJ and Marsh RE (1969) Crystal structure of trimesic acid (benzene-1,3,5-tricarboxylic acid). *Acta Crystallographica Section B-Structural Crystallography and Crystal Chemistry* B 25: 5–19.
181. Kolotuchin SV, Thiessen PA, Fenlon EE, Wilson SR, Loweth CJ, and Zimmerman SC (1999) Self-assembly of 1,3,5-benzenetricarboxylic (trimesic) acid and its analogues. *Chemistry A European Journal* 5: 2537–2547.
182. Griessl S, Lackinger M, Edelwirth M, Hietschold M, and Heckl WM (2002) Self-assembled two-dimensional molecular host–guest architectures from trimesic acid. *Single Molecules* 3: 25–31.
183. Dmitriev A, Lin N, Weckesser J, Barth JV, and Kern K (2002) Supramolecular assemblies of trimesic acid on a Cu(100) surface. *Journal of Physical Chemistry B* 106: 6907–6912.
184. Lackinger M, Griessl S, Heckl WA, Hietschold M, and Flynn GW (2005) Self-assembly of trimesic acid at the liquid–solid interface – a study of solvent-induced polymorphism. *Langmuir* 21: 4984–4988.
185. Lei SB, Tahara K, De Schryver FC, Van der Auweraer M, Tobe Y, and De Feyter S (2008) One building block, two different supramolecular surface-confined patterns: Concentration in control at the solid–liquid interface. *Angewandte Chemie International Edition* 47: 2964–2968.
186. Ye YC, Sun W, Wang YF, et al. (2007) A unified model: Self-assembly of trimesic acid on gold. *Journal of Physical Chemistry C* 111: 10138–10141.
187. MacLeod JM, Ivasenko O, Perepichka DF, and Rosei F (2007) Stabilization of exotic minority phases in a multicomponent self-assembled molecular network. *Nanotechnology* 18: 424031.
188. Kampschulte L, Lackinger M, Maier AK, et al. (2006) Solvent induced polymorphism in supramolecular 1,3,5-benzenetribenzoic acid monolayers. *Journal of Physical Chemistry B* 110: 10829–10836.
189. Ruben M, Payer D, Landa A, et al. (2006) 2D supramolecular assemblies of benzene-1,3,5-triyl-tribenzoic acid: Temperature-induced phase transformations and hierarchical organization with macrocyclic molecules. *Journal of the American Chemical Society* 128: 15644–15651.
190. Kampschulte L, Griessl S, Heckl WM, and Lackinger M (2005) Mediated coadsorption at the liquid–solid interface: Stabilization through hydrogen bonds. *Journal of Physical Chemistry B* 109: 14074–14078.
191. Theobald JA, Oxtoby NS, Phillips MA, Champness NR, and Beton PH (2003) Controlling molecular deposition and layer structure with supramolecular surface assemblies. *Nature* 424: 1029–1031.
192. Nath KG, Ivasenko O, Miwa JA, et al. (2006) Rational modulation of the periodicity in linear hydrogen-bonded assemblies of trimesic acid on surfaces. *Journal of the American Chemical Society* 128: 4212–4213.
193. Nath KG, Ivasenko O, MacLeod JM, et al. (2007) Crystal engineering in two dimensions: An approach to molecular nanopatterning. *Journal of Physical Chemistry C* 111: 16996–17007.
194. Gambardella P, Stepanow S, Dmitriev A, et al. (2009) Supramolecular control of the magnetic anisotropy in two-dimensional high-spin Fe arrays at a metal interface. *Nature Materials* 8: 189–193.
195. Stepanow S, Lin N, and Barth JV (2008) Modular assembly of low-dimensional coordination architectures on metal surfaces. *Journal of Physics-Condensed Matter* 20: 15.
196. Barth JV (2009) Fresh perspectives for surface coordination chemistry. *Surface Science* 603: 1533–1541.
197. Lin N, Stepanow S, Ruben M, and Barth JV (2009) Surface-confined supramolecular coordination chemistry. In: *Templates in Chemistry III.*, vol. 287, pp. 1–44. Berlin: Springer.
198. Tait SL, Langner A, Lin N, et al. (2007) One-dimensional self-assembled molecular chains on Cu(100): Interplay between surface-assisted coordination chemistry and substrate commensurability. *Journal of Physical Chemistry C* 111: 10982–10987.
199. Stepanow S, Lin N, Payer D, et al. (2007) Surface-assisted assembly of 2D metal–organic networks that exhibit unusual threefold coordination symmetry. *Angewandte Chemie International Edition* 46: 710–713.

200. Schlickum U, Decker R, Klappenberger F, *et al.* (2007) Metal–organic honeycomb nanomeshes with tunable cavity size. *Nano Letters* 7: 3813–3817.
201. Lin N, Stepanow S, Vidal F, Barth JV, and Kern K (2005) Manipulating 2D metal–organic networks via ligand control. *Chemical Communications* 2005: 1681–1683.
202. Lin N, Stepanow S, Vidal F, *et al.* (2006) Surface-assisted coordination chemistry and self-assembly. *Dalton Transactions* 2006: 2794–2800.
203. Stepanow S, Lin N, Barth JV, and Kern K (2006) Surface-template assembly of two-dimensional metal–organic coordination networks. *Journal of Physical Chemistry B* 110: 23472–23477.
204. Classen T, Fratesi G, Costantini G, *et al.* (2005) Templated growth of metal–organic coordination chains at surfaces. *Angewandte Chemie International Edition* 44: 6142–6145.
205. Langner A, Tait SL, Lin N, Chandrasekar R, Ruben M, and Kern K (2008) Ordering and stabilization of metal–organic coordination chains by hierarchical assembly through hydrogen bonding at a surface. *Angewandte Chemie International Edition* 47: 8835–8838.
206. Furukawa M, Tanaka H, and Kawai T (2000) Formation mechanism of low-dimensional superstructure of adenine molecules and its control by chemical modification: A low-temperature scanning tunneling microscopy study. *Surface Science* 445: 1–10.
207. Berner S, Brunner M, Ramoino L, Suzuki H, Guntherodt HJ, and Jung TA (2001) Time evolution analysis of a 2D solid–gas equilibrium: A model system for molecular adsorption and diffusion. *Chemical Physics Letters* 348: 175–181.
208. Kroto HW, Heath JR, Obrien SC, Curl RF, and Smalley RE (1985) C-60 – Buckminsterfullerene. *Nature* 318: 162–163.
209. Taylor R and Walton DRM (1993) The chemistry of fullerenes. *Nature* 363: 685–693.
210. Allemand PM, Khemani KC, Koch A, *et al.* (1991) Organic molecular soft ferromagnetism in a fullerene-C_{60}. *Science* 253: 301–303.
211. Prato M (1997) [60] Fullerene chemistry for materials science applications. *Journal of Materials Chemistry* 7: 1097–1109.
212. Bethune DS, Johnson RD, Salem JR, Devries MS, and Yannoni CS (1993) Atoms in carbon cages – the structure and properties of endohedral fullerenes. *Nature* 366: 123–128.
213. Saunders M, Cross RJ, JimenezVazquez HA, Shimshi R, and Khong A (1996) Noble gas atoms inside fullerenes. *Science* 271: 1693–1697.
214. Shinohara H (2000) Endohedral metallofullerenes. *Reports on Progress in Physics* 63: 843–892.
215. Bingel C (1993) Cyclopropylation of fullerenes. *Chemische Berichte-Recueil* 126: 1957–1959.
216. Hummelen JC, Knight BW, Lepeq F, Wudl F, Yao J, and Wilkins CL (1995) Preparation and characterization of fulleroid and methanofullerene derivatives. *Journal of Organic Chemistry* 60: 532–538.
217. Bosi S, Da Ros T, Spalluto G, and Prato M (2003) Fullerene derivatives: An attractive tool for biological applications. *European Journal of Medicinal Chemistry* 38: 913–923.
218. Guldi DM (2000) Fullerenes: Three dimensional electron acceptor materials. *Chemical Communications* 2000: 321–327.
219. Diederich F and Gomez-Lopez M (1999) Supramolecular fullerene chemistry. *Chemical Society Reviews* 28: 263–277.
220. Stepanow S, Lingenfelder M, Dmitriev A, *et al.* (2004) Steering molecular organization and host–guest interactions using two-dimensional nanoporous coordination systems. *Nature Materials* 3: 229–233.
221. Fasel R, Agostino RG, Aebi P, and Schlapbach L (1999) Unusual molecular orientation and frozen librational motion of C-60 on Cu(110). *Physical Review B* 60: 4517–4520.
222. Li M, Deng K, Lei SB, *et al.* (2008) Site-selective fabrication of two-dimensional fullerene arrays by using a supramolecular template at the liquid–solid interface. *Angewandte Chemie International Edition* 47: 6717–6721.
223. Meier C, Landfester K, and Ziener U (2008) Topological selectivity in a supramolecular self-assembled host–guest network at the solid–liquid interface. *Journal of Physical Chemistry C* 112: 15236–15240.
224. Spillmann H, Kiebele A, Stohr M, *et al.* (2006) A two-dimensional porphyrin-based porous network featuring communicating cavities for the templated complexation of fullerenes. *Advanced Materials* 18: 275–279.
225. Pan C, Sampson MP, Chai Y, Hauge RH, and Margrave JL (1991) Heats of sublimation from a polycrystalline mixture of C_{60} and C_{70}. *Journal of Physical Chemistry* 95: 2944–2946.
226. Repp J, Moresco F, Meyer G, Rieder K-H, Hyldgaard P, and Persson M (2000) Substrate mediated long-range oscillatory interaction between adatoms: Cu /Cu(111). *Physical Review Letters* 85: 2981.
227. Kiebele A, Bonifazi D, Cheng FY, *et al.* (2006) Adsorption and dynamics of long-range interacting fullerenes in a flexible, two-dimensional, nanoporous porphyrin network. *Chemphyschem* 7: 1462–1470.
228. Ishikawa Y, Ohira A, Sakata M, Hirayama C, and Kunitake M (2002) A two-dimensional molecular network structure of trimesic acid prepared by adsorption-induced self-organization. *Chemical Communications* 2002: 2652–2653.
229. Griessl SJH, Lackinger M, Jamitzky F, Markert T, Hietschold M, and Heckl WM (2004) Room-temperature scanning tunneling microscopy manipulation of single C-60 molecules at the liquid–solid interface: Playing nanosoccer. *Journal of Physical Chemistry B* 108: 11556–11560.
230. Gade LH, Galka CH, Hellmann KW, *et al.* (2002) Tetraaminoperylenes: Their efficient synthesis and physical properties. *Chemistry – A European Journal* 8: 3732–3746.
231. Stohr M, Wahl M, Galka CH, Riehm T, Jung TA, and Gade LH (2005) Controlling molecular assembly in two dimensions: The concentration dependence of thermally induced 2D aggregation of molecules on a metal surface. *Angewandte Chemie International Edition* 44: 7394–7398.
232. Stohr M, Wahl M, Spillmann H, Gade LH, and Jung TA (2007) Lateral manipulation for the positioning of molecular guests within the confinements of a highly stable self-assembled organic surface network. *Small* 3: 1336–1340.
233. Mena-Osteritz E and Bauerle P (2006) Complexation of C-60 on a cyclothiophene monolayer template. *Advanced Materials* 18: 243–246.
234. Pan GB, Cheng XH, Hoger S, and Freyland W (2006) 2D supramolecular structures of a shape-persistent macrocycle and co-deposition with fullerene on HOPG. *Journal of the American Chemical Society* 128: 4218–4219.
235. Otsubo T, Aso Y, and Takimiya K (2002) Functional oligothiophenes as advanced molecular electronic materials. *Journal of Materials Chemistry* 12: 2565–2575.
236. Mena-Osteritz E (2002) Superstructures of self-organizing thiophenes. *Advanced Materials* 14: 609–616.

237. Griessl SJH, Lackinger M, Jamitzky F, Markert T, Hietschold M, and Heckl WA (2004) Incorporation and manipulation of coronene in an organic template structure. *Langmuir* 20: 9403–9407.
238. Walzer K, Sternberg M, and Hietschold M (1998) Formation and characterization of coronene monolayers on HOPG(0001) and MoS2(0001): A combined STM/STS and tight-binding study. *Surface Science* 415: 376–384.
239. Lackinger M, Griessl S, Heckl WM, and Hietschold M (2002) STM and STS of coronene on HOPG(0001) in UHV – adsorption of the smallest possible graphite flakes on graphite. *Analytical and Bioanalytical Chemistry* 374: 685–687.
240. Chernichenko KY, Sumerin VV, Shpanchenko RV, Balenkova ES, and Nenajdenko VG (2006) "Sulflower": A new form of carbon sulfide. *Angewandte Chemie International Edition* 45: 7367–7370.
241. Ivasenko O, MacLeod JM, Chernichenko KY, et al. (2009) Supramolecular assembly of heterocirculenes in 2D and 3D. *Chemical Communications* 10: 1121–1280.
242. Blunt M, Lin X, Gimenez-Lopez MD, Schroder M, Champness NR, and Beton PH (2008) Directing two-dimensional molecular crystallization using guest templates. *Chemical Communications* 2008: 2304–2306.
243. Stepanow S, Lin N, Barth JV, and Kern K (2006) Non-covalent binding of fullerenes and biomolecules at surface-supported metallosupramolecular receptors. *Chemical Communications* 2006: 2153–2155.
244. Spillmann H, Dmitriev A, Lin N, Messina P, Barth JV, and Kern K (2003) Hierarchical assembly of two-dimensional homochiral nanocavity arrays. *Journal of the American Chemical Society* 125: 10725–10728.
245. Perepichka DF and Rosei F (2009) Extending polymer conjugation into the second dimension. *Science* 323: 216–217.
246. Sakamoto J, van Heijst J, Lukin O, and Schluter AD (2009) Two-dimensional polymers: Just a dream of synthetic chemists? *Angewandte Chemie International Edition* 48: 1030–1069.
247. Gourdon A (2008) On-surface covalent coupling in ultrahigh vacuum. *Angewandte Chemie International Edition* 47: 6950–6953.
248. Barner J, Mallwitz F, Shu LJ, Schluter AD, and Rabe JP (2003) Covalent connection of two individual polymer chains on a surface: An elementary step towards molecular nanoconstructions. *Angewandte Chemie International Edition* 42: 1932–1935.
249. Grim PCM, De Feyter S, Gesquiere A, et al. (1997) Submolecularly resolved polymerization of diacetylene molecules on the graphite surface observed with scanning tunneling microscopy. *Angewandte Chemie International Edition* 36: 2601–2603.
250. Wegner G (1969) Topochemical reactions of monomers with conjugated triple bonds. 1. Polymerization of 2.4-hexadiyn-1.6-diols deivative in crystalline state. *Zeitschrift Fur Naturforschung Part B-Chemie Biochemie Biophysik Biologie Und Verwandten Gebiete* B 24: 824.
251. Wegner G (1972) Topochemical reactions of monomers with conjugated triple bonds. *Makromolekulare Chemie* 154: 35–48.
252. Lio A, Reichert A, Nagy JO, Salmeron M, and Charych DH (1996) Atomic force microscope study of chromatic transitions in polydiacetylene thin films. *Journal of Vacuum Science and Technology B* 14: 1481–1485.
253. Batchelder DN, Evans SD, Freeman TL, Haussling L, Ringsdorf H, and Wolf H (1994) Self-assembled monolayers containing polydiacetylenes. *Journal of the American Chemical Society* 116: 1050–1053.
254. Enkelmann V (1984) Structural aspects of the topochemical polymerization of diacetylenes. *Advances in Polymer Science* 63: 91–136.
255. Okawa Y and Aono M (2001) Nanoscale control of chain polymerization. *Nature* 409: 683–684.
256. Okawa Y and Aono M (2001) Linear chain polymerization initiated by a scanning tunneling microscope tip at designated positions. *Journal of Chemical Physics* 115: 2317–2322.
257. Neumann W and Sixl H (1981) The mechanism of the low-termperature polymerization reaction in diacetylene crystals. *Chemical Physics* 58: 303–312.
258. Albrecht TR, Dovek MM, Kirk MD, Lang CA, Quate CF, and Smith DPE (1989) Nanometer-scale hole formation on graphite using a scanning tunneling microscope. *Applied Physics Letters* 55: 1727–1729.
259. Sullivan SP, Schmeders A, Mbugua SK, and Beebe TP (2005) Controlled polymerization of substituted diacetylene self-organized monolayers confined in molecule corrals. *Langmuir* 21: 1322–1327.
260. Takajo D, Okawa Y, Hasegawa T, and Aono M (2007) Chain polymerization of diacetylene compound multilayer films on the topmost surface initiated by a scanning tunneling microscope tip. *Langmuir* 23: 5247–5250.
261. Nishio S, I-i D, Matsuda H, Yoshidome M, Uji-i H, and Fukumura H (2005) Formation of molecular wires by nanospace polymerization of a diacetylene derivative induced with a scanning tunneling microscope at a solid–liquid interface. *Japanese Journal of Applied Physics* 44: 5417–5420.
262. Miura A, De Feyter S, Abdel-Mottaleb MMS, et al. (2003) Light- and STM-tip-induced formation of one-dimensional and two-dimensional organic nanostructures. *Langmuir* 19: 6474–6482.
263. Takami T, Ozaki H, Kasuga M, et al. (1997) Periodic structure of a single sheet of a clothlike macromolecule (atomic cloth) studied by scanning tunneling microscopy. *Angewandte Chemie International Edition* 36: 2755–2757.
264. Sakaguchi H, Matsumura H, and Gong H (2004) Electrochemical epitaxial polymerization of single-molecular wires. *Nature Materials* 3: 551–557.
265. Sakaguchi H, Matsumura H, Gong H, and Abouelwafa AM (2005) Direct visualization of the formation of single-molecule conjugated copolymers. *Science* 310: 1002–1006.
266. Matena M, Riehm T, Stohr M, Jung TA, and Gade LH (2008) Transforming surface coordination polymers into covalent surface polymers: Linked polycondensed aromatics through oligomerization of N-heterocyclic carbene intermediates. *Angewandte Chemie International Edition* 47: 2414–2417.
267. Ullmann F and Bielecki J (1901) Synthesis in the Biphenyl series. (I. Announcement). *Berichte Der Deutschen Chemischen Gesellschaft* 34: 2174–2185.
268. Hassan J, Sevignon M, Gozzi C, Schulz E, and Lemaire M (2002) Aryl–aryl bond formation one century after the discovery of the Ullmann reaction. *Chemical Reviews* 102: 1359–1469.
269. Xi M and Bent BE (1992) Iodobenzene on Cu(111) – formation and coupling of adsorbed phenyl groups. *Surface Science* 278: 19–32.
270. Xi M and Bent BE (1993) Mechanisms of the Ullmann reaction in adsorbed monolayers. *Journal of the American Chemical Society* 115: 7426–7433.
271. McCarty GS and Weiss PS (2004) Formation and manipulation of protopolymer chains. *Journal of the American Chemical Society* 126: 16772–16776.

272. Lipton-Duffin JA, Ivasenko O, Perepichka DF, and Rosei F (2009) Synthesis of polyphenylene molecular wires by surface-confined polymerization. *Small* 5: 592–597.
273. Grill L, Dyer M, Lafferentz L, Persson M, Peters MV, and Hecht S (2007) Nano-architectures by covalent assembly of molecular building blocks. *Nature Nanotechnology* 2: 687–691.
274. Champness NR (2007) Building with molecules. *Nature Nanotechnology* 2: 671–672.
275. Veld MI, Iavicoli P, Haq S, Amabilino DB, and Raval R (2008) Unique intermolecular reaction of simple porphyrins at a metal surface gives covalent nanostructures. *Chemical Communications* 2008: 1536–1538.
276. Weigelt S, Busse C, Bombis C, et al. (2007) Covalent interlinking of an aldehyde and an amine on a Au(111) surface in ultrahigh vacuum. *Angewandte Chemie International Edition* 46: 9227–9230.
277. Weigelt S, Busse C, Bombis C, et al. (2008) Surface synthesis of 2D branched polymer nanostructures. *Angewandte Chemie International Edition* 47: 4406–4410.
278. Treier M, Fasel R, Champness NR, Argent S, and Richardson NV (2009) Molecular imaging of polyimide formation. *Physical Chemistry Chemical Physics* 11: 1209–1214.
279. Treier M, Richardson NV, and Fasel R (2008) Fabrication of surface-supported low-dimensional polyimide networks. *Journal of the American Chemical Society* 130: 14054–14055.
280. Zwaneveld NAA, Pawlak R, Abel M, et al. (2008) Organized formation of 2D extended covalent organic frameworks at surfaces. *Journal of the American Chemical Society* 130: 6678–6679.
281. Takami K, Mizuno J, Akai-Kasaya M, Saito A, Aono M, and Kuwahara Y (2004) Conductivity measurement of polydiacetylene thin films by double-tip scanning tunneling microscopy. *Journal of Physical Chemistry B* 108: 16353–16356.
282. Akai-Kasaya M, Shimizu K, Watanabe Y, Saito A, Aono M, and Kuwahara Y (2003) Electronic structure of a polydiacetylene nanowire fabricated on highly ordered pyrolytic graphite. *Physical Review Letters* 91: 255501.
283. Takami K, Kuwahara Y, Ishii T, Akai-Kasaya M, Saito A, and Aono M (2005) Significant increase in conductivity of polydiacetylene thin film induced by iodine doping. *Surface Science* 591: L273–L279.
284. Wang QH and Hersam MC (2008) Orthogonal self-assembly of interconnected one-dimensional inorganic and organic nanostructures on the Si(100) surface. *Journal of the American Chemical Society* 130: 12896–12897.
285. MacLeod JM, Miwa RH, Srivastava GP, and McLean AB (2005) The electronic origin of contrast reversal in bias-dependent STM images of nanolines. *Surface Science* 576: 116–122.
286. Harikumar KR, Lim TB, McNab IR, et al. (2008) Dipole-directed assembly of lines of 1,5-dichloropentane on silicon substrates by displacement of surface charge. *Nature Nanotechnology* 3: 222–228.

2 Bio-Mediated Assembly of Ordered Nanoparticle Superstructures

W L Cheng, S J Tan, M J Campolongo, M R Hartman, J S Kahn, and D Luo, Cornell University, Ithaca, NY, USA

© 2010 Elsevier B.V. All rights reserved.

2.1 Introduction

One of the central goals of nanoscience is to build small structures for advanced materials design, high-performance nanodevices and miniaturized electronics. Inorganic nanoparticles are particularly attractive building blocks for such purposes owing to their unique optical, electronic, magnetic, and catalytic properties [1–6], many of which can be tailored simply by tuning size, shape, and surface functionality of nanoparticles without changing their material composition. Thus far, significant progress has been made using wet chemistry strategies to synthesize high-quality nanoparticles from a variety of inorganic materials, including gold, silver, iron oxide, and semiconductors [1–6]. Manipulating the synthesis conditions allows for rational control of nanoparticle morphology and provides a means to tailor material properties in the process. Due to their unique physical properties, nanoparticles are often described as artificial atoms [7–9]. Synthetic advances allow for precise control over structural parameters affecting nanoparticle formation, enabling the properties of these artificial atoms to be tailored accordingly. Hence, a nanoparticle periodic table, in which the nanoparticle elements can be obtained through various wet chemistry syntheses, has been proposed by Murray and coworkers [10].

In addition to engineering size, shape, and surface functionality of the individual nanoparticle atoms, assembly of these artificial atoms into well-defined highly ordered structures affords additional control over material properties. In particular, the collective properties of highly ordered nanoparticle assemblies can be different from those of materials in the bulk phase, isolated nanoparticles, and disordered nanoparticle assemblies [9,11–18]. For example, ordered silver nanoparticle arrays show coherent vibrational modes that only appear in highly ordered nanoparticle superlattices [19], while synergistic effects in superlattices can lead to enhanced p-type conductivity [20].

The collective properties of these nanoparticle assemblies originate from the electromagnetic coupling among nanoparticles [21]. Despite recent progress in generating highly ordered nanoparticle superlattices, the design of new materials by using the elements in the nanoparticle periodic table is still in its infancy. The full potential of designing nanoparticle-based superstructures cannot be achieved until we are capable of manipulating the nanoparticles in a fashion analogous to the way chemists manipulate atoms into versatile architectures. In the nanoparticle-based world, spatially defined stoichiometric nanoparticle assemblies [22] (nanoparticle molecules) and extended arrays of nanoparticles (nanoparticle superlattices) will lay the foundation for next-generation materials and optoelectronic devices.

A critical step in building the nanoparticle molecular world is to control nanoparticle binding interactions, particularly the vectorial and robust bonds that can spatially coordinate nanoparticles. This turns out to be a challenge as nonspecific van der Waals and electrostatic forces often dominate the interparticle potential [23–29]. One way to manipulate the interparticle potential is to coat nanoparticles with organic ligands that act as a soft steric layer to counteract the strong van der Waals attractions between nanoparticles. For this purpose, organic ligands that have been typically used include surfactant molecules and biomolecules. Biomolecules are particularly appealing due to their chemical and physical characteristics: (1) bio-interactions are highly specific (e.g., Watson–Crick base-pairing and antigen–antibody interactions); (2) biomolecular length and topology can be rationally tuned by available technologies (e.g., DNA molecular length can be customized from the nanometer to micrometer scale; DNA topology can also be engineered into linear, branched, or more complex shapes [30,31]); and (3) biomolecules are inherently adaptive to specific environments [32].

Several excellent review papers have described the bio-based assembly of nanoparticles [12,33–37]. Herein, we specifically focus on the use of biomolecules for building highly ordered assemblies. In particular, we will discuss wet chemistry synthesis and bioconjugation of single-crystalline nanoparticles, interaction energetics between biofunctionalized nanoparticles, and the current progress in building the nanoparticle molecular world. We expect our contribution to supplement earlier reviews [34–38] and inspire innovation in nanobiotechnology toward the discovery of novel materials and development of smart nanoparticle-based devices.

2.2 Synthesis and Biofunctionalization of Nanoparticles

2.2.1 Wet Chemistry Synthesis of Nanoparticles

Research interest in nanoparticle superlattices is largely driven by the unique collective properties arising from synergistic coupling effects between nanoparticles. The control over size, shape, and interparticle spacing is critical for regulating these collective properties for a variety of applications in nanoscience and nanotechnology. Wet chemistry synthesis, which is emerging as an inexpensive route to synthesize single-crystalline nanoparticles in large quantities, is the foundation for fabrication of high-quality superlattice structures. The synthesis and modular assembly of nanoparticles allows us to exploit their unique properties, which can lead to novel applications in catalysis, electronics, photonics, as well as chemical and biological sensing. So far, significant progress has been achieved in synthesizing nanoparticles with sophisticated control over their structural parameters. For example, nanometer-sized metallic spheres, rods, wires, cubes, and other complex shapes have been reported via wet chemistry approaches. In this chapter, we briefly summarize some of the representative wet chemistry methods used to synthesize highly crystalline nanoparticles for the purpose of bioconjugation and subsequent assembly of nanoparticle superlattices.

Typical spherical colloidal nanoparticle synthesis revolves around the dynamic solvation of inorganic cores with surfactant molecules [39]. In order to achieve solutions that can maintain distinct nanoparticle crystals within a broad possible range of sizes, it is necessary to choose chemicals and conditions that allow for nucleation and continued, controlled growth from these nucleated crystals [39–43]. This growth is driven by factors including the advantageous kinetic breakdown of molecular precursors into functional chemical monomers, the temporal absorption and desorption of surfactant molecules on the inorganic surface to allow for controlled growth depending on face-specific surface energies, and aggregation reduction by the presence of these surfactants.

Initial efforts in this field focused on various methods of spherical nanoparticle formation: copolymer stabilized solution [44], arrested precipitation [45], inverse micelles [46], and hot injection into a coordinating solvent [40]. Micelles are able to provide a partitioned microenvironment, enabling confined interaction between molecular precursors that results in nanoparticle formation. Hot injection uses a supersaturated solution of molecular precursors to induce nucleation while controlling growth based on a carefully tuned temperature profile and an optimal surfactant to solvent ratio; this allows the surfactant to surround the growing crystal with a dynamic protective monolayer. Current methods are typically based on the principles underlying these procedures, with a few notable exceptions including vapor-phase synthesis [47] and laser ablation [48]. The synthesis and application of semiconducting nanoparticles, such as TiO_2 and CdSe, have also been thoroughly explored [42].

When considering the synthesis of nanoparticles for biological applications, one would expect water to be the ideal solvent because it is inexpensive, non-toxic, and amenable to biofunctionalization [49,50]. However, it does not always offer the stable, tailored environment that a carefully chosen organic solvent would provide. Earlier efforts reported difficulties in producing smaller nanoparticles, which is in part due to the stability of nanoparticles in buffered environments [51]. In addition, there is an increased likelihood of Ostwald ripening, in which particles below a critical size dissolve into monomers and adsorb onto the larger nanoparticles already present in solution [52,53].

Various methods have been employed to produce nanoparticles in the aqueous phase. Seed-mediated growth has shown great success in producing monodisperse, single-crystalline particles in aqueous solution [54,55]. In addition, particles can be transferred between organic and aqueous phases via ligand exchange [56] or through the introduction of a surfactant bilayer [57,58]. These methods can alter the polarity of the particle's protective layer to be compatible with the desired solvent. For example, oleic acid-stabilized particles in hexane can be transferred into the aqueous phase by using α-cyclodextrin to create a surfactant bilayer [57]. In general, aliphatic ligands that are common in organic solvents must be replaced or modified by more hydrophilic molecules.

Due to shape-dependent properties, significant efforts have been directed toward fabrication of nanoparticles beyond spherical morphologies. As opposed to a nanosphere, a nanorod possesses different surface energies and strains on its crystal faces, allowing for increased functionality from reduced symmetry [59,60]. Multiple plasmon bands, notably the longitudinal and transverse bands, cause surface plasmon resonance modes to differ over the surface of the crystal, allowing for anisotropic light absorption and scattering [55]. These differences contribute to surface-enhanced Raman scattering (SERS) and surface-enhanced fluorescence. In general, many of these characteristics are shape dependent and are only present in asymmetric materials.

The synthesis of anisotropic nanoparticles can be achieved through seed-mediated growth, whereby growth occurs epitaxially on seed particles [2,40,54,55,61,62]. The synthesis techniques introduced here are mainly intended for face-centered cubic (f.c.c.) metals such as gold and silver, which are widely used as model systems in nanoparticle assemblies. Seeds can be generated by reducing metal salts in water at room temperature with a strong reducing agent such as sodium borohydride [55]. The nanorod growth occurs with the addition of more metal salt, a weak reducing agent, and a surfactant (or surfactant mixture) that directs structure. Murphy and others provided evidence that the surfactant forms a bilayer on selective faces of the growing crystal, favoring adsorption to the lower-energy longitudinal faces and directing metal ions to the ends of the rod [63,64]. A schematic of how this synthesis mechanism can lead to anisotropic structures is shown in **Figure 1** [39]. In addition, one would expect smaller seeds to yield nanorods with higher aspect ratios, as the diameter of the original seed directly influences the final width of the nanorod. Although it is clear that both of these parameters significantly affect the final shape and size, the synthesis of anisotropic nanoparticles may also be impacted by other parameters, including temperature, reactant concentration, and reactant molar ratios [65].

Significant research efforts have been focused on the photochemical synthesis of nanoparticle prisms and triangles [65–68]. Although common chemical reduction methods and surfactant face-blocking have typically been used to direct growth in the production of plate-like morphologies, visible light can also be used to specifically target growth on different crystal faces of a seed particle based on the different surface plasmon resonances (SPRs). By matching wavelengths with the dipole and quadrupole SPR modes of a silver nanoparticle, the corresponding surfaces are excited, and reduced atoms are deposited preferentially onto these faces [69]. Longer exposure times or higher surfactant concentrations directly

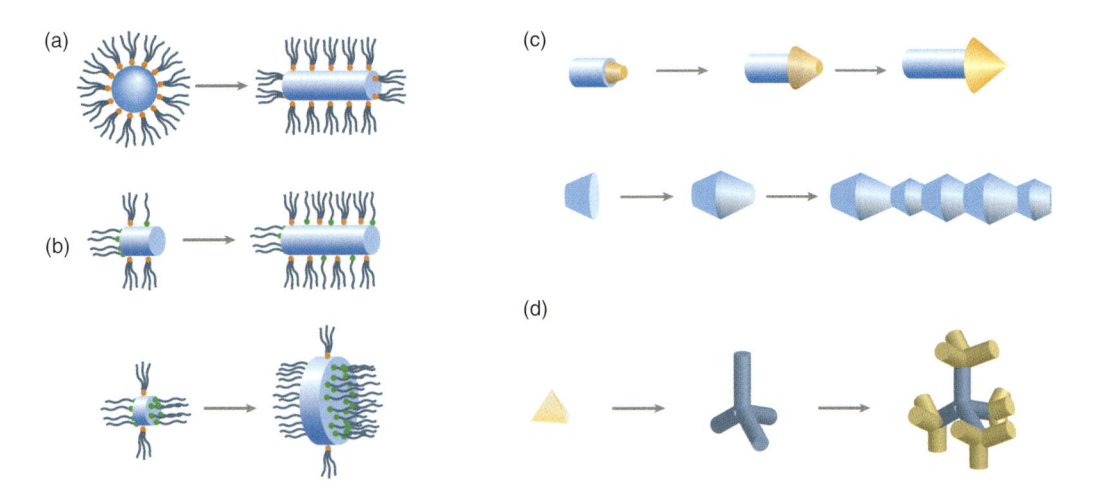

Figure 1 Anistropic synthesis of nanoparticles. (a) Certain faces of the nanoparticle possess higher energy than others; high-energy faces grow more quickly than low-energy faces. (b) Kinetic shape control through selective adhesion. Organic molecules selectively bind to certain faces, further regulating nanoparticle growth. (c) Sequential elimination of a high-energy facet. (d) Controlled branching of colloidal nanoparticles. Reproduced with permission from Yin Y and Alivisatos AP (2004) Colloidal nanocrystal synthesis and the organic–inorganic interface. *Nature* 437: 664–670.

result in increased edge length on the exposed face [65]. Other techniques using ultraviolet light, laser ablation, radiolysis, and biologically-mediated systems have been shown to produce similar nanoprisms, but so far little control over the resulting platelet's thickness has been shown. However, the photochemical control of edge lengths over a large range (from <40 nm to hundreds of nanometers) allows for the production of nanoparticles with multiple plasmon resonances. Further efforts in manipulating nanoparticle growth have yielded structures such as cubes [70], polyhedra [71], and cages [72]. For example, Yang and coworkers used a procedure that was modified from the synthesis of silver nanocubes to produce polyhedral shapes using polyvinylpyrrolidone (PVP) as a stabilizing agent while reducing a metal salt by a diol solvent at a high temperature (~180°C) [71]. Another attempt at creating particles with distinct optical properties and true three-dimensional (3D) anisotropy was pursued by Hafner and co-workers [73] in the synthesis of star-shaped gold nanoparticles. Although this structure was not produced in high yields, the nanostars present an opportunity to detect orientations in three dimensions by making use of its multidirectional scattering.

2.2.2 Functionalization of Nanoparticles with Biomolecules

Surface functionalization of nanoparticles is essential in imparting specificity and control to the interactions of nanoparticles and significantly affects the assembly and organization of higher-ordered nanoparticles. The functionalization of nanoparticles with biomolecules has been widely investigated for its importance in interfacing inorganic nanoparticles with biology for biotechnological applications. The techniques typically used for functionalization include electrostatic binding, specific affinity interactions, and covalent binding [1,33,74–76]. In particular, there are several spatial modes in which the nanoparticle can be functionalized: N:1 functionalization, in which multiple ligands are attached onto the nanoparticle without any specific spatial location; 1:1 functionalization, in which a single ligand is conjugated onto the nanoparticle; and anisotropic functionalization, in which ligands are bound to a specific site on the surface of the nanoparticle. These will be discussed in detail in the following sections.

2.2.2.1 N:1 functionalized nanoparticles

Electrostatic adsorption of biomolecules onto nanoparticles is a relatively straightforward nonspecific binding technique that typically yields N:1 functionalized nanoparticles. Positively charged biomolecules (e.g., peptides or proteins) readily bind onto nanoparticles stabilized with negatively charged molecules (e.g., citrate, tartrate, or oleic acid). For example, citrate-stabilized gold nanoparticles were functionalized with immunoglobin G (IgG) at pH 6.0, which is below the isoelectric point of the protein [77]. As a result, the positively charged side chains of the protein were able to bind to the negatively charged citrate groups on the nanoparticles. Using this method, it is also possible to selectively bind specific proteins with different isoelectric points onto negatively charged nanoparticle surfaces simply by changing the pH of the solution [78]. Bovine serum albumin (BSA) is another protein that can be attached onto nanoparticles. The available lysine residues on its surface allow for convenient chemical conjugation to other biological molecules such as peptides [79]. In addition, enzymes, such as horseradish peroxidase (HRP) can be adsorbed onto gold nanoparticle surfaces for electrochemical studies [80].

Nucleic acids can also be attached onto nanoparticles via electrostatic interactions. The negative charge of the phosphate backbone in double-stranded DNA (dsDNA), which is relatively rigid, prevents it from adsorbing onto the negatively charged citrate-capped surface of gold nanoparticles [81]. However, single-stranded DNA (ssDNA) is sufficiently flexible and able to expose its bases and conceal the phosphate backbone, allowing it to adsorb onto these nanoparticles. In contrast, dsDNA is able to bind onto the Cd^{2+}-rich surface of CdS nanoparticles via electrostatic interactions [82].

Covalent coupling of biomolecular ligands onto nanoparticles is a reliable and robust route to add both function and stability to nanoparticles in solution. Unlike nonspecific electrostatic interactions, covalent binding can occur with relatively specific chemistry between functional groups and surfaces. The chemisorption of thiol derivatives onto the surfaces of metal nanoparticles is one of the most widely used strategies for covalent functionalization. Functional proteins can covalently bind onto gold nanoparticles by means of the thiol groups in cysteine residues, forming stable Au–S bonds [83]. For cases in which there are no available groups on the protein surface, thiol modifications can be incorporated into proteins by introducing cysteine residues via genetic engineering [84]. Alternatively, thiol groups can also be chemically added to proteins using reagents such as 2-iminothiolane (Traut's reagent) [85] or N-succinimidyl 3-(2-pyridyldithio)-propionate (SPDP) [86]. In addition, thiolated DNA has been attached onto gold nanoparticles for the organization of higher-ordered structures [87,88].

Nanoparticles can also be primed with chemical groups that can covalently bind to biomolecules. For example, the amino acid cysteine can bind onto the surface of gold nanoparticles via its thiol linkage, surrounding the crystals with amine and carboxyl groups that can be used for further conjugation or polymerization [89]. Additionally, disteaoryl-phospholipid-coated magnetic nanoparticles with a maleimide functional group on one end can be functionalized with thiol-terminated DNA by reaction of maleimide with the thiol to form a stable thiol–ether linkage [90]. Glucose oxidase (GOD) was also covalently anchored onto gold nanoparticles coated with carboxyl-terminated alkanethiols through an 1-ethyl-3-(3-dimethyl aminopropyl) carbodiimide hydrochloride (EDC)/N-hydroxysuccinimide (NHS) coupling reaction that resulted in the formation of an amide bond between GOD and the nanoparticles [91].

Semiconductor nanoparticles, such as quantum dots, that are synthesized from organic routes typically lack aqueous solubility and compatibility. Hence, it is often necessary for these nanoparticles to undergo some form of surface modification before they can be directly functionalized with biomolecules. A detailed coverage of this topic is discussed in the review of quantum dot bioconjugates by Medintz et al. [5].

The functionalization of nanoparticles can also occur through specific affinity interactions between two biomolecules. Nanoparticles can be primed with biotin, which enables site-specific binding between biotin and streptavidin – one of the most widely used conjugation techniques due to its extremely high association constant [92,93].

2.2.2.2 1:1 functionalized nanoparticles

Biomolecule ligands can also attach onto nanoparticle surfaces with a 1:1 ratio, which affords greater spatial specificity and control in terms of ligand–ligand recognition. For example, 1:1 nanoparticle monoconjugates allow for the self-assembly of nanoparticle molecules as well as nanoparticle polymer chains [87,94–98].

Several strategies have been developed to obtain 1:1 conjugates. One approach used small 1.4 nm nanoparticles, allowing for attachment of only a single N-propylmaleimide substituent as a result of surface area restrictions. Hence, thiol-terminated ssDNA can then be conjugated onto the nanoparticle in a 1:1 ratio via thiol–maleimide chemistry [87]. In a similar fashion, lysine can be monoconjugated onto 2 nm nanoparticles through a solid-phase reaction [99]. For larger nanoparticles, careful control of the biomolecule-to-nanoparticle ratio is required for obtaining 1:1 monoconjugates. For example, 5 and 10 nm gold nanoparticles functionalized with single strands of DNA were obtained by optimizing the DNA:Au ratio, and subsequently isolated using gel electrophoresis [100]. There is limited resolution for ssDNA lengths below 50 bases, but this was overcome by hybridizing a longer ssDNA strand for reversible length extension. Alternatively, short ssDNA strands conjugated onto gold nanoparticles were obtained upon enzymatic cleavage of longer strands by restriction endonucleases [101]. Anion-exchange high-performance liquid chromatography (AE-HPLC) was also used to isolate monofunctionalized short DNA conjugates (<40 bases) on much larger nanoparticles (>20 nm) [102].

2.2.2.3 Anisotropically functionalized nanoparticles

Anistropically functionalized particles containing functional ligands on site-specific regions of the nanoparticle surface can provide a greater level of control in self-assembly processes that involve interactions and molecular recognition between surface ligands. Several approaches to obtaining larger (>100 nm) anisotropically functionalized particles have already been reviewed [103] and can be extended to smaller nanoparticles. The methodologies typically involve some form of geometric restriction that either prevents or limits ligand binding to specific sites on the nanoparticle surface.

One strategy is to immobilize nanoparticles functionalized with two different ssDNA sequences (A and B) onto a substrate containing a complementary sequence to one of the ssDNA sequences (A′) [104]. As a result of the immobilization, only one hemisphere is exposed and made available for further functionalization via hybridization of a linker strand of the other complementary ssDNA sequence (B′). With this strategy, Janus (two-faced) nanoparticle dimers and clusters can be readily obtained with relatively high throughput [104].

Another strategy is to use microparticles functionalized with ssDNA (sequence C_1C_2) as a geometric restriction template [105]. Nanoparticles functionalized with shorter strands of ssDNA (sequence C_1') and an extension strand (sequence C_2') can then hybridize onto the microparticles, resulting in duplex linkers (sequence C_1C_2-$C_1'C_2'$) between the nanoparticles and the microparticle. The strands C_1' and C_2' are ligated and subsequently released from the microparticle via thermal melting. This leads to formation of assymetrically functionalized nanoparticles containing extended ssDNA strands at a specific site.

One additional strategy involved a linker strand (sequence D_1D_2) connecting ssDNA-functionalized nanoparticles (sequence D_1) and ssDNA-functionalized microparticles (sequence D_2) [106]. In this approach, sequence D_2 is designed to be shorter than D_1 and the anisotropically functionalized nanoparticles are released based on the difference in thermal melting properties of each sequence.

Anisotropic functionalization can also be incorporated into template-based fabrication procedures. For example, DNA was attached onto the inner surfaces of hollow gold nanopyramids while their outer surfaces were embedded in a template during the fabrication process [107]. The outer surfaces were functionalized with an alkanethiol layer during a wet etch to remove the template.

2.3 Interactions between Biofunctionalized Nanoparticles

The elegance of biomolecule-conjugated nanoparticles arises from the ability to manipulate nanoparticles based purely on specific biorecognition interactions [108]. Among different biomolecules, DNA has drawn much attention for directing nanoparticle assembly due to its unique physical and chemical characteristics [30].

Since Watson and Crick elucidated its molecular structure in the early 1950s, DNA has begun to gain recognition as an incredibly versatile generic material. From a chemical perspective, DNA can be engineered with extreme ease; a multitude of biochemical techniques can be used to efficiently process DNA with angstrom-level precision [31]. There exist thousands of biochemically characterized enzymes that can manipulate DNA [109], surpassing that of any other known polymer, natural, or synthetic. Enzymatic cleavage and ligation allows for the synthesis of nucleic acids with tunable length that are completely monodisperse. The four bases in ssDNA act as sidechains that can be positioned precisely. Each base can bind with its complement, which allows two complementary strands to recognize and bind to each other forming the dsDNA duplex (hybridization). Furthermore, ssDNA is also capable of self-annealing, resulting in tertiary structures that can bind specifically to other macromolecules. From a mechanical perspective, DNA is quite robust. dsDNA is relatively rigid below its persistence length of about 50 nm [110], while it can have enormous aspect ratios, spanning several hundred micrometers despite its diameter of approximately 2 nm.

Due to the aforementioned properties, DNA has been predominantly used over other biomolecules for directing the assembly of nanoparticles. To rationally program nanomaterials synthesis, a critical step is the understanding of fundamental interactions between DNA-functionalized nanoparticles, including both specific and nonspecific interactions.

2.3.1 Specific Chemical Bonding Interactions

The interaction energetics among DNA-capped nanoparticles cannot be simply estimated by means of the same thermodynamic treatment that applies to free DNA molecules. Multiple DNA strands confined within a small volume often demonstrate a melting behavior that departs from what is ordinarily observed for free DNA strands. For example, the melting transitions of DNA associated with nanoparticles are substantially sharper than those of free DNA in solution [88,111].

The experimental observations by Jin *et al.* [111] show that the interparticle potential between DNA-capped nanoparticles is dependent on the size of the nanoparticles, surface density of DNA, dielectric constant of the surrounding medium, concentration of linker strands, and the position of the nanoparticles with respect to one another within the aggregate. In particular, a sharper melting transition was observed for denser coatings of DNA strands and larger nanoparticle sizes. The melting temperature was shown to increase linearly with DNA linker length, which was also later shown to be true of DNA-based gold nanoparticle networks [112]. Lukatsky and Frenkel [113] proposed a model to account for the experimental observations of DNA-capped nanoparticles [111,114]. They argued that the binding of DNA-coated nanoparticles is analogous to the behavior of quantum particles obeying fractional statistics. With this model, the phase-separation temperature of the nanoparticles increases with DNA surface number density and the dissociation temperature increases logarithmically with the salt concentration. More recently, Dreyfus *et al.* [115] developed a general model in which quantitative control is demonstrated over the dissociation temperature and the sharpness of the dissociation curve. This model accounted for an entropy cost relating to the reduced configurational freedom of tethered DNA strands together with hybridization-mediated attractions.

Recent work has also highlighted the influence of nanoparticle size on the surface density of immobilized oligonucleotides [116–118]. A higher density of thiol-terminated oligonucleotides can be immobilized on the nanoparticle surface as the nanoparticle diameter decreases. In particular, 10 nm nanoparticles, for which the density reaches its limit, were shown to have a threefold higher density than 200 nm nanoparticles [116]. This occurs because smaller nanoparticles have a higher surface curvature that reduces steric hindrance due to crowding of the oligonucleotides as they extend radially outward from the surface. As the particle size increases, the density approaches a flat surface limit that is comparable to oligonucleotide monolayers on a planar surface. In addition to base-pairing interactions between complementary oligos on different nanoparticles, Mirkin proposed a second type of hybridization that arises from surface curvature called the 'slipping' interaction (**Figure 2**) [119]. Terminal sequences may encounter difficulties hybridizing via base-pair interactions as the curvature prevents them from reaching each other. However, these sequences are still capable of interacting through nonspecific base-pairing, which helps to improve the stability of the aggregates.

Schatz and coworkers [120] provided an excellent overview of various techniques for modeling the self-assembly of complex DNA nanostructures, as well as other biomolecules.

2.3.2 Nonspecific Physical Bonding Interactions

Biomolecule-mediated superlattice formation is by no means limited to ligand-specific interactions. Several nonspecific physical interactions govern the behavior of colloidal particles including (1) attractive van der Waals forces between nanoparticle surfaces, (2) repulsive steric interactions between surface ligands, (3) electrostatic interactions between charged nanoparticle surfaces

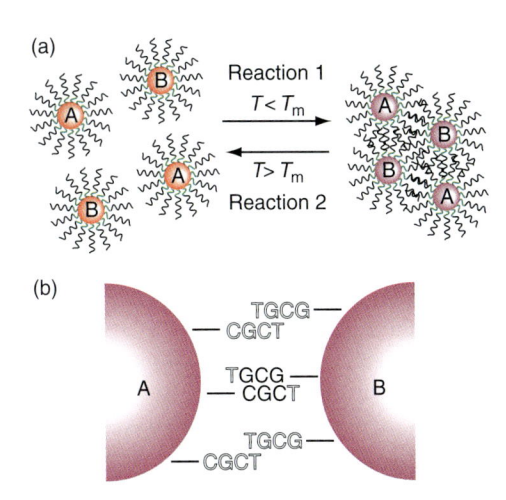

Figure 2 Aggregation and melting is controlled by adjusting the system to be below and above the melting temperature, respectively. Particle aggregates are stabilized by primary hybridization between complementary strands on adjacent particles. However, the particle radius introduces curvature that prevents complementary regions from reaching each other, though noncomplementary interactions can still occur and add stability to the system.

described by DLVO theory (named for Derjaguin and Landau, Verwey and Overbeek) [121,122], and (4) attractive and repulsive forces between charged ligands. For nonpolar solvents, the first two interactions generally dominate. For aqueous solvents, DLVO theory takes into account the effects of van der Waals attractive forces but also the electrostatic repulsion that results from counterions that gather at charged surfaces. These attractive forces arise from counterion layers that gather at the surface and, consequently, are highly sensitive to even small fluctuations in the electrolytic environment [123].

In 1937, Hamaker [124] developed a theory to describe the London and van der Waals adhesive forces that are experienced between two small particles. At very small interparticle separations, the Hamaker potential strongly governs the formation and stabilization of aggregates. The Hamaker potential is given by

$$V_H = -\frac{A}{6}\left[\frac{2R_1 R_2}{C^2 - (R_1 + R_2)^2} + \frac{2R_1 R_2}{C^2 - (R_1 - R_2)^2} + \ln\frac{C^2 - (R_1 + R_2)^2}{C^2 - (R_1 - R_2)^2}\right] \quad (1)$$

In this equation, A is the Hamaker constant (which is 1.95 eV for gold–gold interaction), and C is the center-to-center distance between particles, which can also be expressed as $C = S - R_1 - R_2$, where S is the shortest separation between the particles. **Equation 1** behaves differently at its two limits (**Figure 3**) [21]. Consider two particles of equal radius, $R = R_1 = R_2$. At short separations in which $S \ll 2R$, the potential scales as S^{-1}. At large separations in which $S \gg 2R$, the potential scales as S^{-6}, falling off at the same rate as the well-known van der Waals attractive potential.

The earliest superlattice formations were reported by Bentzon et al. [125] in which ordered 3D arrays were observed after drying colloidal solutions of iron oxide particles. Subsequent works involved the use of ligand-stabilized particles, such as alkylthiol and citrate ligands [126,127]. Although ligand chemistry and structure are important factors for achieving order, Murthy showed that monodispersity of nanoparticles is also required for long-range order [128,129]. At the length scale achievable with alkyl ligands, the Hamaker potential alone is sufficient for modeling systems consisting of polydisperse, short-ligand-stabilized particles in nonpolar solvents, as demonstrated by Ohara et al. [130] by means of simulation. Ligands that help disperse nanoparticles, however, can also contribute to the interparticle potential by introducing repulsive steric interactions [14,123]. The total potential energy can be expressed as

$$V_{\text{total}} = V_{\text{H}} + V_{\text{steric}} \qquad (2)$$

The steric term is given as

$$V_{\text{steric}} = \frac{100 R \delta^2}{(C-2R)\pi\sigma^3} k_{\text{B}} T e^{-\pi(C-2R)/\delta} \qquad (3)$$

The surface ligands have an effective height above the nanoparticle surface, δ (called the brush height), and each ligand occupies an area on the surface (called the footprint) with diameter σ. Unlike hard-sphere systems, the steric forces that arise from the ligand monolayer effectively make the metallic particles behave as soft spheres, for which we may find it convenient to define a unitless softness parameter, $\chi = \delta/R$. For short ligands, a balance will exist between van der Waals attraction and steric repulsion (**Figure 4(a)**). However, as the ligand length increases, the metallic cores are no longer close enough for van der Waals forces to play a role in stabilization (**Figure 4(b)**).

In general, the addition of longer linkers pushes the particles far enough out of this range such that the core–core van der Waals interactions are negligible. In this case, the stability of aggregates must be maintained entirely by the interactions between ligand molecules [15,131,132], which are typically more complicated and are often investigated through simulations [131,133]. Short-range nonspecific interactions between interdigitated and bundled ligands include hydrophobic forces and electrostatic interactions between charged polymer chains. Such forces are the dominant factors that govern the formation and stability of hard-sphere/soft-corona systems. For example, nonspecific physical interactions between non-base-pairing DNA molecules has been shown to lead to highly ordered nanoparticle arrays using a drying-mediated self-assembly strategy [132].

A unique advantage of biomolecules compared to surfactant molecules is wide range length controllability. For example, DNA molecular length can be tuned from several nanometers to tens of micrometers. In comparison, surfactants that are used for

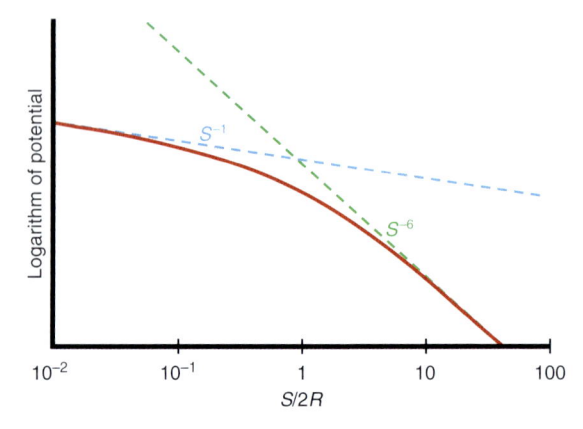

Figure 3 The Hamaker potential behaves differently at its limits. For small separations, the potential is inversely proportional to the separation, and at larger distances it falls off as S^{-6} similar to the van der Waals attractive potential.

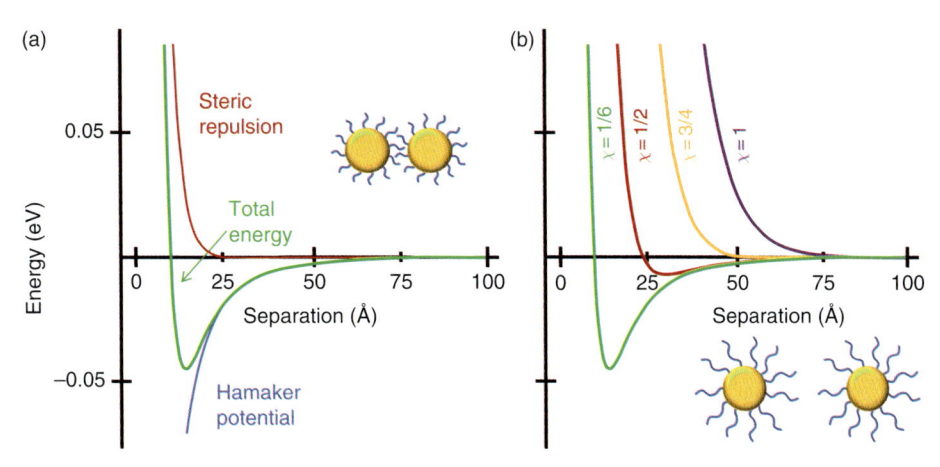

Figure 4 The potential between two ligand-stabilized particles in the absence of electrostatic forces results from a balance of attractive forces with repulsive steric forces. (a) For alkylthiol-stabilized gold nanoparticles ($A = 1.95$ eV, brush height $\delta = 10$ Å, footprint diameter $\sigma = 4.3$ Å [14], diameter $d = 60$ Å, $T = 300$ K), the well depth exceeds the thermal energy, forming kinetically stable aggregates ($V_{min} > 1.5 k_B T$ at 300 K). (b) As the ligand length increases with respect to the particle radius, the particles are pushed far enough away that attractive forces between the metallic cores becomes negligible.

coating nanoparticles are typically much shorter in length due to synthetic limitations. Hence, biomolecules can lead to much thicker corona layers around inorganic nanoparticles to create softer spheres. Although physical interactions among these soft nanoparticle sytems have yet to be fully understood, it has been demonstrated that nonspecific interactions among biomolecules can regulate self-assembly of hard inorganic nanoparticles into highly ordered arrays [132]. Another advantage of using biomolecules to coat nanoparticles is that the mechanical properties of the soft corona can be tailored. For instance, coronas comprised of dsDNA are expected to be stiffer than those of ssDNA due to its much longer persistence length. Coronas from rigid DNA motifs, such as Y-shaped DNA [30] or dendrimer-like DNA molecules [134], can potentially increase the stiffness due to reduced conformational flexibility.

2.4 Assembly of Ordered Nanoparticle Superstructures

2.4.1 Nanoparticle Molecules

The programmable nature of DNA that arises from its biological relevance facilitates the directed organization of nanoparticle molecules (assemblies containing a discrete number of nanoparticle building blocks) in a controlled and modular fashion.

Alivisatos and coworkers were the first to demonstrate that DNA could be used to precisely engineer nanoparticles into nanoparticle molecules by virtue of Watson–Crick base pairing. In this seminal work, small 1.4 nm gold nanoparticles monofunctionalized with a single 18-base strand of thiol-modified ssDNA self-assembled into homodimers or homotrimers upon addition of a complementary ssDNA template [22]. The versatility of this approach was explored in a subsequent report in which heterodimers and heterotrimers were also assembled using complementary ssDNA linkers, resulting in a variety of molecular spatial arrangements that include collinear and triangular configurations (**Figure 5**) [94].

One of the challenges in engineering DNA-based molecular assemblies lies in tailoring the exact amount of DNA conjugated onto each particles for various lengths of DNA and sizes of nanoparticles. Zanchet *et al.* [100] described the successful isolation of 5 and 10 nm gold nanoparticles conjugated with discrete numbers of ssDNA strands via gel electrophoresis. However, electrophoretic separation is limited to ssDNA lengths of at least 50 bases, below which the difference in mobility becomes relatively insignificant. One workaround for this issue is to hybridize an extension strand to elongate the ssDNA conjugate during electrophoretic isolation [100]. The resulting purified conjugates allow for enhanced control and an improved yield in the assembly of dimeric and trimeric nanoparticles [135]. Alternatively, restriction enzymes – one of

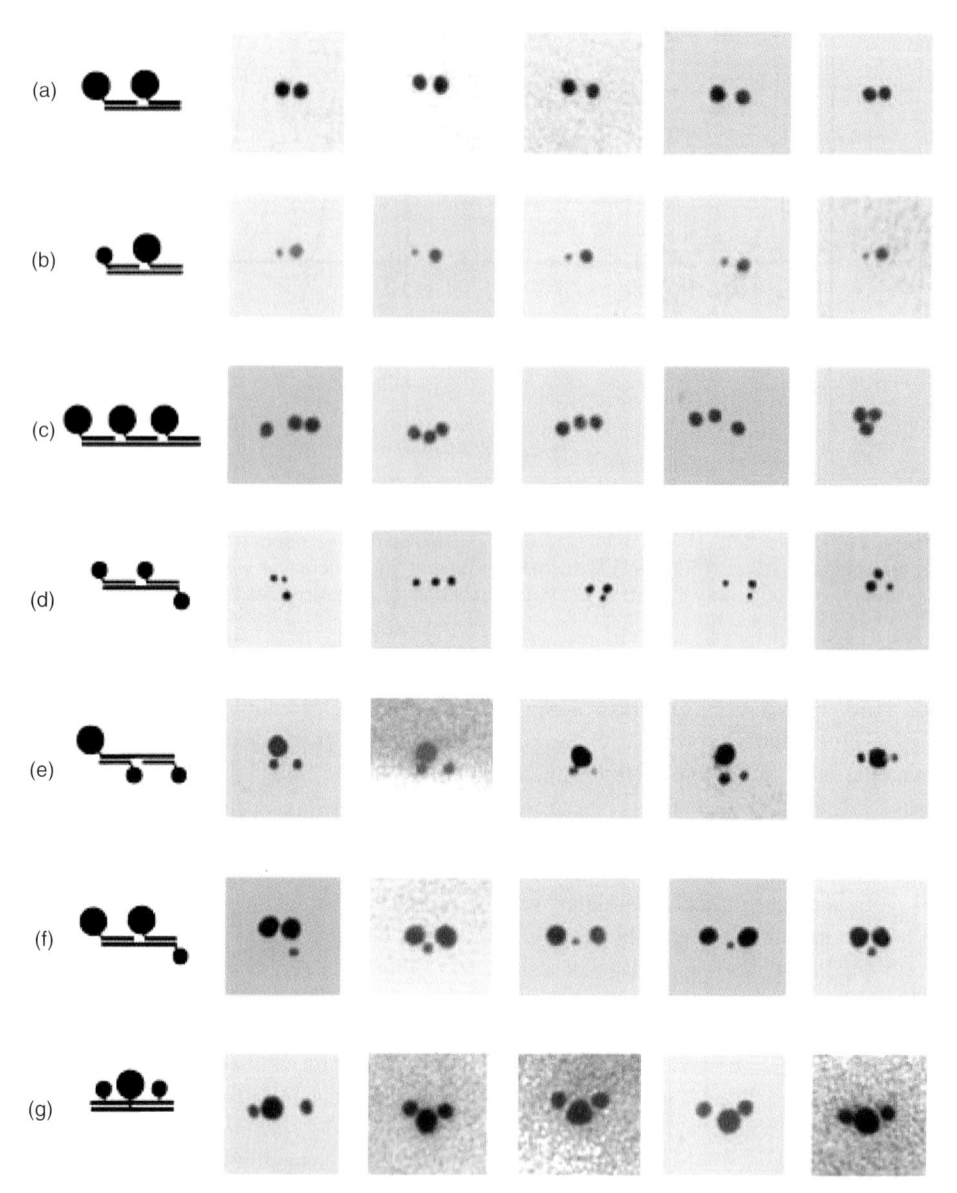

Figure 5 Schematic illustrations and representative TEM images for specific nanocrystal molecules. (a) 10 nm homodimer, (b) 5/10 nm heterodimers, (c) 10 nm homotrimer, (d) 5 nm homotrimer, (e) 10/5/5 nm heterotrimer, (f) 5/10/10 nm heterotrimer, (g) 5/10/5 nm heterotrimer. Reproduced with permission from Alivisatos AP, Johnson KP, Peng et al. (1996) Organization of 'nanocrystal molecules' using DNA. *Nature* 382: 609–611.

the versatile biological tools available for manipulating DNA sequences – could be used to obtain short ssDNA on gold nanoparticles. In the work reported by Qin and Yung [101], 10 nm nanoparticles were first capped with a specific number of long dsDNA, which were subsequently cleaved through restriction enzyme digestion, resulting in a specific number of short DNA strands (<20 bases) attached onto each nanoparticle. The conditions affecting the formation of well-defined nanoassemblies – dimers, trimers, and tetramers – using these short-DNA–nanoparticle conjugates were also explored in detail [136]. Discrete large nanoparticles (20 nm) conjugated with short ssDNA strands (<40 bases) and passivated with polyethylene glycol (PEG) can also be isolated using AE-HPLC [102]. These large nanoparticle–short ssDNA conjugates are ideal for plasmon coupling studies, as surface plasmon intensity increases with particle diameter but decreases with interparticle distance [137].

Claridge and coworkers [138] utilized enzymatic ligation to create discrete multi-nanoparticle building blocks for self-assembly into more complex structures. ssDNA–nanoparticle monoconjugates were hybridized to a complementary DNA strand and were enzymatically ligated. Upon removal of the template strand, nanoparticle dimers connected by ssDNA were obtained. This approach can be extended to fabricate nanoparticle trimers, tetramers, and even more complex assemblies. Branched DNA scaffolds were also used as templates for the formation of multi-nanoparticle assemblies [95].

Mirkin and coworkers [135] also demonstrated the versatility of using DNA base-pair interactions to direct the assembly of binary nanoparticle materials. By functionalizing complementary ssDNA onto gold nanoparticles of different sizes, they could organize nanoparticle satellite structures, in which 13 nm gold satellite nanoparticles were connected to a 50 nm core particle via DNA duplexes.

In a work that combined the use of DNA recognition with synthetic molecular linkers, Aldaye and Sleiman [97] sequentially assembled gold nanoparticles into a hexagonal configuration by using cyclic ssDNA as an addressable template. Gold nanoparticles were conjugated onto the ends of a pair of short ssDNA strands linked by rigid 120° synthetic vertices. This facilitates a step-by-step cyclic assembly of hexagonal nanoparticle molecules by sequential addition of ssDNA–nanoparticle conjugates that are partially complementary to the ssDNA hexagonal template (**Figure 6**). The use of the synthetic vertices provided the necessary additional rigidity to the system that resulted in the formation of discrete, well-defined hexagonal assemblies.

Aldaye and Sleiman further incorporated greater modularity into the assembly of nanoparticle molecules by creating dynamic DNA templates that could be actively modified to obtain tailor-made nanoparticle molecules (**Figure 7**) [98]. They used ssDNA as a template with distinct sections onto which DNA–nanoparticle conjugates containing the complementary sequence could hybridize. As a result, they obtained well-defined nanoparticle triangles and squares when cyclic ssDNA was used as the template, and organized linear assemblies when linear ssDNA was used as the template. Due to the modular nature of this assembly, they were also able to obtain heterotrimeric structures that incorporated multiple nanoparticles in a variety of configurations. Interestingly, they also demonstrated structural switching within the same template. For example, by adding a fraction of nanoparticles conjugates with a shorter ssDNA strand that hybridizes onto the square template to create a hairpin loop on one side, they were able to effectively obtain trapezoidal and rectangular nanoparticle molecular configurations out of a single template. In addition, they showed that these molecules possessed write/erase capabilities, in which any nanoparticle in the assembly could be readily replaced with a different one. To do this, the nanoparticles were conjugated to a 40-base ssDNA strand such that they hybridized to the template with a 20-base overhang. These could then be easily removed upon addition of an eraser strand

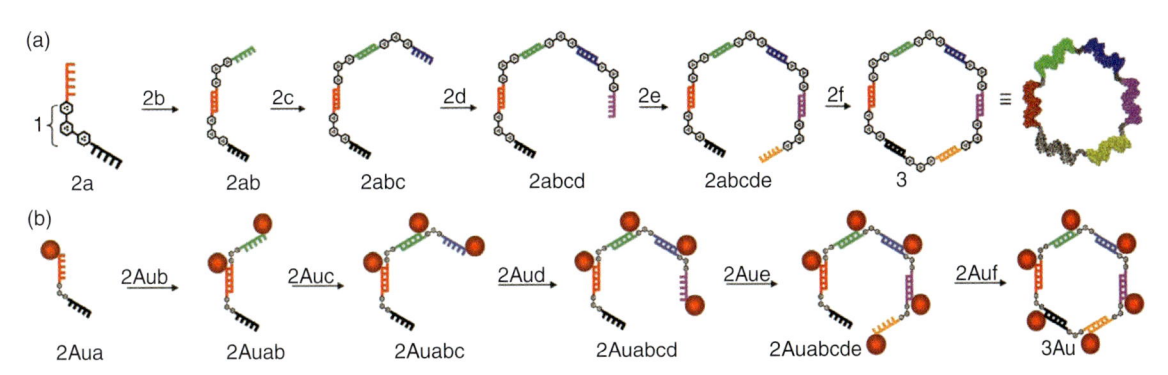

Figure 6 Self-assembly of a DNA hexagon with gold nanoparticles at the vertices. (a) Assembly of a hexagon from ssDNA components with 120° synthetic vertices. (b) Assembly of hexagonal structure with gold nanoparticles functionalized at one end of each ssDNA component. Reproduced with permission from Aldaye FA and Sleiman HF (2006) Sequential self-assembly of a DNA hexagon as a template for the organization of gold nanoparticles. *Angewandte Chemie International Edition* 45: 2204–2209.

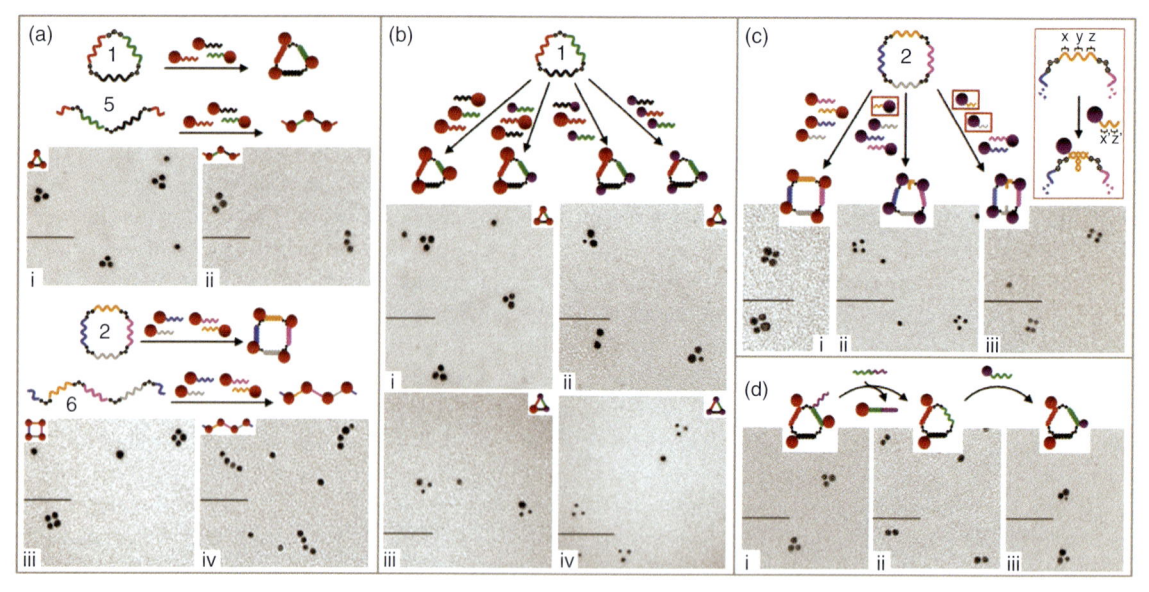

Figure 7 (a) Templates **1** and **2** assemble ssDNA–nanoparticle monoconjugates into triangles (i) and squares (iii), respectively. Templates **5** and **6** lead to linear assemblies of **3** (ii) or **4** particles (iv). (b) Heterotrimers assembled onto template **1** in a triangular configuration containing (i) three large (15 nm, red), (ii) two large/one small (5 nm, purple), (iii) two small/one large, and (iv) three small particles. (c) Template **2** assembles nanoparticle monoconjugates into (i) squares, (ii) trapezoids, and (iii) rectangles. The monoconjugates can be designed to hybridize to the ends of one side, creating a loop that shortens the template's arm. (d) Write/erase functionality in **1** by (i) writing three ssDNA-Au monoconjugates (15 nm, red) onto a triangular ssDNA template, (ii) removing a specific particle upon addition of an eraser strand, and (iii) rewriting the empty arm with a smaller 5 nm ssDNA–Au monoconjugate. Reproduced with permission from Aldaye FA and Sleiman HF (2007) Dynamic DNA templates for discrete gold nanoparticles assemblies: Control of geometry, modularity, write/erase and structural switching. *Journal of the American Chemical Society* 129: 4130–4131.

complementary to the entire length of ssDNA, and a replacement ssDNA–nanoparticle conjugate could then be added to the assembly.

Kotov and colleagues [139] exploited the biotechnological tools available for mass replication of DNA to DNA–nanoparticles and reported the use of the polymerase chain reaction (PCR) to program the assembly and growth of nanoparticle molecules. PCR was performed on the surface of the nanoparticles to amplify the linker strands, creating assemblies of nanoparticles. They observed that the complexity of the product correlated to the the density of primers conjugated onto the surface of the gold nanoparticles. At a low concentration of primers, dimeric and trimeric nanoparticle molecules were formed. As expected, the number of nanoparticles in each assembly increased with the PCR cycle number. These nanoparticle assemblies also displayed chirality, which could give rise to unique optical properties.

Mirkin and coworkers [105,106] also used gold nanoparticles asymmetrically functionalized with oligonucleotides to create unique nanoparticle molecules. These anisotropic building blocks facilitated the directional assembly of assymetric structures such as 'cat paws' as well as sophisticated core-satellite and dendrimer-like structures (**Figure 8**). In addition, Maye *et al.* [104] were able to generate high yields of discrete multimeric Janus nanoparticles using a high throughput method for designing anisotropic DNA-conjugated nanoparticles in a stepwise manner. Gold nanoparticles conjugated with two different ssDNA sequences were hybridized onto a substrate functionalized with a complementary ssDNA sequence. A second complementary ssDNA sequence hybridized onto the exposed side of the nanoparticles to create anisotropically functionalized nanoparticles as a result of the geometric restriction from binding to the surface. Additional conjugations to other nanoparticles can be made on the exposed nanoparticle surface or to the anisotropic nanoparticles upon release from the substrate. Free ssDNA strands complementary to the substrate were eventually added to the system to release the anisotropic nanoparticles into solution. This approach was

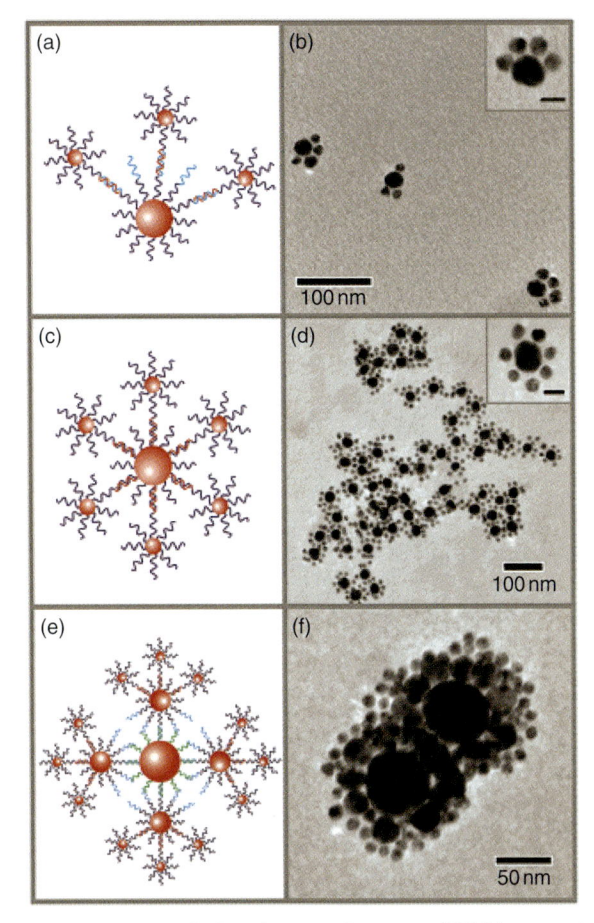

Figure 8 Schematics and representative transmission electron microscopy (TEM) images showing the directional assembly of anisotropically functionalized AuNPs into (a, b) cat paw, (c, d) core-satellite, and (e, f) dendrimer-like structures [105]. Scale bars of insets = 20 nm. Reproduced with permission from Xu X, Rosi NL, Wang Y, Huo F, and Mirkin CA (2006) Asymmetric functionalization of gold nanoparticles with oligonucleotides. *Journal of the American Chemical Society* 128: 9286–9287.

reported to produce a significantly high yield of Janus nanoparticles and multimeric clusters and could represent a step forward towards large-scale DNA-based nanoparticle molecules.

In general, nanoparticle molecules afford an excellent amount of control and sensitivity for detection and sensing. For example, Alivisatos and coworkers [140] proposed the use of metallic nanoparticle pairs linked by DNA as a 'plasmon ruler' that can measure the separation distance between nanoparticles up to 70 nm for more than 3000 s. This allowed them to observe the DNA-mediated assembly of nanoparticle pairs in real time, and also investigate the kinetics of single DNA hybridization events.

Sebba *et al.* [141] have also shown that plasmon coupling in the core-satellite nanoparticle assemblies can be specifically modulated using reconfigurable DNA linker strands designed with hairpin loops. The system can switch between a hairpin or an extended configuration upon addition or removal of a bridging strand that hybridizes to the hairpin sequence in the DNA linker. The interparticle spacing between the satellite and core particles changes with each configuration, resulting in a detectable shift in plasmon coupling. Further investigation into these core-satellite assemblies suggested that the detection of plasmon coupling was resistant to minor variations in particle properties as well.

In addition, significant plasmon enhancements have been observed in dimeric nanoparticle molecules [143]. In the approach of Bidault *et al.* [143], 5, 8, and 18 nm gold nanoparticles were grouped into 'molecules'. This method involved the use of 3' and 5'-thiolated DNA, which enabled subnanometer interparticle spacings to be achieved upon DNA hybridization. The small interparticle distance was critical for attaining strong plasmon coupling enhancement, which allowed the assembly to act as a nanoscale plasmonic lens for sensing applications.

2.4.2 Nanoparticle Superlattices

2.4.2.1 Programmable DNA-based assembly

- One-dimensional ordered assemblies

  The linearity of DNA can also be exploited for the templated assembly of nanoparticle chains (nanoparticle polymers). Nonspecific electrostatic interactions between DNA molecules and other functional materials provide a straightforward means toward direct synthesis of one-dimensional (1D) materials [144–147]. On the other hand, highly specific Watson–Crick base-pairing forces can be used to assemble nanoparticles into well-defined 1D structures. Mao's and Simmel's groups took advantage of long, repetitive DNA single strands obtained by an elegant biochemical technique called rolling-circle amplification (RCA) as templates to organize nanoparticle polymers [96,148]. RCA is a variation of standard PCR, in which a single-stranded circular DNA is used as a template for a DNA polymerase. The RCA polymerase can copy the circular template without stopping, leading to a very long ssDNA strand with repeated sequences. Mao and coworkers [96] synthesized DNA–nanoparticle monoconjugates, which were then directly hybridized to the recurring segments on the long ssDNA strand, resulting in a micrometer–long nanoparticle polymer (**Figure 9**). In Simmel's [148] strategy, biotinylated short DNA strands were attached onto the RCA product to create a periodic arrangement of biotin sites. Subsequently, streptavidin-coated gold nanoparticles were anchored specifically to the biotin sites, allowing for the organization of 1D nanoparticle assemblies.

  Li et al. [149] demonstrated a route to fabricate much stiffer nanoparticle polymers. In their efforts, triple-crossover DNA (TX-DNA) molecules were used to form a rigid template. Biotin binding sites can then be selectively distributed uniformly on either one side or both sides of the template. Using biotin–streptavidin binding interactions, nanoparticles can be organized into a single- or double-layered nanoparticle polymer.

  Yan and coworkers [150] reported the formation of complex 3D geometric architectures of

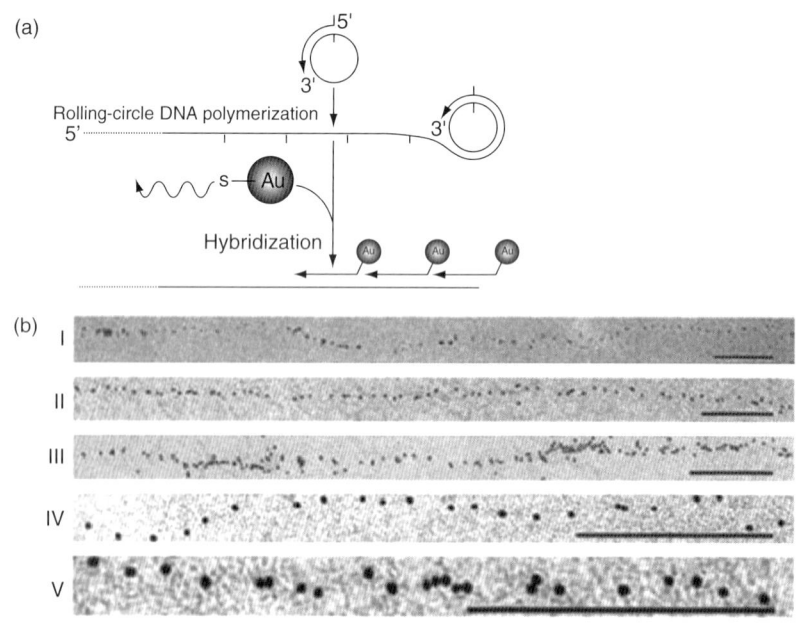

Figure 9 Synthesis of linear arrays of gold nanoparticles through RCA [96]. (a) Scheme showing the ssDNA circular template amplified to produce long ssDNA with periodic sequence repeats. Gold nanoparticles with ssDNA complementary to these repeats hybridize in a periodic fashion to the linear RCA product. (b) TEM images of gold nanoparticles anchored onto the RCA product. (IV) and (V) are close-up views of (I) and (II). Scale = 200 nm. Reproduced with permission from Deng Z, Tian Y, Lee S-H, Ribbe AE, and Mao C (2005) DNA-encoded self-assembly of gold nanoparticles into one-dimensional arrays. *Angewandte Chemie International Edition* 44: 3582–3585.

nanoparticles using a DNA-based strategy. Four different double-crossover DNA (DX-DNA) tiles were assembled into a 2D array containing periodic linear arrangements of nanoparticles (**Figure 10**). The degree of electrostatic and steric repulsion between gold nanoparticles can be influenced by varying the sizes of the nanoparticles and incorporating stem loops in the tiles. As a result, these arrays can be directed to assemble into a variety of tubular nanoparticle structures, including stacked rings, spiral tubes, and double helices. Electron tomography reconstructions revealed a resemblance of these structures to carbon nanotubes. This work demonstrated the possibility of rationally controlled assembly of nanoparticles with nanoscale precision, which may lead to integration of functional nanocomponents into miniaturized devices.

- **Two-dimensional ordered assemblies**

DNA can also be used for organizing nanoparticles into extended two-dimensional (2D) arrays [151]. Rational design of DNA sequences can lead

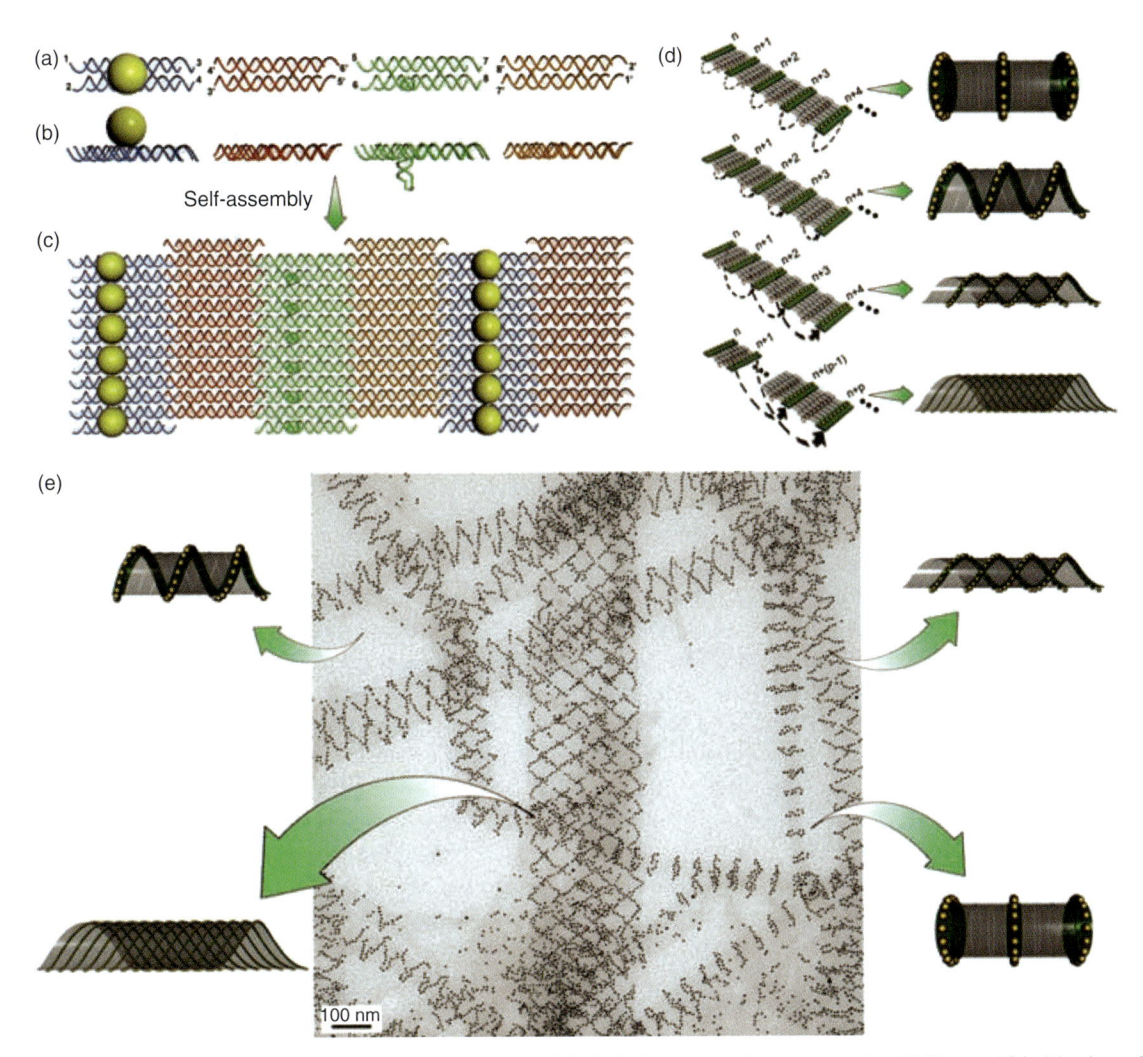

Figure 10 ssDNA self-assembles into tubular structures labeled with 5 nm gold nanoparticles. (a) Top and (b) side view of DX-DNA building blocks. Blue subunit carries a 5 nm gold nanoparticle and green subunit carries a stem loop. (c) The subunits were designed to assemble into a 2D tiled array. (d) Due to steric and electrostatic repulsion of gold nanoparticles, these 2D arrays fold into tubular structures with nanoparticles arranged into repeating rings or helices. (e) TEM image displaying various forms of nanoparticles-labeled tubular structures [150]. Reproduced with permission from Sharma J, Chhabra R, Cheng A, Brownell J, Liu Y, and Yan H (2009) Control of self-assembly of DNA tubules through integration of gold nanoparticles. *Science* 323: 112–116.

to the formation of periodic DNA nanostructures, which can act as templates for programming the placement of nanoparticles. In this approach, a rigid DNA template can be obtained using DX-DNA (pioneered by Seeman and coworkers [152]), which is strong enough to preserve a regular structure. The resulting arrays can be designed to include sticky ends that hybridize to DNA-conjugated nanoparticles or other functionalized moieties, producing 2D periodic patterns. With this strategy, lattice formation is entirely controlled by the specific recognition between sticky ends on adjacent DNA tiles.

Several different DX-DNA arrays with different architectures have been successfully demonstrated, differing primarily in the structure of the DNA repeating unit. For example, in one approach, oligonucleotides were designed with specific sequences to form DNA tiles (DX-DNA subunits) that could hybridize with each other to form a tiled array [153,154]. The DNA tiles formed a 2D scaffold with periodic sticky ends that served as attachment points for DNA-conjugated gold nanoparticles, resulting in regular patterns [155]. The binary arrangement of nanoparticles could also be patterned by this tile-based strategy, resulting in patterns such as alternating rows of 5 and 10 nm gold nanoparticles [156]. In addition to gold nanoparticles, this particular DX-DNA scaffold has been extended to the patterning of many different moieties including proteins [157,158] and quantum dots [159].

An alternate DX-DNA scaffold is the tensegrity triangle motif. This architecture has been assembled using both dsDNA [160] and DX-DNA motifs [161]. In the DX-DNA case, two triangular motifs (termed 3D-DX triangles) were used to assemble a 2D rhombic lattice. These arrays were produced in a multistep process in which the two types of triangular subunits were synthesized separately, and then combined and assembled. Gold nanoparticles were conjugated to the triangular subunits before assembly of the final pattern. Each triangular subunit was composed of DX-DNA, with two attachment points involved in the 2D lattice and one attachment point used for attaching gold nanoparticles. Since two types of triangles were used, it was possible to organize two different sizes of gold nanoparticles in an alternating fashion.

The four-armed DX-DNA structure represents another variation of DX-DNA architecture that has been used for the construction of ordered 2D arrays. In this system, nine oligonucleotides formed a cross-shaped structure that self-assembled into a 2D nanogrid with periodic square cavities [162].

ssDNA-functionalized gold nanoparticles could be annealed to sticky ends within the cross-shaped subunits, or they could be directly conjugated to one of the component oligonucleotides before the subunit was formed [163]. Variations on the scaffold structure were also demonstrated; for instance, two complementary types of cross-shaped DNA building blocks were used to create 'chessboard' arrangements of DNA tiles (**Figure 11**) and the DNA sequence design allowed for control over the size and orientation of the tiles within the assembly [164]. The placement of the nanoparticles either above or below the plane of the DNA scaffold could also be controlled. This DX-DNA structure has been extended to patterning nanoparticle assemblies based on the principles of sequence symmetry [165]. More complex nanoparticle arrays could also be templated from 10- and 16-tile DNA arrays [166,167].

For systems in which DNA-conjugated nanoparticles hybridize with 2D DNA scaffolds, N:1 functionalized gold nanoparticles can cross-hybridize with multiple sticky ends. This prevents a one-to-one correspondence between nanoparticles and attachment sites and interferes with the fidelity of the resulting pattern. Cross-hybridization can be minimized with sufficiently large spacing between attachment points but this introduces an undesirable design constraint for particular systems, such as in plasmonic applications. Another solution to this problem is to use 1:1 functionalized nanoparticles, which can stoichiometrically bind to the attachment points. However, these monoconjugates tend to aggregate under high salt conditions due to the lower surface density of protecting ligands. Sharma *et al.* [163] have developed a strategy that combines the advantages of both 1:1 and N:1 functionalized nanoparticles. Gold nanoparticles were functionalized with a small amount of DNA and the gold nanoparticles with only one DNA ligand were purified and isolated. Nonspecific DNA was then used to coat the remaining surface of the gold nanoparticles in order to limit their aggregation at high salt concentrations. These monovalent DNA-conjugated gold nanoparticles

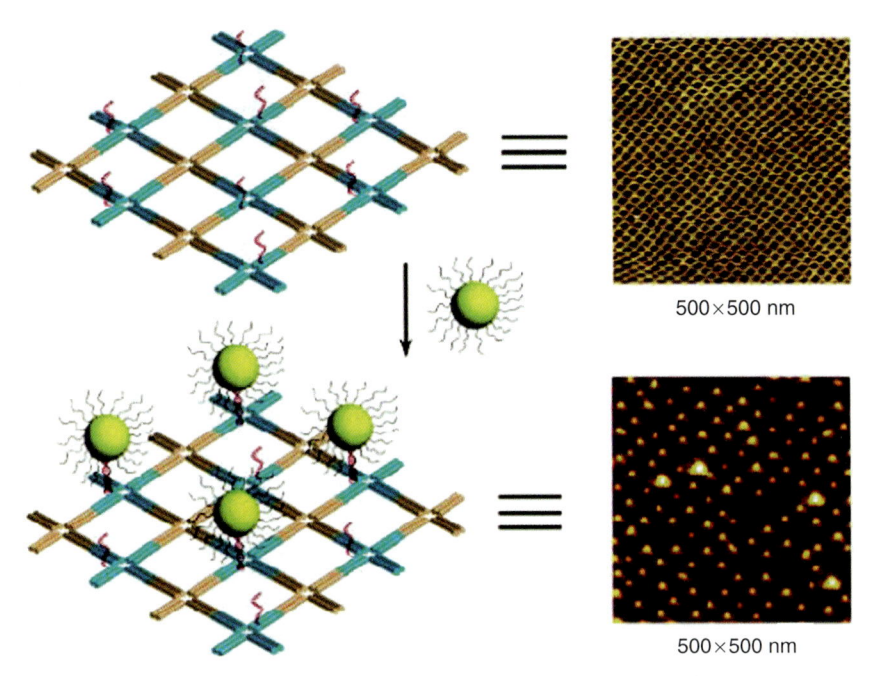

Figure 11 Two-dimensional periodic 'nanogrid' from cross-shaped DX-DNA subunits. Scheme and atomic force microscopy (AFM) images of fully assembled nanogrids before (top) and after (bottom) hybridization with DNA-coated 5 nm gold nanoparticles. The edge-to-edge spacing between nanoparticles is approximately 18 nm [164]. Reproduced with permission from Zhang JP, Liu Y, Ke YG, and Yan H (2006) Periodic square-like gold nanoparticle arrays templated by self-assembled 2D DNA nanogrids on a surface. *Nano Letters* 6: 248–251.

could then be arrayed onto a DNA template with a more precise one-to-one correspondence between nanoparticle and sticky end.

Recent progress in DNA nanotechnology has led to a variety of sophisticated 2D DNA arrays, which could potentially be used to template the assembly of more complex nanoparticle materials. For example, Rothemund [168] demonstrated the feasibility of DNA origami, in which complicated patterns can be obtained from long, single-stranded phage DNA and short ssDNA 'staple' strands in an efficient, single-step approach. Other DNA structures have included star [169], hexagon [170,171], and parallelogram subunits [172], as well as a DNA box [173].

Achieving complex nanostructures from branched DNA building blocks is still in many ways an art form. Understanding the interactions between DNA building blocks is of the utmost importance for directed self-assembly. A precise balance between internal stress and flexibility is essential to achieve the long-range order observed in extended 2D DNA arrays [174,175], though more junction flexibility is necessary for the formation of 3D structures [176]. In addition to physical and structural characteristics of individual building blocks, the interplay between building blocks must also be understood. Starr and Sciortino [177] have investigated self-assembly of four-armed branched DNA molecules, showing that gelation is highly dependent on temperature and diffusivity. Through these and similar studies, it may be possible to optimize DNA subunits such that control over macroscale transitions can be exercised. Further work with molecular modeling may elucidate the conditions necessary for periodic 3D DNA networks, which could potentially be used as scaffolds for 3D assemblies of macromolecules.

- **Three-dimensional ordered assemblies**
  In 1996, Mirkin described the reversible self-assembly of gold nanoparticles via specific DNA base-pairing interactions. Although the structures obtained were better classified as amorphous polymers, this effort not only represented a starting point for designing 3D ordered structures but has also led to a plethora of diagnostic applications. In Mirkin's seminal report, two batches of gold nanoparticles were labeled with two different sequences (A or B) of 5′-thiolated

oligonucleotides. The sequences A and B were both semicomplementary to a target DNA linker, with sticky end sequences A' and B'. When this target DNA was added to the solution, both semicomplementary oligonucleotides hybridized to the target strand to form stable duplexes. Through this one-pot DNA hybridization, the gold nanoparticles assembled into 3D nanoparticle polymeric aggregates. This aggregation was both temperature sensitive and reversible; if the mixture was heated beyond the melting point of the hybridized DNA, then the DNA-conjugated gold nanoparticles re-dispersed into solution.

The assembly of gold nanoparticles accompanied a reversible color change from red (absorbance at 520 nm) to blue (absorbance around 600–700 nm), a plasmon coupling effect resulting from the change in interparticle distance [137]. This striking red to blue color change associated with the aggregation of gold nanoparticles quickly led to applications in colorimetric sensing of target DNA oligonucleotides [178,179]. Besides target DNA detection, the sensitivity of DNA-mediated gold nanoparticle aggregation has also been applied toward studying aspects of self-assembly. For instance, the role of DNA hybridization in nanoparticle aggregation was explored by introducing noncomplementary ssDNA into a DNA-mediated self-assembly system [180]. The noncomplementary ssDNA, which did not base-pair with any other DNA in the system, was conjugated onto the surface of gold nanoparticles along with different ratios of complementary ssDNA. As the proportion of noncomplementary oligonucleotides was increased, both the average size and the melting temperature, T_m, of the nanoparticles aggregates decreased. Apparently, the presence of noncomplementary DNA ligands introduced repulsive forces without contributing any favorable base-pairing interactions, controlling the relative contributions of the forces involved in aggregation. The noncomplementary DNA used in this study could be replaced with additional sets of complementary DNA to form specific linkages with other components.

Despite significant attempts on the utilization of DNA for the assembly of 3D nanoparticle aggregates, highly ordered 3D structures were not obtained until recently. In 2008, Gang and coworkers [181] and Mirkin and coworkers [182] reported the first highly ordered 3D arrays of gold nanoparticles by a DNA-based route in conjunction with temperature control.

In the approach of Gang and coworkers [181] two batches of ssDNA-conjugated nanoparticles (A and B), each with sequences complementary to the other, were mixed together to induce hybridization. A flexible region was included in the form of a spacer between the nanoparticle and the overlapping hybridized region. The result of this hybridization was an ordered, 3D nanoparticle array as verified with small-angle X-ray scattering (SAXS). The crystal structure was shown to be body-centered cubic (b.c.c.), CsCl-type superlattices, which were favored due to attractive forces between binary components A and B and the repulsive forces between negatively charged DNA ligands.

In a similar approach, Mirkin and coworkers [182] used a dsDNA linker to induce the crystallization of ssDNA-conjugated nanoparticles. The crystalline structure was reversible over multiple heating and cooling cycles and long-range order was demonstrated for several lengths of dsDNA linkers and two sizes of gold nanoparticles (10 and 15 nm). The design of the double-stranded linker determined whether the system was single-component or binary (**Figure 12**). When a linker composed of two identical, partially self-complementary ssDNA strands was used, the system crystallized into an f.c.c. configuration. When two different, nonpalindromic, complementary ssDNA strands were used, a b.c.c. configuration was observed. This control over the self-assembled crystal structure, which arose entirely from specific base-pairing interactions, demonstrates the potential of DNA ligands for controlled self-assembly.

In these systems, DNA plays the role of 'assembler' while the nanoparticles compose the bulk of the crystalline aggregate and give rise to its properties. With different DNA sequences corresponding to different sizes, shapes, and species of nanoparticles, these approaches may lead to versatile techniques for assembling a variety of hybrid 3D crystalline assemblies.

2.4.2.2 Nonspecific DNA-based assembly
The DNA-based approach can also be coupled with drying-mediated self-assembly to obtain stable nanoparticle superlattices in the dehydrated state. In a drying-mediated self-assembly strategy, the organization of nanoparticles is based on entropic effects. The weak

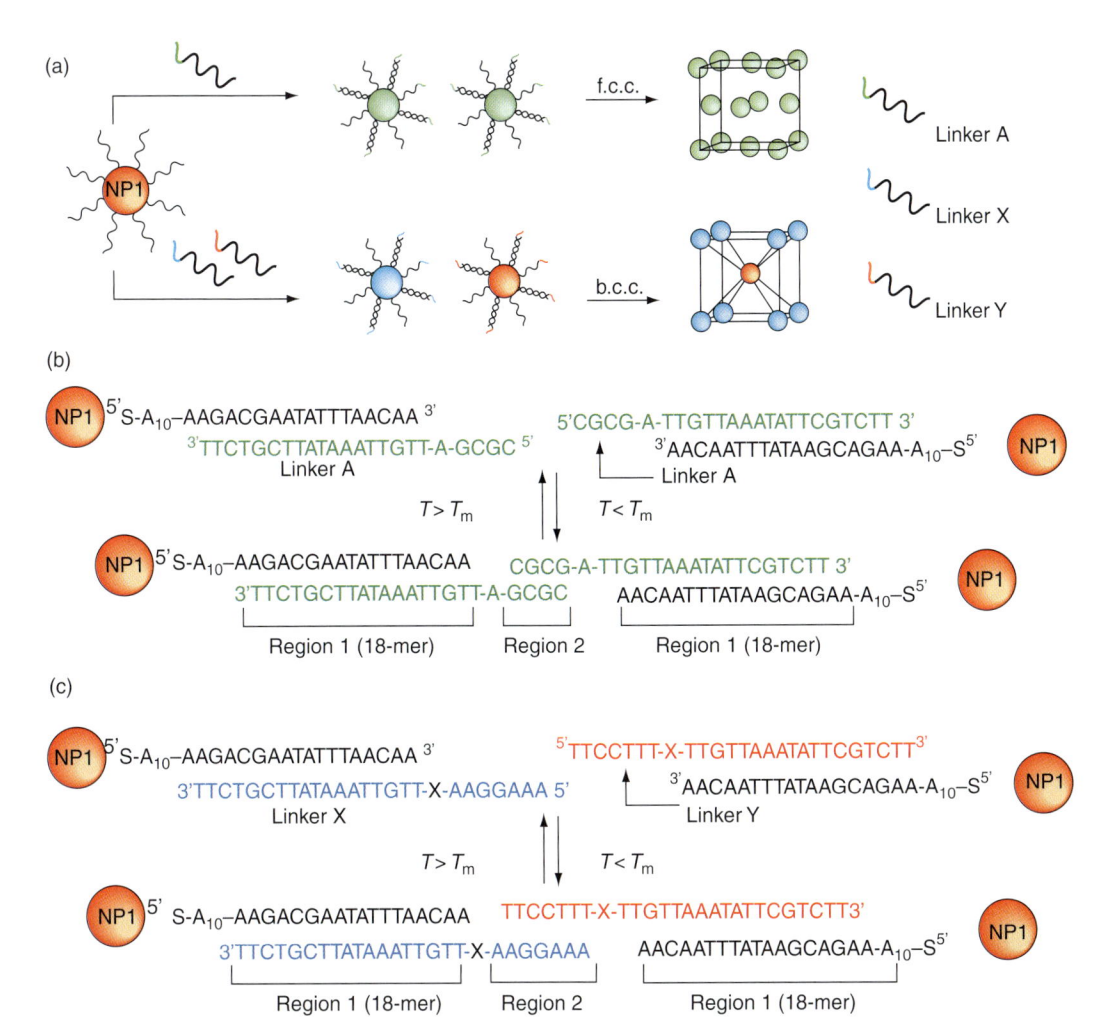

Figure 12 Scheme demonstrating the assembly of DNA-capped gold nanoparticles into f.c.c. and b.c.c. lattices. Whether the assembly becomes f.c.c. or b.c.c. is entirely determined by the sequence design of the DNA linker. (a) A linker with self-complementary sequences leads to an f.c.c. crystalline structure, whereas a linker composed of two noncomplementary ssDNA sequences gives rise to a b.c.c. structure. (b) Details of the self-complementary linker (linker A). (c) Details of the noncomplementary ssDNA linkers (linkers X and Y) [182]. Reproduced with permission from Park SY, Lytton-Jean AKR, Lee B, Weigand S, Schatz GC, and Mirkin CA (2008) DNA-programmable nanoparticle crystallization. *Nature* 451: 553–556.

attractive forces between nanoparticles shielded in solution become more prominent upon volume reduction due to solvent evaporation. This process induces the gradual packing of nanoparticles into aggregates, allowing superlattices to spontaneously assemble through minimization of the free energy of the system [24].

The application of nanoparticle superlattices in optoelectronic devices typically requires methods capable of patterning them into desired structures while maintaining a high degree of internal order. However, drying-mediated self-assembly is a far-from-equilibrium process [9,11–18] and its statistical nature makes it challenging to pattern nanoparticle superlattices using existing methods. Although superlattices can form on a surface, the overall structures are difficult to control and local structures are kinetically trapped. For instance, capillary flow induced by a non-uniform evaporation field [183] and fluid fluctuations during late-stage drying [24] often lead to irregular features such as isolated islands, worm-like domains, ring-like structures, and cellular networks [184,185].

To overcome this challenge, we demonstrated that nanoparticle superlattices can be patterned into versatile structures using spatial confinement of a microdroplet on a substrate [186]. This simple approach involves the use of a polydimethylsiloxane

(PDMS) mold to spatially regulate the drying of a droplet containing ssDNA–nanoparticle conjugates. Using this methodology, nanoscale features can be obtained from a micrometer-sized mold. In particular, we were able to fabricate 12 nm-thick nanoparticle wires using a 5 μm-diameter mold. We also obtained single-particle-thickness superlattice microdisks (2D), submicrometer-sized 'supra-crystals' (3D), and double-superlattice corrals (**Figure 13**). Moreover, these superlattices can be addressed onto micropatterned electrode arrays, a step towards integrating nanoparticle superlattices into large-area electronic nanodevices.

Despite the success of drying-mediated assembly in generating nanoparticle superlattices that are stable in the dried state, the limited molecular length of commonly used ligands prevents us from widely tuning interparticle spacing. Taking advantage of both DNA's wide-ranging length controllability and monodispersity, our group has used gold nanoparticles and the nonspecific interactions of DNA as a model system for the drying-mediated self-assembly of biofunctionalized nanoparticles. In our efforts, we used DNA as a generic ligand with nonspecific ligand–ligand interactions rather than exploiting the typical specific recognition capabilities of Watson–Crick base pairing. Nanoparticle superstructures assembled via DNA base-pairing interactions need to be stabilized in aqueous buffered environments and tend to collapse upon removal of water. These solution-based structures are also difficult to pattern into well-defined structures, posing difficulties for integration into solid-state optoelectronic devices. By relying on the nonspecific interactions of DNA for assembly, we circumvented the issues of buffer stability and structural collapse from solvent evaporation, allowing us to apply drying-mediated techniques to DNA–nanoparticle conjugates for the assembly of nanoparticle superlattices.

The drying-mediated assembly of DNA–nanoparticle conjugates within the spatial confinement of a microhole allows us to fabricate suspended sheets of nanoparticle superlattices, driven by the nonspecific interactions of the ssDNA corona and volume reduction from solvent evaporation. This single-step, rapid self-organization process is completed within a few minutes and could generate free-standing superlattices of single-particle thickness (a thickness of only 12 nm), exhibiting paper-like mechanical properties (**Figure 14**). Using the unique tools available for controlling DNA length over a wide range, we were able to obtain highly ordered nanoparticle superlattices with edge-to-edge interparticle spacing over a much larger regime (~1.0 to ~20.0 nm) than has been achieved for alkyl molecular ligands (~1.2 to ~2.3 nm) [132,187]. This enables us to control the interparticle spacing and modulate the mechanical properties, as revealed from AFM indentation studies and plasmon shifts over a wide range in these superlattices.

This tunable regime of DNA ligand length also allows us to reveal the fundamental dynamics

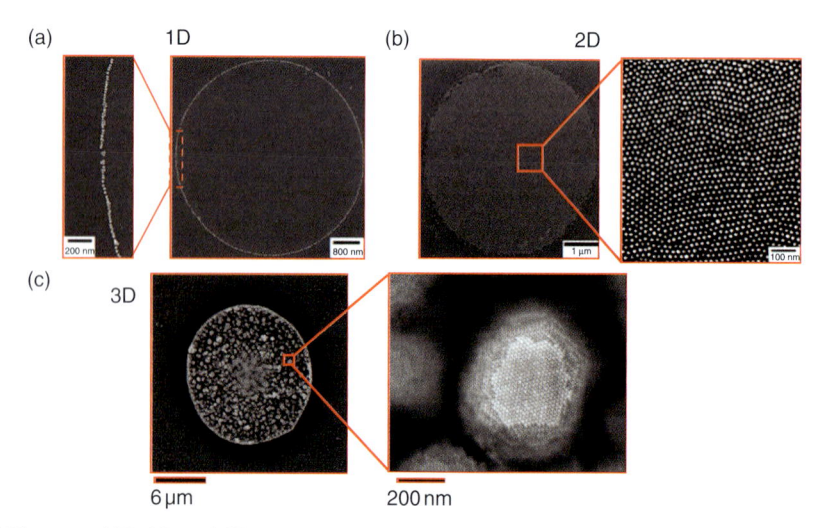

Figure 13 SEM images of 1D, 2D, and 3D ordered assemblies of 12 nm gold nanoparticles assembled through nonspecific DNA physical interactions within a drying droplet. (a) Linear arrangement of nanoparticles; (b) monolayer of gold nanoparticles; and (c) 'supra-crystals' of gold nanoparticles [186]. Reproduced with permission from Cheng W, Park N, Walter MT, Hartman MR, and Luo D (2008) Nanopatterning self-assembled nanoparticle superlattices by moulding microdroplets. *Nature Nanotechnology* 3: 682–690.

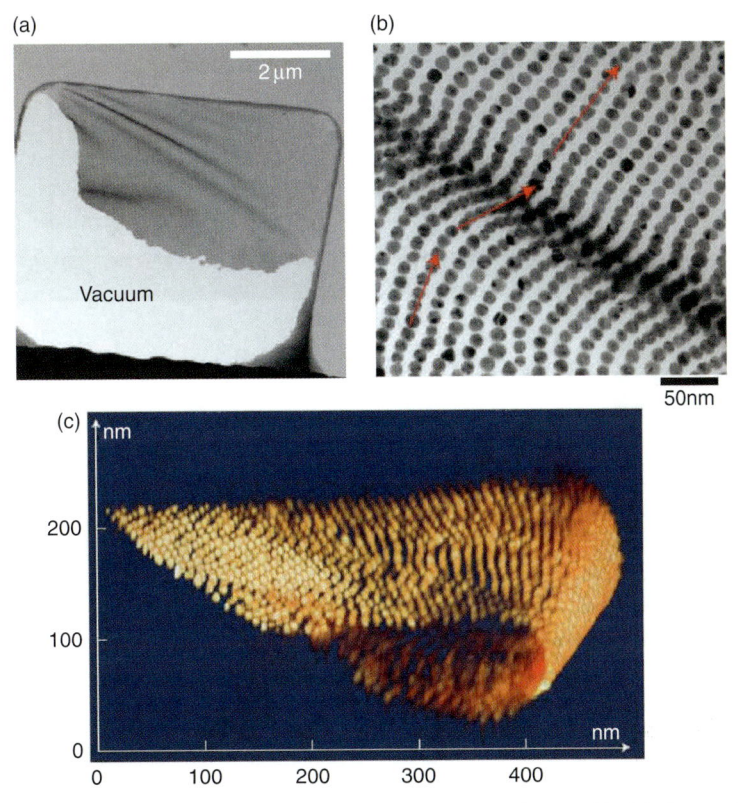

Figure 14 Free-standing superlattice sheet of 12 nm gold nanoparticles self-assembled during contact line deposition from a drying droplet through nonspecific DNA physical interactions. (a) TEM image showing a crumpled superlattice sheet over a hole in the carbon substrate. (b) Close-up shows alignment of nanoparticles even when sheet is crumpled. (c) Three-dimensional scanning tunneling electron microscopy (3D STEM) tomography reconstruction of a folded sheet [132]. Reproduced with permission from Cheng W, Campolongo MJ, Cha JJ, et al. (2009) Free-standing nanoparticle superlattice sheets controlled by DNA. *Nature Materials*, doi: 10.1038/nmat2440.

underlying the crystallization of soft corona/hard core nanoparticles (unpublished work), and also to tailor the electromagnetic and mechanical properties of nanoparticle superlattices [132]. Futhermore, by controlling the drying process of a water droplet containing the soft corona/hard core nanoparticles, we obtained 3D nanoparticle superlattices (unpublished work).

2.4.3 Assembly of Nanoparticles by Other Biomolecules

While DNA has proven to be a versatile and programmable molecule for obtaining and controlling nanoparticle assemblies, there are a myriad of other biomolecules that have been used as well. Polypeptides that specifically bind onto a variety of unmodified inorganic nanoparticle surfaces can be genetically engineered through directed evolution and screened using phage and cell-surface display techniques [188]. In addition, these designer peptides could possess unique capabilities, such as facilitating the synthesis of gold nanoparticles [189]. Rosi and coworkers [190] used such a peptide and modified it with an alkyl chain to create an amphiphilic system that leads to the formation of gold nanoparticles and self-assembly into ordered double-helical structures resembling 'nanoribbons' (**Figure 15**). Proteins can also assist the organization of nanoparticles via specific recognition and strong affinity binding, such as antigen–antibody and biotin–streptavidin interactions. For example, gold nanorods functionalized with biotin assemble linearly in an end-to-end fashion upon addition of streptavidin [191].

Bacterial surface layer (S-layer) proteins provide another means for template-based ordered assembly of nanoparticles as well. S-layer proteins self-assemble into 2D crystalline structures on a variety of

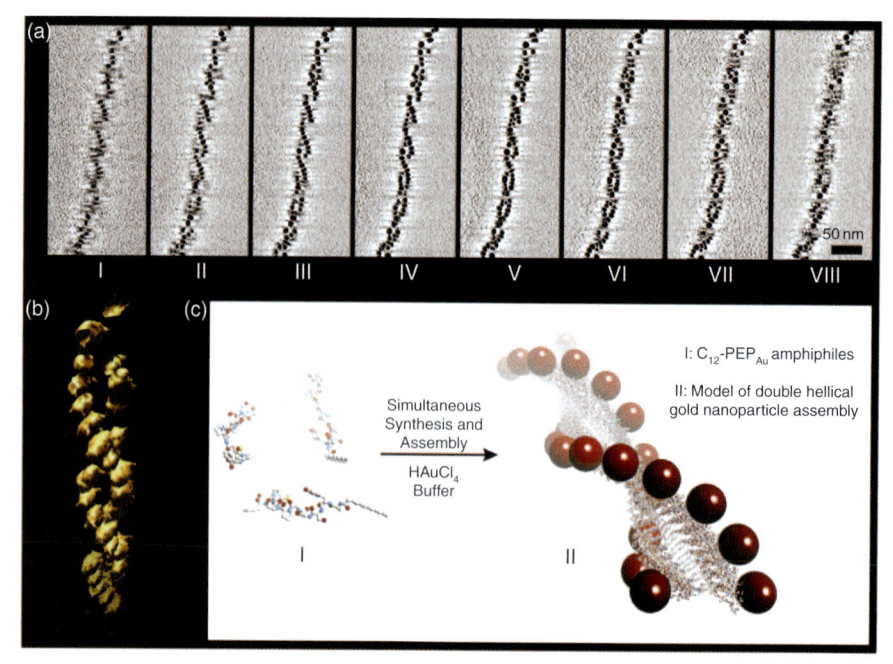

Figure 15 Structure of a gold nanoparticle double helix assembled from peptide amphiphiles as observed through electron tomography (a) and the corresponding 3D surface rendering of the tomographic volume (b) illustrate the left-handedness of these helices. Schematic representation of the formation of these helices is shown in (c) [190]. Reproduced with permission from Chen C-L, Zhang P, and Rosi NL (2008) A new peptide-based method for the design and synthesis of nanoparticle superstructures: Construction of highly ordered gold nanoparticle double helices. *Journal of the American Chemical Society* 130: 13555–13557.

substrates and are excellent for templating the growth and assembly of nanoparticles [192,193]. In particular, Tang *et al.* [194] used crystalline monolayers of S-layer proteins to template the assembly of highly ordered gold nanoparticles. Genetically engineered chaperonin protein subunits were designed to assemble into 2D periodic templates containing repeated units of hollow ring stuctures surrounded by exposed thiol groups (**Figure 16**) [195]. These templates can organize nanoparticles in a size-selective manner determined by the size of each ring, which in turn can be controlled through genetic modification. This demonstrates the potential of genetically engineered biomolecules in patterning and organizing nanoparticles for future optoelectronic devices in the nanoscale.

Genetically engineered viruses can also be exploited as a means of directing the organization of nanoparticles and nanowires. For example, a tetraglutamate-modified M13 virus was used to nucleate cobalt nanowires in an ordered virus scaffold. In the same work, cationic nanoparticles were also assembled onto the negatively charged virus scaffold via electrostatic interactions [196].

2.5 Characterization

Building well-defined nanoparticle architectures requires an assortment of tools to characterize nanoparticle building blocks, bioconjugates, the dynamic self-assembly process, and the resulting ordered nanoparticle assemblies.

2.5.1 Microscopy

Various microscopic tools enable direct observation of the size, shape, and crystalline phase of an individual nanoparticle, as well as structural information about nanoparticle assemblies [40,140,197–200]. These tools typically include TEM, SEM, and light microscopy. The electron microscopy techniques allow for direct visualization of nanoparticle shape and internal atomic structure [197,198], whereas light microscopy provides the ability to compare shapes and sizes of different particles based on their light scattering properties in dark field (**Figure 17**) [140,199].

TEM represents one of the most widely used characterization techniques for nanoparticles and their resultant structures. High-resolution TEM

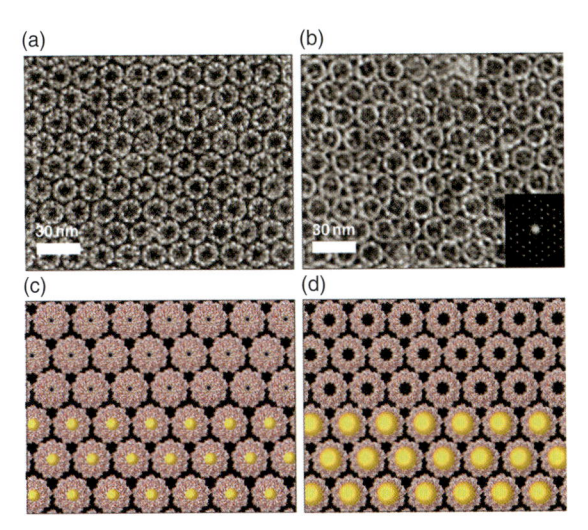

Figure 16 Two-dimensional crystals of genetically engineered chaperonins for the templated assembly of quantum dot arrays [195]. (a) TEM image of a 2D crystal of the beta chaperonin variant with exposed cysteine substitutions at the pores. (b) The 2D crystal of the loopless chaperonin variant shows an increase in pore size from 3 to 9 nm. Inset illustrates the ordering of the crystal in a fast Fourier transform. (c) Graphical representation of small 5 nm nanoparticles assembled within the 3 nm pores in the genetically engineered crystals corresponding to the image in (a), and (d) of larger 10 nm particles assembled within the 9 nm pores corresponding to (b). Reproduced with permission from McMillan RA, Paavola CD, Howard J, Chan SL, Zaluzec NJ, and Trent JD (2002) Ordered nanoparticle arrays formed on engineered chaperonin protein templates. *Nature Materials* 1: 247–252.

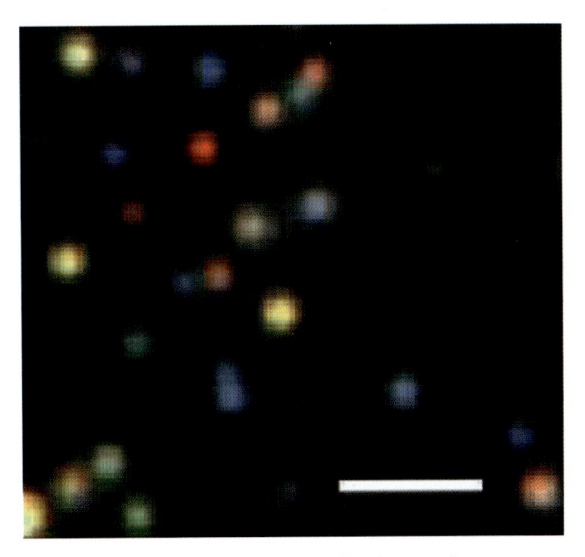

Figure 17 Dark-field light scattering image of a heterogeneous mixture of silver nanoparticles of different shapes [40]. Scale = 2 μm. Reproduced with permission from Oren dorff CJ, Sau TK, and Murphy CJ (2006) Shape-dependent plasmon-resonant gold nanoparticles. *Small* 2: 636–639.

enables visualization of individual atoms and sub-nanometer surface details. In addition, the ability to obtain diffraction patterns allows for determination of face orientation and lattice structure (including interfaces within polycrystalline structures) [201]. A limitaion of TEM imaging is that thin samples need to be deposited on the electron beam-transparent surface. Complementary to TEM, SEM imaging reveals surface structures of thick nanoparticle samples by detecting back-scattered electrons. A major advantage of the SEM is that it requires minimal sample preparation when compared to the TEM, although its imaging resolution is lower [40,198,200]. Electron microscopy imaging can also be combined with X-ray detection to obtain the energy-dispersive X-ray spectroscopy (EDX) of individual nanoparticles [197,198,201,202]. The added functionality afforded by EDX allows for elemental analysis and chemical characterization of the nanoparticle sample [201,202]. Hence, electron microscopy can provide comprehensive information by combining high-resolution imaging with surface elemental mapping, making it an indispensible technique for characterizing nanosized objects [197].

2.5.2 Gel Electrophoresis

Gel electrophoresis is a well-established molecular biology technique typically used for the separation of biomolecules such as DNA and proteins [203,204]. It is based on the electrokinetic phenomena in which charged particles or molecules migrate relative to a fluid in the presence of an electric field [205]. The differential electrophoretic mobility of molecules in a gel matrix arising from the relative disparities in size, shape, and charge allows for the distinct separation of molecules in solution.

In addition to biomolecules, nanoparticles of various shapes and sizes can also be differentiated based on the principles of gel electrophoresis as well [206]. This approach can also be extended to the isolation and identification of DNA–nanoparticle conjugates [135,207,208].

Although the absolute effective diameters of nanoparticle conjugates cannot be accurately determined using gel electrophoresis alone, it is a powerful technique for analyzing the attachment of biomolecular ligands onto nanoparticles with a high sensitivity of detection for the relative increase in effective diameters [207]. In fact, gel electrophoresis is even able to resolve DNA–nanoparticle monoconjugates. For example, Alivisatos and coworkers [100] demonstrated that gold nanoparticles conjugated to a specific number of 100-base ssDNA strands could be identified and separated via gel electrophoresis (**Figure 18**). In addition, gel electrophoresis was used to characterize gold nanoparticles capped with various lengths of ssDNA, ranging from 5 to 90 nm (**Figure 19**) [132]. Nanoparticle molecules, such as dimers and trimers, could be isolated based on their differences in electrophoretic mobility as well [135].

2.5.3 Optical Spectroscopy

Nanoparticles exhibit optical properties which can be characterized via near-field and far-field spectroscopic tools [209,210]. Attachment of biomolecules onto nanoparticles affords a way to manipulate these optical properties by altering dielectric environments surrounding nanoparticles [211] or tailoring interparticle spacing [88,114,132,212]. In particular, size- and shape-dependent plasmonic properties of metallic nanoparticles can be easily characterized by UV–visible spectroscopy [137,213–216]. The modes and intensities of surface plasmon resonance can be well resolved, and can be correlated with structures [209,210,217] and quantum size effects [218] of nanoparticles.

Bioconjugated nanoparticles can also be characterized by monitoring the characteristic absorbance of both biomolecules and nanoparticles. For example, DNA-capped gold nanoparticles exhibit two peaks at 260 and ~530 nm in the UV–visible spectroscopic window, corresponding to DNA absorbance and surface plasmon resonance of gold nanoparticles, respectively. The number of DNA molecules per

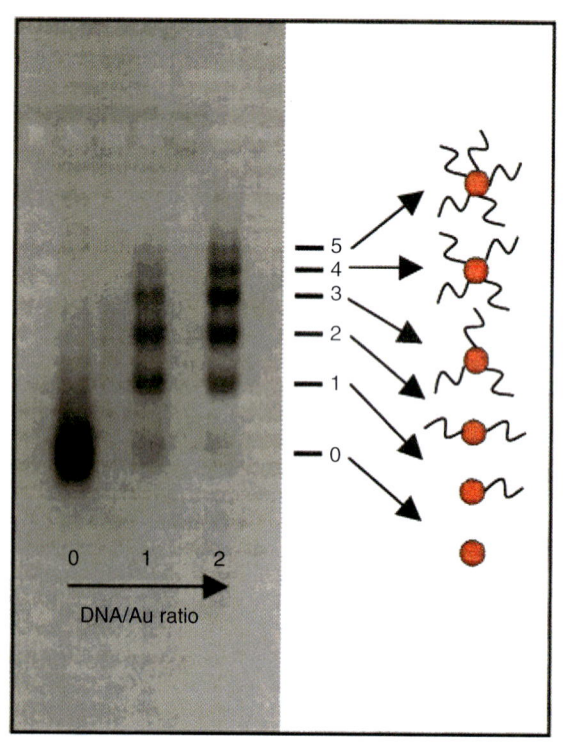

Figure 18 Electophoretic mobility of ssDNA–gold nanoparticle conjugates. The first lane on the left corresponds to a single band of only 5 nm nanoparticles. Discrete bands (**0** to **5**) appear upon addition of DNA to the gold nanoparticles (middle lane). When the amount of DNA is doubled, the intensity of the bands increases. Each discrete band directly corresponds to a specific number of DNA strands attached on each particle, as depicted by the illustration on the right [100]. Reproduced with permission from Zanchet D, Micheel CM, Parak WJ, Gerion D, and Alivisatos AP (2001) Electrophoretic isolation of discrete Au nanocrystal/DNA conjugates. *Nano Letters* 1: 32–35.

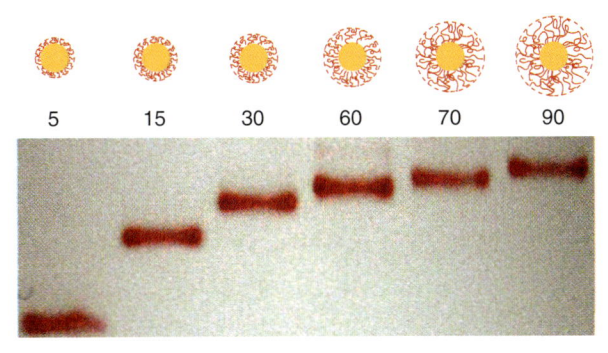

Figure 19 Electrophoretic mobility of DNA–nanoparticle conjugates with varying lengths of ssDNA (5–90 nm) attached onto the surface of gold nanoparticles [132].

nanoparticle can be derived based on Beer's law and the extinction coefficients of DNA and nanoparticles.

In addition to characterizing individual nanoparticles and their bioconjugation, spectroscopic tools can also be used for monitoring biological binding events for sensing applications [219,220]. The change in plasmon resonance wavelength of individual DNA-capped nanoparticles depends on the length of the DNA due to the sensitivity of the plasmon band to surrounding dielectric environments [221]. An average wavelength shift of approximately 1.24 nm was observed per DNA base pair by Alivisatos and coworkers. In an earlier effort, Haes and Van Duyne [222] demonstrated a triangular silver nanoparticles-based plasmonic biosensor for detecting streptavidin. They found that the surface plasmon resonance spectrum was unexpectedly sensitive to nanoparticle size, shape, and local external dielectric environment.

Furthermore, the configurational changes in dimeric plasmonic nanoparticle molecules can be precisely measured by UV–visible spectroscopy. Plasmonic behaviors within metallic nanoparticle molecules is strongly dependent on the interparticle distance which governs the plasmon coupling. In a report by Alivisatos' group [140], kinetics of ssDNA hybridization events in the dimeric nanoparticle molecules can be monitored in real-time by following plasmon peak shifts (**Figure 20**). Such shifts can not only reveal the configurational changes in nanoparticle molecules, but can also be used for diagnostic sensing applications. One of the advantages of plasmon-based sensing is that it is free from bleaching and blinking phenomena that dominate fluorescence-based sensing strategies and also allows for monitoring greater distances than Förster resonance energy transfer.

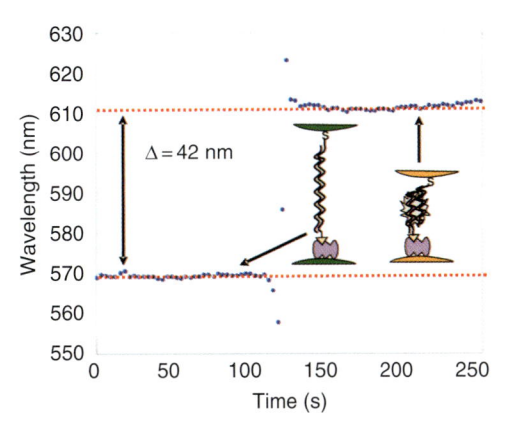

Figure 20 Spectral position as a function of time shows a sudden redshift of 42 nm as a result of interparticle distance change from addition of a DNA binding dendrimer [140]. Reproduced with permission from Sonnichsen C, Reinhard BM, Liphardt J, and Alivisatos AP (2005) A molecular ruler based on plasmon coupling of single gold and silver nanoparticles. *Nature Biotechnology* 23: 741–745.

2.5.4 Small-Angle X-ray Scattering

A powerful method for revealing the crystal structure of ordered assemblies of nanoparticles is SAXS, which can reveal structural information for materials with features in the nanometer range. In particular, high-energy synchrotron-based SAXS gives high-resolution scattering patterns from very short exposure times [223]. In this process, the incident monochromatic X-ray beam strikes the sample and scatters. For ordered structures, this produces diffraction rings that correspond to Bragg planes within the assembled crystallites. The scattered X-rays can be gathered by a detector, resulting in 2D images of concentric rings that are subsequently integrated to produce plots of intensity versus scattering angle

(**Figure 21**). The position and relative intensity of the resultant diffraction peaks enables the elucidation of the type of crystal structure (e.g., f.c.c. or b.c.c.) and degree of ordering in the sample. When samples contain single-crystalline domains, intense spots in the powder rings can appear in the 2D detector image. These spots correspond to individual crystallites, whose size can be estimated by the Scherrer formula [224]

$$t = \frac{K\lambda}{B\cos\theta}$$

in which t is the crystallite thickness, K the Scherrer constant, λ the X-ray wavelength, B the peak breadth (full-width at half maximum of the peak), and θ is the incident angle. For systems with significant complexity, an accurate match between theoretical and observed crystal structure may require simulation by software [225,226].

As a complementary technique to electron microscopy, SAXS can also obtain structural information from ordered nanoparticle assemblies. SAXS is particularly appropriate for characterizing biological systems since most biological samples require an aqueous environment to maintain their native structure: conditions that are incompatible with electron microscopy. Synchrotron-based SAXS can characterize nanoscale structures in real time, in aqueous environments, making the technique uniquely well suited for characterizing biomolecule mediated assemblies.

2.5.5 Atomic Force Microscopy

AFM, invented in 1986 by Binnig *et al.* [227], is a powerful technique for characterizing nanoscale structures, and is particularly well suited for imaging and performing force measurements on nanoparticle superlattices. Unlike electron microscopy, AFM imaging requires physical contact of the sample surface by a probe and is limited to a scanning area on the order of several hundred micrometers, making it a relatively low throughput technique. However, it is capable of providing accurate topographical information, which is useful for identifying minimally thin structures, such as DNA lattices and crystallite layers

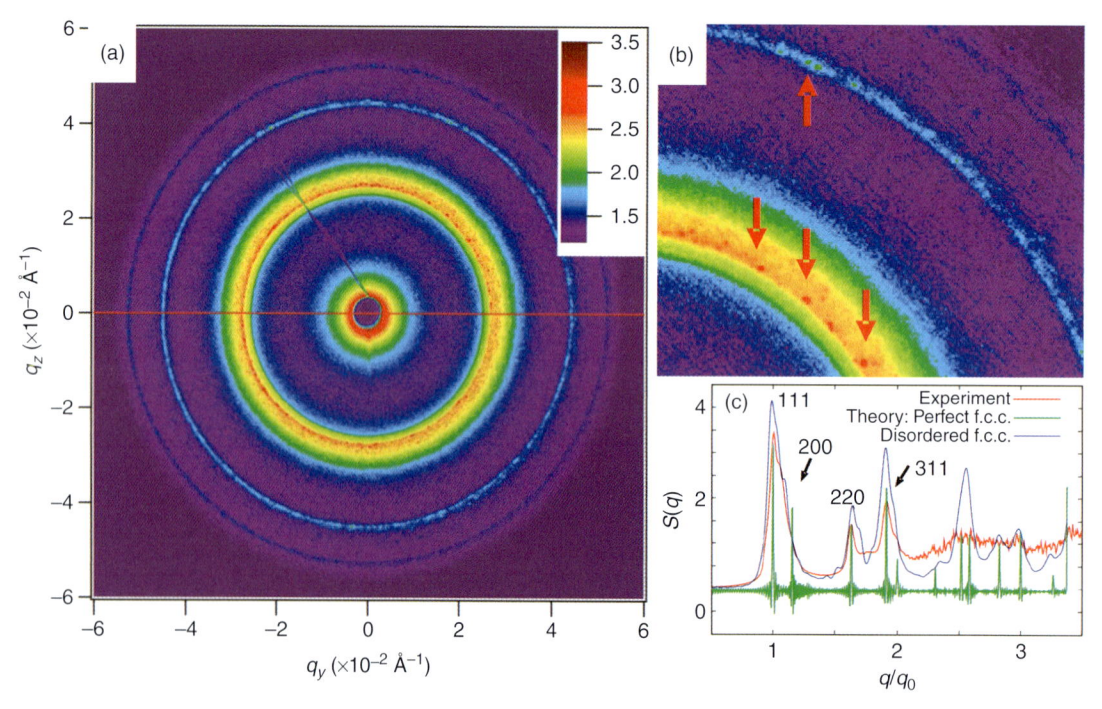

Figure 21 SAXS pattern from 3D DNA-mediated crystallization of gold nanoparticles. (a) Two-dimensional detector image displaying diffraction pattern from single-component system with self-complementary linker. Pseudocoloring corresponds to logarithmic intensity scale. (b) Magnified 2D detector image showing individual spots (red arrows) from single-crystalline domains. (c) Integrated intensity corresponding to f.c.c. crystal structure. The *x*-axis is normalized with respect to the first peak (at 2.76 Å) [182]. Reproduced with permission from Park SY, Lytton-Jean AKR, Lee B, Weigand S, Schatz GC, and Mirkin CA (2008) DNA-programmable nanoparticle crystallization. *Nature* 451: 553–556.

[228]. In addition, AFM is also capable of directly observing DNA, which is not achievable in electron microscopy without the addition of a negative stain. Imaging also does not require extreme environmental conditions but can be performed in both air and liquid, which is practical for observing biological structures in a more stable, physiological environment.

Another novel feature of AFM is that it has enabled force spectroscopy of soft materials. In the 1990s, force spectroscopy became a popular method for microrheology studies of biological specimens [229], including cells [230,231], organelles [232,233], and biomolecules [229,234]. Force measurements are carried out by indenting with a probe while recording the indentation force versus sample displacement, from which elastic properties of the sample can be extracted. In order to obtain the elastic modulus, data were typically fitted to the Hertz model [235], which describes the contact interaction between two elastically deformable spheres.

Free-standing thin films have proven to be ideal systems for investigation by force spectroscopy [236,237]. The free-standing configuration facilitates such measurements by allowing for direct measurement of elastic response without complication from supporting substrates. Theoretical models describing the indentation of suspended, circular thin films [238,239] have made it possible to estimate mechanical characteristics, such as the elastic modulus and tensile strength, for thin crystalline materials. Lee *et al.* [240] have used this approach to measure the elastic modulus of suspended single layers of graphene, which were in good agreement with experimental data. Mueggenburg *et al.* [15] measured the Young's moduli of free-standing alkylthiol-capped nanoparticle superlattices, with values ranging from 3 to 39 GPa. In a similar fashion, force measurements were performed on free-standing DNA-capped nanoparticle superlattices (**Figure 22**) [132]. Measurements of the spring constant for different DNA spacer lengths revealed a trend of increasing stiffness as the spacer length was decreased, indicating that a level of control could be exercised over the mechanical properties. Furthermore, estimates suggested that these sheets possess elastic moduli on the order of a few gigapaseal, which is comparable to their alkylthiol-capped counterparts.

2.6 Summary and Outlook

The discovery of new materials and control over their properties have played a critical role in shaping our world. There is an increasing demand for adaptive high-performance materials with precisely regulated electronic, photonic, and magnetic properties, which will lead to technological breakthroughs in efficient energy-harvesting and sustainable environmental systems. The materials in current optoelectronic devices are predominantly hard materials that lack adaptive capabilities. In contrast, the machinery of life is comprised of soft building blocks such as nucleic acids,

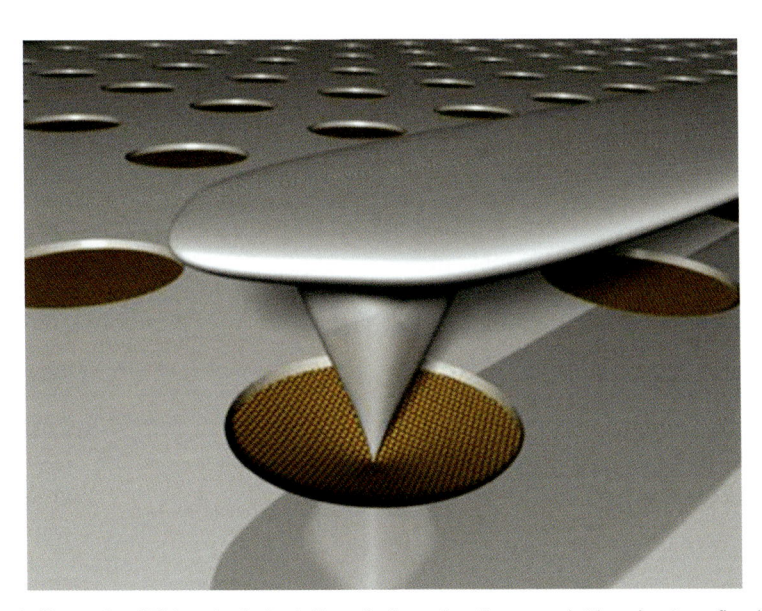

Figure 22 Schematic illustrating AFM probe indentation of a free-standing superlattice sheet confined to a microhole [132].

proteins, carbohydrates, and lipids. With these building blocks, nature has evolved elaborate biocircuitry systems which are self-organized, highly ordered, self-sustained and adaptive to various environments. The combination of nanoparticles with soft biological building blocks affords a powerful approach to creating nanobiomaterials, which may allow us to integrate the optoelectronic functions of nanoparticles with biological machinery.

One of the most important goals for fabricating nanoparticle-based devices is to organize them into well-defined structures in an inexpensive way. Chemists have developed an array of wet chemical synthesis techniques to create large quantities of low-cost nanoparticle building blocks with enhanced size- and shape-dependent properties. Despite these significant advances, it remains a challenge to rationally assemble nanoparticles into well-defined nanoarchitectures that meet device requirements. In nature, biology-based assemblies exploit the unique specific recognition capabilities of biomolecules and the molecular biology toolbox that has evolved through time to circumvent this issue. Applying the specific and nonspecific interactions of these biomolecular building blocks can facilitate the assembly of highly ordered nanoparticle arrays, which have great potential for photonics, electronics, sensors, spintronics, and energy-harvesting systems in the nanoscale.

This review has summarized recent progress in the fabrication of highly ordered nanoparticle assemblies via the bio-based strategy. In particular, DNA has been heavily used for this purpose due to the inherently unique characteristics of DNA molecules [37]. Despite its success, the bio-based approach to real-world devices is still in its nascent state. One limitation to the integration of bio-based nanoparticle assemblies into solid-state devices is the instability of biospecific interactions in the dehydrated state. Using nonspecific DNA ligand interactions in a drying-mediated self-assembly process could potentially provide a solution with the caveat of lower specificity. Further development could lead to methods that maintain the programmable ordered structures in the dehydrated state for their integration into optoelectronic devices. Our increased understanding of the mechanisms of biological interactions, as well as the governing dynamics behind controlled nanoparticle synthesis, drive the discovery of new materials and properties at this unique crossroads of two exciting fields of nanotechnology and biotechnology.

References

1. Daniel MC and Astruc D (2004) Gold nanoparticles: Assembly, supramolecular chemistry, quantum-size-related properties, and applications toward biology, catalysis, and nanotechnology. *Chemical Reviews* 104: 293–346.
2. Pérez-Juste J, Pastoriza-Santos I, Liz-Marzán LM, and Mulvaney P (2005) Gold nanorods: Synthesis, characterization and applications. *Coordination Chemistry Reviews* 249: 1870–1901.
3. Lu AH, Salabas EL, and Schuth F (2007) Magnetic nanoparticles: Synthesis, protection, functionalization, and application. *Angewandte Chemie International Edition* 46: 1222–1244.
4. Shipway AN, Katz E, and Willner I (2000) Nanoparticle arrays on surfaces for electronic, optical, and sensor applications. *ChemPhysChem* 1: 18–52.
5. Medintz IL, Uyeda HT, Goldman ER, and Mattoussi H (2005) Quantum dot bioconjugates for imaging, labelling and sensing. *Nature Materials* 4: 435–446.
6. Grieve K, Mulvaney P, and Grieser F (2000) Synthesis and electronic properties of semiconductor nanoparticles/quantum dots. *Current Opinion in Colloid and Interface Science* 5: 168–172.
7. Alivisatos AP (1996) Semiconductor clusters, nanocrystals, and quantum dots. *Science* 271: 933–937.
8. Banin U, Cao YW, Katz D, and Millo O (1999) Identification of atomic-like electronic states in indium arsenide nanocrystal quantum dots. *Nature* 400: 542–544.
9. Collier CP, Saykally RJ, Shiang JJ, Henrichs SE, and Heath JR (1997) Reversible tuning of silver quantum dot monolayers through the metal-insulator transition. *Science* 277: 1978–1981.
10. Shevchenko EV, Talapin DV, Murray CB, and O'Brien S (2006) Structural characterization of self-assembled multifunctional binary nanoparticle superlattices. *Journal of the American Chemical Society* 128: 3620–3637.
11. Bigioni TP, Lin XM, Nguyen TT, Corwin EI, Witten TA, and Jaeger HM (2006) Kinetically driven self assembly of highly ordered nanoparticle monolayers. *Nature Materials* 5: 265–270.
12. Kalsin AM, Fialkowski M, Paszewski M, Smoukov SK, Bishop KJM, and Grzybowsk BA (2006) Electrostatic self-assembly of binary nanoparticle crystals with a diamond-like lattice. *Science* 312: 420–424.
13. Kiely CJ, Fink J, Brust M, Bethell D, and Schiffrin DJ (1998) Spontaneous ordering of bimodal ensembles of nanoscopic gold clusters. *Nature* 396: 444–446.
14. Korgel BA, Fullam S, Connolly S, and Fitzmaurice D (1998) Assembly and self-organization of silver nanocrystal superlattices: Ordered "soft spheres". *Journal of Physical Chemistry B* 102: 8379–8388.
15. Mueggenburg KE, Lin XM, Goldsmith RH, and Jaeger HM (2007) Elastic membranes of close-packed nanoparticle arrays. *Nature Materials* 6: 656–660.
16. Murray CB, Kagan CR, and Bawendi MG (1995) Self-organization of CdSe nanocrystallines into 3-dimensional quantum-dot superlattices. *Science* 270: 1335–1338.
17. Pileni MP (2001) Nanocrystal self-assemblies: Fabrication and collective properties. *Journal of Physical Chemistry B* 105: 3358–3371.
18. Shevchenko EV, Talapin DV, Kotov NA, O'Brien S, and Murray CB (2006) Structural diversity in binary nanoparticle superlattices. *Nature* 439: 55–59.
19. Courty A, Mermet A, Albouy PA, Duval E, and Pileni MP (2005) Vibrational coherence of self-organized silver

nanocrystals in f.c.c. supra-crystals. *Nature Materials* 4: 395–398.
20. Urban JJ, Talapin DV, Shevchenko EV, Kagan CR, and Murray CB (2007) Synergism in binary nanocrystal superlattices leads to enhanced p-type conductivity in self-assembled PbTe/Ag-2 Te thin films. *Nature Materials* 6: 115–121.
21. Collier CP, Vossmeyer T, and Heath JR (1998) Nanocrystal superlattices. *Annual Review of Physical Chemistry* 49: 371–404.
22. Alivisatos AP, Johnsson KP, Peng X, et al. (1996) Organization of 'nanocrystal molecules' using DNA. *Nature* 382: 609–611.
23. Ohara PC, Leff DV, Heath JR, and Gelbart WM (1995) Crystallization of opals from polydisperse nanoparticles. *Physical Review Letters* 75: 3466–3469.
24. Rabani E, Reichman DR, Geissler PL, and Brus LE (2003) Drying-mediated self-assembly of nanoparticles. *Nature* 426: 271–274.
25. Kalsin AM, et al. (2006) Electrostatic self-assembly of binary nano particle crystals with a diamond-like-lattice. *Science* 312: 420–424.
26. Chandler D (2005) Interfaces and the driving force of hydrophobic assembly. *Nature* 437: 640–647.
27. Qin Y and Fichthorn KA (2003) Molecular-dynamics simulation of forces between nanoparticles in a Lennard-Jones liquid. *Journal of Chemical Physics* 119: 9745.
28. Kolny J, Kornowski A, and Weller H (2002) Self-organization of cadmium sulfide and gold nanoparticles by electrostatic interaction. *Nano Letters* 2: 361–364.
29. Lalatonne Y, Richardi J, and Pileni MP (2004) Van der Waals versus dipolar forces controlling mesoscopic organizations of magnetic nanocrystals. *Nature Materials* 3: 121–125.
30. Seeman NC (2003) DNA in a material world. *Nature* 421: 427–431.
31. Luo D (2003) The road from biology to materials. *Materials Today* 6: 38–43.
32. Liu J and Lu Y (2006) Smart nanomaterials responsive to multiple chemical stimuli with controllable cooperativity. *Advanced Materials* 18: 1667–1671.
33. Katz E and Willner I (2004) Integrated nanoparticle–biomolecule hybrid systems: Synthesis, properties, and applications. *Angewandte Chemie International Edition* 43: 6042–6108.
34. You CC, Verma A, and Rotello VM (2006) Engineering the nanoparticle–biomacromolecule interface. *Soft Matter* 2: 190–204.
35. Niemeyer CM and Simon U (2005) DNA-based assembly of metal nanoparticles. *European Journal of Inorganic Chemistry* 18: 3641–3655.
36. Ofir Y, Samanta B, and Rotello VM (2008) Polymer and biopolymer mediated self-assembly of gold nanoparticles. *Chemical Society Reviews* 37: 1814–1823.
37. Kwon YW, Lee CH, Choi DH, and Jin JI (2009) Materials science of DNA. *Journal of Materials Chemistry* 19: 1353–1380.
38. Katz E and Willner I (2004) Integrated nanoparticle–biomolecule hybrid systems: Synthesis, properties, and applications. *Angewandte Chemie International Edition* 43: 6042–6108.
39. Yin Y and Alivisatos AP (2004) Colloidal nanocrystal synthesis and the organic–inorganic interface. *Nature* 437: 664–670.
40. Casavola M, Buonsanti R, Caputo G, and Cozzoli PD (2008) Colloidal strategies for preparing oxide-based hybrid nanocrystals. *European Journal of Inorganic Chemistry* 2008: 837–854.

41. Sugimoto T (1987) Preparation of monodispersed colloidal particles. *Advances in Colloid and Interface Science* 28: 65–108.
42. Murray CB, Norris DJ, and Bawendi MG (1993) Synthesis and characterization of nearly monodisperse CdE (E = sulfur, selenium, tellurium) semiconductor nanocrystallites. *Journal of the American Chemical Society* 115: 8706–8715.
43. Manna L, Scher EC, and Alivisatos AP (1999) Synthesis of soluble and processable rod-, arrow-, teardrop-, and tetrapod-shaped CdSe nanocrystals. *Langmuir* 15: 2002.
44. Rossetti R, Nakahara S, and Brus LE (1983) Quantum size effects in the redox potentials, resonance Raman spectra, and electronic spectra of CdS crystallites in aqueous solution. *Journal of Chemical Physics* 79: 1086.
45. Chestnoy N, Hull R, and Brus LE (1986) Higher excited electronic states in clusters of ZnSe, CdSe, and ZnS: Spin-orbit, vibronic, and relaxation phenomena. *Journal of Chemical Physics* 85: 2237.
46. Steigerwald ML, Alivisatos AP, Gibson JM, et al. (1988) Surface derivatization and isolation of semiconductor cluster molecules. *Journal of the American Chemical Society* 110: 3046–3050.
47. Swihart MT (2003) Vapor-phase synthesis of nanoparticles. *Current Opinion in Colloid and Interface Science* 8: 127–133.
48. Mafune F, Kohno J, Takeda Y, Kondow T, and Sawabe H (2001) Formation of gold nanoparticles by laser ablation in aqueous solution of surfactant. *Journal of Physical Chemistry B* 105: 5114–5120.
49. Aslam M, Fu L, Su M, Vijayamohanan K, and Dravid VP (2004) Novel one-step synthesis of amine-stabilized aqueous colloidal gold nanoparticles. *Journal of Materials Chemistry* 14: 1795–1797.
50. Shon YS and Cutler E (2004) Aqueous synthesis of alkanethiolate-protected Ag nanoparticles using bunte salts. *Langmuir* 20: 6626–6630.
51. Sun S and Zeng H (2002) Size-controlled synthesis of magnetite nanoparticles. *Journal of the American Chemical Society* 124: 8204–8205.
52. Ostwald W (1897) Studien über die Bildung und Umwandlung fester Körper. *Zeitschrift für Physikalische Chemie* 22: 289–330.
53. Wagner C (1961) Theorie Der Alterung Von Niederschlagen Durch Umlosen (Ostwald-Reifung). *Zeitschrift Fur Elektrochemie* 65: 581–591.
54. Johnson CJ, Dujardin E, Davis SA, Murphy CJ, and Mann S (2002) Growth and form of gold nanorods prepared by seed-mediated, surfactant-directed synthesis. *Journal of Materials Chemistry* 12: 1765–1770.
55. Murphy CJ, Sau TK, Gole AM, et al. (2005) Anisotropic metal nanoparticles: Synthesis, assembly, and optical applications. *Journal of Physical Chemistry B* 109: 13857–13870.
56. Song Y and Murray RW (2000) Dynamics and extent of ligand exchange depend on electronic charge of metal nanoparticles. *Science* 291: 2390–2392.
57. Wang Y, Wong JF, Teng X, Lin XZ, and Yang H (2003) "Pulling" nanoparticles into water: Phase transfer of oleic acid stabilized monodisperse nanoparticles into aqueous solutions of [alpha]-cyclodextrin. *Nano Letters* 3: 1555–1559.
58. Swami A, Kumar A, and Sastry M (2003) Formation of water-dispersible gold nanoparticles using a technique based on surface-bound interdigitated bilayers. *Langmuir* 19: 1168–1172.
59. Goldstein RV and Morozov NF (2007) Mechanics of deformation and fracture of nanomaterials and nanotechnology. *Physical Mesomechanics* 10: 235–246.

60. Cervellino A, Giannini C, and Guagliardi A (2003) Determination of nanoparticle structure type, size and strain distribution from X-ray data for monatomic fcc-derived non-crystallographic nanoclusters. *Journal of Applied Crystallography* 36: 1148–1158.
61. Gole A and Murphy CJ (2004) Seed-mediated synthesis of gold nanorods: Role of the size and nature of the seed. *Chemistry of Materials* 16: 3633–3640.
62. Burda C, Chen XB, Narayanan R, and El-Sayed MA (2005) Chemistry and properties of nanocrystals of different shapes. *Chemical Reviews* 105: 1025–1102.
63. Nikoobakht B and El-Sayed MA (2001) Evidence for bilayer assembly of cationic surfactants on the surface of gold nanorods. *Langmuir* 17: 6368–6374.
64. Sau TK and Murphy CJ (2005) Self-assembly patterns formed upon solvent evaporation of aqueous cetyltrimethylammonium bromide-coated gold nanoparticles of various shapes. *Langmuir* 21: 2923–2929.
65. Millstone JE, Hurst SJ, Metraux GS, Cutler JI, and Mirkin CA (2009) Colloidal gold and silver triangular nanoprisms. *Small* 5: 646–664.
66. Xue C, Millstone JE, Li S, and Mirkin CA (2007) Plasmon-driven synthesis of triangular core–shell nanoprisms from gold seeds. *Angewandte Chemie International Edition* 46: 8436.
67. Xue C, Métraux GS, Millstone JE, and Mirkin CA (2008) Mechanistic study of photomediated triangular silver nanoprism growth. *Journal of the American Chemical Society* 130: 8337–8344.
68. Wu X, Redmond PL, Liu H, Chen Y, Steigerwald M, and Brus L (2008) Photovoltage mechanism for room light conversion of citrate stabilized silver nanocrystal seeds to large nanoprisms. *Journal of the American Chemical Society* 130: 9500–9506.
69. Jin R, Cao YC, Hao E, Métraux GS, Schatz GC, and Mirkin CA (2003) Controlling anisotropic nanoparticle growth through plasmon excitation. *Nature* 425: 487–490.
70. Sun Y and Xia Y (2002) Shape-controlled synthesis of gold and silver nanoparticles. *Science* 298: 2176.
71. Tao A, Sinsermsuksakul P, and Yang P (2006) Polyhedral silver nanocrystals with distinct scattering signatures. *Angewandte Chemie International Edition* 45: 4597–4601.
72. Skrabalak SE, Chen J, Sun Y, et al. (2008) Gold nanocages: Synthesis, properties, and applications. *Accounts of Chemical Research* 41: 1587–1595.
73. Nehl CL, Liao H, and Hafner JH (2006) Optical properties of star-shaped gold nanoparticles. *Nano Letters* 6: 683–688.
74. Caruso F (2001) Nanoengineering of particle surfaces. *Advanced Materials* 13: 11–22.
75. Gittins DI and Caruso F (2001) Tailoring the polyelectrolyte coating of metal nanoparticles. *Journal of Physical Chemistry B* 105: 6846–6852.
76. Michalet X, Pinaud FF, Bentolila LA, et al. (2005) Quantum dots for live cells, in vivo imaging, and diagnostics. *Science* 307: 538–544.
77. Ho K-C, Tsai P-J, Lin Y-S, and Chen Y-C (2004) Using biofunctionalized nanoparticles to probe pathogenic bacteria. *Analytical Chemistry* 76: 7162–7168.
78. Teng C-H, Ho K-C, Lin Y-S, and Chen Y-C (2004) Gold nanoparticles as selective and concentrating probes for samples in MALDI MS analysis. *Analytical Chemistry* 76: 4337–4342.
79. Tkachenko AG, Xie H, Coleman D, et al. (2003) Multifunctional gold nanoparticle–peptide complexes for nuclear targeting. *Journal of the American Chemical Society* 125: 4700–4701.
80. Yi X, Huang-Xian J, and Hong-Yuan C (2000) Direct electrochemistry of horseradish peroxidase immobilized on a colloid/cysteamine-modified gold electrode. *Analytical Biochemistry* 278: 22–28.
81. Li H and Rothberg L (2004) Colorimetric detection of DNA sequences based on electrostatic interactions with unmodified gold nanoparticles. *Proceedings of the National Academy of Sciences of the United States of America* 101: 14036–14039.
82. Mahtab R, Rogers JP, and Murphy CJ (1995) Protein-sized quantum dot luminescence can distinguish between "straight", "bent", and "kinked" oligonucleotides. *Journal of the American Chemical Society* 117: 9099–9100.
83. Sasaki YC, Yasuda K, Suzuki Y, et al. (1997) Two-dimensional arrangement of a functional protein by cysteine–gold interaction: Enzyme activity and characterization of a protein monolayer on a gold substrate. *Biophysical Journal* 72: 1842–1848.
84. Hong HG, Jiang M, Sligar SG, and Bohn PW (1994) Cysteine-specific surface tethering of genetically engineered cytochromes for fabrication of metalloprotein nanostructures. *Langmuir* 10: 153–158.
85. Jue R, Lambert JM, Pierce LR, and Traut RR (1978) Addition of sulfhydryl groups of *Escherichia coli* ribosomes by protein modification with 2-iminothiolane (methyl 4-mercaptobutyrimidate). *Biochemistry* 17: 5399–5406.
86. Carlsson J, Drevin H, and Axén R (1978) Protein thiolation and reversible protein–protein conjugation. N-Succinimidyl 3-(2-pyridyldithio)propionate, a new heterobifunctional reagent. *Biochemical Journal* 173: 723–737.
87. Alivisatos AP, Johnsson KP, Peng X, et al. (1996) Organization of 'nanocrystal molecules' using DNA. *Nature* 382: 609–611.
88. Mirkin CA, Letsinger RL, Mucic RC, and Storhoff JJ (1996) A DNA-based method for rationally assembling nanoparticles into macroscopic materials. *Nature* 382: 607–609.
89. Naka K, Itoh H, Tampo Y, and Chujo Y (2003) Effect of gold nanoparticles as a support for the oligomerization of l-cysteine in an aqueous solution. *Langmuir* 19: 5546–5549.
90. Grancharov SG, Zeng H, Sun S, et al. (2005) Bio-functionalization of monodisperse magnetic nanoparticles and their use as biomolecular labels in a magnetic tunnel junction based sensor. *Journal of Physical Chemistry B* 109: 13030–13035.
91. Li D, He Q, Cui Y, Duan L, and Li J (2007) Immobilization of glucose oxidase onto gold nanoparticles with enhanced thermostability. *Biochemical and Biophysical Research Communications* 355: 488–493.
92. Weber PC, Ohlendorf DH, Wendoloski JJ, and Salemme FR (1989) Structural origins of high-affinity biotin binding to streptavidin. *Science* 243: 85–88.
93. Gref R, Couvreur P, Barratt G, and Mysiakine E (2003) Surface-engineered nanoparticles for multiple ligand coupling. *Biomaterials* 24: 4529–4537.
94. Loweth CJ, Caldwell WB, Peng X, Alivisatos AP, and Schultz PG (1999) DNA-based assembly of gold nanocrystals. *Angewandte Chemie International Edition* 38: 1808–1812.
95. Claridge SA, Goh SL, Fréchet JMJ, Williams SC, Micheel CM, and Alivisatos AP (2005) Directed assembly of discrete gold nanoparticle groupings using branched DNA scaffolds. *Chemistry of Materials* 17: 1628–1635.
96. Deng Z, Tian Y, Lee S-H, Ribbe AE, and Mao C (2005) DNA-encoded self-assembly of gold nanoparticles into one-dimensional arrays. *Angewandte Chemie International Edition* 44: 3582–3585.
97. Aldaye FA and Sleiman HF (2006) Sequential self-assembly of a DNA hexagon as a template for the

organization of gold nanoparticles. *Angewandte Chemie International Edition* 45: 2204–2209.
98. Aldaye FA and Sleiman HF (2007) Dynamic DNA templates for discrete gold nanoparticle assemblies: Control of geometry, modularity, write/erase and structural switching. *Journal of the American Chemical Society* 129: 4130–4131.
99. Aili D, Enander K, Baltzer L, and Lie B (2007) Synthetic *de novo* designed polypeptides for control of nanoparticle assembly and biosensing. *Biochemical Society Transactions* 35: 532–534.
100. Zanchet D, Micheel CM, Parak WJ, Gerion D, and Alivisatos AP (2001) Electrophoretic isolation of discrete Au nanocrystal/DNA conjugates. *Nano Letters* 1: 32–35.
101. Qin WJ and Yung LYL (2005) Nanoparticle-DNA conjugates bearing a specific number of short DNA strands by enzymatic manipulation of nanoparticle-bound DNA. *Langmuir* 21: 11330–11334.
102. Claridge SA, Liang HW, Basu SR, Fréchet JMJ, and Alivisatos AP (2008) Isolation of discrete nanoparticle-DNA conjugates for plasmonic applications. *Nano Letters* 8: 1202–1206.
103. Perro A, Reculusa S, Ravaine S, Bourgeat-Lami E, and Duguet E (2005) Design and synthesis of Janus micro- and nanoparticles. *Journal of Materials Chemistry* 15: 3745–3760.
104. Maye MM, Nykypanchuk D, Cuisinier M, van der Lelie D, and Gang O (2009) Stepwise surface encoding for high-throughput assembly of nanoclusters. *Nature Materials* 8: 388–391.
105. Xu X, Rosi NL, Wang Y, Huo F, and Mirkin CA (2006) Asymmetric functionalization of gold nanoparticles with oligonucleotides. *Journal of the American Chemical Society* 128: 9286–9287.
106. Huo F, Lytton-Jean AKR, and Mirkin CA (2006) Asymmetric functionalization of nanoparticles based on thermally addressable DNA interconnects. *Advanced Materials* 18: 2304–2306.
107. Hasan W, Lee J, Henzie J, and Odom TW (2007) Selective functionalization and spectral identification of gold nanopyramids. *Journal of Physical Chemistry C* 111: 17176–17179.
108. Tkachenko AV (2002) Morphological diversity of DNA-colloidal self-assembly. *Physical Review Letters* 89: 148303.
109. Roberts RJ, Vincze T, Posfai J, and Macelis D (2005) REBASE – restriction enzymes and DNA methyltransferases. *Nucleic Acids Research* 33: D230–D232.
110. Hagerman PJ (1988) Flexibility of DNA. *Annual Review of Biophysics and Biophysical Chemistry* 17: 265–286.
111. Jin RC, Wu GS, Li Z, Mirkin CA, and Schatz GC (2003) What controls the melting properties of DNA-linked gold nanoparticle assemblies? *Journal of the American Chemical Society* 125: 1643–1654.
112. Park S-J, Lazarides AA, Storhoff JJ, Pesce L, and Mirkin CA (2004) The structural characterization of oligonucleotide-modified gold nanoparticle networks formed by DNA hybridization. *Journal of Physical Chemistry B* 108: 12375–12380.
113. Lukatsky DB and Frenkel D (2004) Phase behavior and selectivity of DNA-linked nanoparticle assemblies. *Physical Review Letters* 92: 068302.1–068302.4.
114. Taton TA, Mirkin CA, and Letsinger RL (2000) Scanometric DNA array detection with nanoparticle probes. *Science* 289: 1757–1760.
115. Dreyfus R, Leunissen ME, Sha R, et al. (2009) Simple quantitative model for the reversible association of DNA coated colloids. *Physical Review Letters* 102: 048301.
116. Hill HD, Millstone JE, Banholzer MJ, and Mirkin CA (2009) The role radius of curvature plays in thiolated oligonucleotide loading on gold nanoparticles. *ACS Nano* 3: 418–424.
117. Hurst SJ, Lytton-Jean AKR, and Mirkin CA (2006) Maximizing DNA loading on a range of gold nanoparticle sizes. *Analytical Chemistry* 78: 8313.
118. Kira A, Kim H, and Yasuda K (2009) Contribution of nanoscale curvature to number density of immobilized DNA on gold nanoparticles. *Langmuir* 25: 1285–1288.
119. Hill HD, Hurst SJ, and Mirkin CA (2009) Curvature-induced base pair "slipping" effects in DNA-nanoparticle hybridization. *Nano Letters* 9: 317–321.
120. McCullagh M, Prytkova T, Tonzani S, Winter ND, and Schatz GC (2008) Modeling self-assembly processes driven by nonbonded interactions in soft materials. *Journal of Physical Chemistry B* 112: 10388–10398.
121. Verwey EJW and Overbeek J (1948) *Theory of the Stability of Lyophobic Colloids*. New York: Elsevier.
122. Derjaguin BV and Landau L (1941) Theory of stability of highly charged lyophobic sols and adhesion of highly charged particles in solutions of electrolytes. *Acta Physicochima URSS* 14: 633–652.
123. Prasad BLV, Sorensen CM, and Klabunde KJ (2008) Gold nanoparticle superlattices. *Chemical Society Reviews* 37: 1871–1883.
124. Hamaker HC (1937) The London – van der Waals attraction between spherical particles. *Physica IV* 10: 1058–1070.
125. Bentzon MD, Van Wonterghem J, Mørup S, Thölén A, and Koch CJW (1989) Ordered aggregates of ultrafine iron oxide particles: "Super crystals". *Philosophical Magazine B* 60: 169–178.
126. Giersig M and Mulvaney P (1993) Formation of ordered two-dimensional gold colloid lattices by electrophoretic deposition. *Journal of Physical Chemistry* 97: 6334–6336.
127. Giersig M and Mulvaney P (1993) Preparation of ordered colloid monolayers by electrophoretic deposition. *Langmuir* 9: 3408–3413.
128. Murthy S, Bigioni TP, Wang ZL, Khoury JT, and Whetten RL (1997) Liquid-phase synthesis of thiol-derivatized silver nanocrystals. *Materials Letters* 30: 321–325.
129. Murthy S, Wang ZL, and Whetten RL (1997) Thin films of thiol-derivatized gold nanocrystals. *Philosophical Magazine Letters* 75: 321–327.
130. Ohara PC, Leff DV, Heath JR, and Gelbart WM (1995) Crystallization of opals from polydisperse nanoparticles. *Physical Review Letters* 75: 3466–3470.
131. Luedtke WD and Landman U (1996) Structure, dynamics, and thermodynamics of passivated gold nanocrystallites and their assemblies. *Journal of Physical Chemistry* 100: 13323–13329.
132. Cheng W, Campolongo MJ, Cha JJ, et al. (2009) Free-standing nanoparticle superlattice sheets controlled by DNA. *Nature Materials* 8: 519–525.
133. Zhang ZL, Horsch MA, Lamm MH, and Glotzer SC (2003) Tethered nano building blocks: Toward a conceptual framework for nanoparticle self-assembly. *Nano Letters* 3: 1341–1346.
134. Li YG, Tseng YD, Kwon SY, et al. (2004) Controlled assembly of dendrimer-like DNA. *Nature Materials* 3: 38–42.
135. Mucic RC, Storhoff JJ, Mirkin CA, and Letsinger RL (1998) DNA-Directed synthesis of binary nanoparticle network materials. *Journal of the American Chemical Society* 120: 12674–12675.
136. Qin WJ and Yung LYL (2008) Well-defined nanoassemblies using gold nanoparticles bearing specific number of DNA strands. *Bioconjugate Chemistry* 19: 385–390.

137. Jain PK, Huang WY, and El-Sayed MA (2007) On the universal scaling behavior of the distance decay of plasmon coupling in metal nanoparticle pairs: A plasmon ruler equation. *Nano Letters* 7: 2080–2088.
138. Claridge SA, Mastroianni AJ, Au YB, et al. (2008) Enzymatic ligation creates discrete multinanoparticle building blocks for self-assembly. *Journal of the American Chemical Society* 130: 9598–9605.
139. Chen W, Bian A, Agarwal A, et al. (2009) Nanoparticle superstructures made by polymerase chain reaction: Collective interactions of nanoparticles and a new principle for chiral materials. *Nano Letters* 9: 2153–2159.
140. Sonnichsen C, Reinhard BM, Liphardt J, and Alivisatos AP (2005) A molecular ruler based on plasmon coupling of single gold and silver nanoparticles. *Nature Biotechnology* 23: 741–745.
141. Sebba DS, Mock JJ, Smith DR, LaBean TH, and Lazarides AA (2008) Reconfigurable core–satellite nanoassemblies as molecularly-driven plasmonic switches. *Nano Letters* 8: 1803–1808.
142. Sebba DS and Lazarides AA (2008) Robust detection of plasmon coupling in core–satellite nanoassemblies linked by DNA. *Journal of Physical Chemistry C* 112: 18331–18339.
143. Bidault S, Garcia de Abajo FJ, and Polman A (2008) Plasmon-based nanolenses assembled on a well-defined DNA template. *Journal of the American Chemical Society* 130: 2750–2751.
144. Wang GL and Murray RW (2004) Controlled assembly of monolayer-protected gold clusters by dissolved DNA. *Nano Letters* 4: 95–101.
145. Patolsky F, Weizmann Y, Lioubashevski O, and Willner I (2002) Au-nanoparticle nanowires based on DNA and polylysine templates. *Angewandte Chemie International Edition* 41: 2323–2327.
146. Nakao H, Shiigi H, Yamamoto Y, et al. (2003) Highly ordered assemblies of Au nanoparticles organized on DNA. *Nano Letters* 3: 1391–1394.
147. Warner MG and Hutchison JE (2003) Linear assemblies of nanoparticles electrostatically organized on DNA scaffolds. *Nature Materials* 2: 272–277.
148. Beyer S, Nickels P, and Simmel FC (2005) Periodic DNA nanotemplates synthesized by rolling circle amplification. *Nano Letters* 5: 719–722.
149. Li H, Park SH, Reif JH, LaBean TH, and Yan H (2004) DNA-templated self-assembly of protein and nanoparticle linear arrays. *Journal of the American Chemical Society* 126: 418–419.
150. Sharma J, Chhabra R, Cheng A, Brownell J, Liu Y, and Yan H (2009) Control of self-assembly of DNA tubules through integration of gold nanoparticles. *Science* 323: 112–116.
151. Lin C, Liu Y, Rinker S, and Yan H (2006) DNA tile based self-assembly: Building complex nanoarchitectures. *ChemPhysChem* 7: 1641–1647.
152. Winfree E, Liu F, Wenzler LA, and Seeman NC (1998) Design and self-assembly of two-dimensional DNA crystals. *Nature* 394: 539–544.
153. Liu F, Sha R, and Seeman NC (1999) Modifying the surface features of two-dimensional DNA crystals. *Journal of the American Chemical Society* 121: 917–922.
154. Xiao S, Liu F, Rosen AE, et al. (2002) Selfassembly of metallic nanoparticle arrays by DNA scaffolding. *Journal of Nanoparticle Research* 4: 313–317.
155. Le JD, Pinto Y, Seeman NC, Musier-Forsyth K, Taton TA, and Kiehl RA (2004) DNA-templated self-assembly of metallic nanocomponent arrays on a surface. *Nano Letters* 4: 2343–2347.
156. Pinto YY, Seeman NC, Musier-Forsyth K, Taton TA, and Kiehl RA (2005) Sequence-encoded self-assembly of multiple-nanocomponent arrays by 2D DNA scaffolding. *Nano Letters* 5: 2399–2402.
157. Chhabra R, Sharma J, Ke Y, et al. (2007) Spatially addressable multiprotein nanoarrays templated by aptamer-tagged DNA nanoarchitectures. *Journal of the American Chemical Society* 129: 10304–10305.
158. Williams BAR, Lund K, Liu Y, Yan H, and Chaput JC (2007) Self-assembled peptide nanoarrays: An approach to studying protein–protein interactions. *Angewandte Chemie International Edition* 46: 3051–3054.
159. Sharma J, Ke Y, Lin C, et al. (2008) DNA-tile-directed self-assembly of quantum dots into two-dimensional nanopatterns. *Angewandte Chemie International Edition* 47: 5157–5159.
160. Liu D, Wang M, Deng Z, Walulu R, and Mao C (2004) Tensegrity: Construction of rigid DNA triangles with flexible four-arm DNA junctions. *Journal of the American Chemical Society* 126: 2324–2325.
161. Zheng JW, Constantinou PE, Micheel C, Alivisatos AP, Kiehl RA, and Seeman NC (2006) Two-dimensional nanoparticle arrays show the organizational power of robust DNA motifs. *Nano Letters* 6: 1502–1504.
162. Yan H, Park SH, Finkelstein G, Reif JH, and LaBean TH (2003) DNA-templated self-assembly of protein arrays and highly conductive nanowires. *Science* 301: 1882–1884.
163. Sharma J, Chhabra R, Liu Y, Ke Y, and Yan H (2006) DNA-templated self-assembly of two-dimensional and periodical gold nanoparticle arrays. *Angewandte Chemie International Edition* 45: 730–735.
164. Zhang JP, Liu Y, Ke YG, and Yan H (2006) Periodic square-like gold nanoparticle arrays templated by self-assembled 2D DNA nanogrids on a surface. *Nano Letters* 6: 248–251.
165. He Y, Tian Y, Chen Y, Deng Z, Ribbe AE, and Mao C (2005) Sequence symmetry as a tool for designing DNA nanostructures. *Angewandte Chemie International Edition* 44: 6694.
166. Park SH, Yin P, Liu Y, Reif JH, LaBean TH, and Yan H (2005) Programmable DNA self-assemblies for nanoscale organization of ligands and proteins. *Nano Letters* 5: 729–733.
167. Lund K, Liu Y, Lindsay S, and Yan H (2005) Self-assembling a molecular pegboard. *Journal of the American Chemical Society* 127: 17606–17607.
168. Rothemund PWK (2006) Folding DNA to create nanoscale shapes and patterns. *Nature* 440: 297–302.
169. He Y, Tian Y, Ribbe AE, and Mao C (2006) Highly connected two-dimensional crystals of DNA six-point-stars. *Journal of the American Chemical Society* 128: 15978–15979.
170. Chelyapov N, Brun Y, Gopalkrishnan M, Reishus D, Shaw B, and Adleman L (2004) DNA triangles and self-assembled hexagonal tilings. *Journal of the American Chemical Society* 126: 13924–13925.
171. Ding B, Sha R, and Seeman NC (2004) Pseudohexagonal 2D DNA crystals from double crossover cohesion. *Journal of the American Chemical Society* 126: 10230–10231.
172. Mao C, Sun W, and Seeman NC (1999) Designed two-dimensional DNA Holliday junction arrays visualized by atomic force microscopy. *Journal of the American Chemical Society* 121: 5437–5443.
173. Andersen ES, Dong M, Nielsen MM, et al. (2009) Self-assembly of a nanoscale DNA box with a controllable lid. *Nature* 459: 73–76.
174. He Y and Mao C (2006) Balancing flexibility and stress in DNA nanostructures. *Chemical Communications* 2006: 968–969.

175. Jaeger L and Chworos A (2006) The architectonics of programmable RNA and DNA nanostructures. *Current Opinion in Structural Biology* 16: 531–543.
176. Zhang C, Su M, He Y, et al. (2008) Conformational flexibility facilitates self-assembly of complex DNA nanostructures. *Proceedings of the National Academy of Sciences of the United States of America* 105: 10665.
177. Starr FW and Sciortino F (2006) Model for assembly and gelation of four-armed DNA dendrimers. *Journal of Physics: Condensed Matter* 18: L347–L353.
178. Elghanian R, Storhoff JJ, Mucic RC, Letsinger RL, and Mirkin CA (1997) Selective colorimetric detection of polynucleotides based on the distance-dependent optical properties of gold nanoparticles. *Science* 277: 1078–1081.
179. Storhoff JJ, Elghanian R, Mucic RC, Mirkin CA, and Letsinger RL (1998) One-pot colorimetric differentiation of polynucleotides with single base imperfections using gold nanoparticle probes. *Journal of the American Chemical Society* 120: 1959–1964.
180. Maye MM, Nykypanchuk D, van der Lelie D, and Gang O (2007) DNA-regulated micro- and nanoparticle assembly. *Small* 3: 1678.
181. Nykypanchuk D, Maye MM, van der Lelie D, and Gang O (2008) DNA-guided crystallization of colloidal nanoparticles. *Nature* 451: 549–552.
182. Park SY, Lytton-Jean AKR, Lee B, Weigand S, Schatz GC, and Mirkin CA (2008) DNA-programmable nanoparticle crystallization. *Nature* 451: 553–556.
183. Deegan RD, Olgica B, Dupont TF, Greg H, Nagel SR, and Thomas AW (1997) Capillary flow as the cause of ring stains from dried liquid drops. *Nature* 389: 827–829.
184. Martin CP, Blunt MO, Paulic-Vaujour E, et al. (2007) Controlling pattern formation in nanoparticle assemblies via directed solvent dewetting. *Physical Review Letters* 99: 116103.
185. Huang JX, Fan R, Connor S, and Yang PD (2007) One-step patterning of aligned nanowire arrays by programmed dip coating. *Angewandte Chemie International Edition* 46: 2414–2417.
186. Cheng W, Park N, Walter MT, Hartman MR, and Luo D (2008) Nanopatterning self-assembled nanoparticle superlattices by moulding microdroplets. *Nature Nanotechnology* 3: 682–690.
187. Martin JE, Wilcoxon JP, Odinek J, and Provencio P (2000) Control of the interparticle spacing in gold nanoparticle superlattices. *Journal of Physical Chemistry B* 104: 9475–9486.
188. Sarikaya M, Tamerler C, Jen AKY, Schulten K, and Baneyx F (2003) Molecular biomimetics: Nanotechnology through biology. *Nature Materials* 2: 577–585.
189. Slocik JM, Stone MO, and Naik RR (2005) Synthesis of gold nanoparticles using multifunctional peptides13. *Small* 1: 1048–1052.
190. Chen C-L, Zhang P, and Rosi NL (2008) A new peptide-based method for the design and synthesis of nanoparticle superstructures: Construction of highly ordered gold nanoparticle double helices. *Journal of the American Chemical Society* 130: 13555–13557.
191. Caswell KK, Wilson JN, Bunz UHF, and Murphy CJ (2003) Preferential end-to-end assembly of gold nanorods by biotin–streptavidin connectors. *Journal of the American Chemical Society* 125: 13914–13915.
192. Sotiropoulou S, Sierra-Sastre Y, Mark SS, and Batt CA (2008) Biotemplated nanostructured materials. *Chemistry of Materials* 20: 821–834.
193. Sleytr UB, Messner P, Pum D, and Sara M (1999) Crystalline bacterial cell surface layers (S layers): From supramolecular cell structure to biomimetics and nanotechnology. *Angewandte Chemie International Edition* 38: 1034–1054.
194. Tang J, Badelt-Lichtblau H, Ebner A, et al. (2008) Fabrication of highly ordered gold nanoparticle arrays templated by crystalline lattices of bacterial S-layer protein. *ChemPhysChem* 9: 2317–2320.
195. McMillan RA, Paavola CD, Howard J, Chan SL, Zaluzec NJ, and Trent JD (2002) Ordered nanoparticle arrays formed on engineered chaperonin protein templates. *Nature Materials* 1: 247–252.
196. Yoo PJ, Nam KT, Qi J, et al. (2006) Spontaneous assembly of viruses on multilayered polymer surfaces. *Nature Materials* 5: 234–240.
197. Wang ZL (2000) Transmission electron microscopy of shape-controlled nanocrystals and their assemblies. *Journal of Physical Chemistry B* 104: 1153–1175.
198. Grabar KC, Brown KR, Keating CD, Stranick SJ, Tang S, and Natan MJ (1997) Nanoscale characterization of gold colloid monolayers: A comparison of four techniques. *Analytical Chemistry* 69: 471–477.
199. Orendorff CJ, Sau TK, and Murphy CJ (2006) Shape-dependent plasmon-resonant gold nanoparticles. *Small* 2: 636–639.
200. Casavola M, Grillo V, Carlino E, et al. (2007) Topologically controlled growth of magnetic-metal-functionalized semiconductor oxide nanorods. *Nano Letters* 7: 1386–1395.
201. Tanaka M, Takeguchi M, and Furuya K (2008) X-ray analysis and mapping by wavelength dispersive X-ray spectroscopy in an electron microscope. *Ultramicroscopy* 108: 1427–1431.
202. Dabbousi BO, Rodriguez-Viejo J, Mikulec FV, et al. (1997) (CdSe) ZnS core–shell quantum dots: Synthesis and characterization of a size series of highly luminescent nanocrystallites. *Journal of Physical Chemistry B* 101: 9463–9475.
203. Serwer P (1983) Agarose gels: Properties and use for electrophoresis. *Electrophoresis* 4: 227–231.
204. Tietz D (1987) Gel electrophoresis of intact subcellular particles. *Journal of Chromatography* 418: 305.
205. Viovy JL (2000) Electrophoresis of DNA and other polyelectrolytes: Physical mechanisms. *Reviews of Modern Physics* 72: 813–872.
206. Hanauer M, Pierrat S, Zins I, Lotz A, and Sonnichsen C (2007) Separation of nanoparticles by gel electrophoresis according to size and shape. *Nano Letters* 7: 2881–2885.
207. Pellegrino T, Sperling RA, Alivisatos AP, and Parak WJ (2006) Gel electrophoresis of gold-DNA nano-conjugates. *Journal of Biomedicine and Biotechnology* 2007, doi:10.1155/2007/26796.
208. Sandstrom P and Akerman B (2004) Electrophoretic properties of DNA-modified colloidal gold nanoparticles. *Langmuir* 20: 4182–4186.
209. El-Sayed MA (2004) Small is different: Shape-, size-, and composition-dependent properties of some colloidal semiconductor nanocrystals. *Accounts of Chemical Research* 37: 326–333.
210. El-Sayed MA (2001) Some interesting properties of metals confined in time and nanometer space of different shapes. *Accounts of Chemical Research* 34: 257–264.
211. Cheng WL, Dong SJ, and Wang EK (2003) Synthesis and self-assembly of cetyltrimethylammonium bromide-capped gold nanoparticles. *Langmuir* 19: 9434–9439.
212. Haynes CL, McFarland AD, Zhao L, Van Duyne RP, and Schatz GC (2003) Nanoparticle optics: The importance of radiative dipole coupling in two-dimensional nanoparticle arrays. *Journal of Physical Chemistry B* 107: 7337–7342.
213. Huang WY, Qian W, Jain PK, and El-Sayed MA (2007) The effect of plasmon field on the coherent lattice phonon

oscillation in electron-beam fabricated gold nanoparticle pairs. *Nano Letters* 7: 3227–3234.
214. Jain PK and El-Sayed MA (2007) Universal scaling of plasmon coupling in metal nanostructures: Extension from particle pairs to nanoshells. *Nano Letters* 7: 2854–2858.
215. Jiang CY, Markutsya S, and Tsukruk VV (2004) Collective and individual plasmon resonances in nanoparticle films obtained by spin-assisted layer-by-layer assembly. *Langmuir* 20: 882–890.
216. Maier SA, Kik PG, and Atwater HA (2002) Observation of coupled plasmon-polariton modes in Au nanoparticle chain waveguides of different lengths: Estimation of waveguide loss. *Applied Physics Letters* 81: 1714–1716.
217. Wei QH, Su KH, Durant S, and Zhang X (2004) Plasmon resonance of finite one-dimensional Au nanoparticle chains. *Nano Letters* 4: 1067–1071.
218. Ramakrishna G, Varnavski O, Kim J, Lee D, and Goodson T (2008) Quantum-sized gold clusters as efficient two-photon absorbers. *Journal of the American Chemical Society* 130: 5032–5033.
219. Anker JN, Hall WP, Lyandres O, Shah NC, Zhao J, and Van Duyne RP (2008) Biosensing with plasmonic nanosensors. *Nature Materials* 7: 442–453.
220. Alivisatos P (2004) The use of nanocrystals in biological detection. *Nature Biotechnology* 22: 47–52.
221. Liu GL, Yin Y, Kunchakarra S, et al. (2006) A nanoplasmonic molecular ruler for measuring nuclease activity and DNA footprinting. *Nature Nanotechnology* 1: 47–52.
222. Haes AJ and Van Duyne RP (2002) A nanoscale optical biosensor: Sensitivity and selectivity of an approach based on the localized surface plasmon resonance spectroscopy of triangular silver nanoparticles. *Journal of the American Chemical Society* 124: 10596–10604.
223. Forster S, Timmann A, Konrad M, Schellbach C, and Meyer A (2005) Scattering curves of ordered mesoscopic materials. *Journal of Physical Chemistry B* 109: 1347–1360.
224. Warren BE (1990) *X-ray Diffraction*. New York: Courier Dover.
225. Förster S, Timman A, Schellbach C, et al. (2007) Order causes secondary Bragg peaks in soft materials. *Nature Materials* 6: 888–893.
226. Forster S, Konrad M, and Lindner P (2005) Shear thinning and orientational ordering of wormlike micelles. *Physical Review Letters* 94: 017803.
227. Binnig G, Quate CF, and Gerber CH (1986) Atomic force microscope. *Physical Review Letters* 56: 930–933.

228. Novoselov KS, Jiang D, Schedin F, et al. (2005) Two-dimensional atomic crystals. *Proceedings of the National Academy of Sciences of the United States of America* 102: 10451–10453.
229. Vinckier A, Dumortier C, Engelborghs Y, and Hellemans L (1996) Dynamical and mechanical study of immobilized microtubules with atomic force microscopy. *Journal of Vacuum Science and Technology B: Microelectronics and Nanometer Structures* 14: 1427.
230. Weisenhorn AL, Khorsandi M, Kasas S, Gotzos V, and Butt HJ (1993) Deformation and height anomaly of soft surfaces studied with an AFM. *Nanotechnology* 4: 106–1106.
231. Radmacher M, Fritz M, Kacher CM, Cleveland JP, and Hansma PK (1996) Measuring the viscoelastic properties of human platelets with the atomic force microscope. *Biophysical Journal* 70: 556–567.
232. Parpura V and Fernandez JM (1996) Atomic force microscopy study of the secretory granule lumen. *Biophysical Journal* 71: 2356–2366.
233. Laney DE, Garcia RA, Parsons SM, and Hansma HG (1997) Changes in the elastic properties of cholinergic synaptic vesicles as measured by atomic force microscopy. *Biophysical Journal* 72: 806–813.
234. Noy A, Vezenov DV, Kayyem JF, Maade TJ, and Lieber CM (1997) Stretching and breaking duplex DNA by chemical force microscopy. *Chemistry and Biology* 4: 519–527.
235. Johnson KL, Kendall K, and Roberts AD (1971) Surface energy and the contact of elastic solids. *Proceedings of the Royal Society of London. Series A, Mathematical and Physical Sciences* 324: 301–313.
236. Jiang C, Markutsya S, Pikus Y, and Tsukruk VV (2004) Freely suspended nanocomposite membranes as highly sensitive sensors. *Nature Materials* 3: 721–728.
237. Jiang C and Tsukruk VV (2006) Freestanding nanostructures via layer-by-layer assembly. *Advanced Materials* 18: 829–840.
238. Komaragiri U, Begley MR, and Simmonds JG (2005) The mechanical response of freestanding circular elastic films under point and pressure loads. *Journal of Applied Mechanics* 72: 203.
239. Wan KT, Guo S, and Dillard DA (2003) A theoretical and numerical study of a thin clamped circular film under an external load in the presence of a tensile residual stress. *Thin Solid Films* 425: 150–162.
240. Lee C, Wei XD, Kysar JW, and Hone J (2008) Measurement of the elastic properties and intrinsic strength of monolayer graphene. *Science* 321: 385–388.

3 Chiral Molecules on Surfaces

C J Baddeley, University of St. Andrews, St. Andrews, UK
G Held, University of Reading, Reading, UK

© 2010 Elsevier B.V. All rights reserved.

3.1 Introduction

Since the mid-1990s, the number of investigations of chirality at surfaces has increased dramatically. Developments in techniques for characterizing the adsorption site of relatively complex molecular adsorbates have enabled considerable insight into the bonding of, for example, amino acids to metal surfaces. Similarly, advances in the technique of scanning tunneling microscopy (STM) have been crucial in enabling the visualization of single chiral molecules, clusters, and extended arrays. As such, dramatic advances have been achieved in the fundamental understanding of the interactions of chiral molecules with surfaces and the phenomena of chiral amplification and chiral recognition. These issues are of considerable technological importance, for example, in the development of heterogeneous catalysts for the production of chiral pharmaceuticals and in the design of biosensors. In addition, the understanding of chirality at surfaces may be a key to unraveling the complexities of the origin of life.

3.1.1 Definition of Chirality

The word chirality is derived from the Greek $\chi\varepsilon\iota\rho$ (cheir) meaning 'hand'. It is the geometric property of an object that distinguishes a right hand from a left hand. Lord Kelvin provided a definition of chirality in his 1884 Baltimore Lectures "I call any geometrical figure or group of points "chiral" and say it has "chirality", if its image in a plane mirror, ideally realised, cannot be brought into coincidence with itself." For an isolated object, for example, a molecule, the above statement can be interpreted as being equivalent to requiring that the object possesses neither a mirror plane of symmetry nor a point of symmetry (center of inversion). If a molecule possesses either one of these symmetry elements, it can be superimposed on its mirror image and is therefore achiral. A chiral molecule and its mirror image are referred to as being a pair of enantiomers. Many organic molecules possess the property of chirality.

Chiral centers are most commonly associated with the tetrahedral co-ordination of four different substituents. However, there are many examples of other rigid structures that have chiral properties where a significant barrier exists to conformational change within the molecule.

3.1.2 Nomenclature of Chirality – the R,S Convention

Most of the physical properties (e.g., boiling and melting points, density, refractive index, etc.) of two enantiomers are identical. Importantly, however, the two enantiomers interact differently with polarized light. When plane polarized light interacts with a sample of chiral molecules, there is a measurable net rotation of the plane of polarization. Such molecules are said to be optically active. If the chiral compound causes the plane of polarization to rotate in a clockwise (positive) direction as viewed by an observer facing the beam, the compound is said to be dextrorotatory. An anticlockwise (negative) rotation is caused by a levorotatory compound. Dextrorotatory chiral compounds are often given the label D or (+) while levorotatory compounds are denoted L or (−).

In this chapter, we will use an alternative convention which labels chiral molecules according to their absolute stereochemistry. The R,S convention or Cahn–Ingold–Prelog system was first introduced by Robert S. Cahn and Sir Christopher K. Ingold (University College, London) in 1951 and later modified with Vlado Prelog (Swiss Federal Institute of Technology) [1]. Essentially, the four atomic substituents at a stereocenter are identified and assigned a priority (1 (highest), 2, 3, 4 (lowest)) by atomic mass. If two atomic substituents are the same, their priority is defined by working outward along the chain of atoms until a point of difference is reached. Using the same considerations of atomic mass, the priority is then assigned at the first point of difference. For example, a —CH_2—CH_3 substituent has a higher priority than a —CH_3 substituent. Once the priority has been assigned around the

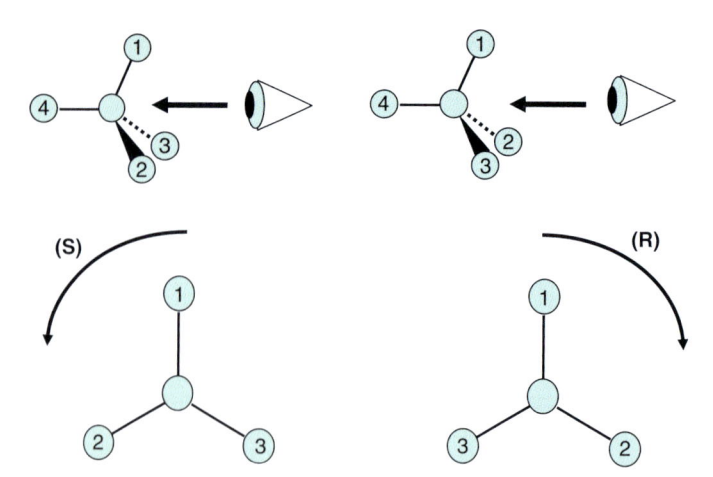

Figure 1 Schematic diagram explaining the Cahn–Ingold–Prelog convention for determining the absolute stereochemistry of a chiral molecule.

stereocenter, the tetrahedral arrangement is viewed along the bond between the central atom and the lowest priority (4) substituent (often a C—H bond) from the opposite side to the substituent (**Figure 1**). If the three other substituents are arranged such that the path from 1 to 2 to 3 involves a clockwise rotation, the stereocenter is labeled R (Latin *rectus*, meaning right). By contrast, if the path involves an anti-clockwise rotation, the stereocenter is labeled S (Latin *sinister*, meaning left). It is important to note that the absolute stereochemistry cannot be predicted from the L or D labels and vice versa.

In nature, a remarkable, and so far unexplained, fact is that the amino acid building blocks of all proteins are exclusively left-handed and that the sugars contained within the double-helix structure of DNA are exclusively right handed. The consequences of the chirality of living organisms are far reaching. The human sense of smell, for example, is able to distinguish between pure R-limonene (smelling of oranges) and S-limonene (smelling of lemons). More significantly, two enantiomeric forms of an organic molecule can have different physiological effects on the human body. In many cases, one enantiomer is the active component while the opposite enantiomer has no effect (e.g., ibuprofen, where the S-enantiomer is active). However, often the two enantiomers have dramatically different effects. For example, S-methamphetamine is a psychostimulant while R-methamphthetamine is the active ingredient in many nasal decongestants (**Figure 2**).

In the pharmaceutical industry, about half of all of the new drugs being tested require the production of exclusively one enantiomeric product.

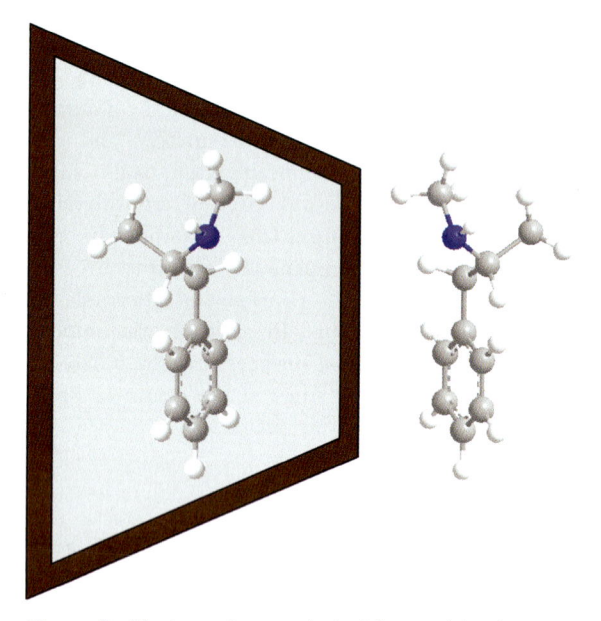

Figure 2 The two mirror-equivalent forms of the drug methamphetamine. On the right is shown the S-form of the molecule. On the left is the R-enantiomer.

Thermodynamically, this is a challenging problem since the two isolated enantiomers have identical Gibbs energies, the reaction from pro-chiral reagent to product should, therefore, result in a 50:50 (racemic) mixture at equilibrium. To skew the reaction pathway to form one product with close to 100% enantioselectivity is nontrivial. Knowles [2], Noyori [3], and Sharpless [4] were awarded the Nobel Chemistry Prize in 2001 for the development of enantioselective homogeneous catalysts capable of producing chiral molecules on an industrial scale.

Typically, these catalysts consist of organometallic complexes with chiral ligands. Access to the metal center by the reagent is strongly sterically influenced by the chiral ligands resulting in preferential formation of one enantiomeric product. There are many potential advantages to using heterogeneous catalysts, not least the ease of separation of the catalyst from the products. However, despite extensive research over several decades, relatively few successful catalysts have been synthesized on a laboratory scale, and the impact in industrial catalysis is essentially negligible. One of the primary motivations behind surface science studies of chirality at surfaces is to understand the surface chemistry underpinning chiral catalysis and to develop methodologies for the rational design of chiral catalysts. Similarly, those interested in issues related to the origin of life are investigating the possibility that surfaces were responsible for the initial seeding of the chiral building blocks of life and that, presumably via some chiral amplification effects, this led to the overwhelming dominance of left-handed amino acids and right-handed sugars in biological systems on Earth. As such, the surface chemistry of chiral solids, chiral amplification and chiral recognition are all important sub-topics of chiral surface science. STM has proved to be the single most important tool of researchers in this field.

This chapter aims to provide insight into the current state of knowledge about adsorption and chemistry of chiral molecules at surfaces. Since the molecule–surface interaction plays a more crucial role in the chemistry and geometric arrangements of smaller molecules, we concentrate on these rather than large organic molecules whose supramolecular arrangements are the subject of other contributions to this volume. For additional information we also refer to a number of recent reviews of certain aspects of this subject [5–9].

3.2 Surface Chirality following Molecular Adsorption

3.2.1 Achiral Molecules on Achiral Surfaces

In many cases the presence of a surface reduces the symmetry of an adsorbed molecule with respect to the gas phase. Therefore, a molecule which is achiral in the gas phase may adsorb on an achiral surface to produce a chiral adsorbate–substrate complex. In the formal definition of chirality proposed by Lord Kelvin, a species, which in its most stable conformation has no mirror plane or center of symmetry, is formally chiral. However, if there exists a low-energy pathway to the enantiomer, for example, by a low-frequency vibrational mode, then, this would not normally be considered to be chiral in the context of chemistry. However, if adsorption of such a species raises the frequency of the vibration substantially, then the energy barrier between the two 'enantiomers' may become chemically significant such that the adsorbed molecule is meaningfully described as chiral. An example of this is the case of the de-protonated glycine species adsorbed on copper surfaces. A glycinate anion, although lacking any mirror plane or center of symmetry, is nevertheless readily converted into its enantiomer principally by a rotation around the C—N bond, with an energy barrier of $\sim$35 kJ mol^{-1}, which might readily be overcome at room temperature, such that glycine or glycinate are not generally considered chiral. However, on Cu, for example, adsorption takes place through both O atoms and the N atom in a 'three-point' interaction with the copper surface with each atom in an approximately atop site [10, 11]. This inhibits the interconversion of enantiomers, and the surface-induced chirality leads to two distinct species on the surface related by mirror symmetry [12]. A (3 × 2) unit cell is formed containing one molecule of each enantiomer giving rise to a heterochiral arrangement with the two enantiomers being related by glide symmetry as shown in **Figure 3**. This proposal based on low-energy electron diffraction (LEED), STM, and infrared (IR) data [10] has been confirmed by photoelectron diffraction (PhD) [11] and by density functional theory (DFT) calculations [13].

Consider the catalytically important pro-chiral reagent, methylacetoacetate (MAA). In the gas phase, in its most symmetric form, this molecule possesses one mirror plane (the molecular plane), that is, it has C$_s$ symmetry. If, for example, MAA adsorbs with the molecular plane parallel to the surface, the mirror symmetry of the molecule is destroyed by the formation of the adsorbate–metal interaction (**Figure 4**). As such, the combined adsorbate–substrate system will be locally chiral, that is, chirality is induced by the adsorption process. The molecule can then exist in two enantiomeric forms, although necessarily as a racemic mixture in the absence of any other influences that might lead to a preference of one rather than the other. If the

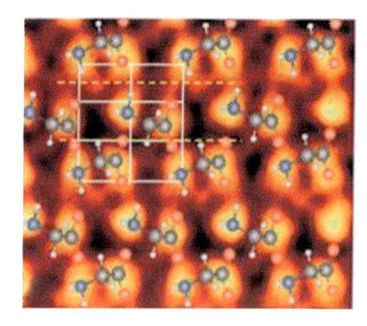

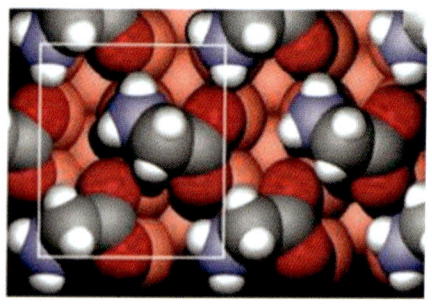

Figure 3 The left-hand panel shows a molecular model of the glycinate/Cu{110} structure with both enantiomers present in the heterochiral (3 × 2) unit cell, superimposed on an STM image of this surface. Adapted with permission from Chen Q, Frankel DJ, and Richardson NV (2002) *Surface Science* 497: 37. The right-hand panel shows the confirmation of this structure calculated by DFT, clearly indicating the atop adsorption sites occupied by the N and both O atoms in this system. Reproduced from Rankin RB and Sholl DS (2004) *Surface Science* 548: 301.

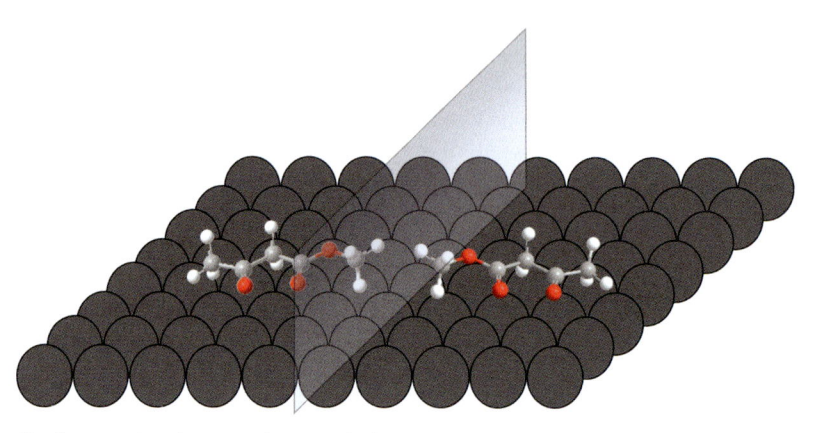

Figure 4 Schematic diagram showing two mirror-equivalent adsorbate–substrate complexes following the adsorption of the pro-chiral reagent MAA on an achiral surface. The two complexes are nonsuperimposable and, therefore, chiral.

molecule adsorbs with the molecular plane perpendicular to the surface plane, the adsorbate–substrate system will be locally chiral if the mirror plane of the adsorbate is not coincident with the mirror plane of the substrate. As will be explained in detail below, the ability of a pro-chiral molecule to adsorb in a preferred chiral geometry is highly significant in enantioselective hydrogenation catalysis because the attack of H atoms from the side of the molecular plane attached to the surface and the transition state so formed determine the chirality of the hydrogenation product.

3.2.2 Chiral Molecules on Achiral Surfaces

An isolated molecule, which is chiral in the gas phase, will necessarily be chiral on adsorption if the basic structure and conformation of the molecule are retained. Adsorption of the opposite gas-phase enantiomer is necessary to generate the mirror-image adsorbate system. There is a large body of information available on the adsorption of chiral molecules on achiral metal surfaces, which has been reviewed recently by Barlow and Raval [6]. The best-studied chiral molecules are carboxylic acids in particular amino acids and tartaric acid, which have been identified as chiral modifiers for heterogeneous catalysts in asymmetric hydrogenation reactions of β-ketoesters [14–16]. Adsorption studies have concentrated mainly on the {110} and {100} surfaces of Cu and Ni, but also Pd, Pt, and Au surfaces have been studied.

With the exception of gold, the chemisorption bond usually involves a de-protonated carboxylate group with each oxygen atom forming a bond with a surface metal atom.

3.2.2.1 Adsorption without substrate modification

We now discuss some examples of adsorption of chiral molecules on different surfaces.

1. *Tartaric acid*

One of the most heavily studied examples of the adsorption of a simple chiral molecule on an achiral metal surface involves the adsorption of tartaric acid onto Cu{110}. Tartaric acid (H_2TA; HOOC.CHOH.CHOH.COOH) (see **Figure 5**) can exist in the (R,R), (S,S), and (R,S) forms. The initial work was motivated by a desire to understand why (R,R)-tartaric acid is the most successful chiral modifier in the Ni-catalyzed enantioselective hydrogenation of β-ketoesters. Work from the catalysis community had proposed that ordered, nanoporous two-dimensional (2D) arrays of chiral molecules may be important in defining the active site for chiral catalytic reactions [15, 17]. The shape of the chiral nanopores could favor the adsorption of a reactant molecule in a geometry favoring the formation of one enantiomeric product. Alternatively, it was proposed that a direct interaction between a pro-chiral reagent and a single chiral modifier may be sufficient to direct the reaction along one enantiomeric route [14]. Hence, it was important to investigate how tartaric acid binds to a metal surface and the extent to which it forms ordered 2D arrays. The two carboxylic acid groups can each form bonds with metal surfaces. When adsorbed on Cu{110} this molecule displays a rich phase diagram of ordered structures (see **Figure 6** and Refs. [6,18]). On the basis of IR spectroscopy (RAIRS), it could be shown that adsorption at room temperature leads to a μ_2 mono-tartrate (HTA) adsorption complex with the two oxygen atoms of the de-protonated carboxylate group forming bonds to surface Cu atoms [18]. The intact carboxylic acid (COOH) group is not interacting with the metal surface. Transformation into a μ_4 bi-tartrate (TA) adsorption complex has a higher activation barrier and takes place at higher temperatures, between 350 and 405 K. Fasel *et al.* carried out a detailed XPD characterization of the adsorption geometry of (R,R)- and (S,S)-tartaric acid in the (9 0; ±1 2) phase (for a detailed description of the matrix notation, see the review article by Barlow and Raval [6]) and concluded that individual TA species were adsorbed with the planes defined by the two carboxylate OCO planes of each TA species being distorted away from the ⟨−110⟩ azimuth. The distortion observed by (R,R)-TA was exactly mirrored for (S,S)-TA [19]. This species is only observed at low coverage, where enough Cu sites are available to form four bonds per molecule [18]. At higher coverage the conversion is blocked and a mono-tartrate overlayer with a different long-range order is observed, which consists of hydrogen-bonded tartrate dimers (see **Figure 6**).

The adsorption of tartaric acid on Cu{111} has been studied by STM in equilibrium with aqueous solution by Yan *et al.* [20]. Based on the STM images, the authors suggest a bi-tartrate adsorption complex, similar to Cu{110} with hydrogen-bonded pairs of molecules forming a (4×4) superstructure.

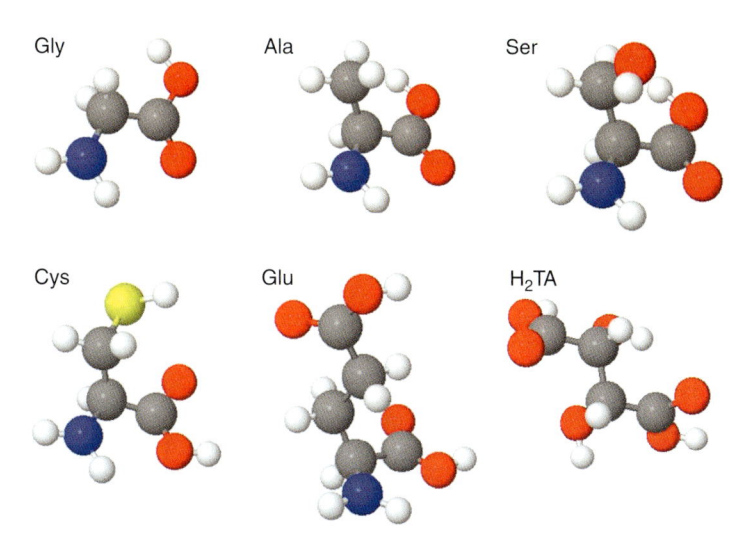

Figure 5 Schematic diagram showing the gas-phase molecular structures of various simple chiral carboxylic acids: glycine (Gly), alanine (Ala), serine (Ser), cysteine (Cys), glutamic acid (Glu), and tartaric acid (H_2TA).

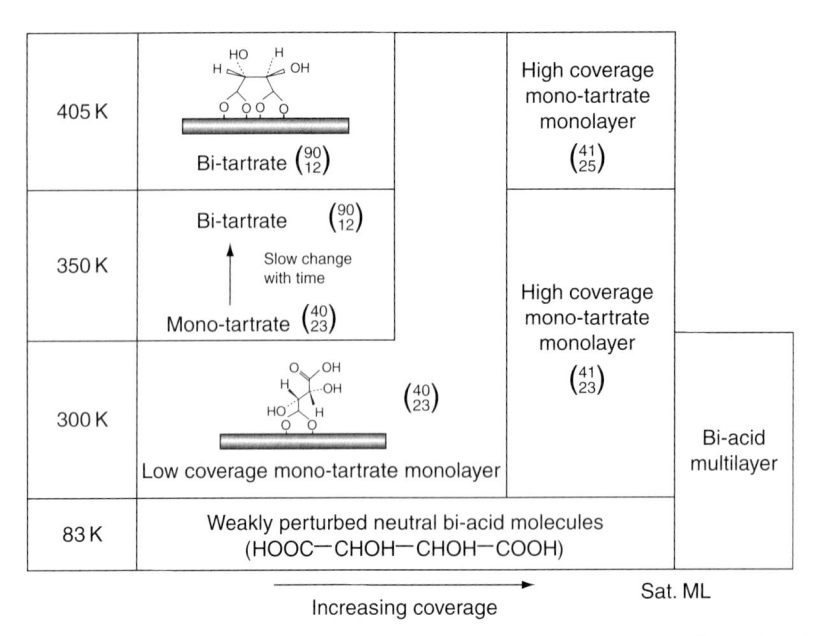

Figure 6 Adsorption phase diagram summarizing the ordered structures and molecular conformation of adsorbed species following the adsorption of (R,R)-tartaric acid on Cu{110}. Reproduced with permission from Lorenzo MO, Haq S, Bertrams T, Murray P, Raval R, and Baddeley CJ (1999) *Journal of Physical Chemistry B* 103: 10661.

Tartaric acid on Ni{110} shows a similar phase diagram as far as the local adsorption structures are concerned, only the activation barrier for bi-tartrate formation is lower. Thus, this transition takes place already at room temperature for low coverages and is observed even for saturation coverage after annealing 380 K. The fundamental difference between adsorption on Cu{110} and Ni{110} is that no long-range ordering is observed on Ni. This is in line with observations for other organic molecules [21, 22]. The bond to Ni atoms is more localized, which reduces the lateral mobility of molecules and, hence, the ability to form long-range ordered overlayers. DFT calculations and STM images for this system reveal perturbations in the bi-tartrate adsorbate, which induce a chiral relaxation of atoms at the Ni{110} surface. Humblot *et al.* [23] explain these relaxations by the need to accommodate the large chiral footprint of bi-tartrate–Ni$_4$, thus providing a mechanism of transferring chirality from the molecule into the metal surface.

On Ni{111}, tartaric acid adsorption produces two distinct ordered adlayer structures, each stabilized by intermolecular hydrogen-bonds. The thermal stability increases with increasing coverage [24]. Using STM and IR spectroscopy, Jones and Baddeley identified mono-tartrate and bi-tartrate species after adsorption at room temperature. In the latter case, the individual carboxylate groups can bond either via a mono-dentate or a bi-dentate geometry. The former leads to pseudo-1D growth and the latter to ordered 2D adlayers. Mono-tartrate surfaces are relatively stable at 300 K. It appears that intermolecular hydrogen bonding kinetically hinders the conversion to bi-tartrate at 300 K while this conversion is facile at 350 K.

STM and LEED have been widely used to investigate how substrate-mediated interactions and intermolecular hydrogen bonding influence the growth of 1-D and 2-D clusters and long range ordered structures. In many cases, if chiral molecules form ordered structures on metal surfaces, the adsorbate forms an oblique unit cell such that the ordered adsorbate structure itself is chiral. In this case, the surface possesses both local chirality (determined by the molecule-surface complex) and global chirality (determined by the chirality of the ordered adsorbate domains).

On Cu{110}, a range of ordered structures were identified with STM following (R,R)-tartaric acid adsorption [18] as functions of tartaric acid coverage and temperature. These are summarized in **Figure 6**. At 300 K and above, tartaric acid adsorption occurred via de-protonation of either one or both –COOH functionalities to produce mono-tartrate (HTA) or bi-tartrate (TA) species. Some of the ordered structures, such as (4 0; 2 3) (or c(4×6)), would be indistinguishable from that produced by (S,S)-tartaric acid because the unit cell has a centered-rectangular structure.

One adlayer phase, a (9 0; 1 2) structure with an oblique unit cell, was particularly significant from the point of view of surface chirality. This structure was observed exclusively, with no evidence being found for the mirror image structure (9 0 ; −1 2). By contrast, the adsorption of (S,S)-tartaric acid gave only the (9 0; −1 2) structure under similar preparation conditions (**Figure** 7) [25]. In these structures, tartaric acid is adsorbed across the troughs of the Cu{110} surface in the doubly de-protonated bi-tartrate form. Barbosa and Sautet [26] used DFT calculations to examine the preference by one enantiomer to form one of the two mirror equivalent domains. It was found that there is a $\sim$10 kJ mol^{-1} preference for one ordered arrangement over the other. The energetic preference is believed to be derived from an optimization of intramolecular H-bonding interactions involving the two −OH groups at the chiral centers and was not believed to be related to intermolecular H-bonding interactions, because adjacent molecular species are too far apart for any significant H-bonding interactions. An interesting feature of the (9 0; ±1 2) structure is the tendency for clusters of three molecular features to be observed in the STM images. It is implicitly assumed in the proposed structural models that the adsorption site of each TA species is essentially equivalent. If this was the case, then it is not obvious why TA species from clusters of three species separated by channels in the surface. Under certain tip conditions, the three features of the cluster appear to give different z-contrast. This may suggest that the three species are in slightly different adsorption sites and that it is energetically more favorable to have an 'empty' channel between rows of clusters than to accommodate an additional TA species in a less favorable adsorption site. There is some evidence from STM images of the Ni{110}/tartaric acid system that the

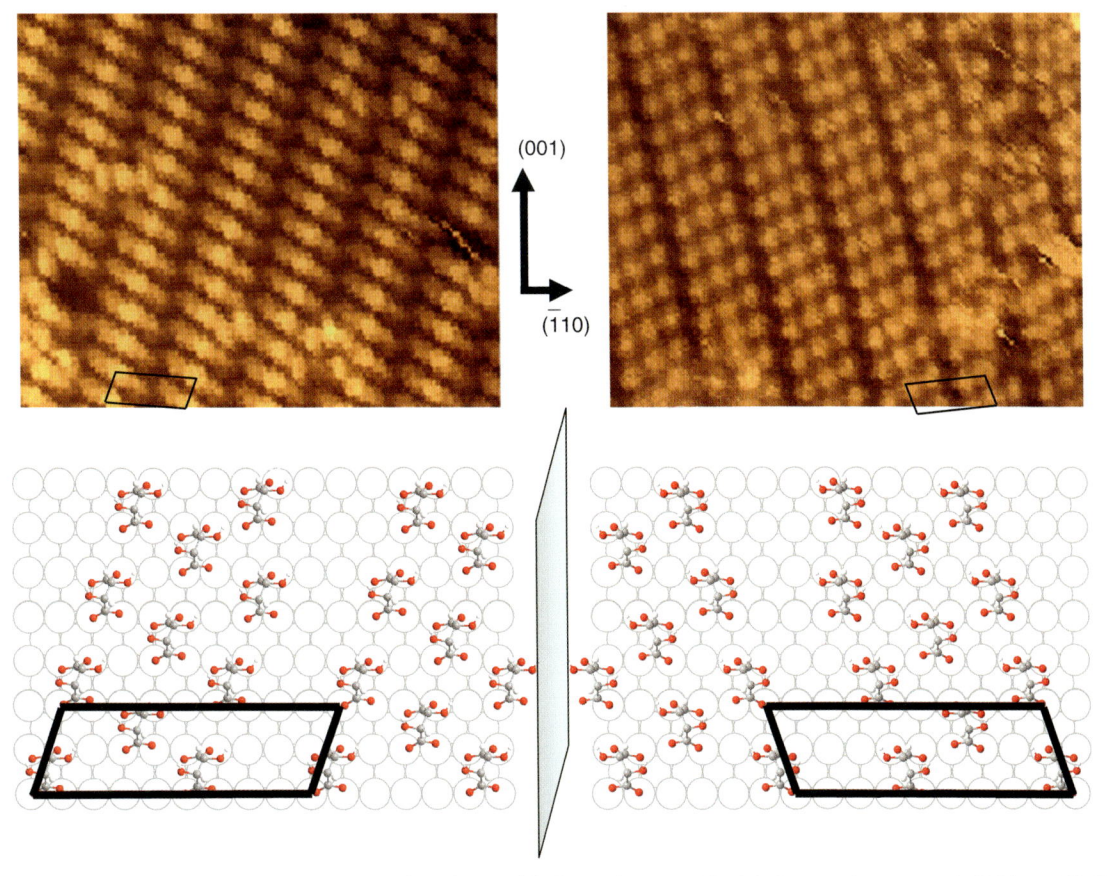

Figure 7 STM images (13.5 nm × 11.5 nm) of the (9 0; 1 2) (left) and (9 0; −1 2) (right) phases of respectively (R,R)- and (S,S)-tartaric acid on Cu{110}. Adapted from Lorenzo MO, Baddeley CJ, Muryn C, and Raval R (2000) *Nature* 404: 376.

TA species influence the electronic structure of the underlying Ni in the vicinity of the adsorbed TA species perhaps via some local restructuring of Ni atoms [21, 27]. The formation of clusters and channels in the Cu{110} experiments may be related to a release of strain in the surface copper atoms. This proposed mechanism is supported by a combined DFT and kinetic Monte Carlo study by Hermse et al. [28].

2. Alanine and other amino acids on Cu surfaces

In terms of surface science studies, alanine (CH$_3$—CHNH$_2$—COOH, see **Figure 5**) is by far the best-studied chiral amino acid. Adsorption on Cu{110} has been investigated by a variety of experimental surface science techniques, such as RAIRS, STM, LEED, X-ray photoelectron spectroscopy (XPS), PhD and near-edge X-ray absorption fine-structure spectroscopy (NEXAFS) [29–35], alongside theoretical model calculations [35–38]. Based on the experimental and theoretical evidence, it is well established that this molecule chemisorbs as alaninate and prefers a μ_3 adsorption complex with bonds through the two O atoms and the N atom, as shown in **Figure 8**. This adsorption geometry is found for the chemisorbed layer at all coverages after annealing to 470 K. For lower annealing temperatures, an additional μ_2 minority species is found at high coverages. The mismatch between the triangular molecular footprint of the μ_3 adsorption complex and the positions of the Cu atoms in the substrate surface leads to distortions in the adsorbate and substrate, which in turn affect the lateral interaction between the molecules and, hence, the long-range order of the adsorption system. The molecular orientation and the positions of the oxygen and nitrogen atoms with respect to the substrate have been characterized experimentally by PhD [34] and NEXAFS [35]. The experimental data are compatible with a tilt of the O-O axis with respect to the close-packed Cu-Cu rows of the {110} surface, which leaves the O atoms somewhat off the atop position with respect to the Cu atoms (see **Figure 8**).

As discussed above for the case for (R,R)-tartaric acid adsorption on Cu{110}, the adsorption of S-alanine on Cu{110} results in a number of discrete 2D phases [31, 32]. At relatively high coverage and adsorption temperature, a 'pseudo (3×2)' phase with an unusually distorted diffraction pattern is observed with two molecules per unit cell (analogous to the structure observed for glycine/Cu{110}). It is interesting to note that, in both the glycine and alanine cases, glide symmetry is observed in the (3×2)

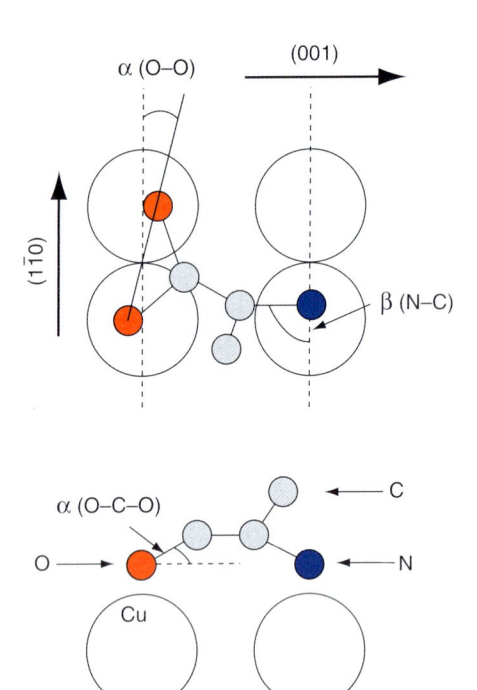

Figure 8 Schematic representation of the adsorption complex of alanine on Cu{110}. Reproduced with permission from Jones G, Jones LB, Thibault-Starzyk F, et al. (2006) Surface Science 600: 1924.

diffraction pattern. The glide symmetry is explained by the presence of a glide plane parallel to the ⟨001⟩ direction. In the case of glycine, this can easily be explained by the presence of two mirror-equivalent adsorption geometries in each unit cell giving rise to a 'heterochiral' structure. For alanine, it is impossible to create a molecularly heterochiral structure, so the observed glide plane must arise from the dominant scatterers in the structure, that is, either the carboxylate groups and/or the N-atoms (each of which are bound directly to the surface) and/or reconstructed surface Cu atoms [32]. At lower adsorption temperature, a globally chiral phase (2 −2; 5 3) is observed. This gives extensive ordered domains of either hexamer or octamer clusters of alaninate species in a globally chiral structure. The clusters are separated from each other by troughs in a manner analogous to the globally chiral (9 0; 1 2) structure observed for (R,R)-tartaric acid on Cu{110} [25].

The local adsorption geometries of glycine and phenylglycine on Cu{110} were studied by a variety of methods, indicating a very similar pattern of adsorption bonds through their triangular O—O—N footprint, each atom bonding to a single surface Cu atom [11, 12, 30, 36, 37, 39–42].

According to DFT model calculations by Rankin and Sholl and STM experiments by Zhao *et al.* and Iwai *et al.*, the same general bonding scheme also applies when alanine or glycine adsorb on Cu{100} [36, 37, 43–45]. In this case, however, the square surface lattice forces the molecule into an even more constrained geometry than on Cu{110}, which leads to interesting surface reconstructions that will be discussed below.

The triangular O—O—N adsorption footprint is also found for adsorption of amino acids with larger side groups on Cu{110}, such as serine, norvaline, methionine, and lysine [6]. Depending on the length and reactivity of the side group, a fourth bond may be formed, as it seems the case for lysine [46]. There is, however, a lack of accurate geometric information, either experimental or theoretical for larger amino acids.

3. *Amino acids on Pt group metals*

The adsorption of amino acids on Pt group metals, such as Pd and Pt, is more complicated than on Cu and Ni. Experimental studies of glycine and alanine adsorbed on Pd{111} and Pt{111} consistently report the presence of zwitterionic (NH_3^+—RCH—COO^-) species in addition de-hydrogenated or anionic (NH_2—RCH—COO^-) forms [47–50]. In a theoretical study by James and Sholl, the zwitterions of glycine and alanine on Pd{111} were found to be 0.37 and 0.46 eV, respectively, less stable than the anionic form [51], whereas the experimental studies find zwitterions as the majority species. The adsorption geometry for the anionic form is tridentate, similar to those discussed before, whereas the zwitterions only form strong bonds with Pd atoms through the carboxylate group; the N atom appears to be involved in a weak hydrogen bond.

4. *Cysteine on gold*

A very different type of adsorption complex was found by Kühnle *et al.* for cysteine (HS—CH_2—CH(NH_2)—COOH) on Au{110} by combining STM results and DFT calculations [52, 53]. Here substrate bonds are formed though the amino groups and the de-protonated S atoms. Equal enantiomers form dimers via hydrogen bonds between the carboxylic acid groups. The preference of the sulfur atoms to form bonds with low-coordinated Au atoms causes a substantial reconstruction of the substrate, leading to the displacement of four Au atoms from the close-packed rows (see **Figure 9**). On the flat Au{110} surface, the adsorption of racemic cysteine leads to the formation of (D,D) and (L,L) dimers, which can be discriminated by their inclination angles with respect to the close-packed Au rows along the [1$\bar{1}$0] direction. At low coverage, equal amounts of the two dimer types are found on each {110} terrace. Chiral kink sites at terrace boundaries, however, provide a template for more stable adsorption (**Figure 9**); (D,D) dimers are more stable on R kink sites). These sites act as nucleation centers for enantiopure islands at high adsorbate densities and, hence, lead to enantioselection even within the terraces [52, 54]. This is enantioselectivity on the nanometer scale, which can only be observed by microscopic techniques such as STM; on the macroscopic scale, the adsorbate layer is still racemic.

5. *Chiral alcohols and carboxylic acids*

The adsorption geometries of a number of other chiral molecules, which are potential chiral modifiers in catalytic processes, have been studied theoretically and experimentally. The most stable adsorption geometries of small chiral alcohols involve bonds of their de-protonated oxygen atoms with high-coordinated

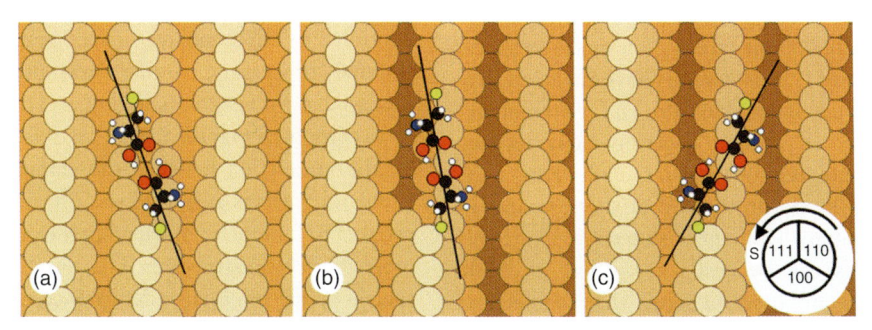

Figure 9 (a) Adsorption geometry of a D-cysteine dimer on a four-atom long vacancy structure showing a 20° rotation compared to the [1–10] direction. (b) Adsorption geometry of a D-cysteine dimer at an S kink site, showing a 10° rotation. (c) Adsorption geometry of an L-cysteine dimer at an S kink site, showing a 30° rotation. Reproduced with permission from Kühnle A, Linderoth TR, and Besenbacher F (2006) *Journal of the American Chemical Society* 128: 1076.

bridge or hollow sites. Theoretical calculations by Stacchiola et al. show that 2-butanol adsorbs on Pd{111} in a threefold hollow site [55], whereas the O atom of alaninol adsorbed on Cu{100} is bonded to a twofold bridge site [56]. The latter molecule forms a second bond with the Cu surface through its amino group on an atop site.

Like amino acids, the most stable adsorption species of 2-methylbutanoic acid adsorbed on Pd{111} is also de-protonated, 2-methylbutanoate. DFT calculations predict a bi-dentate adsorption geometry with the two oxygen atoms near atop Pd sites [57].

3.2.2.2 Chiral substrate modification

For a face-centered cubic (f.c.c.) crystal, the low index faces (e.g., {111}, {100}, and {110}) are the most thermodynamically stable, having the lowest surface free energies. Chemisorption can lead to large changes in surface free energies. There are many examples where chemisorption of organic molecules on a low-index crystal face results in faceting of a metal surface. A number of factors influence the formation of facets including face-specific adsorption energies, the energy difference between kinks, steps, and terraces, and substrate mediated intermolecular interactions and surface diffusion barriers. Recent studies of organic molecules adsorbed on low-index surfaces have found that high-index facets can be formed with complex organic molecular adsorbates containing electronegative elements, such as O and N atoms in their functional groups. In these systems, the energy gain, which drives the morphology change, could originate from the molecule–substrate interactions and substrate mediated, interadsorbate interactions, which stabilize the steps and kinks of the substrates.

Organic molecules with carboxylic acid functionalities commonly exhibit faceting on metal surfaces. For example, STM investigations have revealed that formic acid [58], benzoic acid [59], and p-aminobenzoic acid [60] all exhibit faceting behavior on Cu{110}. As pointed out earlier, carboxylic acid groups tend to be de-protonated at room temperature. A preferential alignment of step edges along the [112] directions can be easily identified for both formate and acetate. It seems likely that the driving force for the formation of this orientation of step edge is the ordering of the molecular species into $c(2\times 2)$ arrangements [61], where the adsorbate molecules are aligned along ⟨112⟩-type surface azimuthal directions. Surface structures formed by the adsorption of benzoic acid are much more complicated [59]. Benzoate species can adopt either flat-lying or upright geometries and may form several different periodic structures, depending on the coverage and annealing temperature. The related molecule p-aminobenzoic acid also displays extensive faceting on the Cu{110} surface [60]. In these cases, it is possible to identify two symmetrically equivalent (11 13 1) facets giving the characteristic sawtooth arrangement of facets. The fact that similar facets are observed both for benzoic and p-aminobenzoic acid leads to the conclusion that the formation of facets is directed by the flat-lying carboxylate units. In the case of formate and acetate, where vibrational spectroscopy reveals upright carboxylate units, step bunching is not observed, thus leading to the proposal that the adsorbate mediated step–step interaction required for step bunching is at best only weakly attractive when the carboxylate is perpendicular to the surface [61].

A surface reconstruction of similar type was observed for R-3-methyl cyclohexanone adsorption on the achiral stepped Cu{533} surface [62]. The step edges, which run along the [01$\bar{1}$] direction on the clean surface, are modified into a zigzag shape upon adsorption with kinked steps running along the [$\bar{1}$3$\bar{2}$] and [$\bar{1}$2̄3] directions. Although a chiral modification would be expected in principle, the STM images do not show any significant net chirality for this adsorbate.

On the chiral Cu{643} surface, the chiral molecule 3-methyl cyclohexanone forms a well-ordered adsorption structure with one molecule per unit cell. Achiral cyclohexanone, however, induces a reconstruction that consists of {11 7 5} and {532} facets [63].

Pascual et al. [64] investigated the adsorption of the pro-chiral carboxylic acid 4-[trans-2-(pyrid-4-yl-vinyl)]benzoic acid (PVBA) on Ag{110}. Following exposure to submonolayer coverages of PVBA and thermal processing, similar sawtooth facets were observed as for benzoic acid on Cu{110} (**Figure 10**). It was proposed that the formation of facets was driven by the interaction between the carboxylate and the {100} microfacets at step edges. The microfacets then act as chiral templates nucleating the growth of supramolecular PVBA structures. The chirality of the PVBA species at the microfacet determined the structure of the first four assembled rows of molecules.

An interesting effect that may provide some insight into chiral modifications of achiral surfaces in general has been discovered by Zhao et al. Upon

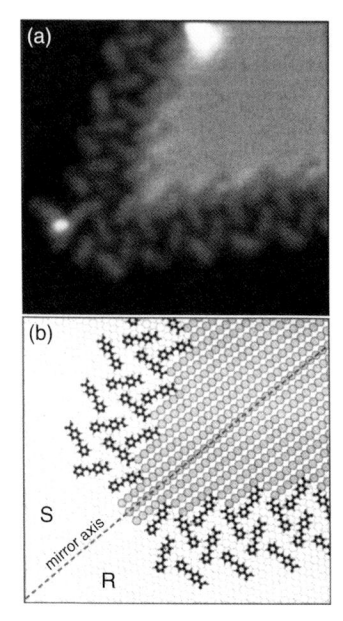

Figure 10 (a) STM image (10 nm × 10 nm, tip bias +0.52 V, tunneling current 0.5 nA) of a PVBA-induced sawtooth blade in a restructured Ag(110) surface terrace. (b) Structural model of the chiral kink arrangements induced by lateral interaction of molecular carboxylate endgroups with Ag {100} microfacets. Reproduced with permission from Pascual JI, Barth JV, Ceballos G, et al. (2004) *Journal of Chemical Physics* 120: 11367.

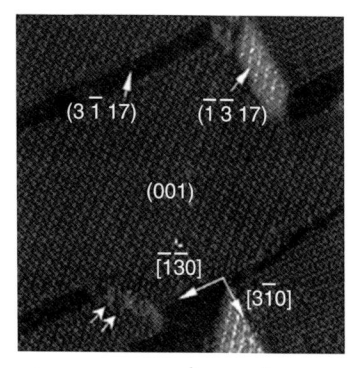

Figure 11 STM image (550 Å × 550 Å, 1.0 V, 0.6 nA) acquired from the L-lysine/Cu(001) surface annealed at 430 K, showing coexistence of {001} terraces and {3 1 17} facets formed through bunching of ⟨310⟩ faceted steps. Reproduced with permission from Zhao XY (2000) *Journal of the American Chemical Society* 122: 12584.

adsorption of glycine, alanine, and lysine on Cu{100}, the adsorbate molecules induce the formation of chiral {3 1 17} facets [43, 44, 65]. Achiral glycine and the smallest chiral amino acid alanine induce all eight possible facets of the {3 1 17} family (four of these facets are rotationally equivalent to each other as are the other four, the two sets of rotationally equivalent facets are related by reflectional symmetry to each other and are denoted R- and S-type facets). In contrast, lysine (NH$_2$-(CH$_2$)$_4$—CH(NH$_2$)—COOH) leads to the formation of only one enantiomorph. After prolonged annealing of the S-lysine covered Cu{100} surface to 430 K, only {3 1 17}^{R} facets are found (**Figure 11**). In order to explain this faceting, Zhao *et al.* assumed the same μ_3 adsorption complex as found for glycine on Cu{100} or Cu{110}.

The {3 1 17} surface has small {100} facets, which are big enough to allow the adsorption of two glycine or alanine molecules per surface unit cell in the preferred bonding configuration, but there is only room for one lysine molecule per unit cell (**Figure 11**). DFT calculations by Rankin and Sholl show that the most stable adsorption site for glycine and alanine involves a bond between the amino group and a kink Cu atom of the substrate [66, 67]. A clear preference for faceting in terms of surface energies is only found for dense layers with two molecules per unit cell. The calculations show very little enantioselectivity with respect to the chirality of the reconstructed surface. The calculated surface energy differences between R and S facets for enantiopure molecular layers are of the order of 0.01 J m^{-2} (the combination S/S is preferred over R/S), which explains the lack of enantioselectivity in the experimental data.

Surface facetting may be significant in chiral heterogeneous catalysis – particularly in the Ni/β-ketoester system. The adsorption of tartaric acid and glutamic acid onto Ni is known to be corrosive, and it is also established that modifiers are leached into solution during both the modification and catalytic reaction [14]. The preferential formation of chiral step–kink arrangements by corrosive adsorption could lead to catalytically active and enantioselective sites at step–kinks with no requirement for the chiral modifier to be present on the surface.

3.2.3 Chiral Amplification and Recognition

3.2.3.1 Chiral amplification in two dimensions

In Section 3.2.1, we discussed the phenomenon of adsorbate induced chirality, whereby the adsorption of achiral species (e.g., glycine) results in the formation of two mirror-equivalent domains on the surface. It has recently been shown that the presence of relatively small mole fractions of chiral dopants can result in the exclusive formation of one of the two mirror-equivalent domains of the achiral species. For example, succinic acid (HOOC.CH$_2$.CH$_2$.COOH), an achiral molecule, forms two mirror-equivalent domains (9 0; −2 2) and (9 0; 2 2) on Cu{110} [22]. The doubly de-protonated succinate species are bound via both carboxylate groups to the Cu surface − the mirror relationship between the two domains is thought to arise due to the twist of the carbon backbone of succinate with respect to the [001] surface direction.

When as little as 2 mol% (R,R)-tartaric acid is co-adsorbed with succinic acid, LEED beams associated with the (9 0; −2 2) structure are extinguished. The opposite behavior is observed when the dopant is (S,S)-tartaric acid [68]. This behavior is analogous to the "sergeants and soldiers" principle observed for helical polyisocyanate co-polymers [69]. The mechanism for this effect is proposed to be substrate mediated. Succinate species are unable to form intermolecular H bonds, so a chiral footprint imposed on the surface by a tartrate species is thought to control the adsorption geometry of the surrounding complex creating an effect that is amplified over 30–50 molecules in a given domain [68].

3.2.3.2 Chiral recognition

A simple example of chiral recognition occurs when one enantiomer of a chiral entity displays a stronger interaction with one enantiomer of a different chiral entity than with its mirror image. This is the key to the significance of chirality in biology and underpins the need to develop chiral products in the pharmaceutical and agrochemical industries. Studying chiral recognition processes at surfaces has considerable relevance in the development of enantioselective catalysts, biosensors, and biocompatible materials.

The differences in adsorption energy of chiral molecules at chiral step–kink sites (Section 3.2.4) are a manifestation of chiral recognition in the surface context. Similarly, the growth of a single molecule type into chiral clusters and 1D and 2D arrays is an example of chiral recognition.

An interesting example of chiral recognition at surfaces comes from the work of Chen and Richardson on the interaction of S-phenylglycine with adenine/Cu{110}. Adsorption of adenine on a Cu{110} surface gives rise to flat-lying molecules that form a racemic mixture of homochiral chains made up of adenine dimers, whose direction on the Cu{110} substrate is correlated with their chirality [70], as shown in the left panel of **Figure 12**.

Subsequent adsorption of one enantiomer of phenylglycine leads to an intermolecular recognition process that favors the decoration of chains running in the 1,2 direction by S-phenylglycine (right-hand panel of **Figure 12**), whereas R-phenylglycine decorates the mirror image 1,-2 adenine chains [71]. The origin of the strong interaction between the anionic amino acid species, and the nucleic acid base is electrostatic favoring the close approach of the carboxylate functionality of phenylglycine to the nitrogen of the adenine's amine group, which lies on the periphery of the chain [42]. Chiral recognition occurs because there is also a repulsive interaction between the amine groups of the two molecules and this is less for the favored enantiomer than for the other [42].

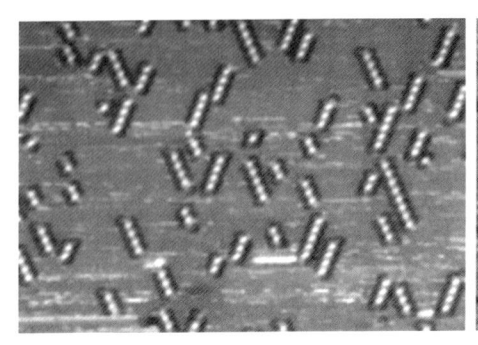

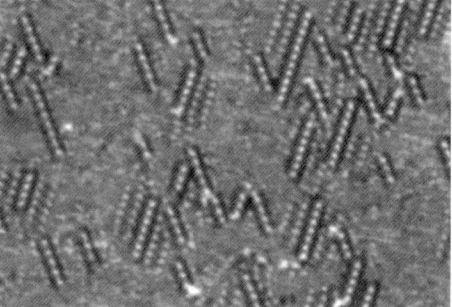

Figure 12 The left-hand STM image shows homochiral adenine rows aligned in low symmetry but mirror image azimuths on a Cu(110) surface. On the right, adenine rows in the (1,2) direction are decorated with double rows of S-phenylglycine molecules, while no such interaction occurs with (1,−2) rows. Adapted with permission from Chen Q and Richardson NV (2003) *Nature Materials* 2: 324.

3.2.4 Chiral Molecules on Chiral Surfaces

Chiral metal and mineral surfaces show enantioselective behavior with regard to the adsorption and reactions of organic molecules [5, 72–81]. Examples are surfaces of naturally occurring crystals of some common minerals, such as α-quartz (chiral bulk symmetry) or calcite (achiral bulk symmetry), and high-Miller-index surfaces of metal single crystals with achiral bulk lattice symmetry, which are normally not found at metal crystallites in their equilibrium shape but can be created by cutting a single crystal along the respective plane or by chiral modification (see Section 3.2.2.2). Intrinsically chiral surfaces have no mirror symmetry and cannot be superimposed onto their mirror images. Therefore, one can expect enantiomeric differences in the adsorption and reaction behavior of chiral molecules. The first reports by Schwab and Rudolph about enantioselective reactions catalyzed by enantiopure α-quartz particles date back to the 1930s [72]. A number of enantioselective effects have been reported since in connection with the adsorption of small chiral molecules on intrinsically chiral metal and mineral surfaces under well-defined conditions [74, 76, 78, 82].

In recent years, single-crystal model systems involving intrinsically chiral surfaces have been studied with surface-science methods in some detail, mainly by STM, photoemission spectroscopies (XPS, PhD, NEXAFS), IR spectroscopy, electrochemical methods, and DFT (see [76, 78–81], and references therein). Owing to the choice of experimental methods and the complexity of these systems, there is relatively little known to date about the exact adsorption geometries on such surfaces, although structural information is crucial for the understanding of the stereoselective behavior of these systems. Of particular interest are the key factors determining adsorption geometries and, thus, the stereochemistry of adsorbed molecules, either the shape of the actual adsorption site or the lateral interaction between neighboring molecules (chiral modifiers or reactants).

3.2.4.1 Chiral substrate geometries

The most abundant naturally occurring intrinsically chiral surfaces are those of α-quartz and calcite. There are examples in the recent literature showing that even unmodified quartz or calcite surfaces can lead to enantioselective adsorption and/or reactions [5, 74, 77]. Because α-quartz has a chiral bulk structure with Si-O-Si chains forming either clockwise or counter-clockwise spirals, every surface is intrinsically chiral. Natural quartz crystals are, however, often strongly twinned internally. The largest facets of naturally occurring α-quartz crystallites are terminated by $(10\bar{1}0)$ (m), $(10\bar{1}1)$ (r), and $(01\bar{1}1)$ (z) surfaces, which are shown in **Figure 13** alongside the macroscopic shapes of left- and right-handed quartz crystals. Detailed investigations into the structure of quartz surfaces have concentrated on theoretical modeling of the {0001} surface, which is not one of the commonly found crystal faces. It was found that the clean unreconstructed surface has a

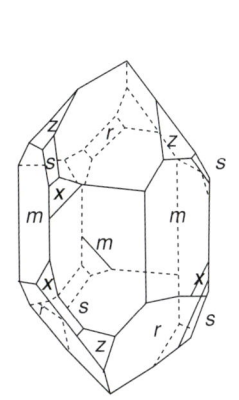

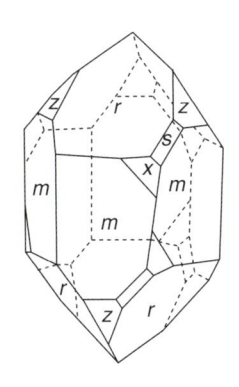

Figure 13 External morphology of natural left-handed (left) and right-handed (middle) quartz crystals often serves to distinguish left- versus right-handed crystals. The image on the right-hand side shows the arrangement of atoms in the m and r surface planes. Adapted with permission from Hazen RM and Sholl DS (2003) *Nature Materials* 2: 367.

very high surface energy, whereas the fully hydroxylated reconstructed surface is much more stable. The reconstruction leads to the formation of six-membered rings of hydroxylated oxygen atoms parallel to the surface plane [83]. The adsorption of methanol (CH_3OH) and methylamine (CH_3NH_2) was studied by Han et al. [84]. Although they are not chiral, these molecules contain functional groups that are common in many chiral adsorbates discussed above. For both molecules, the surface bond occurs through hydrogen bonds involving the amino and alcohol groups, respectively, which makes quartz a much more gentle substrate than the metal surfaces discussed so far.

Calcite has an achiral bulk lattice but its common [21$\bar{3}$1] trigonal scalenohedral (dog-tooth) crystal form has chiral (3$\bar{1}$21) and (2$\bar{1}$31) surface planes, which are opposite enantiomorphs. Recent experiments by Hazen et al. found some degree of enantioselective adsorption behavior of aspartic acid enantiomers on these surfaces [77] of the order 1% ee.

In recent studies, the main emphasis of intrinsically chiral substrates has been on high-Miller-index kinked surfaces of metal single crystals with cubic bulk lattice symmetry, such as Pt or Cu. The motivation for these studies lies in the widespread use of transition metals as industrial catalysts. If the three Miller indices h, k, l are different from each other and from zero, these surfaces have no mirror symmetry and cannot, therefore, be superimposed onto their mirror images, which makes them chiral substrates. Some examples of intrinsically chiral f.c.c. metal surfaces are shown in **Figure 14**. It has been found that such surfaces show enantioselectivity with respect to the adsorption and reactions of chiral molecules [76, 78, 82, 85].

In principle, the specific Miller indices (hkl) unambiguously define the chirality of a surface. For crystals with cubic symmetry, it is easy to verify that a change in sign of one index, (hkl) → ($\bar{h}kl$), or a permutation of two indices, (hkl) → (khl), leads to the opposite chirality. The notation (hkl) using round brackets defines one specific surface plane with surface normal parallel to the [hkl] direction (in the case

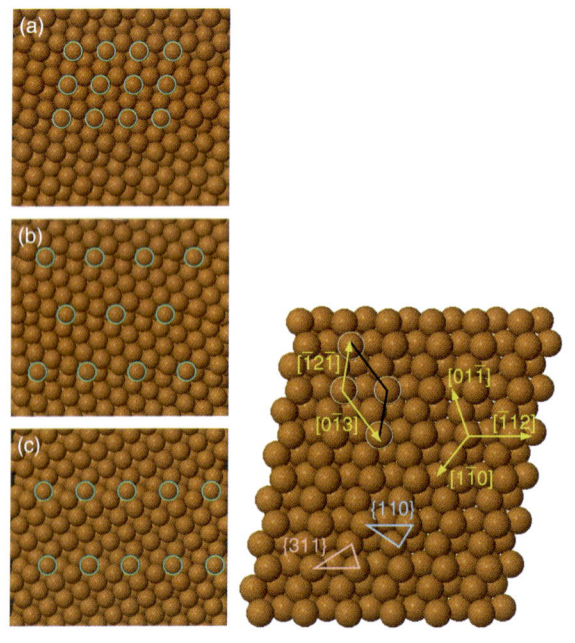

Figure 14 (Left) Examples of chiral f.c.c. surfaces: (a) {531}, (b) {643}, and (c) {17 3 1}. The sixfold coordinated kink atoms are marked by circles. With permission from Held G and Gladys MJ (2008) *Topics in Catalysis* 48: 128.. (Right) Arrangement of atoms in the bulk-terminated Cu{531}^{R} surface. The surface unit cell and the most relevant crystallographic directions are indicated at the top of the figure; the {311} and {110} microfacets are indicated at the bottom. Reproduced with permission from Gladys MJ, Stevens AV, Scott NR, Jones G, Batchelor D, and Held G (2007) *Journal of Physical Chemistry C* 111: 8331.

of cubic crystal lattice); the curly brackets {hkl} define a family of surface planes, which are related to the (hkl) surface by the symmetry operations of the bulk lattice, which, for cubic crystal symmetries, also includes mirror operations. Hence, there are two enantiomorphs within each family of {hkl} surfaces. Gellman and coworkers [73] have introduced the notation {hkl}$^{R/S}$ in order to distinguish between them in analogy to chiral molecules, which was later generalized by Attard [76]. The latter notation is based on the existence of kink atoms, which are surrounded by microfacets of different atomic densities (e.g., {111}, {100}, and {110} for surfaces of f.c.c. crystals). Following these facets in order of decreasing density describes either a clockwise or anticlockwise rotation, which defines the surface chirality as 'R' or 'S', respectively. More recently Jenkins and Pratt suggested an alternative approach, which is based on the rotation of high-symmetry crystallographic directions around the surface normal [7, 86]. This notation, using 'D' and 'L' instead of 'R' and 'S', is also applicable to crystal structures (such as base-centered cubic (b.c.c.)) where the chiral surfaces do not have kink atoms. For f.c.c. crystals the 'R' surface always corresponds to the 'D' surface and 'S' to 'L'; such a statement of equivalence does not hold generally for other crystal structures, where the 'D'/'L' notation is preferred. Ref. [7] gives a very comprehensive overview of chiral metal surfaces.

For f.c.c. transition metals, such as Cu or Pt, {531} surfaces have the smallest unit cell of all possible chiral surfaces and are, therefore, well-suited for ab initio modeling and LEED studies. The surface unit cell also has the right proportions to accommodate small amino acids such as alanine or glycine [81, 85].

Recent combined quantitative LEED and DFT studies of Pt{531} and Cu{531} [87–89] showed that these surfaces do not reconstruct and have arrangements of atoms very similar to bulk termination, except for very strong inward relaxation of the kink atoms in the topmost layer. On the other hand, theoretical and experimental studies also showed that step–kinked Pt surfaces are thermally relatively unstable [87, 90, 91] and tend to roughen. This instability is due to the low (sixfold) coordination of the kink atoms in the topmost surface layer and the short-range bonding between Pt atoms.

Copper surfaces are less affected by roughening or faceting because the long-range interaction between substrate atoms beyond the nearest neighbors is more important for Cu than for Pt atoms [89]. The structures of Cu{531} and Cu{532} have been studied theoretically and experimentally (only Cu{531}). No indications of large-scale reconstructions were found, but large relaxations of the under-coordinated atoms [89, 92]. STM studies of the clean chiral Cu{643} and Cu{5 8 90} surfaces [63, 93] show high mobility of the Cu surface atoms but, on average, a regular arrangement of kinks and steps for {643}.

3.2.4.2 Adsorption of chiral molecules on chiral surfaces

A number of studies have reported that medium-sized chiral molecules show enantioselective behavior on chiral substrate surfaces [62, 67, 78, 79, 81, 82, 85, 94–96].

Only a small number of very recent experimental studies, however, were able to determine certain important aspects of adsorption geometries, including the molecular orientation and adsorption sites. These results are mainly based on the application of modern synchrotron radiation techniques, such as XPD [33, 97] and NEXAFS [35, 85, 98] and STM [44, 52]. For the interpretation of these data DFT, which is now capable of routinely dealing with complex chiral adsorption structures, has proved to be an invaluable tool [35, 66, 97].

At present the available structural information is too patchy to develop a general picture of key factors causing enantioselectivity on intrinsically chiral surfaces but some common general features can be identified from the examples discussed below.

1. *Small amino acids on Cu{531}* (*glycine, alanine, serine, cysteine, methionine*)

Cu{531} is the only intrinsically chiral surface for which ordered molecular superstructures have been reported so far. A series of studies of small amino acids adsorbed on this surface showed long-range ordered overlayers for glycine, alanine, and serine, which all are comparable, in size, to the substrate unit cell [81, 85, 99]. The adsorption geometry of the p(1 × 4) overlayer of alanine on Cu{531} was characterized by Gladys *et al.* in detail using NEXAFS and XPS [81, 85]. The chemical shifts of the C1s, N1s, and O1s core level signal in XPS are essentially the same as for alanine on Cu{110} at low coverage, which indicates that the molecules assume the same μ_3 alaninate adsorption complex as shown in **Figure 8**. There are two adsorption sites on the {531} surface that can match the triangular footprint, namely {110} and {311} facets (see **Figure 14**).

The azimuthal orientation and, hence, the adsorption site can be determined through the angular dependence of the carbon or oxygen K-edge NEXAFS spectra. The half-filled π orbital in the carboxylate (O—C—O) group of alaninate gives rise to a 1s→π^* absorption resonance for photon energies around 288 eV (C1s→π^*) and 533 eV (O1s→π^*), respectively. According to the dipole selection rule, the cross section of this resonance is at a maximum when the polarization vector, E, of the exciting photon is perpendicular to the O—C—O triangle and goes to zero when E lies in the plane of the triangle. The angular dependence is $\cos^2 \gamma$, where γ is the angle between E and the surface normal of the O—C—O plane [100]. Hence, the orientation of the carboxylate group can be determined by fitting the experimental angular dependence of the 1s→π^* resonance, which had been used previously to determine the orientations of formate, glycinate and alaninate on Cu{110} [35, 101, 102].

Selected NEXAFS spectra and the angular dependence of the C1s→π^* resonance for saturated layers of R and S alaninate on Cu{531}^{R} are shown in **Figure 15**. The difference in the angular dependence for the two enantiomers indicates a certain degree of enentioselectivity with regard to local adsorption geometries. The fact that the signal is well above the zero level for all angles indicates that the layer consists of molecules with two different azimuthal orientations. By fitting the data, their azimuthal orientation could be determined, which is consistent with adsorption on {311} and {110} microfacets, as shown in **Figure 15**. DFT model calculations for glycinate do not indicate a particular preference for one of the two microfacets but strongly favor an adsorption complex with two first-layer (kink) Cu atoms and one second-layer atom involved , which is the case in the p(1 × 4) overlayer geometry depicted schematically in **Figure 15**.

In this geometry, the molecules on {110} microfacets can form strong hydrogen bonds between their amine groups and the oxygen atoms of neighboring molecules with O—H distances close to the ideal value of 1.5 Å [38] (hydrogen atoms are not shown in the figure).

The angular dependence of the C1s→π^* resonance is very different for dilute alaninate layers. There is a clear preference for the adsorption sites on the {311} microfacets and no more enantiomeric difference [81]. This indicates that the adsorption energy for isolated alaninate is higher on these facets and the difference in adsorption energies between {311} and {110} must be higher than the possible gain from forming hydrogen bonds, which is not possible if only one facet is occupied.

Interestingly, glycine adsorbed on the Cu{531} does not show this behavior. From low coverage to saturation adsorption sites on both facets are occupied in equal numbers [81]. Apparently, the methyl side group in alanine has a significant effect on the substrate bond, most likely through a repulsive interaction with the Cu surface atoms. Enantiomeric difference in the adsorption geometry is the result of intermolecular interactions, either attractive (through hydrogen bonds) or repulsive (through the methyl side group). The importance of hydrogen bonds for the arrangement of adsorbed amino acids has been pointed out by several authors before [38, 103].

A recent XPS and NEXAFS study comparing the adsorption geometries of the two sulfur-containing amino acids cysteine (HS—CH$_2$—CH(NH$_2$)—COOH) and methionine (CH$_3$—S—(CH$_2$)$_2$—CH(NH$_2$)—COOH) on Cu{531} indicates that these molecules also form surface bonds through the de-protonated carboxylate and the amino groups [98]. Cysteine forms an additional fourth bond through the thiolate side group, which is also de-protonated. The methyl endgroup of methionine does not interact strongly with the metal surface.

The adsorption of serine (HO—CH$_2$—CH(NH$_2$)—COOH) (which is structurally very similar to cysteine) on Cu{531} was also studied recently by Eralp *et al.* [99]. This molecule forms four surface bonds (through the oxygen atoms of the carboxylate group, the amino group and the OH group) at low coverage and three at saturation (carboxylate and amino groups). The azimuthal dependence of the NEXAFS spectra shows large differences between the two enantiomers. This suggests that enantiomeric differences in the adsorption geometries of amino acids become more pronounced with increasing length of the side group. This is in line with the suggestion by Easson and Steadman that stereoselectivity requires the involvement of three side groups of the asymmetric carbon atom [104].

2. Glycine and alanine on Cu{3 1 17}

The adsorption of glycine and alanine on Cu{3 1 17} was studied theoretically by Rankin and Sholl [66, 67]. Zhao *et al.* found that {3 1 17} facets were formed upon adsorption of amino acids on Cu{100} [43, 65], suggesting that this surface orientation

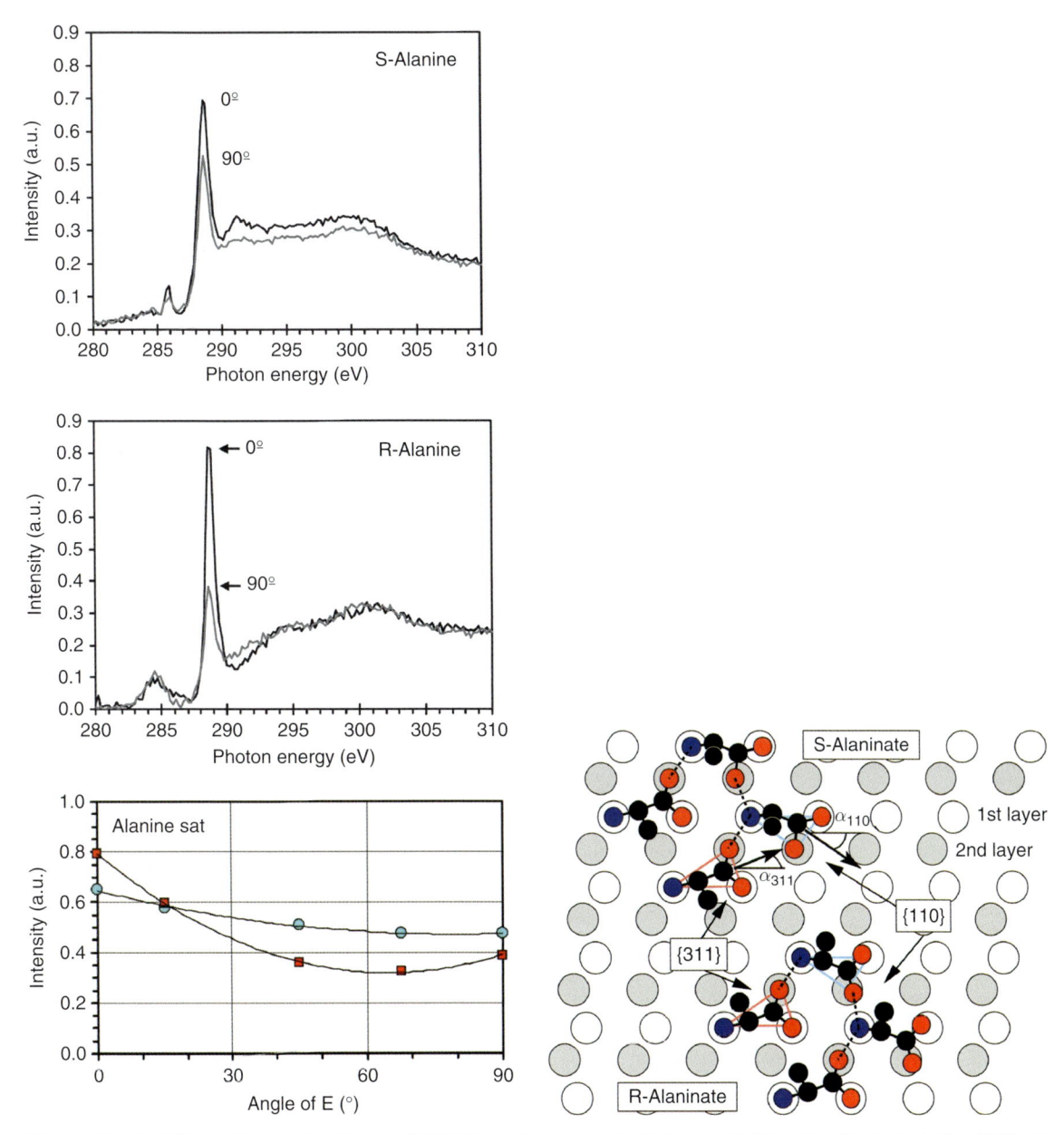

Figure 15 (Left) Normalized carbon K-edge NEXAFS spectra recorded with horizontal (0°) and vertical polarization (90°) of E for (S)- (top) and (R)-alanine (middle) on Cu{531}^{R}. (Bottom) Angular variation of the π–resonance for (R)- and (S)-alanine. (Right) Proposed adsorption geometry for alaninate on Cu{531}. Reproduced with permission from Gladys MJ, Stevens AV, Scott NR, Jones G, Batchelor D, and Held G (2007) *Journal of Physical Chemistry C* 111: 8331.

provides a better match for these molecules. Unlike {531}, {3 1 17} surfaces have relatively wide terraces with {100} orientation that can accommodate up to two small amino acids per unit cell. The most stable adsorption site for a dilute layer of glycine and both alanine enantiomers involves the kink Cu atoms and two atoms in the lower {100} facet, which together form a {311} microfacet, with the familiar triangular bonding pattern (through the carboxylate and amino groups). The {311} microfacet was also found to be the most stable adsorption site for alanine on Cu{531}. The additional molecules at saturation coverage adsorb on the {100} terrace in a geometry, which is essentially the same as on Cu{100}. The combination of preferred {311} adsorption sites and a network of hydrogen bonds in these dense layers leads to a reduction in the total surface energy of 0.03 J m^{-2} (Ala) and 0.07 J m^{-2} (Gly), respectively, for adsorption on Cu{3 1 17} as compared to Cu{100}, which explains the facetting observed in the experiments.

3. *Cysteine on Au{17 11 9}*

In contrast to Cu{531}, intermolecular interactions do not appear to be responsible for the enantiospecific adsorption geometries observed for cysteine on Au{17 11 9}. In a combined XPD and DFT study Greber *et al.* [97] found that cysteine forms bonds with the gold substrate only through the amino group and the de-protonated sulfur atom, the latter bond involves two Au atoms. Unlike on Au{110} (see above, [52]) no hydrogen-bonded dimers are formed, nor is the carboxylic acid group involved in the substrate bond. According to Ref. [97], it shows even a slightly repulsive interaction with the surface. The enantioselective adsorption behavior is due to the fact that the preferred bonding partner for the kink Au atom is the amino group for the R-enantiomer but with the thiolate group for the S-enantiomer (see **Figure 16**). It is noteworthy, that on the basis of energy differences calculated by DFT alone such behavior was not expected [105].

4. *Other chiral molecules*

In order to systematically determine the criteria governing enantioselective behavior, Bhatia and Sholl theoretically examined the energy differences between the enantiomers of a number of small chiral molecules on the chiral Cu surfaces {874}, {432}, and {821} [79, 80]. The list of molecules includes fluoroaminomethoxy (FAM), aminoethoxy (AE), propylene oxide (PO), epichlorohydrin (EPO), 3-methylcyclohexanone (3-MCHO), aminomethoxythiolate (AMT), and aminoethoxythiolate (AET).

As can be seen from **Figure 17**, this list includes examples that are expected to lead to mono-dentate, bi-dentate, and tri-dentate binding on Cu surface. As a general trend, the number of surface bonds is as predicted, and kink atoms are involved in the most stable adsorption sites of all molecules for low coverages (one molecule or less per kink atom). The authors found significant enantiospecific energy differences for surfaces with large {111} terraces, {874} and {432}, but none for {821}, which has large {100} terraces. This finding cannot, however, be generalized as there is experimental evidence for the fact that such differences exist for larger molecules [65]. Unfortunately, there is very little overlap with experimental data for the systems studied in this work. 3-Methyl cyclohexanone (3-MCHO) has been studied by STM on Cu{643} [63], which also has large {111} terraces and is, therefore, comparable with the {874} and {432} surfaces studied in [80]. The experimental data are compatible with the ketone group bonding to the kink site and the ring lying flat on the {111} terrace maximizing the van der Waals interaction, as shown in **Figure 18**.

3.3 Kinetics of Desorption Processes

3.3.1 Achiral Surfaces

In this section we will concentrate on the desorption and decomposition behavior of alanine and tartaric acids, which are the most important prototypes of chiral carboxylic acids.

Temperature-programmed desorption (TPD) spectra of alanine on Cu{110} show multilayer desorption around room temperature and sharp desorption peaks of the decomposition products

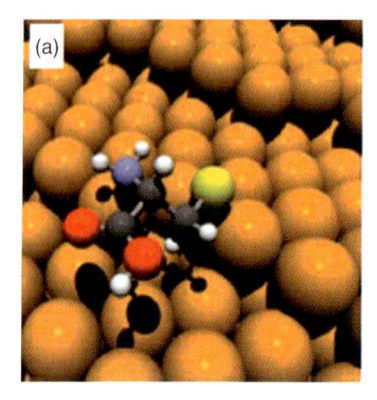

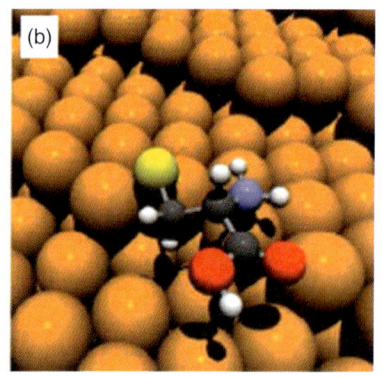

Figure 16 Three-dimensional images of the adsorption structures of (a) D- and (b) L-cysteine. Reproduced with permission from Greber T, Sljivancanin Z, Schillinger R, Wider J, and Hammer B (2006) *Physical Review Letters* 96: 056103.

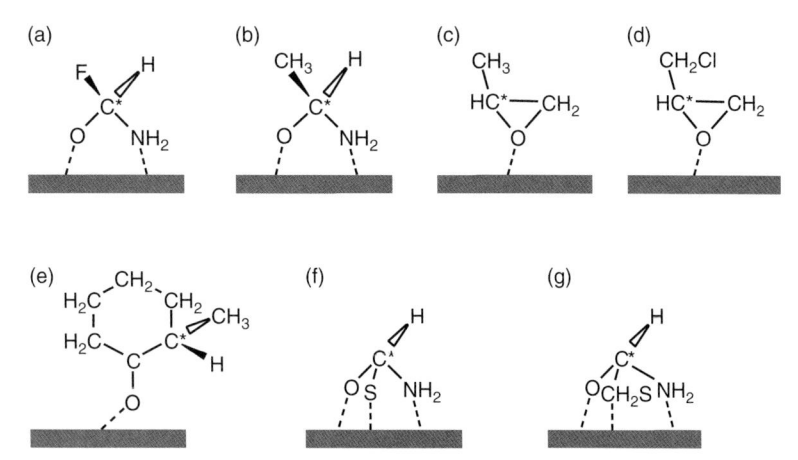

Figure 17 Schematic illustrations of surface bonding of (a) FAM, (b) AE, (c) PO, (d) EPO, (e) 3-MCHO, (f) AMT, and (g) AET. In each case, dashed lines indicate bonds formed with the metal surface and the chiral carbon is indicated with a star. Reproduced with permission from Bhatia B and Sholl DS (2008) *Journal of Chemical Physics* 128: 144709.

Figure 18 (a) STM image of R-3-MCHO on Cu(643) deposited at room temperature. The adsorption structure is ordered along steps. (b) Scaled model representation of the R-3-MCHO overlayer on Cu(643). Reproduced with permission from Zhao XY and Perry SS (2004) *Journal of Molecular Catalysis A: Chemical* 216: 257.

(hydrogen, CO / CNH_2, and CO_2) at 520 K [32]. Temperature-dependent XPS shows that the molecules are intact up to this temperature [35], and only traces of species containing O, C, and N are left on the surface after decomposition. The phase transition between the two long-range ordered phases (2 − 2; 5 1) and the p(3 × 2), which occurs around 450 K, is not accompanied by significant changes in the XPS data. On the Pt-group metal surfaces Pt{111} and Pd{111}, decomposition of glycine and alanine occurs at lower temperatures, starting at around 350 and 390 K, respectively [47, 49, 50]. The distribution of decomposition products on these surfaces suggests that the main dissociation mechanism is breaking the C—C bond of the molecular backbone.

The thermal evolution of tartaric acid (HOOC—CHOH—CHOH—COOH) is very similar on Cu{110} and Ni{110}. On both surfaces, complete decomposition occurs around 450 K, which is accompanied by sharp desorption peaks of hydrogen, CO/CNH_2, and CO_2 at 520 K [21, 22], again indicating that C—C bond fission is the main dissociation mechanism. The comparison with succinic acid (HOOC—CH_2—CH_2—COOH) on Cu{110} shows that the absence of the OH groups in this molecule increases the thermal stability significantly; for succinic acid decomposition occurs at the much higher temperature of 600 K.

On Ni{111}, decomposition of tartaric acid starts at lower temperatures, between 400 and 420 K, depending on the coverage [24]. On the oxidized Ni{111} surface, however, the onset of decomposition is significantly higher, with desorption peaks of the products at 570 and 620 K [106].

3.3.2 Chiral Surfaces

One of the main interests in desorption spectroscopy from chiral substrate surfaces lies in potential enantioselectivity, that is, differences either in height or position of desorption peaks between enantiomers of the same molecule, or between enantiomorphs of the same surface termination.

The earliest systematic studies of enantioselective behavior of intrinsically chiral single-crystal metal surfaces came from a series of electrochemical experiments by Attard and coworkers [75, 76, 107] showing clear differences in the cyclo-voltammograms (CVs) for electro-oxidation of (L)- and (D)-glucose on Pt electrodes terminated by chiral {643}, {431}, {531}, or {321} surfaces. Changing the enantiomer of the reactant molecule from (L) to (D) in these experiments had the same effect as changing the chirality of the electrode surface. The information provided by these measurements is in many ways comparable to TPD: the current density is proportional to the reaction rate at a certain electrochemical potential; peaks at certain potential values can be associated with features on the surface, for example, terraces of certain terminations and step or kink sites. **Figure 19** shows examples of CVs for glucose electro-oxidation on both enantiomorphs for three chiral Pt surfaces ({643}, {321}, and {531}), with increasing kink density. As the kink density increases, the difference between the two CV curves also increases, in particular in the region between 0.35 and 0.40 V (Pd/H), which is associated with the reaction on {111} terraces. However, the total current density (i.e., the reaction rate) decreases as the kink density increases.

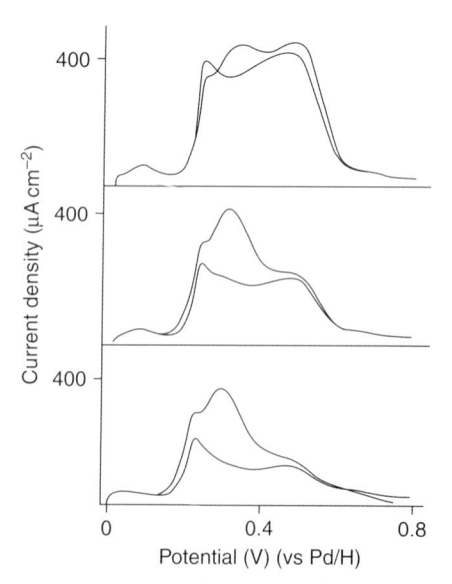

Less pronounced, but still significant, differences were found in temperature-programmed desorption (TPD) experiments performed in ultrahigh vacuum on intrinsically chiral Cu and Ag surfaces by Gellman and coworkers [73, 78, 82, 94–96, 108]. The high-temperature desorption peaks of R- and S-3-MCHO and R- and S-PO on Cu{643} differ by 3.5 and 1.0 K, respectively. The temperature difference for 3-MCHO on Cu{531} is 2.2 K. These small temperature differences can be used to separate the enantiomers in a racemic mixture [78]. The differences in desorption temperatures are equivalent to differences in the adsorption energies of 1.0 kJ mol^{-1} (R-3-MCHO on S substrate is more stable) and 0.3 kJ mol^{-1} (R-PO on R substrate is more stable). The DFT calculations by Bhatia and Sholl lead to a similar adsorption energy difference between the two enantiomers: 2 to 4 kJ mol^{-1} for PO on Cu{874}, Cu{432} and Cu{821} of (R-PO on S substrate is more stable) and 0–2 kJ mol^{-1} for 3-MCHO on Cu{874} and Cu{432} [79, 80].

By comparison between different chiral and achiral surfaces and/or by selectively blocking adsorption sites with iodine, Gellman *et al.* could identify the kink atoms as the origin of the enantioselective adsorption/reactions on these surfaces. Examples are given in **Figure 20**: the achiral hexagonal Cu{111} surface exhibits a single desorption peak at 230 K; the stepped Cu{533} and Cu{221} surfaces have {111} terraces (desorption peak at 230 K) separated by {100} or {110} steps, respectively (345 K); the chiral Cu{643} and Cu{653} surfaces have kinks in addition to terraces and steps and exhibit an additional peaks at 385 K. Only the kink-related peaks at 385 K show enantiomeric differences in the desorption temperature of 3.5 K, as shown on the right-hand side of **Figure 20**. This is in agreement with the DFT studies of Refs [79, 80], where the most stable adsorption sites were found to involve bonds to the kink Cu atoms.

Figure 19 CVs of Pt{643}^{S} and Pt{643}^{R} (top), Pt{321}^{S} and Pt{321}^{R} (middle), and Pt{531}^{S}, and Pt{531}^{R} (bottom) in 0.1 M H$_2$SO$_4$ + 5 × 10^{-3} M glucose. Sweep rate: 50 mV s^{-1}. As surface kink density increases, so too does electrochemical enantioselective excess. Reproduced with permission from Attard GA (2001) *Journal of Physical Chemistry B* 105: 3158.

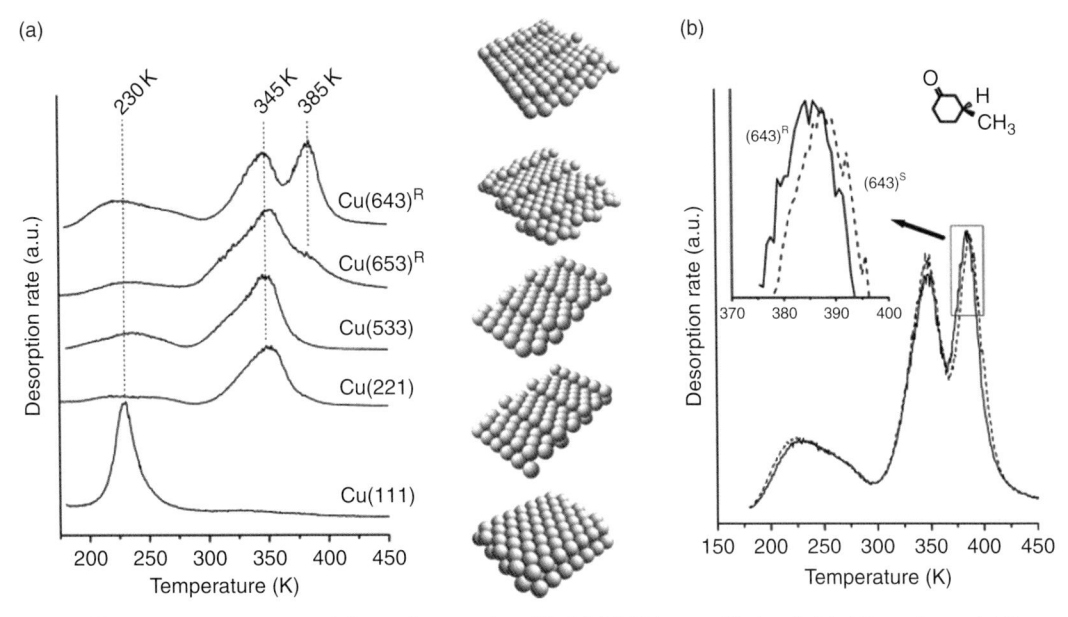

Figure 20 Temperature-programmed desorption spectra of R-3-MCHO from achiral and chiral Cu surfaces. (a) Desorption Cu{111}, Cu{221}, Cu{533}, Cu{653}^{R}, and Cu{643}^{R} surfaces. (b) Desorption of R-3-MCHO from the kinks on the Cu{643}^{R} and Cu{643}^{S} surfaces. The enantiospecific difference in the peak desorption temperatures is 3.5 ± 0.8 K. Reproduced with permission from Horvath JD, Koritnik A, Kamakoti P, Sholl DS, and Gellman AJ (2004) *Journal of the American Chemical Society* 126: 14988.

3.3.3 Effect of Chiral Templating/Modification on Achiral Surfaces

High degrees of enantioselectivity are found in desorption experiments when achiral surfaces are templated by chiral modifiers. Close-packed Pd{111} or Pt{111} surfaces modified by 2-butoxide show a preferential adsorption of PO of the same chirality [109, 110]. On both surfaces, up to 30% (Pt) and 100% (Pd) higher saturation coverages and lower decomposition rates were found for the combination (S/R)-2-butoxide / (S/R)-PO as compared to (S/R)-2-butoxide/(R/S)-PO. A similar degree of enantioselectivity is observed for 2-aminobutanoate on Pd{111}; however, if the amine group in this modifier is replaced by a methyl group, 2-methylbutanoate, all enantioselectivity is lost [57]. There are no indications that these modifiers induce major substrate surface reconstructions. According to DFT calculations, structures of the 2-butyl groups of the 2-butoxide and 2-methylbutanoate species are very similar, implying that conformational changes are not responsible for differences in enantioselectivity. It is also unlikely that differences in the modifier adsorption site play a decisive role [57]. The most likely explanation for this effect is that the 2-butyl group in 2-methylbutanoate species is less rigidly bonded to the surface than in 2-butoxides, allowing it to rotate more freely and, thus, wipe out the enantioselective effect. The effect of the amino group in 2-aminobutanoate species would be to inhibit azimuthal rotation by forming an additional bond to the surface. This points toward a predominantly steric interaction between the modifier and probe molecules. A rigid bond to the substrate, which inhibits azimuthal rotation, is therefore an important factor for the ability to template achiral surfaces [57, 111].

The two types of chirally modified surfaces that have been used successfully to catalyze enantioselective hydrogenation reactions (namely tartaric acid- or alanine-modified Ni and cinchonine/cinchonidine-modified Pt surfaces) [15, 16, 112, 113] cannot be studied by TPD directly because the reactants and products decompose on the relevant surfaces before they desorb.

3.4 Chiral Heterogeneous Catalysis

Despite extensive research over several decades, there remain relatively few examples of chiral heterogeneous catalytic reactions. The two most extensively researched examples are the Pt-catalyzed hydrogenation of α-keto esters (e.g., ethyl pyruvate, often referred to as the Orito reaction [114]) and the Ni-catalyzed hydrogenation of β-keto esters (e.g., methyl acetoacetate) [14, 15, 113]. These systems share a number of common features. In each case, the reagent is close to planar and can exist in either the di-keto or enol tautomeric forms. Adsorption with the molecular plane parallel to the surface results in a chiral adsorbate–surface system at the single molecule level (see Section 3.2.1). The mechanism of the hydrogenation reaction is thought to involve the attack by H atoms on the >C=O group (or C=C of the enol tautomeric form) from beneath the molecular plane. In each case, this attack results in the newly formed chiral center having a C—OH bond pointing away from the metal surface. Thus, the particular enantiotopic face of the prochiral reagent interacting with the metal determines the ultimate chirality of the product. Clearly, on an achiral metal surface, the adsorption energy of each enantioface will be identical and, as expected, the hydrogenation reaction proceeds racemically. This section examines the various ways in which one may achieve the crucial goal of this type of catalytic process – namely to restrict the adsorption geometry of the reagent to just one of the two possible enantioface adsorption states.

The key step in achieving enantioselective behavior is the adsorption, from solution, of chiral modifiers onto the metal surface. In the Pt-catalyzed system, the most successful modifiers are the cinchona alkaloids (e.g., cinchonidine). These modifiers are much larger molecules than those typically used in the Ni-catalyzed system, where α-hydroxy acids (e.g., (R,R)-tartaric acid) and α-amino acids give the best enantioselective performance [14, 15]. All modifiers for the Orito reaction have in common a double aromatic ring system that anchors the molecule firmly on the metal (Pt or Pd) surface [115–117]. This leaves the side group, which acts as the actual active site, free to interact with the reactant molecule. As will be explained below, there are various models to explain how an asymmetric environment is created for the hydrogenation reaction [112, 113].

In the 1990s, there was considerable debate as to the mechanism of chiral promotion. One proposed mechanism involved the formation of 2D porous arrays of chiral modifiers[15, 17]. The chiral shape of the pores could restrict approach to the metal surface by one enantioface while allowing ready access to the active site by the desired enantioface. As such the surface coverage of the reagent would be heavily skewed in favor of one enantioface geometry, and one would expect the reaction to occur enantioselectively. It should be noted that this template model does not require any bonding interaction between the modifier and the reagent. In the Pt system, for which the template model was originally proposed, it was well known that naphthalene (chemically very similar to the quinoline moiety of cinchonidine) formed ordered arrangements on Pt{111} [118]. However, LEED studies of dihydrocinchonidine on Pt{111} [119] and STM studies of the synthetic analogue of cinchonidine, S-1-(naphthyl)ethylamine on Pt{111} [115] failed to identify any ordered overlayer structures. By contrast, the adsorption, onto Cu{110}, of alanine [32] and tartaric acid [18] (two of the most successful modifiers in the Ni-catalyzed reaction) gave a wide range of ordered overlayer structures. It was initially proposed that a template model may operate in the Ni-catalyzed system. However, subsequent work on the more relevant Ni{111} [24] and {110} [23] surfaces revealed that tartaric acid has a much weaker tendency to form ordered structures on these surfaces.

The second proposed mechanism involved a direct one-to-one interaction between modifier and reagent and is often referred to as the active chiral site model [14, 113]. In this case, a direct H-bonding interaction is invoked between the adsorbed modifier and the reagent. In the Pt-catalyzed system, there is very strong evidence for such docking interactions and much of the debate is now over the absolute molecular conformation of the modifier and reactant in the complex. Since cinchonidine is a relatively bulky molecule, it is easier to understand how the reactant geometry is sterically restricted by the chiral modifier. The key interaction is thought to be an N–H–O interaction between the N atom of the quinuclidine moiety of the alkaloid and the carbonyl group of the α-keto ester.

For the Pt/cinchona/α-keto ester system, the main features of the enantioselective behavior can be explained by a model involving the adsorption of the alkaloid modifier via its aromatic component to the Pt surface and an intermolecular H bond between the N of the quinuclidine moiety and the ketone group of the reactant. The geometry of the aromatic ring system is thought to be crucial in optimizing the catalytic ee. Tilting of the plane of the ring system away from the surface is thought to have a detrimental effect and occurs both at high coverage for adsorption at 300 K [120] and upon annealing [121, 122].

Direct evidence of a 1:1 docking interaction has been provided by Lambert and coworkers using ultrahigh-vacuum (UHV) STM [115]. Considerable debate exists over the absolute structure of the modifier–reagent complex. For instance, the structure of the activated complex has been proposed to consist of a protonated cinchonidine interacting with the ketone group, which acts as an H-bond acceptor [113]. Alternatively, Wells and Wilkinson [123] proposed that there is an H-bonding interaction between the half-hydrogenated α-keto ester and the N atom of the quinuclidine moiety. The structure of the docking complex has been the subject of detailed computational analysis. Issues relate to the conformation of the alkaloid and the α-ketoester and the number and type of intermolecular H bonds involved in the stabilizing interaction between modifier and reagent. It has been proposed that the NH$^+$ of the quinuclidine can be involved in a bifurcating H-bonding interaction with both C=O functional groups [124] (**Figure 21**). Alternatively, a secondary H-bonding interaction between the C=O of the ester and the C—H of the aromatic ring has been invoked as a stabilizing factor favoring the adsorption of one enantiotopic face [125]. The rational optimization of the catalytic system is, to an extent, hindered by the lack of detailed understanding as to the structure of this complex.

In the Ni-catalyzed system, similar 1:1 interactions have been invoked. However, in this case, it is much less clear how the chiral modifier is able to restrict which enantiotopic face of the reagent interacts with the metal.

In the Pt/cinchona system, high enantioselectivities are facilitated by the fact that the enantioselective hydrogenation at modified sites has a significantly enhanced rate compared to the activity of the racemic reaction at unmodified sites [15, 113, 126]. In the Ni case, no such rate enhancement is observed and, consequently, the enantioselectivity is thought to be very strongly dependent on modifier coverage. In UHV studies, ordered arrangements of (R,R)-tartaric acid [24] or S-glutamic acid [127] are observed only at high coverage and coexist with dense disordered molecular arrangements. The strong tendency to form intermolecular H-bonding interactions dictates that there is very little available space to accommodate reactant molecules. Indeed, under UHV conditions, the sticking probability of MAA is close to zero on Ni surfaces pre-covered by a saturated monolayer of modifier [128, 129]. This supports the view that the template model is unlikely to be the mechanism of the enantioselective reaction on Ni. At intermediate coverages of modifier, MAA is able to chemisorb and is found to interact directly with the modifier [128, 129]. Interestingly, in the case of tartaric acid, MAA forms a supramolecular H-bonded network with tartaric acid with a 1:1 stoichiometry (**Figure 22**) [129]. From a catalytic viewpoint, the fascinating feature of this network is that each MAA species lies flat and with exclusively one enantioface attached to the surface. This enantioface adsorption geometry would favor the R-product in the hydrogenation reaction – that is, the dominant product observed in the catalytic reaction.

Clearly, it is important to establish the extent to which observations made under UHV conditions are transferable to the behavior of catalytic surfaces prepared and operating at the liquid–solid interface. Baddeley and coworkers have studied the adsorption

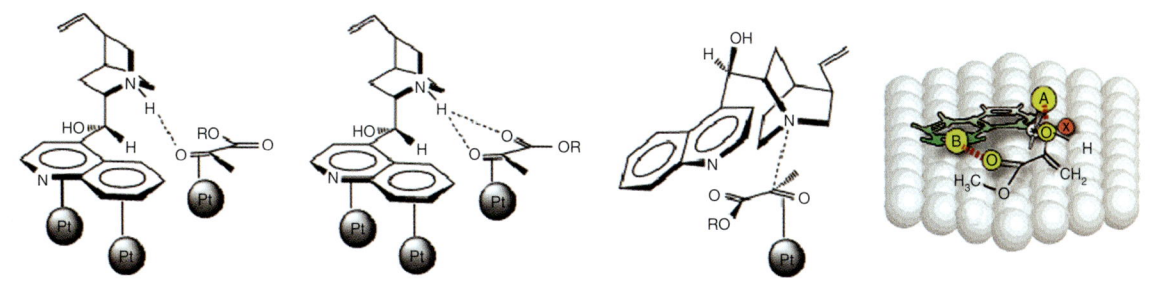

Figure 21 Proposals for the structures of activated complexes on Pt surfaces [124]. >From left to right: the cinchonidine adopts an open(3) conformation involved in an H bond with the α-ketoester in the *trans* configuration; the open(3) conformation involved in a bifurcating H-bonding interaction with the *cis* α-ketoester; and cinchonidine in the closed configuration. Adapted with permission from Studer M, Blaser HU, and Exner C (2003) *Advanced Synthesis and Catalysis* 345: 45. (Right) The proposed C—H—O interaction providing a two-point stabilizing contact between the α-ketoester and the alkaloid. Reproduced with permission from Lavoie S, Laliberte MA, Temprano I, and McBreen PH (2006) *Journal of the American Chemical Society* 128: 7588.

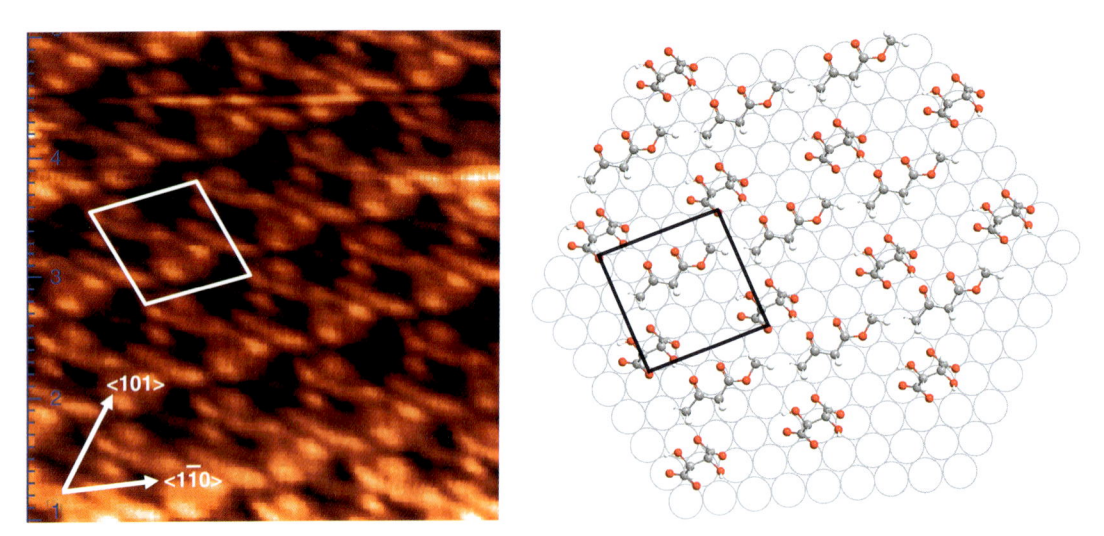

Figure 22 STM image (4 nm × 4 nm) showing the 2D co-crystalline structure consisting of an ordered array of 1:1 H-bonded complexes of (R,R)-tartrate and MAA species on Ni{111} giving a chiral (3 1 | −3 4) structure Adapted with permission from Jones TE and Baddeley CJ (2002) *Surface Science* 519: 237.

of various modifiers (tartaric acid [130], glutamic acid [131], and aspartic acid [132]) on Ni{111} surfaces by reflection absorption infrared spectroscopy (RAIRS) as functions of pH and modification temperature. In addition, they have studied the interaction of MAA with modified surfaces via adsorption from tetrahydrofuran solution. In the preparation of Ni catalysts, the catalyst is modified by treatment in an aqueous solution at a well-defined pH (normally controlled by addition of NaOH to the acidic solution). The catalyst is then washed to remove excess modifier and introduced to a reactor, where it is exposed to a solution of the β-ketoester in an organic solvent (typically MeOH or THF) and a high pressure of hydrogen gas. Several common features emerge from these studies. Adsorption of the modifier at 300 K results in adsorbed species, which are generally characteristic of the dominant solution phase species at the given pH. At low pH, the surface species are generally protonated (e.g., hydrogentartrate, HTA^-, or the zwitterionic or even cationic amino acids). At high pH, TA^{2-} or the anionic/dianionic amino acid species dominate. At higher modification temperature, there is an increased tendency to form nickel salts. When the modified surface is washed, most of the adsorbed modifier is removed from the surface. In the case of tartaric acid (the most effective modifier), the resulting surface maintains a significant modifier coverage. In contrast, washing aspartic acid-modified Ni{111} removes essentially all the modifier from the surface.

When MAA is adsorbed onto modified Ni surfaces, a striking correlation is observed between the tautomeric form of MAA and the pH/modification temperature conditions where catalytic enantioselectivity is optimized. In the case of Ni/tartrate, the catalyst works best at a modification temperature of 350 K and pH 5.0 [133]. Under these conditions, the subsequent adsorption of MAA is primarily in the diketone form [130]. Similarly, in the case of glutamic acid, low modification temperatures and pH 5 are optimal. Under these conditions, the diketone form dominates. Intriguingly, as the modification temperature is increased, glutamic acid modified Ni displays a switch from R-MHB as the dominant product to S-MHB [14]. This correlates closely with a switch in the preferred tautomeric form of MAA from diketone to enol [128, 131]. The preference for diketone versus enol seems to be related to the extent to which the modifier is protonated – for example, in the amino acid case, when the $-NH_3^+$ functionality is present, an H-bonding interaction can be established whereby the diketone accepts the proton from the modifier. In contrast, once the surface species is essentially a nickel amino acid salt, the NH_3^+ functionality is lost, and the dominant tautomeric form of MAA is the enol. It is interesting to note that It has been postulated that the contrasting behavior of Pd and Pt catalysts toward enantioselective hydrogenation of pyruvate esters is related to the tendency of the reactant molecule to exist in the diketone form (Pt) or the enol form (Pd). Modifying such catalysts with cinchonidine results in a tendency to produce R-lactate over Pt catalysts and S- lactate over Pd [17].

The case of Ni/aspartic acid displays interesting behavior. Under conditions where the catalyst operates most enantioselectively [134], the surface coverage of modifier is negligible [132]. This leads to the intriguing possibility that some contribution to the enantioselective behavior of the catalyst comes from unmodified surface sites, presumably in the form of chiral defects. Similar behavior has been proposed for the Pt/cinchona system by Attard and coworkers [135]. In their experiment, step and step–kink sites were selectively blocked by the adsorption of Bi. The enantioselectivity of the Pt/Bi catalysts decreased with increasing population of the defect sites, leading to the conclusion that near step sites contribute to the enantioselective behavior.

In summary, there appear to be several, possibly interrelated, factors determining enantioselective behavior in these hydrogenation reactions. While it seems unlikely that ordered arrangements of modifier define chiral docking sites for reactant molecules, it is possible that 2D supramolecular self-assembly could play a role in forming extended ordered arrangements of modifiers and reagents on metal surfaces. In such assemblies, the driving force for their formation is often the ability to form intermolecular H-bonding interactions. In the case of Pt/cinchona, the relatively large modifiers can envelope the reactant so as to favor adsorption of the reactant via one enantiotopic face. Supramolecular self-assembly (i.e., the ability to interact with several modifiers in an ordered structure) may be required to restrict the adsorption of β-ketoesters to one enantiotopic face in the Ni-catalyzed reaction. In both systems, indirect evidence suggests that chiral defects could contribute to the enantioselectivity of the catalytic reaction. It is patently clear that corrosion and leaching play a role in the Ni-catalyzed system, but it is not yet clear whether chiral metal arrangements can catalyze the enantioselective hydrogenation reaction in the absence of the chiral modifier.

3.5 Conclusions

The adsorption of chiral molecules can modify surfaces in many ways. Local chirality can induce the formation of long-range chiral structures, which can act as templates confining the adsorption geometries of other molecules – chiral or achiral – to certain orientations. Such templates are either based on the arrangement of molecules alone or involve reconstruction and/or faceting of the substrate surface. Stereo-direction can also be achieved by direct local interaction involving chiral molecules. Local chirality can even be induced via the adsorption of achiral molecules onto achiral surfaces due to a reduction of symmetry in the presence of the surface. Finally, intrinsically chiral metal and oxide surfaces can act as templates for enantioselective adsorption and surface reactions without any surface modification.

The ability of STM to probe surfaces on a local scale with atomic/molecular resolution has revolutionized the study of these phenomena. Surfaces, which are globally chiral either due to their intrinsic structure or due to the adsorption of chiral molecules have been shown by STM to establish control over the adsorption behavior of pro-chiral species. A deeper understanding of adsorption and reaction mechanisms requires probes for local bonds and atomic positions, which is provided, to some extent, by spectroscopy and diffraction experiments, such as NEXFAS, XPS, and RAIRS and LEED, PhD, and XPD, respectively. Most of the geometric information available to date at the atomic scale is, however, generated by first-principles DFT calculations.

A better understanding of chiral surface systems could have profound consequences in the development of new generations of heterogeneous chiral catalysts and may eventually make a substantial impact on the pharmaceutical industry. In a more fundamental context, chiral surface modifications or intrinsically chiral surfaces may also have played a decisive role in the origin of homochirality in life on Earth.

References

1. Cahn RS, Ingold CK, and Prelog V (1966) Specification of Molecular Chirality. *Angewandte Chemie-International Edition* 78: 385.
2. Knowles WS (2002) Asymmetric hydrogenations (Nobel lecture). *Angewandte Chemie-International Edition* 41: 1999.
3. Noyori R (2002) Asymmetric catalysis: Science and opportunities (Nobel lecture). *Angewandte Chemie-International Edition* 41: 2008.
4. Sharpless KB (2002) Searching for new reactivity (Nobel lecture). *Angewandte Chemie-International Edition* 41: 2024.
5. Hazen RM and Sholl DS (2003) Chiral selection on inorganic crystalline surfaces. *Nature Materials* 2: 367.
6. Barlow SM and Raval R (2003) Complex organic molecules at metal surfaces: bonding, organisation and chirality. *Surface Science Reports* 50: 201.

7. Jenkins SJ and Pratt SJ (2007) Beyond the surface atlas: A roadmap and gazetteer for surface symmetry and structure. *Surface Science Reports* 62: 373.
8. Ma Z and Zaera F (2006) Organic chemistry on solid surfaces. *Surface Science Reports* 61: 229.
9. James JN and Sholl DS (2008) Theoretical studies of chiral adsorption on solid surfaces. *Current Opinion in Colloid & Interface Science* 13: 60.
10. Barlow SM, Kitching KJ, Haq S, and Richardson NV (1998) A study of glycine adsorption on a Cu{110} surface using reflection absorption infrared spectroscopy. *Surface Science* 401: 322.
11. Booth NA, Woodruff DP, Schaff O, Giessel T, Lindsay R, Baumgartel P, and Bradshaw AM (1998) Determination of the local structure of glycine adsorbed on Cu(110). *Surface Science* 397: 258.
12. Chen Q, Frankel DJ, and Richardson NV (2002) Chemisorption induced chirality: glycine on Cu{110}. *Surface Science* 497: 37.
13. Rankin RB and Sholl DS (2004) Assessment of heterochiral and homochiral glycine adlayers on Cu(110) using density functional theory. *Surface Science* 548: 301.
14. Izumi Y (1983) Modified Raney-Nickel (Mrni) Catalyst - Heterogeneous Enantio- Differentiating (Asymmetric) Catalyst. *Advances in Catalysis* 32: 215.
15. Webb G and Wells PB (1992) Asymmetric Hydrogenation. *Catalysis Today* 12: 319.
16. Baddeley CJ (2003) Fundamental investigations of enantioselective heterogeneous catalysis. *Topics in Catalysis* 25: 17.
17. Sutherland IM, Ibbotson A, Moyes RB, and Wells PB (1990) Enantioselective Hydrogenation .1. Surface Conditions During Methyl Pyruvate Hydrogenation Catalyzed by Cinchonidine- Modified Platinum Silica (Europt-1). *Journal of Catalysis* 125: 77.
18. Lorenzo MO, Haq S, Bertrams T, Murray P, Raval R, and Baddeley CJ (1999) Creating chiral surfaces for enantioselective heterogeneous catalysis: R,R-Tartaric acid on Cu(110). *Journal of Physical Chemistry B* 103: 10661.
19. Fasel R, Wider J, Quitmann C, Ernst KH, and Greber T (2004) Determination of the absolute chirality of adsorbed molecules. *Angewandte Chemie-International Edition* 43: 2853.
20. Yan HJ, Wang D, Han MJ, Wan LJ, and Bai CL (2004) Adsorption and coordination of tartaric acid enantiomers on Cu(111) in aqueous solution. *Langmuir* 20: 7360.
21. Humblot V, Haq S, Muryn C, Hofer WA, and Raval R (2002) From local adsorption stresses to chiral surfaces: (R,R)-tartaric acid on Ni(110). *Journal of the American Chemical Society* 124: 503.
22. Humblot V, Lorenzo MO, Baddeley CJ, Haq S, and Raval R (2004) Local and global chirality at surfaces: Succinic acid versus tartaric acid on Cu(110). *Journal of the American Chemical Society* 126: 6460.
23. Humblot V, Haq S, Muryn C, and Raval R (2004) (R,R)-tartaric acid on Ni(110): The dynamic nature of chiral adsorption motifs. *Journal of Catalysis* 228: 130.
24. Jones TE and Baddeley CJ (2002) A RAIRS, STM and TPD study of the Ni{111}/R,R-tartaric acid system: Modelling the chiral modification of Ni nanoparticles. *Surface Science* 513: 453.
25. Lorenzo MO, Baddeley CJ, Muryn C, and Raval R (2000) Extended surface chirality from supramolecular assemblies of adsorbed chiral molecules. *Nature* 404: 376.
26. Barbosa L and Sautet P (2001) Stability of chiral domains produced by adsorption of tartaric acid isomers on the Cu(110) surface: A periodic density functional theory study. *Journal of the American Chemical Society* 123: 6639.
27. Hofer WA, Humblot V, and Raval R (2004) Conveying chirality onto the electronic structure of achiral metals: (R,R)-tartaric acid on nickel. *Surface Science* 554: 141.
28. Hermse CGM, van Bavel AP, Jansen APJ, Barbosa L, Sautet P, and van Santen RA (2004) Formation of chiral domains for tartaric acid on Cu(110): A combined DFT and kinetic Monte Carlo study. *Journal of Physical Chemistry B* 108: 11035.
29. Williams J, Haq S, and Raval R (1996) The bonding and orientation of the amino acid L-alanine on Cu{110} determined by RAIRS. *Surface Science* 368: 303.
30. Chen Q, Lee CW, Frankel DJ, and Richardson NV (1999) The formation of enantiospecific phases on a Cu{110} surface. *Phys. Chem. Comm.* 2: 41.
31. Barlow SM, Louafi S, Le Roux D, Williams J, Muryn C, Haq S, and Raval R (2004) Supramolecular assembly of strongly chemisorbed size-and shape- defined chiral clusters: S- and R-alanine on Cu(110). *Langmuir* 20: 7171.
32. Barlow SM, Louafi S, Le Roux D, Williams J, Muryn C, Haq S, and Raval R (2005) Polymorphism in supramolecular chiral structures of R- and S-alanine on Cu(110). *Surface Science* 590: 243.
33. Polcik M, Allegretti F, Sayago DI, Nisbet G, Lamont CLA, and Woodruff DP (2004) Circular dichroism in core level photoemission from an adsorbed chiral molecule. *Physical Review Letters* 92.
34. Sayago DI, Polcik M, Nisbet G, Lamont CLA, and Woodruff DP (2005) Local structure determination of a chiral adsorbate: Alanine on Cu(110). *Surface Science* 590: 76.
35. Jones G, Jones LB, Thibault-Starzyk F, Seddon EA, Raval R, Jenkins SJ, and Held G (2006) The local adsorption geometry and electronic structure of alanine on Cu{110}. *Surface Science* 600: 1924.
36. Rankin RB and Sholl DS (2005) Structure of enantiopure and racemic alanine adlayers on Cu(110). *Surface Science* 574: L1.
37. Rankin RB and Sholl DS (2005) Structures of glycine, enantiopure alanine, and racemic alanine adlayers on Cu(110) and Cu(100) surfaces. *Journal of Physical Chemistry B* 109: 16764.
38. Jones G, Jenkins SJ, and King DA (2006) Hydrogen bonds at metal surfaces: Universal scaling and quantification of substrate effects. *Surface Science* 600: L224.
39. Kang JH, Toomes RL, Polcik M, Kittel M, Hoeft JT, Efstathiou V, Woodruff DP, and Bradshaw AM (2003) Structural investigation of glycine on Cu(100) and comparison to glycine on Cu(110). *Journal of Chemical Physics* 118: 6059.
40. Toomes RL, Kang JH, Woodruff DP, Polcik M, Kittel M, and Hoeft JT (2003) Can glycine form homochiral structural domains on low-index copper surfaces?. *Surface Science* 522: L9.
41. Blankenburg S and Schmidt WG (2006) Adsorption of phenylglycine on copper: Density functional calculations. *Physical Review B* 74.
42. Blankenburg S and Schmidt WG (2007) Long-range chiral recognition due to substrate locking and substrate-adsorbate charge transfer. *Physical Review Letters* 99.
43. Zhao XY, Zhao RG, and Yang WS (1999) Adsorption of alanine on Cu(001) studied by scanning tunneling microscopy. *Surface Science* 442: L995.
44. Zhao XY, Gai Z, Zhao RG, Yang WS, and Sakurai T (1999) Adsorption of glycine on Cu(001) and related step faceting and bunching. *Surface Science* 424: L347.

45. Iwai H, Tobisawa M, Emori A, and Egawa C (2005) STM study of D-alanine adsorption on Cu(001). *Surface Science* 574: 214.
46. Humblot V, Methivier C, Raval R, and Pradier CM (2007) Amino acid and peptides on Cu(110) surfaces: Chemical and structural analyses of L-lysine. *Surface Science* 601: 4189.
47. Lofgren P, Krozer A, Lausmaa J, and Kasemo B (1997) Glycine on Pt(111): A TDS and XPS study. *Surface Science* 370: 277.
48. Ernst KH and Christmann K (1989) The interaction of glycine with a platinum (111) Surface. *Surface Science* 224: 277.
49. Gao F, Li ZJ, Wang YL, Burkholder L, and Tysoe WT (2007) Chemistry of Alanine on Pd(111): Temperature-programmed desorption and X-ray photoelectron spectroscopic study. *Surface Science* 601: 3276.
50. Gao F, Li ZJ, Wang YL, Burkholder L, and Tysoe WT (2007) Chemistry of glycine on Pd(111): Temperature-programmed desorption and X-ray photoelectron spectroscopic study. *Journal of Physical Chemistry C* 111: 9981.
51. James JN and Sholl DS (2008) Density Functional Theory studies of dehydrogenated and zwitterionic glycine and alanine on Pd and Cu surfaces. *Journal of Molecular Catalysis a-Chemical* 281: 44.
52. Kuhnle A, Linderoth TR, Hammer B, and Besenbacher F (2002) Chiral recognition in dimerization of adsorbed cysteine observed by scanning tunnelling microscopy. *Nature* 415: 891.
53. Kuhnle A, Linderoth TR, and Besenbacher F (2006) Enantiospecific adsorption of cysteine at chiral kink sites on Au(110)-(1x2). *Journal of the American Chemical Society* 128: 1076.
54. Kuhnle A, Linderoth TR, and Besenbacher F (2003) Self-assembly of monodispersed, chiral nanoclusters of cysteine on the Au(110)-(1 x 2) surface. . *Journal of the American Chemical Society* 125: 14680.
55. Gao F, Kotvis PV, Stacchiola D, and Tysoe WT (2005) Reaction of tributyl phosphate with oxidized iron: Surface chemistry and tribological significance. *Tribology Letters* 18: 377.
56. Irrera S and Costa D (2008) New insight brought by density functional theory on the chemical state of alaninol on Cu(100): Energetics and interpretation of x-ray photoelectron spectroscopy data. *Journal of Chemical Physics* 128.
57. Stacchiola D, Burkholder L, Zheng T, Weinert M, and Tysoe WT (2005) Requirements for the formation of a chiral template. *Journal of Physical Chemistry B* 109: 851.
58. Bowker M, Poulston S, Bennett RA, and Stone P (1998) Gross rearrangement of metal atoms during surface reactions. *Journal of Physics-Condensed Matter* 10: 7713.
59. Chen Q, Perry CC, Frederick BG, Murray PW, Haq S, and Richardson NV (2000) Structural aspects of the low-temperature deprotonation of benzoic acid on Cu(110) surfaces. *Surface Science* 446: 63.
60. Chen Q, Frankel DJ, and Richardson NV (2001) Organic adsorbate induced surface reconstruction: p-aminobenzoic acid on Cu{110}. *Langmuir* 17: 8276.
61. Chen Q and Richardson NV (2003) Surface facetting induced by adsorbates. *Progress in Surface Science* 73: 59.
62. Zhao XY, Perry SS, Horvath JD, and Gellman AJ (2004) Adsorbate induced kink formation in straight step edges on Cu(533) and Cu(221). *Surface Science* 563: 217.
63. Zhao XY and Perry SS (2004) Ordered adsorption of ketones on Cu(643) revealed by scanning tunneling microscopy. *Journal of Molecular Catalysis a-Chemical* 216: 257.
64. Pascual JI, Barth JV, Ceballos G, Trimarchi G, De Vita A, Kern K, and Rust HP (2004) Mesoscopic chiral reshaping of the Ag(110) surface induced by the organic molecule PVBA. *Journal of Chemical Physics* 120: 11367.
65. Zhao XY (2000) Fabricating homochiral facets on Cu(001) with L-lysine. . *Journal of the American Chemical Society* 122: 12584.
66. Rankin RB and Sholl DS (2006) Structures of dense glycine and alanine adlayers on chiral Cu(3,1,17) surfaces. *Langmuir* 22: 8096.
67. Rankin RB and Sholl DS (2006) First-principles studies of chiral step reconstructions of Cu(100) by adsorbed glycine and alanine. *Journal of Chemical Physics* 124.
68. Parschau M, Romer S, and Ernst KH (2004) Induction of homochirality in achiral enantiomorphous monolayers. *Journal of the American Chemical Society* 126: 15398.
69. Green MM, Reidy MP, Johnson RJ, Darling G, Oleary DJ, and Willson G (1989) Macromolecular Stereochemistry - the out-of-Proportion Influence of Optically-Active Co-Monomers on the Conformational Characteristics of Polyisocyanates - the Sergeants and Soldiers Experiment. *Journal of the American Chemical Society* 111: 6452.
70. Chen Q, Frankel DJ, and Richardson NV (2002) Self-assembly of adenine on Cu(110) surfaces. *Langmuir* 18: 3219.
71. Chen Q and Richardson NV (2003) Enantiomeric interactions between nucleic acid bases and amino acids on solid surfaces. *Nature Materials* 2: 324.
72. Schwab GM and Rudolph L (1932) Katalytische Spaltung von Racematen durch Rechts- Linksquarz. *Z. Naturwiss* 20: 363.
73. McFadden CF, Cremer PS, and Gellman AJ (1996) Adsorption of chiral alcohols on "chiral" metal surfaces. *Langmuir* 12: 2483.
74. Soai K, Osanai S, Kadowaki K, Yonekubo S, Shibata T, and Sato I (1999) d- and l-quartz-promoted highly enantioselective synthesis of a chiral organic compound. *Journal of the American Chemical Society* 121: 11235.
75. Attard GA, Ahmadi A, Feliu J, Rodes A, Herrero E, Blais S, and Jerkiewicz G (1999) Temperature effects in the enantiomeric electro-oxidation of D- and L-glucose on Pt{643}(S). *Journal of Physical Chemistry B* 103: 1381.
76. Attard GA (2001) Electrochemical studies of enantioselectivity at chiral metal surfaces. *Journal of Physical Chemistry B* 105: 3158.
77. Hazen RM, Filley TR, and Goodfriend GA (2001) Selective adsorption of L- and D-amino acids on calcite: Implications for biochemical homochirality. *Proceedings of the National Academy of Sciences of the United States of America* 98: 5487.
78. Horvath JD, Koritnik A, Kamakoti P, Sholl DS, and Gellman AJ (2004) Enantioselective separation on a naturally chiral surface. *Journal of the American Chemical Society* 126: 14988.
79. Bhatia B and Sholl DS (2005) Enantiospecific chemisorption of small molecules on intrinsically chiral Cu surfaces. *Angewandte Chemie-International Edition* 44: 7761.
80. Bhatia B and Sholl DS (2008) Characterization of enantiospecific chemisorption on chiral Cu surfaces vicinal to Cu(111) and Cu(100) using density functional theory. *Journal of Chemical Physics* 128: 144709.
81. Held G and Gladys MJ (2008) The chemistry of intrinsically chiral surfaces. *Topics in Catalysis* 48: 128.
82. Horvath JD and Gellman AJ (2001) Enantiospecific desorption of R- and S-propylene oxide from a chiral

Cu(643) surface. *Journal of the American Chemical Society* 123: 7953.
83. Goumans TPM, Wander A, Brown WA, and Catlow CRA (2007) Structure and stability of the (001) alpha-quartz surface. *Physical Chemistry Chemical Physics* 9: 2146.
84. Han JW, James JN, and Sholl DS (2008) First principles calculations of methylamine and methanol adsorption on hydroxylated quartz (0001). *Surface Science* 602: 2478.
85. Gladys MJ, Stevens AV, Scott NR, Jones G, Batchelor D, and Held G (2007) Enantiospecific adsorption of alanine on the chiral Cu{531} surface. *Journal of Physical Chemistry C* 111: 8331.
86. Pratt SJ, Jenkins SJ, and King DA (2005) The symmetry and structure of crystalline surfaces. *Surface Science* 585: L159.
87. Puisto SR, Held G, Ranea V, Jenkins SJ, Mola EE, and King AA (2005) The structure of the chiral Pt{531} surface: A combined LEED and DFT study. *Journal of Physical Chemistry B* 109: 22456.
88. Puisto SR, Held G, and King DA (2005) Energy-dependent cancellation of diffraction spots due to surface roughening. *Physical Review Letters* 95.
89. Jones G, Gladys MJ, Ottal J, Jenkins SJ, and Held G (2009) Surface geometry of Cu{531}. *Physical Review B* 79: 165420.
90. Power TD, Asthagiri A, and Sholl DS (2002) Atomically detailed models of the effect of thermal roughening on the enantiospecificity of naturally chiral platinum surfaces. *Langmuir* 18: 3737.
91. Asthagiri A, Feibelman PJ, and Sholl DS (2002) Thermal fluctuations in the structure of naturally chiral Pt surfaces. *Topics in Catalysis* 18: 193.
92. Mehmood F, Kara A, and Rahman TS (2006) First principles study of the electronic and geometric structure of Cu(532). *Surface Science* 600: 4501.
93. Giesen M and Dieluweit S (2004) Step dynamics and step-step interactions on the chiral Cu(5890) surface. *Journal of Molecular Catalysis a-Chemical* 216: 263.
94. Horvath JD and Gellman AJ (2002) Enantiospecific desorption of chiral compounds from chiral Cu(643) and achiral Cu(111) surfaces. *Journal of the American Chemical Society* 124: 2384.
95. Horvath JD and Gellman AJ (2003) Naturally chiral surfaces. *Topics in Catalysis* 25: 9.
96. Huang Y and Gellman AJ (2008) Enantiospecific adsorption of (R)-3-Methylcyclohexanone on naturally chiral Cu(531)(R&S) surfaces. *Catalysis Letters* 125: 177.
97. Greber T, Sljivancanin Z, Schillinger R, Wider J, and Hammer B (2006) Chiral recognition of organic molecules by atomic kinks on surfaces. *Physical Review Letters* 96: 056103.
98. Thomsen L, Wharmby M, Riley DP, Held G, and Gladys MJ (2009) The adsorption and stability of sulfur containing amino acids on Cu{5 3 1}. *Surface Science* 603: 1253.
99. Eralp T and Shavorskiy A, Held, in preparation.
100. Stohr J (1996) NEXAFS Spectroscopy, 2nd edn. Berlin: Springer.
101. Hasselstrom J, Karis O, Weinelt M, Wassdahl N, Nilsson A, Nyberg M, Pettersson LGM, Samant MG, and Stohr J (1998) The adsorption structure of glycine adsorbed on Cu(110); comparison with formate and acetate/Cu(110). *Surface Science* 407: 221.
102. Nyberg M, Hasselstrom J, Karis O, Wassdahl N, Weinelt M, Nilsson A, and Pettersson LGM (2000) The electronic structure and surface chemistry of glycine adsorbed on Cu(110). *Journal of Chemical Physics* 112: 5420.
103. Nyberg M, Odelius M, Nilsson A, and Pettersson LGM (2003) Hydrogen bonding between adsorbed deprotonated glycine molecules on Cu(110). *Journal of Chemical Physics* 119: 12577.
104. Easson LH and Stedman E (1933) Studies on the relationship between chemical constitution and physiological action. *Journal of Biochemistry* 27: 1257.
105. Sljivancanin Z, Gothelf KV, and Hammer B (2002) Density functional theory study of enantiospecific adsorption at chiral surfaces. *Journal of the American Chemical Society* 124: 14789.
106. Jones TE and Baddeley CJ (2004) An investigation of the adsorption of (R,R)-tartaric acid on oxidised Ni{111} surfaces. *Journal of Molecular Catalysis a-Chemical* 216: 223.
107. Ahmadi A, Attard G, Feliu J, and Rodes A (1999) Surface reactivity at "chiral" platinum surfaces. *Langmuir* 15: 2420.
108. Gellman AJ, Horvath JD, and Buelow MT (2001) Chiral single crystal surface chemistry. *Journal of Molecular Catalysis a-Chemical* 167: 3.
109. Lee I and Zaera F (2005) Enantioselectivity of adsorption sites created by chiral 2-butanol adsorbed on Pt(111) single-crystal surfaces. *Journal of Physical Chemistry B* 109: 12920.
110. Stacchiola D, Burkholder L, and Tysoe WT (2002) Enantioselective chemisorption on a chirally modified surface in ultrahigh vacuum: Adsorption of propylene oxide on 2-butoxide-covered palladium(111). *Journal of the American Chemical Society* 124: 8984.
111. Ma Z and Zaera F (2006) Competitive chemisorption between pairs of cinchona alkaloids and related compounds from solution onto platinum surfaces. *Journal of the American Chemical Society* 128: 16414.
112. Baiker A (1997) Progress in asymmetric heterogeneous catalysis: Design of novel chirally modified platinum metal catalysts. *Journal of Molecular Catalysis a-Chemical* 115: 473.
113. Baiker A (2000) Transition state analogues - a guide for the rational design of enantioselective heterogeneous hydrogenation catalysts. *Journal of Molecular Catalysis a-Chemical* 163: 205.
114. Orito Y, Imai S, and Niwa S (1979) Asymmetric hydrogenation of methyl pyruvate using Pt-C catalyst modified with cinchonidine. *Nippon Kagaku Kaishi*:1118.
115. Bonello JM, Williams FJ, and Lambert RM (2003) Aspects of enantioselective heterogeneous catalysis: structure and reactivity of (S)-(-)-1-(1-naphthyl)ethylamine on Pt{111}. *Journal of the American Chemical Society* 125: 2723.
116. Bonello JM and Lambert RM (2002) The structure and reactivity of quinoline overlayers and the adsorption geometry of lepidine on Pt{111}: model molecules for chiral modifiers in enantioselective hydrogenation. *Surface Science* 498: 212.
117. Bonello JM, Sykes ECH, Lindsay R, Williams FJ, Santra AK, and Lambert RM (2001) Fundamental aspects of enantioselective heterogeneous catalysis: a NEXAFS study of methyl pyruvate and (S)-(-)-1-(1- naphthyl) ethylamine on Pt{111}. *Surface Science* 482: 207.
118. Hallmark VM, Chiang S, Brown JK, and Woll C (1991) Real-space imaging of the molecular organization of Naphtalene on Pt(111). *Physical Review Letters* 66: 48.
119. Carley AF, Rajumon MK, Roberts MW, and Wells PB (1995) XPS and LEED Studies of 10,11-Dihydrocinchonidine Adsorption at Pt(111) - Implications for the Role of Cinchona Alkaloids in Enantioselective Hydrogenation. *Journal of the Chemical Society-Faraday Transactions* 91: 2167.
120. Kubota J and Zaera F (2001) Adsorption geometry of modifiers as key in imparting chirality to platinum

catalysts. *Journal of the American Chemical Society* 123: 11115.
121. Bonello JM, Lindsay R, Santra AK, and Lambert RM (2002) On the orientation of quinoline on Pd{111}: Implications for heterogeneous enantioselective hydrogenation. *Journal of Physical Chemistry B* 106: 2672.
122. Evans T, Woodhead AP, Gutierrez-Sosa A, Thornton G, Hall TJ, Davis AA, Young NA, Wells PB, Oldman RJ, Plashkevych O, Vahtras O, Agren H, and Carravetta V (1999) Orientation of 10,11-dihydrocinchonidine on Pt(111). *Surface Science* 436: L691.
123. Wells PB and Wilkinson AG (1998) Platinum group metals as heterogeneous enantioselective catalysts. *Topics in Catalysis* 5: 39.
124. Burgi T and Baiker A (2000) Model for enantioselective hydrogenation of alpha-ketoesters over chirally modified platinum revisited: Influence of alpha- ketoester conformation. *Journal of Catalysis* 194: 445.
125. Lavoie S, Laliberte MA, Temprano I, and McBreen PH (2006) A generalized two-point H-bonding model for catalytic stereoselective hydrogenation of activated ketones on chirally modified platinum. *Journal of the American Chemical Society* 128: 7588.
126. Studer M, Blaser HU, and Exner C (2003) Enantioselective hydrogenation using heterogeneous modified catalysts: An update. *Advanced Synthesis & Catalysis* 345: 45.
127. Jones TE, Urquhart ME, and Baddeley CJ (2005) An investigation of the influence of temperature on the adsorption of the chiral modifier, (S)-glutamic acid, on Ni{111}. *Surface Science* 587: 69.
128. Jones TE and Baddeley CJ (2006) Investigating the mechanism of chiral surface reactions: The interaction of methylacetoacetate with (S)-glutamic acid modified Ni{111}. *Langmuir* 22: 148.
129. Jones TE and Baddeley CJ (2002) Direct STM evidence of a surface interaction between chiral modifier and pro-chiral reagent: Methylacetoacetate on R,R- tartaric acid modified Ni{111}. *Surface Science* 519: 237.
130. Jones TE and Baddeley CJ (2007) Influence of modification conditions on the interaction of methylacetoacetate with (R,R)-Tartaric acid-modified Ni{111}. *Journal of Physical Chemistry C* 111: 17558.
131. Jones TE, Rekatas AE, and Baddeley CJ (2007) Influence of modification pH and temperature on the interaction of methylacetoacetate with (S)-glutamic acid-modified Ni{111}. *Journal of Physical Chemistry C* 111: 5500.
132. Wilson KE and Baddeley CJ (2009) Understanding the Surface Chemistry of Enantioselective Heterogeneous Reactions: Influence of Modification Variables on the Interaction of Methylacetoacetate with (S)-Aspartic Acid Modified Ni{111}. *Journal of Physical Chemistry C* 113: 10706.
133. Keane MA and Webb G (1992) The Enantioselective Hydrogenation of Methyl Acetoacetate over Supported Nickel-Catalysts .1. The Modification Procedure. *Journal of Catalysis* 136: 1.
134. Izumi Y, Imaida M, Fukuwa H, and Akabori S (1963) Asymmetric Hydrogenation of C=O Double Bond with Modified Raney Nickel. *Bulletin of the Chemical Society of Japan* 42: 2373.
135. Jenkins DJ, Alabdulrahman AMS, Attard GA, Griffin KG, Johnston P, and Wells PB (2005) Enantioselectivity and catalyst morphology: step and terrace site contributions to rate and enantiomeric excess in Pt-catalysed ethyl pyruvate hydrogenation. *Journal of Catalysis* 234: 230.

4 Electron Beam Lithography of Nanostructures

D M Tennant and A R Bleier, Cornell University, Ithaca, NY, USA

© 2010 Elsevier B.V. All rights reserved.

4.1 Basics of Electron Beam Lithography

4.1.1 Introduction

We have all been amazed at some point by a story of a craftsman who works on the repair of tiny watch mechanisms or an artist who carves figures that require a magnifying glass to observe. But what if our aim is to create structures so small that even the best optics cannot bring them into view? This article will describe the ways that researchers in laboratories around the world are creating such structures on a scale down to 1/100 of the wavelength of a visible light photon – more than a hundred thousand times smaller than the watchmaker's gears! While we are not concentrating on the smallest structures ever made (for that see Eigler and Schweizer [1]), they are very nearly so. Likened to the great explorers [2], these nanofabrication experts routinely visit this small-size scale everyday in laboratories at universities and in industry. One of the key instruments in the researcher's tool box to achieve this goal is a very expensive one – the electron beam (or e-beam) lithography system.

In 1985, a Stanford University graduate student, Tom Newman, won a challenge issued in 1959 by Richard Feynman in his famous talk, 'There's Plenty of Room at the Bottom' [3]. To earn the prize, he needed to print a book page in an area 25 000 times smaller than standard print size. At that reduced size, one could, Feynman reasoned, print the *Encyclopedia Britannica* on the head of a pin. Newman claimed the prize by writing the first page of Charles Dickens' *A Tale of Two Cities* in a 6 µm square area [4]. How? By using electron beam lithography (see **Figure 1**).

Electron beam lithography (EBL), as you now gather, is a very precise and high-resolution, direct-write, method for forming patterns in a layer of material. Direct-write lithography distinguishes itself from a wide variety of other methods, in that it allows patterns to be generated from a data file. It is, therefore, a primary method of patterning rather than a replication method. EBL, in its most common form, uses hardware not unlike a scanning electron microscope (SEM) to guide a tiny focused beam of electrons to form a latent image in a thin layer of polymer known as a resist. The result of this exposure is to render the resist either more soluble (called a positive tone resist) or less soluble (negative tone resist) in an appropriate developer solution. The resulting pattern is then transferred via etching or by depositing other materials. By iterating a number of steps of this type, complex structures of very short length scales can be built up.

In the limited space of this article, we will attempt to describe: the basic elements of EBL; discuss why we believe the field has evolved to using high-voltage beams; survey the techniques and materials that are currently popular; and demonstrate the wide impact that the field has had on nanostructure research. There are several earlier book chapters and whole volumes that have been devoted to these various topics, so we will attempt to provide ample references to allow the reader to learn more about selected areas of interest.

4.1.2 E-Beam Tools

4.1.2.1 Elements of an EBL system

Electrons are very light (low-mass) charged particles. This is the root of the love–hate relationship between electrons and lithography engineers. Their low mass and relatively high charge-to-mass ratio allows them to be focused and steered with modest magnetic and electric fields. The low mass also permits them to penetrate reasonably thick layers of material without displacing their heavier atomic cousins and without leaving behind foreign elements. They just enter, deposit their energy, then wander off to some grounded part of the wafer and exit (maybe), leaving behind a squeaky clean, exposed but otherwise unperturbed, substrate. Most lithography is done using ultraviolet light, that is, photons. When we compare electrons to photons as our actinic exposure method, the effects of diffraction again make electrons attractive for lithography. The range of photon wavelengths used in current optical lithography systems is typically 193–436 nm. This gives rise to diffractive artifacts on length scales similar to the wavelength. In contrast, the wavelength of a 100 keV electron is just about 4 pm (picometer), many orders of magnitude smaller than the size of

Figure 1 The first page of Charles Dickens' *A Tale of Two Cities* written inside a 6 μm square by EBL. With permission from Newman TH, Williams KE, and Pease RFW (1987) High resolution patterning system with a single bore objective lens. *Journal of Vacuum Science and Technology B: Microelectronics and Nanometer Structures* 5(1): 88–91.

the patterns being produced! These are among the attributes we like about electrons. Now the dark side – electrons are charged particles with low mass. These same properties also mean that the entire optical system must be held in vacuum to prevent electron collisions with gas molecules.

Any small amount of charging in the optical column will nudge the beam in undesirable (and unexpected) directions. Since the mean free path in solids is rather meager compared to photons (in refractive materials such as glass), only electrostatic or magnetic focusing lenses can be employed. The best magnetic electron lens has been likened to using the bottom of a glass Coke bottle for light optics [5]. And last, electrons undergo both small-angle (inelastic or forward scatter) and large-angle (elastic or backscatter) scattering events, once they enter the resist and substrate. These properties give rise to exposure in the resist well beyond the place where the e-beam was incident. These effects accrue as the beam is moved from dot to dot to form the desired exposure shape. This is illustrated in **Figure 2**. A result of this scattering is unwanted exposure in areas surrounding pattern features known as the 'proximity effect'.

If the incident beam has a Gaussian-shaped intensity cross section, one might expect the energy deposited in a resist layer on a wafer to have a radial dependence represented by the dashed (red) line in **Figure 3**.

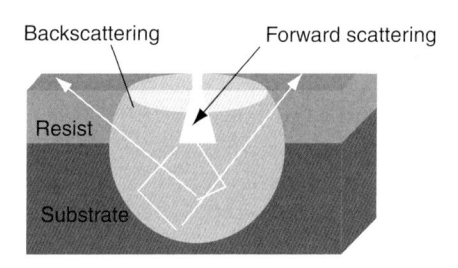

Figure 2 Illustration of the short-range (forward-scatter) and long-range (backscatter) interaction of electrons with resist-coated wafer.

However, due to scattering, there is significant departure from ideal. In effect, the actual exposure is the convolution of the designed pattern and the point spread function (blue line in **Figure 3**). This can require a significant computational effort to correct. Hence, the love–hate relationship. Section 4.3 examines the methods and software solutions for correcting exposures. Mindful of these pros and cons, we push ahead with our description of the basic elements of electron beam exposure systems, and illuminate the evolution of the commercial EBL market.

A simple e-beam column is shown in **Figure 4**. Starting at the top it consists of an electron source (or gun), some probe forming optics, a blanker (that can rapidly turn the beam on and off), a condenser lens to allow changes in the current and corresponding beam diameter, an objective lens to focus the beam on the wafer, and a deflector to scan the

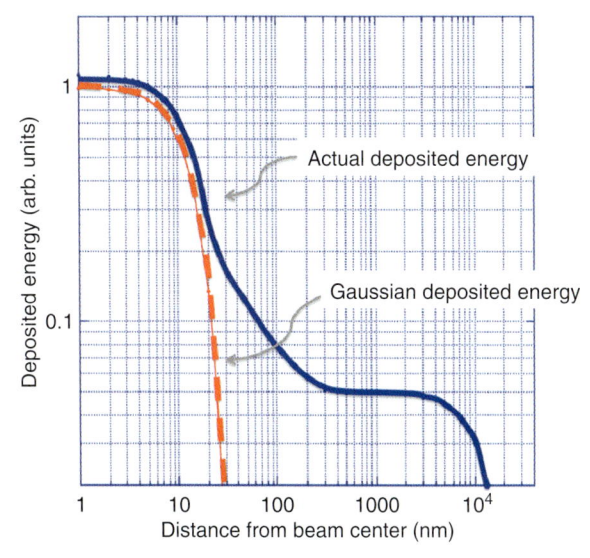

Figure 3 Comparison of a Gaussian exposure (dashed line) and the actual energy deposited in a thin resist on a bulk substrate. The added long-range exposure is due to electron scattering.

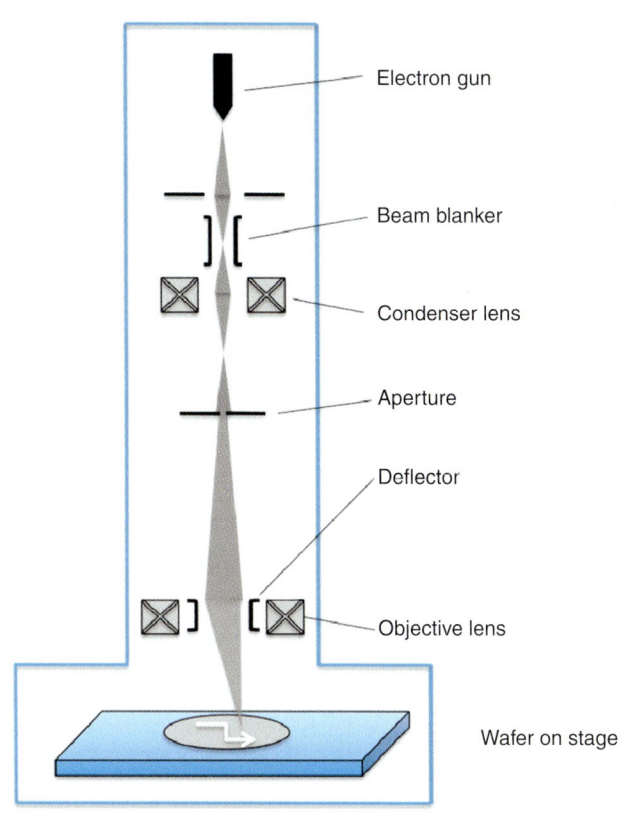

Figure 4 Schematic of a simple electron optical column.

e-beam around within the field. The wafer is placed below on a motorized stage so that it can be accurately positioned and the various calibrations can be performed. More detailed descriptions of the different types of sources, lenses, resolution limits, and the various other components can be found elsewhere [6].

4.1.2.2 Modern electron sources

Ironically, the field emission sources were made practical because the SEM industry realized that extremely low voltage imaging produced images with more surface detail, less charging, and less heating. Yet operating the older thermionic cathodes at low voltage yielded very low current density and, therefore, insufficient signal to form quality images. Rapid development of both cold field emission (CFE) and thermally assisted field emission (TFE) for low voltage SEMs followed, and were soon found to be equally attractive for high voltage applications. The basic idea is simple: a voltage applied to a pointy conductor concentrates the electric field. The Schottky equation describes the lowering of the potential barrier due to the field. Sufficient voltage thereby removes electrons from the tip. The picture gets a bit more complicated because lithography applications pose the added requirement that the current must be extremely stable so that exposure dose is well controlled. Therefore, any flicker in the source would ruin the result. This meant that the thermal field emitter or, more correctly, the Schottky emitter operated in the 'extended Schottky regime', became the gun type most favored for EBL [7]. The most common such tip is ZrO/W that is heated to about 1800 K. The emitter consists of a sharpened crystalline tungsten tip with a (100) emitting plane that selectively receives a surface coating of zirconia by diffusion from a resevoir. The addition of the ZrO layer lowers the work function of the W (100) surface, the high temperature prevents contaminants from adsorbing on the surface and, in the presence of a high electric field, supplies a very stable high brightness source of electrons. Like a fine culinary recipe, the right mixture of electric field, temperature, work function, and crystallinity have been employed to achieve nearly the same brightness observed in CFE sources. The development of the high brightness ZrO/W electron emitter has been a significant engineering achievement.

4.1.2.3 More advanced electron optics

In practice, the electron optics can be much more complex than shown in **Figure 4**. **Figure 5** shows a more complete column, as might be found on a modern dedicated EBL system. The additional elements play a subtle role in improving features and performance. Again working top down, the simple thermionic electron gun is replaced with field emission source to supply a much higher current density.

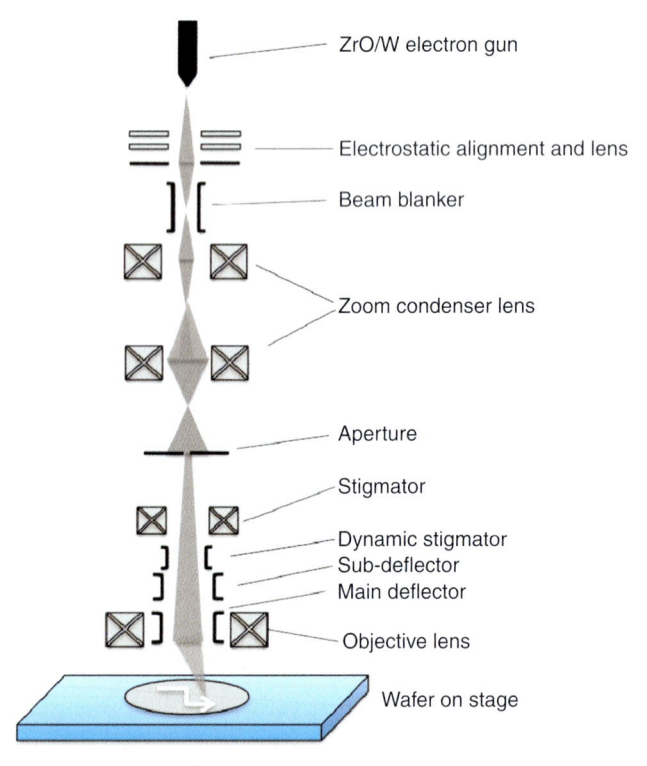

Figure 5 Schematic of a modern electron optical column.

Working further down from the source, we come to the gun alignment part of the column. The alignment electrodes add more knobs to tilt and steer the beam down the axis of the column. Next, we find a zoom condenser that replaces the single lens version to allow the user to change the beam current without appreciably changing the focal plane. The stigmator helps make the beam round, hence producing equal linewidths along various scanned directions. The dynamic focus and stigmation allow for real-time corrections as the system detects height variations during the exposure. The deflector, in practice, consists of multiple stages. This is done to take advantage of the high-speed (sub-field) scanning achievable with digital-to-analog converter (DAC) electronics when the address contains fewer bits (e.g., 12 bits). Then the large-field (main field) scanning is achieved by tiling together sub-fields with a high-precision DAC with larger (and slower) addressing (e.g., 20 bits or greater). Using the DAC sizes above and a 1 nm address grid, we see that the maximum sub-field' size is typically 2^{12} nm or about 4 μm, and the maximum main field is 2^{20} nm or about 1mm. These distances can be increased either by using a coarser address (larger pixel) or by employing larger DACs. Since the electron optics and accompanying electronics typically restrict EBL writing fields to 1 or 2 mm square areas, large patterns are 'stitched' together by precisely moving the specimen stage. The fine control of the stage is tracked using a laser interferometer system that can resolve position changes down to the 0.3–0.6 nm range!

4.1.2.4 Evolution of e-beam writers

EBL began around 1960. It was a direct result of the proliferation of SEMs. Feynman and others around that time thought that, if the microscopists could scan a fine beam over a small area of a surface to magnify an image, then reversing the process could substantially demagnify a pattern to write small features. These ideas were implemented in various university and industrial laboratories around the world. We obtained this account from Prof. R.F.W. Pease [8]:

> The earliest submicron work I have a record of is Mollenstedt and Speidel at U. Tuebingen [Germany] who wrote 'G. Mo. Tub' with 100 nm linewidths on a membrane in a TEM (published in *Physikalische Blatter* in February 1960, the same month that Feynman's talk was published). The idea (EBL) was first put to me by some folk from the Royal Radar Establishment in November 1960. Ken Shoulders at Stanford Research Institute showed a pattern formed by decomposing an organic compound in a TEM in 1961. Oliver Wells and Dick Matta at Westinghouse started a formal program at Westinghouse in 1962 and I believe they presented their work in late 1963 or '64. I formed the free-standing letter 'P' (less than a half-micron high) in my home-built SEM [at Cambridge University] and described that at the European EM conference in July 1963. Alec Broers recognized that the material used for my letter, made with crosslinked polymer contamination, would serve as an excellent ion etch mask and in 1964 at the [then Symposium on Electron and Ion Beam Science and Technology now known as the International Conference on Electron, Ion, and Photon Beam Technology and Nanofabrication or EIPBN] conference showed a beautiful picture of a sub-micron grating of ion-milled gold on silicon. Philip Chang started his [lithography] work on the SEM [at Cambridge] in late 1963 and showed results in 1966 of thermally etched silicon at the International EM conference in Kyoto; his work was in collaboration with the Cambridge Instrument Company. Thornley and Hatzakis at IBM Research showed a transistor with sub-micron features in 1967 (EIPBN, Berkeley) but the first device that worked better because it was smaller was the SAW device that Alec Broers (now at IBM) described in 1969. This got everyone excited and Bell Labs got around to starting a program at the end of 1970 and pulled me ... to Murray Hill to set up a group to develop the materials and processes while Herriott's department built the EBES.

Bell Labs began their electron beam exposure system (EBES) series of EBL systems for their internal research use and then spun it out commercially as the MEBES series produced by ETEC, Inc. These became popular as photomask-making tools for many years. IBM had a similar tool development program that produced several generations of spot-beam and shaped-beam EBL systems. The commercial growth of dedicated EBL systems and vendors was also taking place at this time. The geneology of any business can be complicated. None more so than the that depicted in **Figure 6**, in which the commercial systems of Philips, Cambridge Instruments, and Jenoptik all eventually merged to become Leica and then Vistec, now one of the major suppliers of high-end systems.

Whichever vendor product line one examines, however, we observe certain technology trends. For example, systems developed for making photomasks tend to be shaped-beam tools that operate at 50 kV and use a LaB_6 thermionic source. These tools are

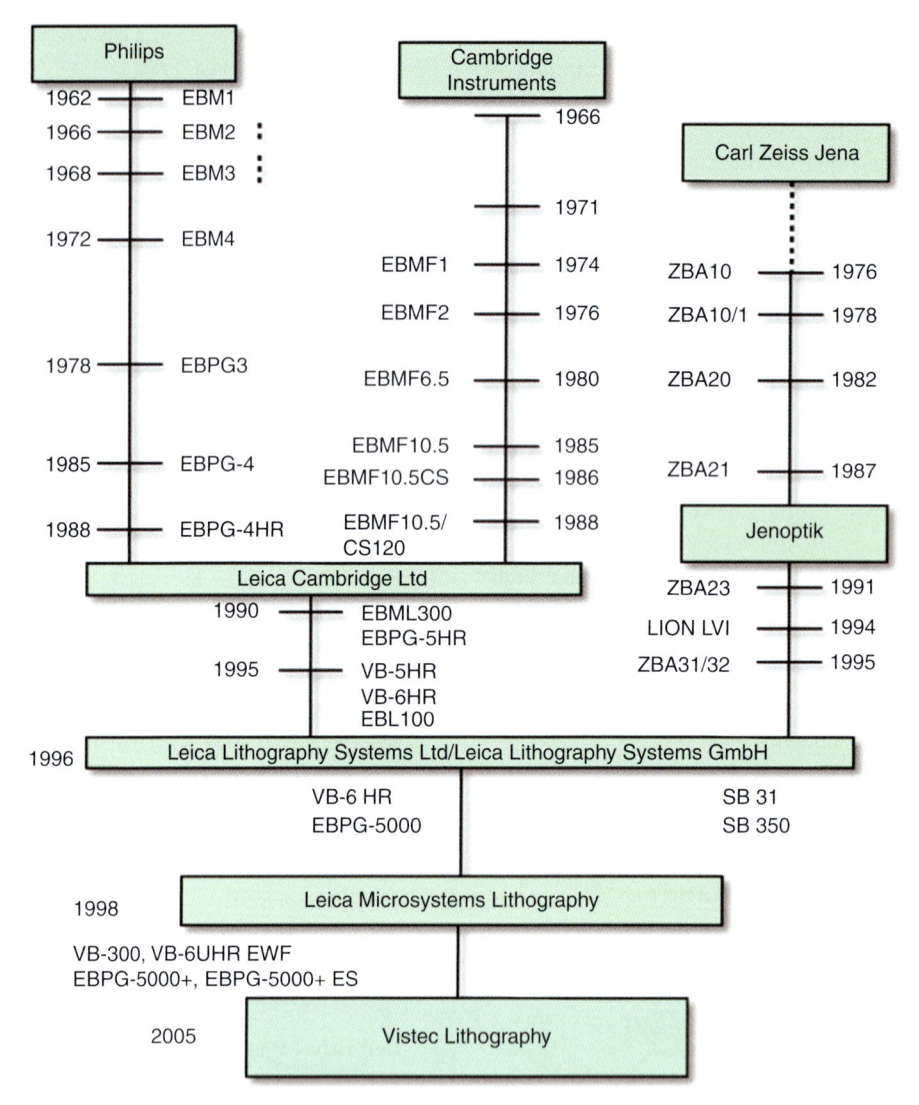

Figure 6 Family tree of Vistec, an EBL system maker. Courtesy of Vistec.

optimized for high throughput to allow economically sensible photomask making. A shaped-beam system exposes a series of rectangles with a single flash of electrons and is, therefore, faster than filling in each rectangle with a series of dots. This is possible, of course, because minimum features on a mask are typically (4×) larger than printed on a wafer. JEOL Ltd., NuFlare, and Vistec all make shaped-beam systems.

The highest resolution is obtained with a spot-beam tool, however. This has caused all high-end instruments to migrate to field emission (and, therefore, higher-brightness) sources with accompanying high-speed deflection systems. Spot-beam systems started out employing accelerating voltages of 20 kV but have evolved to 100 kV.

Much of the impetus for high voltage EBL came from laboratory work in which a high voltage scanning transmission electron microscope (STEM) was adapted for writing. While limited in functionality compared with dedicated instruments, they allowed comparisons of patterning results at various accelerating voltages and yielded experimental values for scattering parameters. Broers [9] demonstrated the dose change experienced when a line was drawn on thin membrane and then across a boundary onto bulk silicon. This was dramatic evidence of the proximity effect in e-beam writing. Mankiewich et al. [10] used an STEM to plot the backscatter range in silicon from 20 to 120 kV. Jones et al. [11] reported the resist sensitivity changes that occur for exposures up to

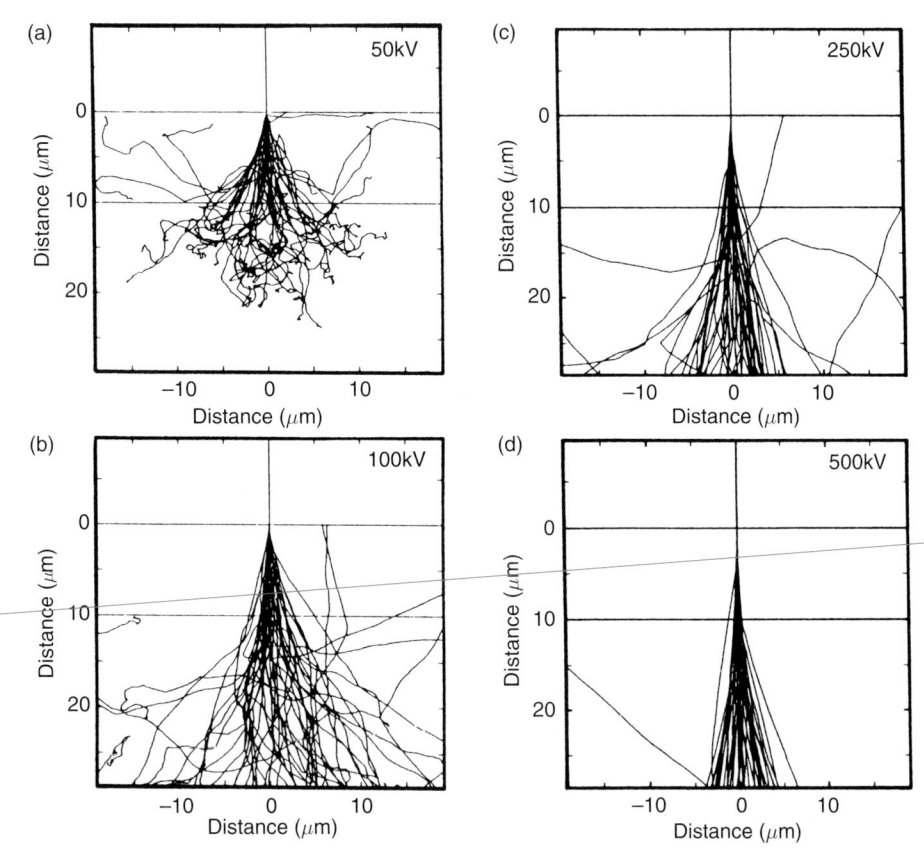

Figure 7 Monte Carlo electron trajectory simulations for 10 μm of resist on a silicon wafer for very high voltage electron lithography. Adapted with permission from Jones G, Blythe S, and Ahmed H (1987) Very high voltage (500 kV) electron beam lithography for thick resists and high resolution. *Journal of Vacuum Science and Technology B* 5(1): 120–123.

500 kV. Simulations of these high-voltage beams in resist are shown in **Figure 7**, revealing almost no broadening of the exposure in the first few microns of the resist layer for voltages greater than 100 kV. This last work also cautions us that at >250 kV can cause damage to the underlying substrate via atomic displacement. So there are practical limits to high-voltage lithography!

There was, for a time, some interest in low voltage (1–3 kV) lithography due to the electron backscatter effects discussed above. At sufficiently low voltage, researchers determined that the scattered electrons were of a short enough range in the substrate that the proximity effect was minimized. Jenoptik offered such a system, called the LION LV1, around 1994. The LION employed a novel technique for making curved structures in which the beam was held at the center of the field and the stage moved continuously beneath it to form smooth, uninterrupted curves. However, for more general-purpose applications, low voltage e-beam columns can limit the minimum spot size that can be attained when larger scanning fields are needed. The low voltage also was accompanied by short penetration range of the beam. This typically made the resist process more complicated, requiring multilayer resist schemes [6]. In general, as researchers sought ever-smaller features (and spaces between features), proximity exposure once again arose as a problem.

Current vendors in the EBL market include JEOL, Vistec, Elionix, and Crestec, all of whom manufacture dedicated e-beam tools. Raith and Nabity that adapt commercially built SEMs for smaller-laboratory use lithography applications.

4.1.2.5 What are the current trends in EBL systems?

Among the high-end systems, high-voltage EBL continues to be the most desirable direction for nanopatterning. Most current e-beam tools operate at 100 kV, while many laboratory systems operate at 50 kV or even 20 kV. The laboratory systems are lower in voltage because the lithography tool is fashioned from add-on electronics to an SEM that has been adapted for lithography. SEMs are optimized for imaging not lithography. As we discussed above, low voltage can have advantages for imaging. Having said that, SEM-based systems can exploit their high resolution to perform some outstanding proof-of-concept research experiments. Yang and Berggren [12] demonstrated modulation of hydrogen silsesquioxane (HSQ) resist down to 5 nm lines by using a rather unorthodox method that involves a salty developer process (**Figure 8**).

These are remarkably fine features that are among the narrowest ever produced on a bulk substrate. It would be difficult to produce such features on any system, no matter what the accelerating voltage, without employing such fortuitous chemistry.

Dedicated EBL systems typically are considerably larger and more expensive than SEM-based systems. **Figure 9** shows a JEOL JBX9300FS EBL system that has been in production for more than 10 years. Systems such as this are geared to long-term stability and automated calibration procedures to allow long unattended operations. We have run single continuous exposures as long as 96 h exposures without interruption on this system. The system shown in the figure operates at 100 kV, contains a high-precision stage with laser interferometric control, a robotic loader for job queuing, and a 50 MHz deflection system for improved writing speed. Given that these features exist in 10-year-old machines, what are the future trends? Here is some analysis that might help predict the features that might grace the next generation of instruments.

Is there considerable benefit to exposure at even higher energies than generally available? For example, why not 150 keV? To help analyze this question, we have run some electron scattering simulations for various accelerating energies from 20 to 200kV (20, 50, 100, 150, 200). **Table 1** summarizes the trends for

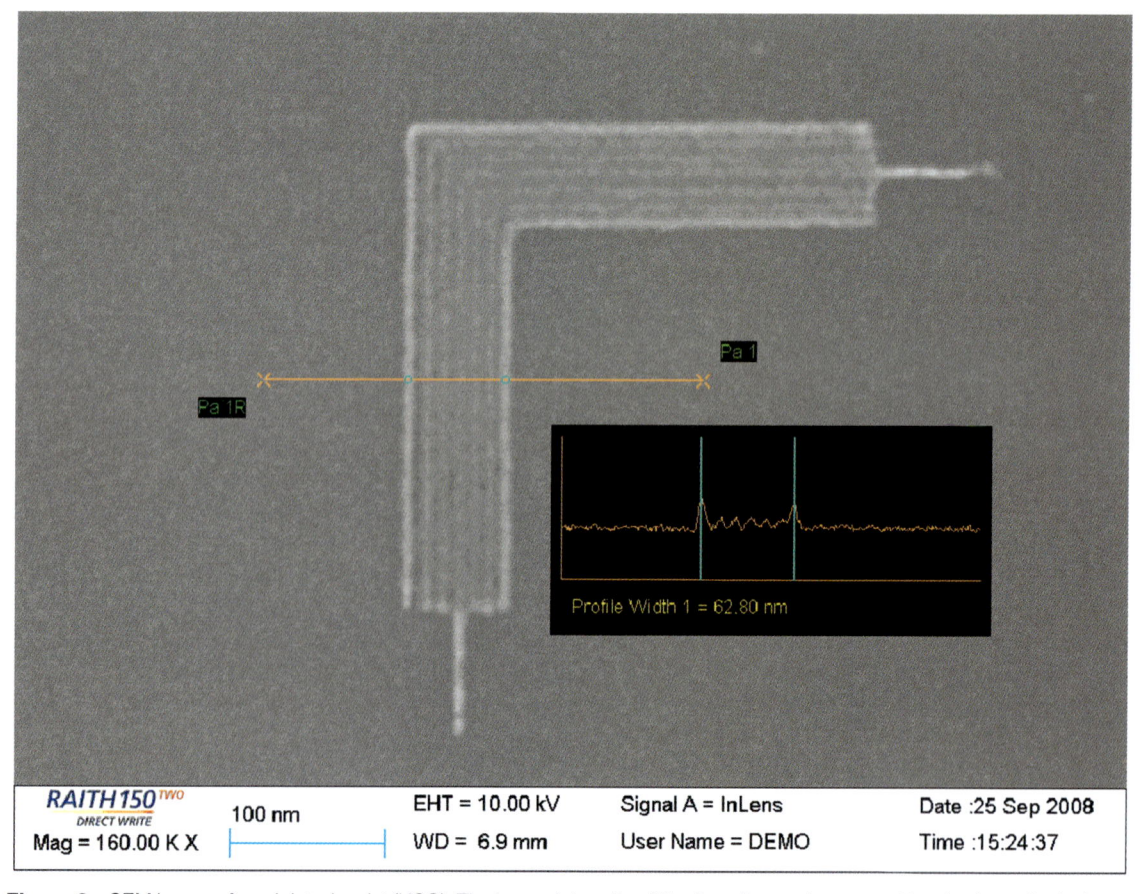

Figure 8 SEM image of modulated resist (HSQ). The image intensity of the 5 nm line and space pattern is shown in the inset. With permission from Yang J and Berggren K (2007) Using high contrast salty development of hydrogen silsequioxane for sub–10-nm half pitch lithography. *Journal of Vacuum Science and Technology B* 25(6): 2025.

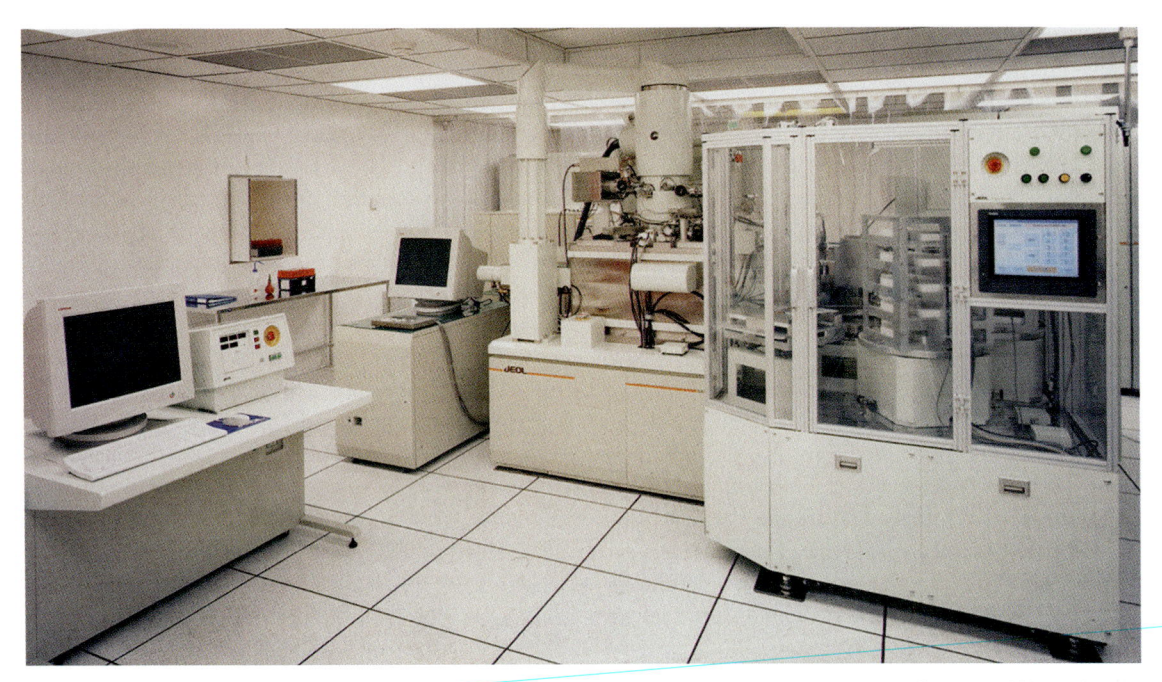

Figure 9 A JEOL JBX 9300FS commercial 100 kV EBL system shown with an optional robotic loader. Courtesy of Alcatel-Lucent.

Table 1 Simulations of point spread function dependence on energy

Material	Energy (keV)	α (μm)	β (μm)	η
GaAs	25	0.0086	1.18	1.13
GaAs	50	0.0078	3.65	0.98
GaAs	100	0.0071	12.78	0.94
GaAs	150	0.0067	25.11	0.94
GaAs	200	0.0066	40.12	0.97
InP	25	0.0087	1.20	1.21
InP	50	0.0078	3.64	1.06
InP	100	0.0071	12.65	1.04
InP	150	0.0068	27.62	0.98
InP	200	0.0067	40.09	1.04
Si	25	0.0086	2.85	0.67
Si	50	0.0077	10.18	0.65
Si	100	0.0070	34.14	0.59
Si	150	0.0067	72.91	0.65
Si	200	0.0065	123.40	0.77

the forward scatter range (α), backscatter range (β), and backscatter yield (η) for three common substrate materials. The data were modeled by using commercial software called SCELETON [13]. The approach uses random scattering events to trace the trajectories of many electron paths that enter a wafer. This method is known as a Monte Carlo simulation.

The traces (2 million electrons for each case) were performed for the 15 cases that appear in the table. We did simulations for three materials, Silicon, GaAs, and InP each with a thin resist layer on top. As we saw in **Figure 1**, the point spread function exhibits a short-range distribution of exposure around the incident electron beam and a long-range exposure due to the scattering in the substrate that causes the backscatter exposure. To represent this analytically, a two-Gaussian fit was obtained for each simulation using another commercial software called BEAMER [14]. This two-Gaussian fit defines the point spread function recorded in **Table 1**. [We note that while for actual proximity correction, better results are obtained with more complicated fits (e.g., four-Gaussian fit), the trends are more difficult to interpret. So for our purpose, we forced the fit to a two-Gaussian model.] The two-function fit allows a simple interpretation of the trends in backscatter range (β), the forward-scatter width (α), and in the ratio (η) of the energy deposited from the backscattered electrons to that of the incident electrons. For all these materials, energies of 100 keV and above give the lowest forward scatter (small α), the lowest backscatter exposure (small η), and the best exposure contrast (large β).

Examining the plots in **Figure 10**, the optimum based on this simple analysis appears to be in the range 100–150 keV, depending on the material. The plot of η shows that silicon has a minimum near 100 kV, while the III–V compounds are minimum near 140–150 kV. This is not the whole story though. The forward scatter and the backscatter both improve monotonically with higher voltage. The higher voltage should provide better minimum linewidth and better contrast due to the larger β and low η. The higher voltage will permit thicker resist exposures. Since many applications use silicon, operating at 125 kV might be a good choice. There are concerns in going to higher voltage. For example, mark detection (MD) signal could well be an issue, requiring either deeper etched alignment marks or thicker metal marks to compensate for the reduced signal. Reduced resist sensitivity also accompanies high-voltage e-beam. The high penetration length in resist means that the energy lost to the resist per unit of path length is less and, therefore, it takes a greater number of electrons to impart the dose required to fully expose the resist. This typically varies as the inverse of the electron energy. So the dose at 100 kV is about twice that at 50 kV. This is mostly offset by the increase in current density from the source that comes with higher voltage. In addition, the simulations of Kyser [15] show that energy deposition radially from the beam is confined to smaller distances as voltage increases. Plots of this trend [16] suggest that improvements likely continue above 100 kV. All things considered, higher beam voltage is likely to be a target specification for future dedicated EBL systems.

One consequence of bringing field emission sources to EBL systems has been that suddenly we are using a few nanoamperes of current where we formerly used 5–10 pA. This huge jump in current density due to the source brightness creates a problem. Resist exposure is the product of the current and exposure time divided by the pixel area. In systems with digital addressing, the ability to move the beam from point to point is limited by the deflector speed of the tool. The need for ever-faster deflection is insatiable. Since 1992, we have advocated scanning speeds of 1 GHz to fully realize the potential of the high-brightness sources. The current state of the art is 50 MHz. Sometimes one is just ahead of one's time. Nevertheless, increasing deflection speed is an important direction for future systems.

4.1.2.6 The high-throughput issue – are multi-beams the answer?

While marvelous tools exist for creating highly imaginative structures for research (see Section 4.2) and

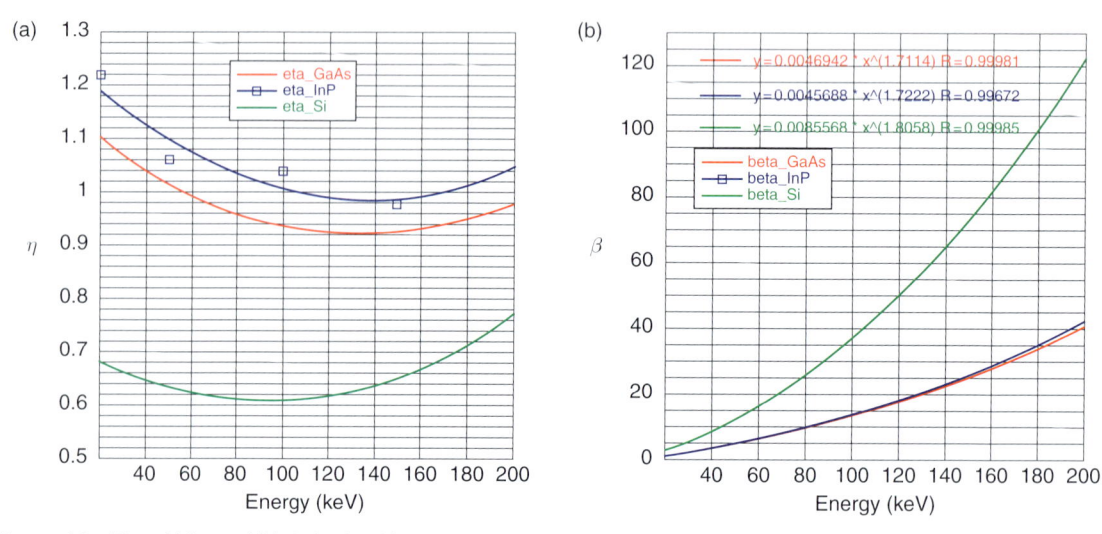

Figure 10 Plot of (a) η and (b) β obtained from simulations for Si, GaAs, and InP over the range 20–200 kV.

for making advanced masks, e-beam systems have not really been a viable option for mainstream integrated circuit manufacturing. This is due to the extraordinary throughput requirements placed on the lithography steps. Typical numbers on present-day optical tools are 100 wafers per hour. Another solution to the throughput issue that has been suggested is the use of multiple beams or other means of achieving parallel writing. A series of projects have been discussed over the past decade or so, but it seems that these systems are just starting to become commercially viable. It will take some time to determine whether these become fully realized, but they are worth mentioning, should the reader choose to follow the progress. Several companies are developing maskless lithography, which includes the multiple e-beam variety. MAPPER, KLA-Tencor, TEL/Multibeam, and others are developing maskless tools, and are motivated by the promise of lowering the cost of processing in lithography by eliminating the photomask in IC production. MAPPER Lithography is a European consortium developing a maskless lithography technology that combines massively parallel electron-beam writing in which 13 000 electron beams are focused on the wafer by electrostatic lens arrays. Each beam is fed by an optical data stream. The throughput goal is ten wafers per hour. The individual e-beams are generated by splitting up a beam from a single electron source. During exposure, the e-beams are deflected over 2 µm perpendicular to the wafer stage movement. Modulated light beams create a modulated e-beam pattern at data rates in the 1–10 GHz range [17]. KLA-Tencor Corp., has also embarked on an e-beam maskless lithography program, based on multi-beam technology, called reflective electron-beam lithography (REBL). Its multi-beam system will consist of 106 beams with throughput goal of 2–10 wafers an hour depending on the litho level. It is a novel approach, in which the electron gun directs electrons into the digital reflective pattern generator in an arrangement analogous to the digital light projector chip used for video images [18]. Tokyo Electron Ltd. (TEL) and Multibeam Systems Inc, are co-developing another multi-beam maskless lithography technology [19,20]. The Multibeam technology incorporates 88 beams in a system based on a high-current, shaped-beam technology. It is also targeting five to ten wafers an hour.

Even with the major improvement these concepts offer, we note that they are only targeting 10% of the customary wafer throughput numbers. This signals that either the optical alternatives (including DUV and EUV) have become either too expensive, or that the technology will not be ready in time to meet the industry roadmap, or both. These outcomes are difficult to forecast and often depend on the decisions made by the major players in the industry. The parallel-beam technology solutions are, for now, an interesting and promising path, but we will withhold any predictions!

4.1.2.7 Resists and processing are keys to success

A small electron beam scanned across a resist-coated sample causes a latent image. Whether that image eventually becomes a useful device, or shapes a new material, or demonstrates some new effect is the purview of nanofabrication. The choice of resist, and the subsequent development and post-processing all determine success or failure. The first step is to choose a resist or a multilayer stack of resists that meet the criteria for the job. Resists seem to come and go as they rise and fall in popularity. The best we can do here is to provide a sampler of currently popular resists and some basic properties. One of the longest-used resists among nanofabricators is a positive tone resist, poly(methylmethacrylate) (PMMA). First reported by Hatzakis [21] many years ago, it is still among the highest-resolution materials in use. PMMA is available in various molecular weight formulations and dissolved in various carrier solvents. The molecular weights are typically in the range 10^4 to 10^6 Da. PMMA is an example of a single-component resist in which the incoming radiation completes the chemical change needed for exposure. There are also multi-component resists in which a process of chemical amplification is used to enhance the sensitivity. In chemically amplified resists, the radiation causes a chemical release, which upon a subsequent heating step, completes the reaction needed for full exposure.

It is desirable to use resists and developers that have highly nonlinear response to exposure dose. An important resist property for high-resolution lithography is the contrast, γ. A high-contrast (positive) resist permits very little change in thickness in the under-exposed regions and completely clears in regions dosed above a threshold value. This nonlinear behavior can help to correct for all manner of dose variations. Contrast is defined by

$$\gamma = \left(\log_{10}(D_{t=0}/D_{t=1})\right)^{-1}$$

where $D_{t=0}$ is the dose at which a large area exposure first clears and $D_{t=1}$ is the dose below which the full thickness of resist is retained. The conventional measure of resist contrast is obtained from a sensitivity curve that plots normalized resist thickness as a function of the $\log_{10}$ of exposure dose. The contrast is the absolute value of the measured slope at the 50% thickness point. High-resolution resists such as PMMA can have contrasts in the range from 6 to 9 depending on process conditions such as developer strength, development time, and pre-exposure bake time and temperature. A value below 2 is considered very low and is usually a poor candidate. Most resists fall in the range from 2 to 10. High-contrast resists tend to produce steeper sidewalls in resist profiles.

The challenge in finding the ideal resist has been to increase the sensitivity (PMMA is rather slow), and improve the etch resistance for better pattern transfer capability (PMMA is a rather poor etch mask) while retaining adequate resolution. **Table 2** shows a short list of resists in common use. They are used in single layers or in combination with other materials to achieve various profiles. **Figure 11** shows five different resist stategies that can be employed. The top (a) is a single-layer resist exposure and development. Using high-voltage lithography, as seen in the simulation in **Figure 7**, the electron beam undergoes minimal forward scattering and, therefore, results in quite vertical sidewall angles. If adequate, a single resist layer can then be used as a mask for either etching the underlying material or as a liftoff mask for deposited material. At times the liftoff process using single-layer resist is not 'clean' at the edges. To remedy this, one can form a bilayer stack. These consist of a 'slow resist' on top and a 'fast resist' on the bottom. In this way a single exposure and development creates either a small (b) or large (c) undercut. The relative sensitivities determine the extent of the undercut [22–25]. This method has been used for decades to improve the liftoff quality for metals, superconductors, and assorted other materials. When etching is to be the next processing step, the resist alone often does not have the etch resistance to meet the goal. A common solution is to use a hard mask as illustrated in (d). Here the resist need only stand up to the etch of thin layer of material chosen for its good etch selectivity properties. Materials such as metals, silicon dioxide, silicon nitride, etc., are often chosen. The hard mask often can withstand an order of magnitude greater etch time than the resist alone. The last example is a trilayer stack (e). By using a thin imaging layer of a known high-resolution resist, the ultimate fine pattern can be produced and one can avoid pattern collapse that often occurs in high-aspect-ratio attempts. Only a thin pattern transfer layer is etched using this imaging resist. The pattern is transferred to a lower, usually thicker, layer using a selective etch. There is great flexibility in choosing the lower layer, however. It is usually a thick polymer that is selected for its good etch resistance or mechanical properties, or some other quality not exhibited by the imaging resist. This is a good way to decouple the imaging properties from the general requirements for the level.

Table 2 is a list of some currently popular resists, their manufacturers, and a suggested base dose and development. It is always a good idea to perform a dose array on your pattern to optimize the process for your particular pattern. The sampling in the table is simply offered as a starting point. Developer conditions and technique can dramatically alter the sensitivity and can play a determining role in your individual results.

4.2 Applications to Nanostructures

We have provided above an introduction to electron beam lithography. We have discussed the basic processes, the interaction of electrons with solid materials, the elements of electron sources and optics, the evolution and trends of dedicated EBL tools, and the common resist methods employed. For more detailed discussion of the various ebeam systems and their data formats see McCords and Rooks [6]. For an in-depth treatment of the electron sources and optics see Swanson and Schwind [7]. For engineering aspects of complex exposure systems see Suzuki's chapter [7a]. For a more detailed treatment of proximity correction and charging effects, see Dobisz and Peckerar [26]. We now turn our attention to a sampling of the many and varied applications that rely on nanofabrication and EBL.

4.2.1 Spintronics

Spintronics exploits the electron spins and the magnetic moments of materials in devices, in addition to using the electronic charge [27]. A current application of spintronics is in the read/write heads of the hard-disk drives used in computers and portable music players. Future applications may include spin-based computation. EBL is used in combination with photolithography to fabricate read and write

Table 2 Common EBL resists with their properties and uses

Resist	Supplier	Tone	Sensitivity ($\mu C/cm^2$) at 100 kV	Developer(s)	Characteristics
PMMA (high molecular weight)	Microchem	Positive	900	Cellosolve Methanol 3:7 or MIBK:IPA1:3 or IPA:Water (conc. varies)	High resolution, single layer or top of Hi/Lo bilayer, liftoff
PMMA (low molecular weight)	Microchem	Positive	800	Cellosolve Methanol 3:7 or MIBK:IPA1:3 or IPA:Water (conc. varies)	High resolution, bottom of Hi/Lo bilayer, liftoff
Copolymer (MMA/MAA)	Microchem	Positive	300	Same as above	Large undercut profile for liftoff when used as bottom layer with PMMA on top
Zep520	Zeon Chemicals	Positive	300	Hexyl Acetate, n-Amyl Acetate, or Xylenes	High resolution, good etch resistance
Zep7000	Zeon Chemicals	Positive	80	3-Pentanone Diethyl Malonate	Fast, good for making masks
NEB31	Sumika Materials	Negative	80	0.26N TMAH	Fast, chemically amplified resist
HSQ	Dow Corning	Negative	1000	0.26N TMAH	High resolution, good etch resistance, flowable oxide
Calixarene	Synthesized	Negative	10 000 or higher	Xylenes, IPA, MIBK	High resolution, good etch resistance

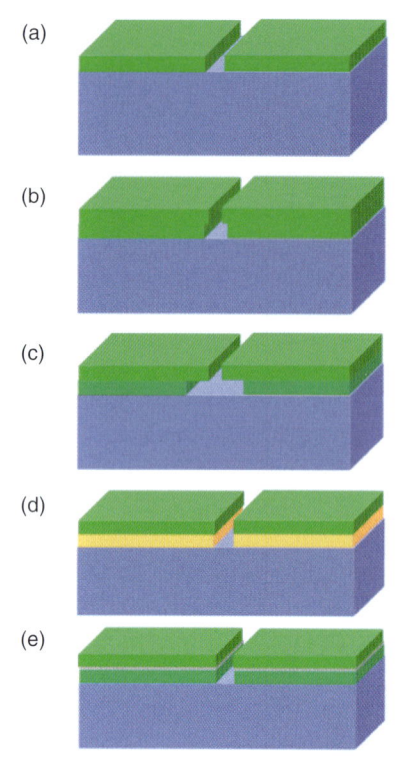

Figure 11 Illustrations of five different resist strategies. (a) Single layer, (b) bilayer with small undercut, (c) bilayer with a large undercut, (d) single layer resist on a hard mask, and (e) a trilayer resist.

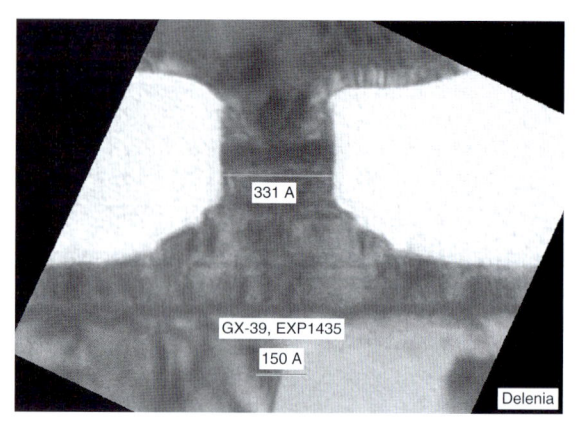

Figure 12 TEM image of a CPP GMR sensor that was defined with EBL. Courtesy of Jordan Katine, Hitachi Global Storage Technologies.

sensors for high-density hard-disk drives, because the size of a bit on the disk has been shrinking steadily. Current-in-plane (CIP) structures using the tunnel junction magneto resistive (TMR) effect have been used widely. Current-perpendicular-to-plane (CPP) architectures using the giant magnetoresistive (GMR) effect are being studied for use in future generations of disk drives [28]. **Figure 12** is an example of a CPP GMR sensor that was defined with EBL. This sensor geometry is being condsidered for future (750 Gbit per square inch density) recording head sensors. Magnetoresistive random access memory (MRAM) devices may replace current memory technologies in the future if they have significant performance and areal density advantages. EBL has been used for the fabrication of prototype devices and for research in the basic physical mechanisms that underly the devices, because they require high-resolution patterning. Some recent examples are as follows: Braganca *et al.* [29] used EBL and ion milling to make elliptical nanomagnets for MRAM studies with a tapered nanopillar geometry that generates a spin current polarized partially out of plane. This geometry allows the use of smaller currents for faster switching of the magnetization of the free layer of the device. Cui *et al.* [30] used similar nanopillars, also created using EBL, to study improvements in nanomagnet switching time and switching power by using microwave-frequency current pulses in combination with square current pulses (**Figure 13**).

4.2.2 Molecular electronics

EBL is useful for studying single molecules and for constructing devices that use single molecules as transistors. **Figure 14** shows a pentacene transistor with 10 nm channel length. The source and drain electrodes together with the guard electrodes were defined with EBL on a Leica VB6 system. To avoid proximity effects [31], a three-step fabrication process was used. The first step was to expose and deposit the alignment marks. The second step exposed the source and upper guard electrodes, followed by e-gun evaporation of 1 nm Ti adhesion layer followed by 30 nm Pt and lift-off. The third step repeated the exposure and metal deposition for drain and lower guard electrodes. This whole process led to channels with lengths of 10 nm and widths between 70 and 230 nm. The guard electrodes were located 20 nm away from the channel area. The metal electrical contact pads were patterned through image reversal and lift-off processes.

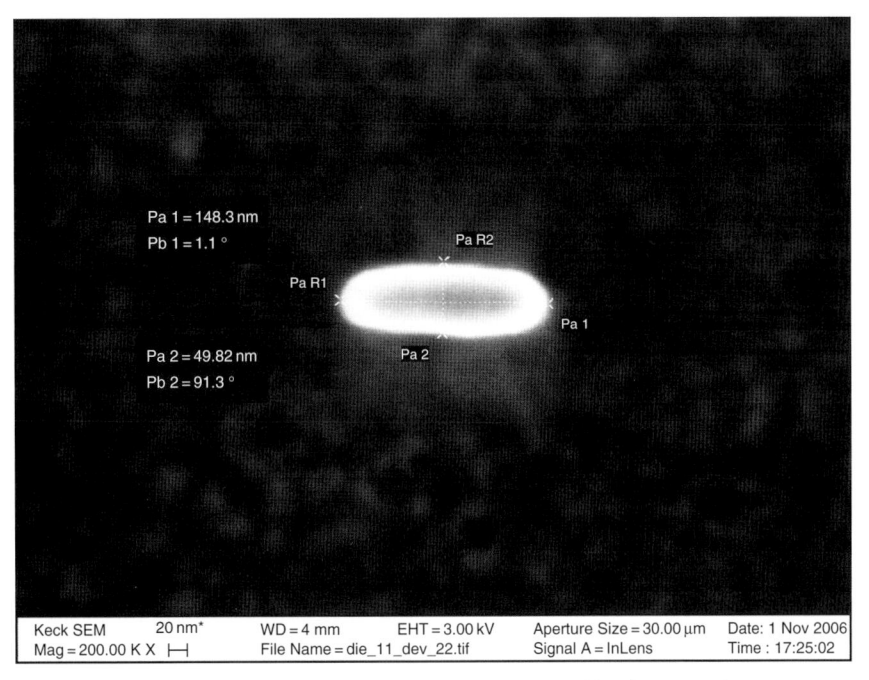

Figure 13 A nanopillar used to study resonant excitation of nanomagnets with microwave-frequency current pulses. Courtesy of Hong-Tao Cui, Cornell University.

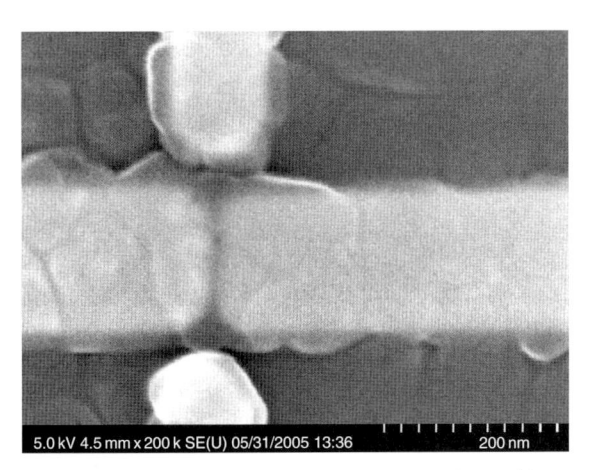

Figure 14 Channel area of a pentacene transistor with 10-nm channel lengths. Source and drain electrodes are on the left and right sides of the image and guard electrodes are on the top and bottom. Courtesy of Yuanjia Zhang and George Malliaras, Cornell University.

4.2.3 Photonic devices

EBL has made a number of advances in micro- and nanophotonics possible [32]. Single-mode waveguides need the smooth sidewalls that EBL produces to reduce light scatter; this is especially important for materials with high indices of refraction. Sub-wavelength features, many with curved geometries, are needed to make distributed Bragg reflectors, tapered couplers, and modulators. All require high resolution and accurate pattern placement. Gratings require precise periodicity to the nearest nanometer to tune them for use with specific wavelengths of light. The interferometrically monitored stage and precise deflection of the beam in EBL systems satisfy these requirements. One example of a device that exploits the capabilities of EBL is optical ring resonators. When coupled with straight waveguides for input and output, they have been used as bandpass filters and as optical switches, by detuning the ring resonance frequency by heating with a pump laser [33]. **Figure 15** is an optical microscope image of a ring resonator written with EBL, with the via regions patterned into resist before etching. This ring is designed to have a p-i-n diode across it for injection of free carriers. **Figure 16** shows several rings of the same type in a zoomed-out image, with the vias etched and metalized, and contacts patterned in resist.

4.2.4 Patterned media for disk drives

Bit-patterned media (BPM) will most likely be required to increase the density of disk drives to 1

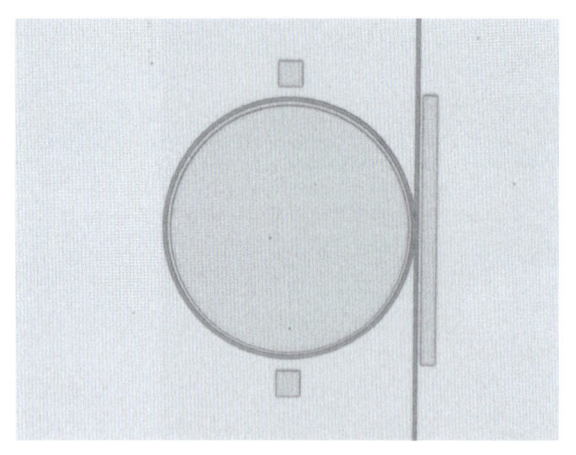

Figure 15 Optical microscope image of a ring resonator written with EBL, and via regions patterned into resist before etching. This ring is designed to have a p-i-n diode across it for injection of free-carriers. Courtesy of Amy Turner, Cornell University.

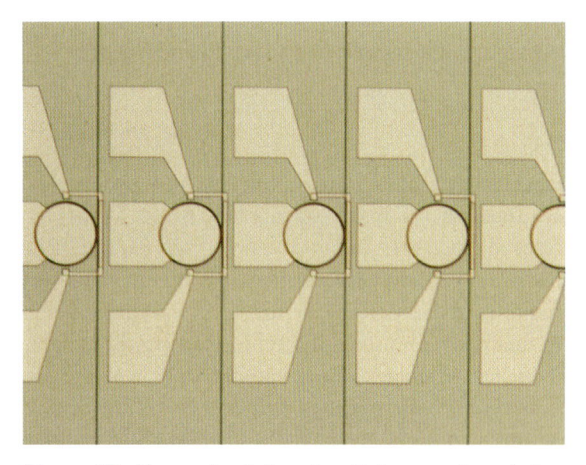

Figure 16 Zoomed-out view of multiple resonators similar to the one in Figure 14. Courtesy of Amy Turner, Cornell University.

Tbit per square inch or higher. Cost-effective manufacturing of BPM might be achieved by self-assembly, ion-beam lithography, or nanoimprinting. For all three methods, EBL would have a role in creating master disks that would be replicated by the less expensive methods [34]. Silicon nitride stencil masks for ion-beam lithography have been made with EBL [35]. **Figure 17** shows high-density patterns made in HSQ [36], similar to patterns which would be needed for nano-imprint masters. Self-assembled block co-polymers can potentially

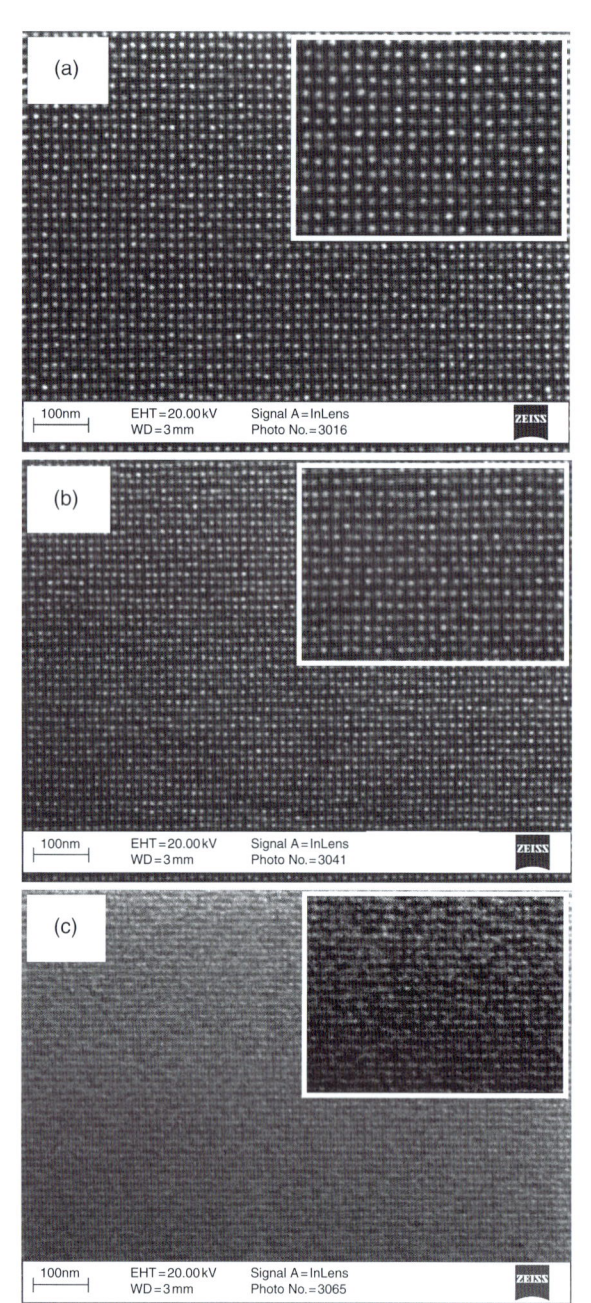

Figure 17 SEM images of 30-nm-thick HSQ resist with pitch of (a) 18 nm (2.0 Tdots per square inch); (b) 15 nm (2.9 Tdots per square inch); (c) 12 nm (4.5 Tdots per square inch). With permission from Yang X, Xiao S, Wu W, et al. (2007) Challenges in 1 teradot/in^2 dot patterning using electron beam lithography for bit-patterned media. *Journal of Vacuum Science and Technology* B 25(6): 2202–2209; reproduced in Terris B (2009) Fabrication challenges for patterned recording media. *Journal of Magnetism and Magnetic Materials* 321(6): 512–517.

produce higher-density patterns than EBL, but they lack the long-range order required for magnetic media. EBL could be used to improve long-range order by writing templates for polymer growth [37].

4.2.5 X-ray Optics

Optics for microscopes that operate at X-ray wavelengths are of various types that can be challenging to fabricate. Most X-ray sources are located at synchrotron beam lines in order to achieve high-brightness, wavelength-selectable, coherent photon flux. The most common focusing optics in use are diffractive lenses in the form of a Fresnel zone plate. These are comprised of hundreds of concentric rings that have diminishing widths as the radius (see **Figure 18**) increases.

To achieve improved performance, the outermost zones (rings) must be made narrower (e.g., 10–20 nm or smaller). For a fixed focal length, this means that the lens must be larger with precise placement of the features (typically ~1/3 the outermost zone), and with increasing aspect ratio to maintain good diffraction efficiency. The rings are typically composed of gold for hard X-rays (energies from 3–30 keV) and nickel for soft X-rays (energies in the few hundreds of eV) and supported on a thin membrane of Si_3N_4 for high transmission. Work in this area is concentrating on developing methods of improving resolution and aspect ratio to achieve sub-10nm imaging [38–41]. An alternative lens design that has been more recently explored is the kinoform refractive lens [42,43]. Bulk materials tend to have a refractive index slightly less than 1 in the X-ray spectrum. Therefore, one can make refractive lenses, but they tend to be rather long to achieve the phase shift needed for focusing. The entrance of the beam for a kinoform lens is 'edge on'. It is, therefore, the cross section of the structure that is seen by the beam. This is also demanding in the etching process that follows the lithography. **Figure 19** shows two such elements fabricated by EBL and deep etching in silicon. The figure on the left is a prism and on the right is one section of a lens. Since the wavelengths employed are extremely short, these elongated lenses require extraordinary pattern placement accuracy in order to maintain long range spatial coherence.

4.2.6 High Frequency Electronics

Electronic circuits that do not require millions of transistors for memory or logic operations, but require very high frequency operation, often make use of III–V semiconductor heterostructure materials such as GaAs/AlGaAs or InGaAs/InAlAs. To maximize the performance, the gate length of the metal gate should be short – an ideal candidate for EBL. Yet the diminished dimension can make the resistance of the gate high. The ideal structure would have a small footprint but a large cross

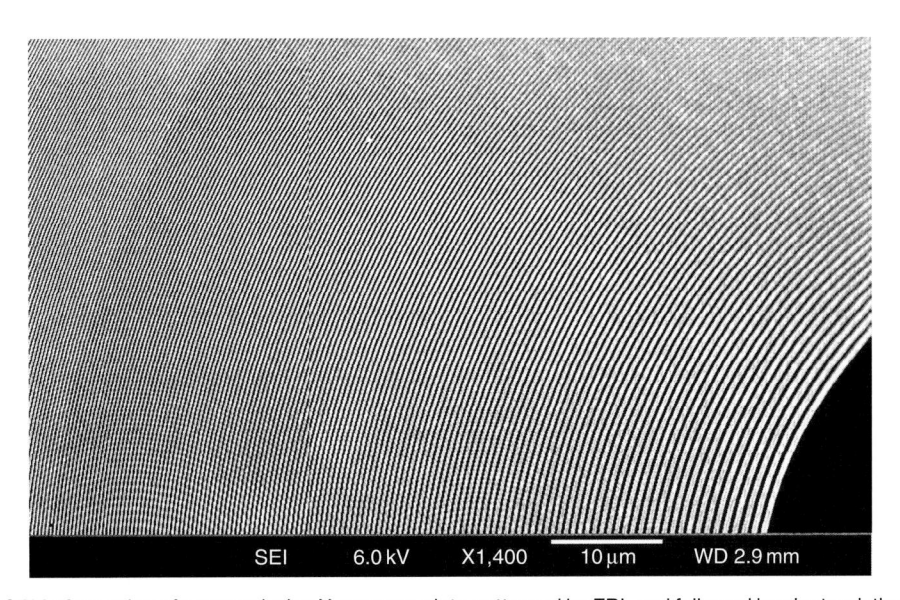

Figure 18 SEM of a portion of a transmission X-ray zone plate patterned by EBL and followed by electroplating of nickel on supporting Si_3N_4 membrane. Courtesy of A. Stein and K. Evans-Lutterodt, Brookhaven National Lab.

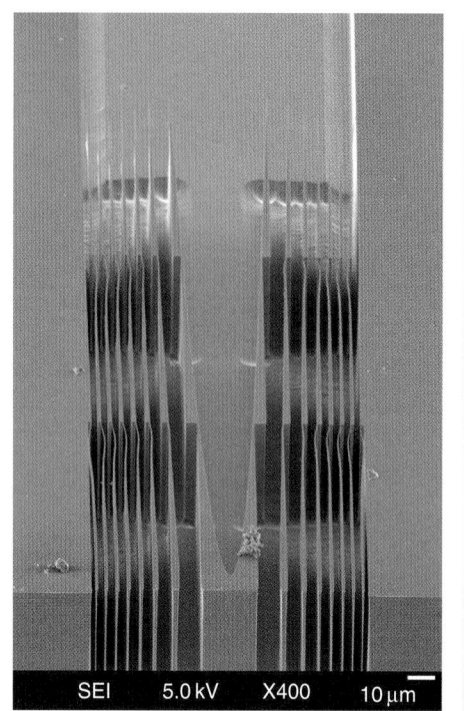

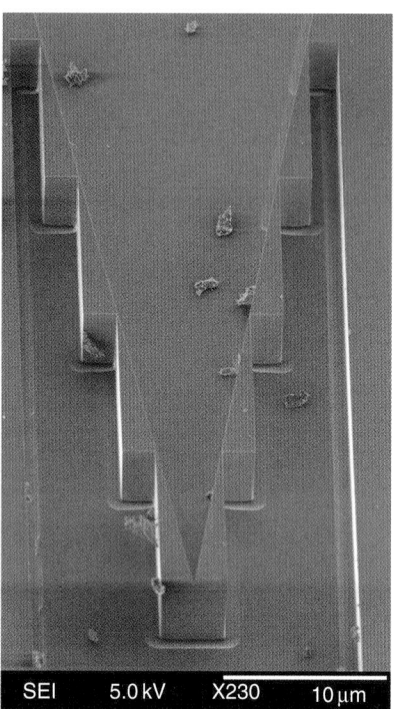

Figure 19 SEM images of (left) a prism and (right) a kinoform lens that act as refractive optical element for X-rays. The vertical deep etch is necessary to provide a good capture of the beam with uniform optical properties. Courtesy of A. Stein and K. Evans-Lutterodt, Brookhaven National Lab.

section to meet both requirements. These 'T-gates' are, therefore, critical for realizing high-quality high-electron-mobility transistor structures (HEMTs). **Figure 20** shows a T-gate made with a Leica VB-6HR electron beam lithography system. A trilayer resist process is typically used for forming T-gate (or mushroom gate) structures. The substrate is coated with a bottom layer of PMMA. The thickness of this layer determines the height of the gate foot. The middle layer is coated with methyl methacrylate/methacrylic acid (MMA) co-polymer, which is thicker than the metal layer that will be evaporated to form the gate. A top layer of PMMA is applied to be thick enough to support overhang of the copolymer. The exposed pattern is a path narrower than the desired gate length, with an overlapping path with the width of the head. The total dose for the gate-defining center line is enough to develop away all three layers, but the dose for the head-defining shape will only develop to the bottom of the copolymer. Three development steps with toluene, methanol:IPA (1:1), and MIBK:IPA (1:1), rinsing with IPA between each step, each cut through a layer of resist. After descum with an O_2 plasma and evaporation of the metal layer, the mushroom-shaped gate results [44]. Since this resist system is rather slow and limited in the relative widths that can be engineered, other T-gate processes are sometimes used [45].

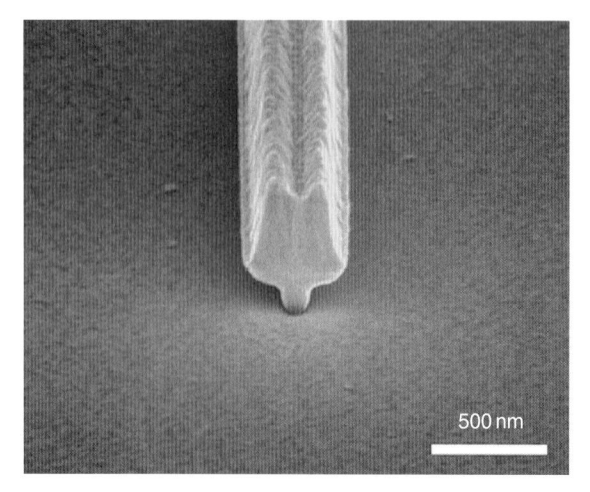

Figure 20 SEM image of a T-gate. Courtesy of Yunju Sun, Cornell University.

4.2.7 Nanotubes, Nanofibers, and Nanowires

Because of their naturally small width and novel electrical and chemical properties, nanotubes and nanowires are of great interest for miniaturization of electronics beyond current lithographically defined dimensions, and also for use as sensors and as models of biological structures. EBL has enabled research in this area because of its flexibility in creating nucleation sites for growth, trenches for isolation of electrodes or suspension of nanotubes, and contacts to nanotubes already in place, with the required small dimensions and placement accuracy.

EBL has been used to create precisely placed nucleation sites for growth of carbon nanofibers, so individual fibers can be characterized and structures can be used to control the functionality of devices made with nanofibers. **Figure 21** shows platinum electrodes made with a Leica VB6-HR system, with gaps varying in size from 40 nm and up, with catalyst dots of nickel made with a second e-beam exposure aligned with the electrodes [46]. **Figure 22** shows the device with the nanofibers grown from the catalyst dots. **Figure 23** shows a nanofiber container that can be used to mimic cells in studies of molecular transport [47]. The square array of nucleation sites was made with a JEOL 6300FS/E at 50 kV. **Figure 24** shows a device used to measure electrical transport through an individual carbon nanotube suspended over a narrow trench. At temperatures below 1 K, this resulted in very clean quantum dots, which

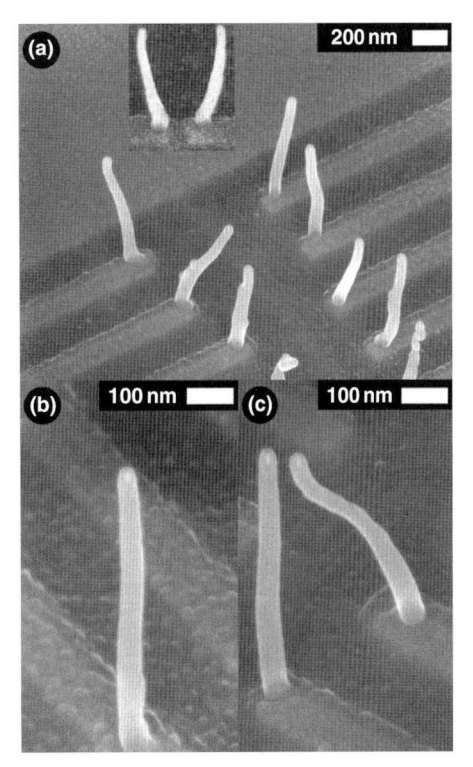

Figure 22 One micron tall carbon nanofibers on tungsten terminals. Panel A and its inset are on the same scale; the inset pictures two nanofibers across a 40-nm gap. Courtesy of Joel Moser and Michael Naughton, Boston College.

were used to show that, in clean nanotubes, the spin and orbital motion of electrons are coupled [48]. The metal contacts on top of the gate electrodes had to be aligned to the trench with high precision, after the wafer was distorted by various processing steps, including etching and thermal oxidation. High-contrast metal alignment marks could not be used, because gold is not compatible with the high-temperature growth of carbon nanotubes. The automatic local alignment system of the JEOL JBX9300FS system allowed the use of low-contrast marks etched into the thin SOI layer.

Nanowires of silicon and other materials can be grown in a bottom-up scheme or patterned by a top-down method such as EBL for greater control over their arrangement for making measurements. Silicon nanowires may be used for scaling logic sizes downward because wraparound gates can be used as an alternative to reducing the gate dielectric thickness. **Figure 25** shows a 300 nm wraparound gate over 28-nm diameter nanowires, defined using EBL, which were used to make the first direct capacitance measurements of silicon nanowires and to determine field

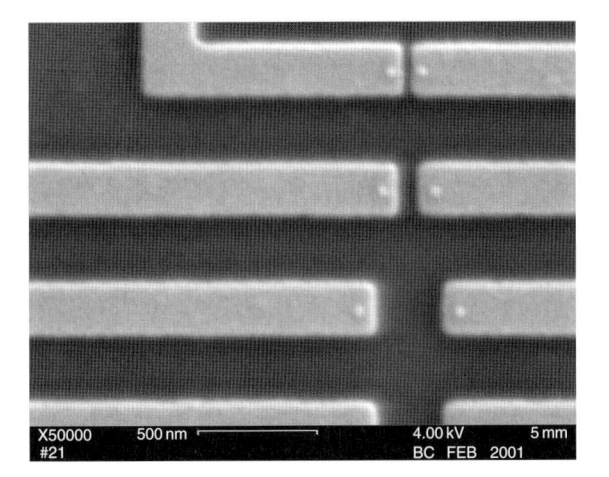

Figure 21 Nickel catalyst dots for growth of carbon nanofibers (CNFs), made with an e-beam exposure aligned to the previously fabricated platinum electrodes. Courtesy of Joel Moser and Michael Naughton, Boston College.

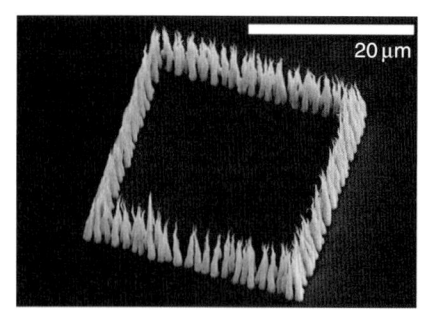

140   Electron Beam Lithography of Nanostructures

Figure 23   A semipermeable, 25-μm square cell defined by deterministically grown fibers. The walls of the cell consist of staggered rows of nanofibers that produce a size-restrictive barrier. A mouth at the bottom of the cell is defined by a single row of fibers for increased permeability. Adapted with permission from Retterer ST, Melechko A, Hensley DK, Simpson ML, and Doktycz MJ (2008) Positional control of catalyst nanoparticles for the synthesis of high density carbon nanofiber arrays. *Carbon* 46(11): 1378–1383.

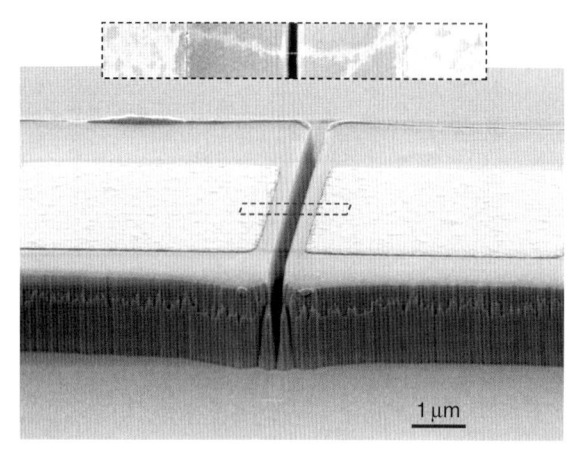

Figure 24   Few-electron carbon nanotube quantum dot device. A single nanotube makes contact to source and drain electrodes, separated by 500 nm, and is gated from below by two gate electrodes. The inset shows the nanotube crossing the deep trench used to isolate gate electrodes. Courtesy of Ferdinand Kuemmeth, Harvard University.

carrier mobilities in undoped-channel SiNW field-effect transistors (FETs) at room temperature [49].

4.3   Proximity Correction: Software and Hardware Solutions

As described above, electrons that are backscattered from the substrate can contribute significantly to the absorbed energy in a resist that is distant from the initial point of exposure to the electron beam. Forward scatter also broadens the exposed volume around the center of the beam, although, as we have shown, this is less problematic for high-energy beams. The scattering leads to increased energy absorption in the resist for large, closely spaced features and relatively less absorption in small, isolated features. This is known as the proximity effect [50, 51] and, if the exposure has uniform dose (in $\mu C\,cm^{-2}$), some features will be overexposed or larger than they should be, and some will be underexposed. A compensation for this effect, to make all features close to the designed size after exposure and development, is called proximity correction.

A relatively simple but effective method of correction is to assign doses to features or segments of features based on size and proximity to other features [52]. This conditional feature assignment can be done manually by placing features in different layers of the GDSII CAD file and assigning each layer to a different hardware clock on the EBL system. After the beam current is measured, the clock frequencies, which are the inverses of the dwell times of the beam on single pixels, are adjusted by the system to give the desired doses during the exposure. This technique has been used, for example, to make small elliptical nanopillars [53] and to make tapers for coupling light from optical fibers to single-mode waveguides on silicon [54]. Additional automation of this process is possible by sorting features into different dose assignments based on height, width, and other criteria with the software that converts from GDSII format to the native file format of the EBL system. A drawback of this semi-automatic process and the manual method is that both require experience and some trial and error to achieve the desired results, and they are not well suited to correcting large, complex patterns.

Better and more automatic results can be obtained by thinking about the correction as an inverse

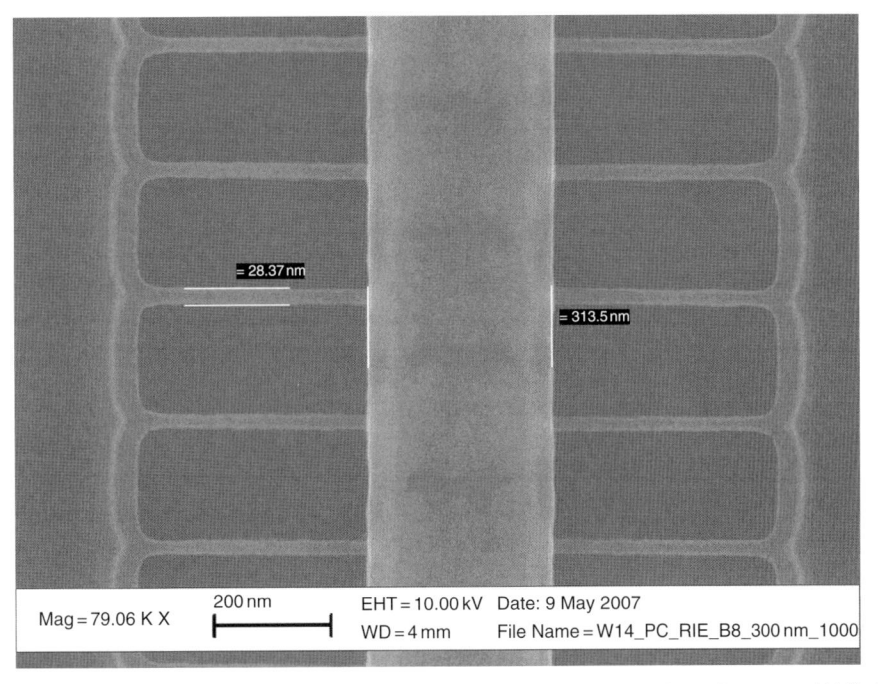

Figure 25 Wraparound gate with 300 nm width, in contact with 28-nm silicon nanowires. Courtesy of Lidija Sekaric, IBM.

problem. We know how to model operations on the input pattern and sample, such as exposure with an electron beam, to do a forward calculation of the resulting pattern of absorbed energy in resist. It is also possible to model the development of the resist, but this may not be necessary depending on the process requirements. In the inverse problem we start with the desired absorbed energy result and then calculate the set of operations on the initial data that give the best possible approximation to that result. The operations include exposure with the electron beam, modulation of the beam by changing the dose given to parts of the pattern, or changing the size and shape of the pattern. The inversion can use analytical methods or may use iterative methods to obtain exposure parameters that give successively closer approximations to the desired results.

A common method to calculate corrected doses for different parts of the pattern is to invert a convolution model of the exposure [55]. The kernel or point spread function (PSF) for the convolution is the absorbed energy in the resist from exposure to an infinitesimally thin beam of electrons. As was done above to compare the voltage effects, simulation of electron scatter and absorption produces a radially symmetric proximity function, $f(r, z)$. Analytical fits to the proximity function as a sum of two or more Gaussian functions have also been used (50, 55). In a two-Gaussian model, a narrow Gaussian with standard deviation α represents forward-scattered electrons and a second broad Gaussian with standard deviation β represents the backscattered electron distribution.

To simplify the calculation, the proximity function at half the depth of the resist can be used in a two-dimensional calculation, which is adequate when the resist is thin and beam energy is sufficiently high that the scatter profile changes slowly with depth. If $D(x, y)$ is the dose given to the resist surface by the EBL system and the PSF is assumed to be spatially invariant, the forward calculation of the absorbed energy in a plane is the convolution

$$E(x,y) = D(x,y) \otimes f(r)$$

The inverse approach is to start with the desired pattern $E_{desired}(x,y)$ and calculate the exposure $D_{corrected}(x,y)$ that gives the desired pattern. Using the property of Fourier transforms that the transform of a convolution of two functions is the product of the transforms of the functions, the previous equation can be solved for D:

$$D_{corrected}(x,y) = F^{-1}\left\{\frac{F\{E_{desired}(x,y)\}}{F\{f(x,y)\}}\right\}$$

Naive application of this formula may result in singularities if the transform of $f(r)$ has zeros, in nonphysical negative doses, and in unreasonably long computation times if the computation grid is too fine, so additional techniques, such as filtering out high spatial frequencies before transforming, may be needed [57]. **Figure 26** illustrates the result of applying this method to a pattern consisting of an array of dots next to an array of lines. Using manual methods, it would not be possible to predict the irregular shape of the isodose contours that produces even energy absorption from the center to the corners of the pattern. A commercially available software system for Monte Carlo modeling of the PSF is SCELETON, from PDF Solutions, Inc. A software package for proximity correction is PROXECCO, also from PDF Solutions, which is integrated into the CATS data conversion software from Synopsys, Inc. Another system that does data conversion and proximity correction is Layout BEAMER, from GenISys GmbH. Single-processor and cluster versions of CATS/PROXECCO and Layout BEAMER are available.

Other methods that have been investigated to perform proximity correction include linear programming [58], constrained optimization [59], correction by dose modulation and shape modification at the same time [60], neural networks [61], hierarchical rule-based schemes [62,63] and methods that take advantage of the periodicity of photonic crystal patterns [64]. To correct features made with shaped-beam direct-write systems, correction by PROXECCO has been followed by rule-based modification of pattern features less than 75 nm in size [65]. Proximity correction methods have been developed to accurately fabricate three-dimensional structures such as blazed gratings by gray scale lithography [66,67]. As critical dimensions of features are reduced below 50 nm it may be necessary to use three-dimensional modeling to get more accurate results than can be obtained by two-dimensional models [68,69].

Commissioning of proximity correction software by measurement of experimental results is required because real EBL systems, substrates, and resists may not be fully described by simple models, and the models do not include effects of post-exposure processing. Data measured by a point exposure distribution method, where the measured diameters of single dots of high-contrast resist, exposed to a wide range of doses, have been used to obtain the parameters of two-Gaussian or Gaussian plus exponential models. The results show significant deviations from the simple two-Gaussian model [70]. For chemically amplified resists in which acid diffusion after exposure and post-exposure baking are not modeled by the Monte Carlo simulations or

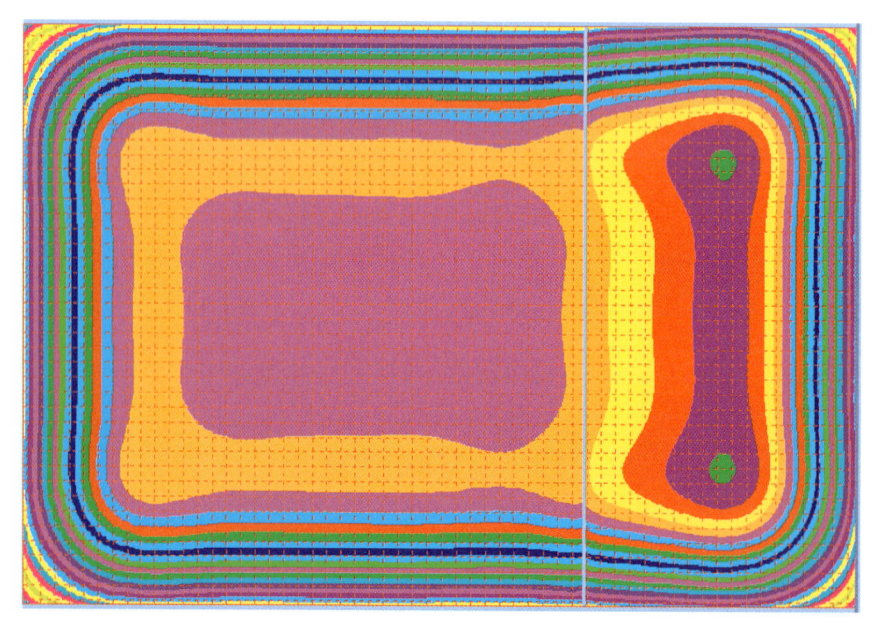

Figure 26 Design with closely spaced small dots on the left and fine lines on the right, proximity-corrected by deconvolution of the proximity function $f(r)$.

multiple-Gaussian approximations to the PSF, effective Gaussian parameters have been obtained by measuring the developed widths of multiple lines [71]. A method that works for thin resists on most solid substrates, without the need for a Monte Carlo calculation for each case, uses a Monte Carlo-derived result for the backscatter component (β) of a two-Gaussian model, and a simple lift-off experiment to find the best value of the forward-scatter component, α [72]. A larger value of α was found to be necessary for chemically amplified resists to compensate for acid diffusion during processing. Another method uses a small set of linewidth measurements to determine the parameters of a three-Gaussian model and to make corrections that take processing effects into account [73].

A few examples of proximity-corrected exposures are now presented. **Figures 27** and **28** show SEM images of the underexposed corners of arrays of dots and lines from an exposure done in the negative resist (XR-1541) without proximity correction. **Figures 29** and **30** show results of applying the corrections shown in **Figure 26**. **Figure 31** shows the proximity-corrected design of a slotted ring resonator next to a slotted waveguide. The dose is reduced in the area where the ring and the waveguide are next to each other, so the size of the waveguides and slots will be maintained; the resulting exposed resist is shown in **Figure 32**.

4.4 Summary

EBL has gone from a curiosity that grew from the advent of SEM in the early 1960s to a commercial mask making industry as early as the 1970s. Simultaneously, researchers quickly recognized the nanolithography capabilities and started relying on adapted SEMs for routine exploration of prototype devices at the nanoscale. In the 1980s, dedicated EBL systems began to be commercialized and popular, but they could be afforded by only the major research laboratories. With the rapid expansion of nanotechnology, the systems, materials, methods, and software for extraordinary levels of resolution and precision are now widely available. We believe that EBL is a unique capability that makes ever-more challenging devices and structures possible and fuels the imagination of researchers worldwide.

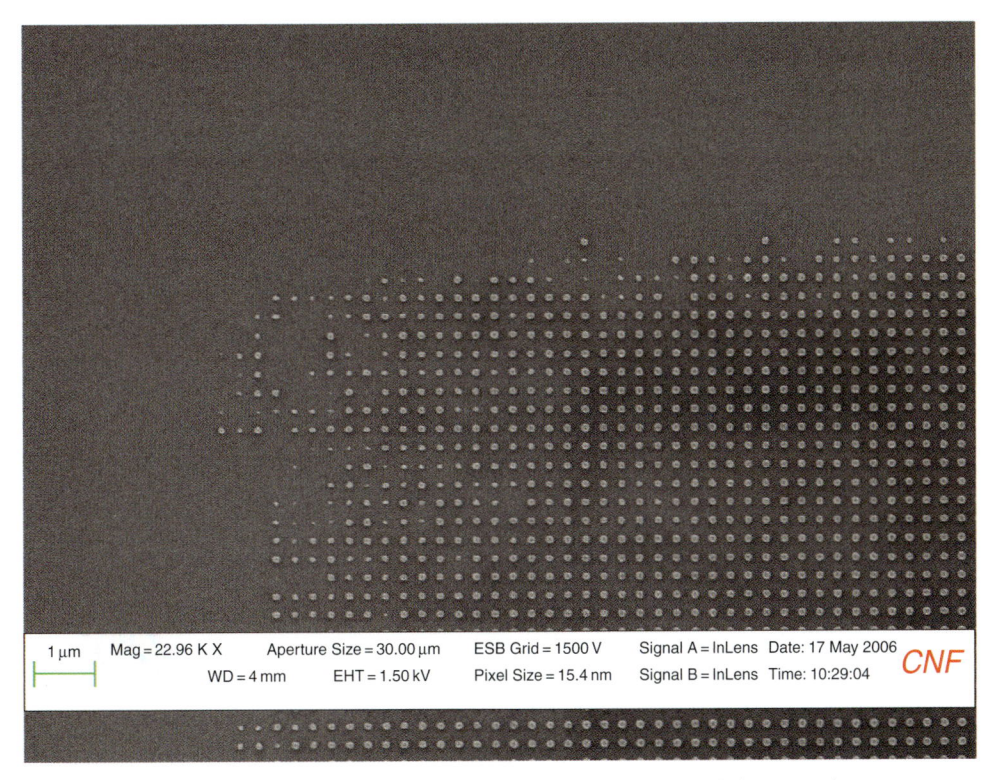

Figure 27 Underexposed dots at the corner of a test pattern exposed without proximity correction.

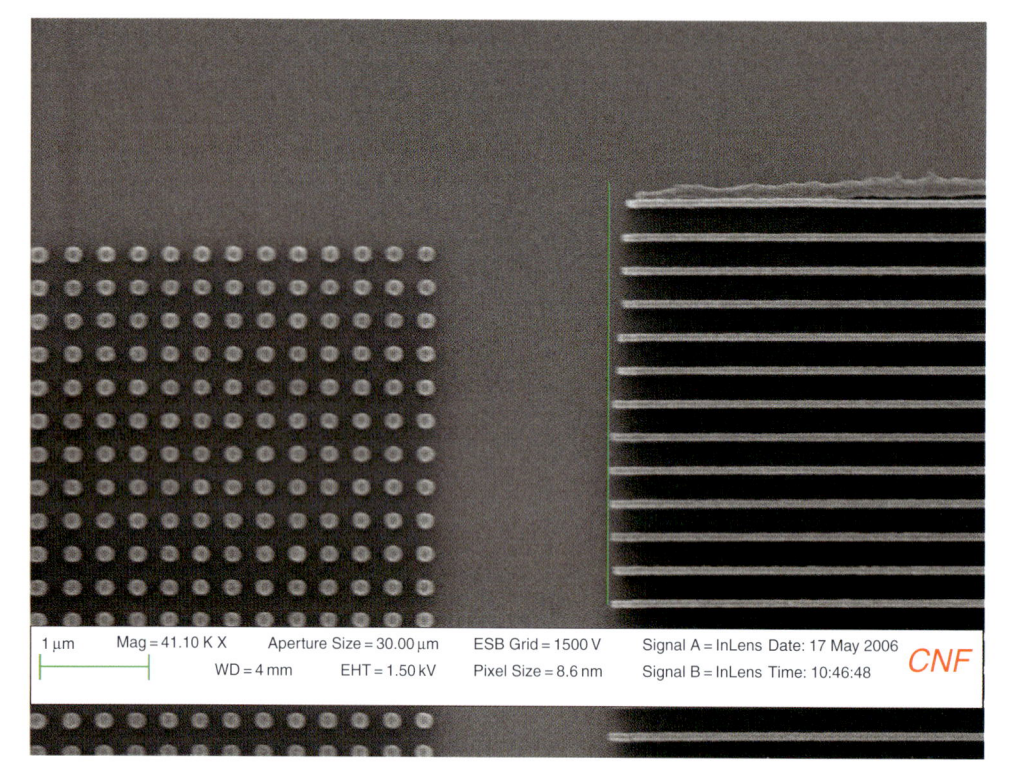

Figure 28 Underexposed line ends at the corner of a test pattern exposed without proximity correction.

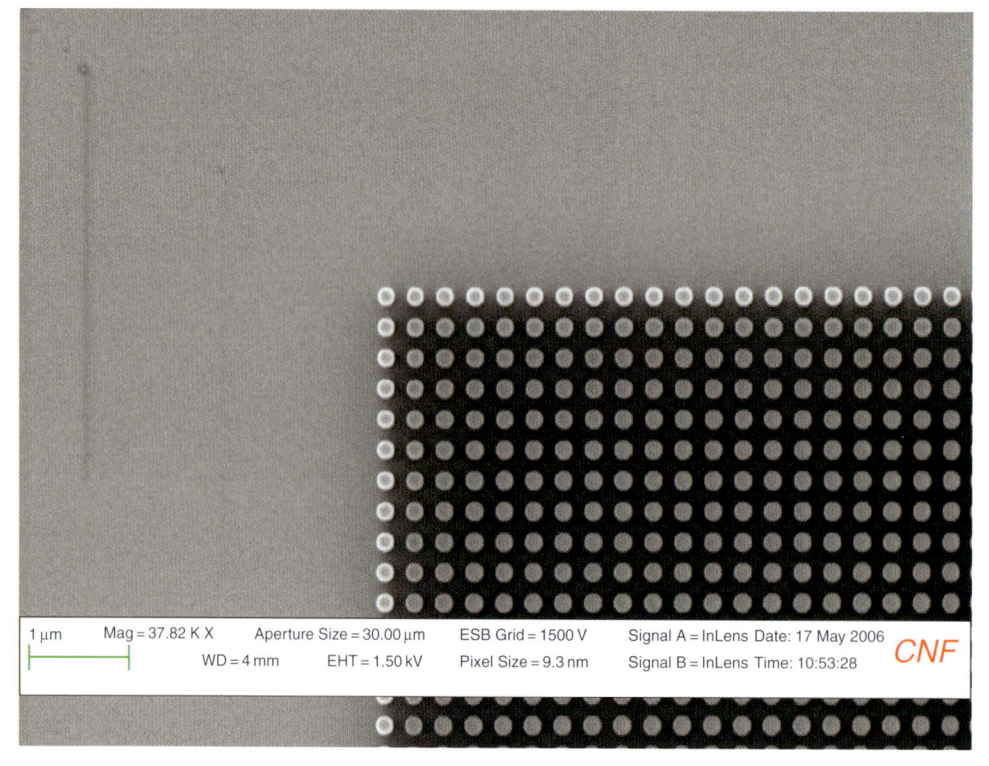

Figure 29 Corner of the proximity-corrected array of dots.

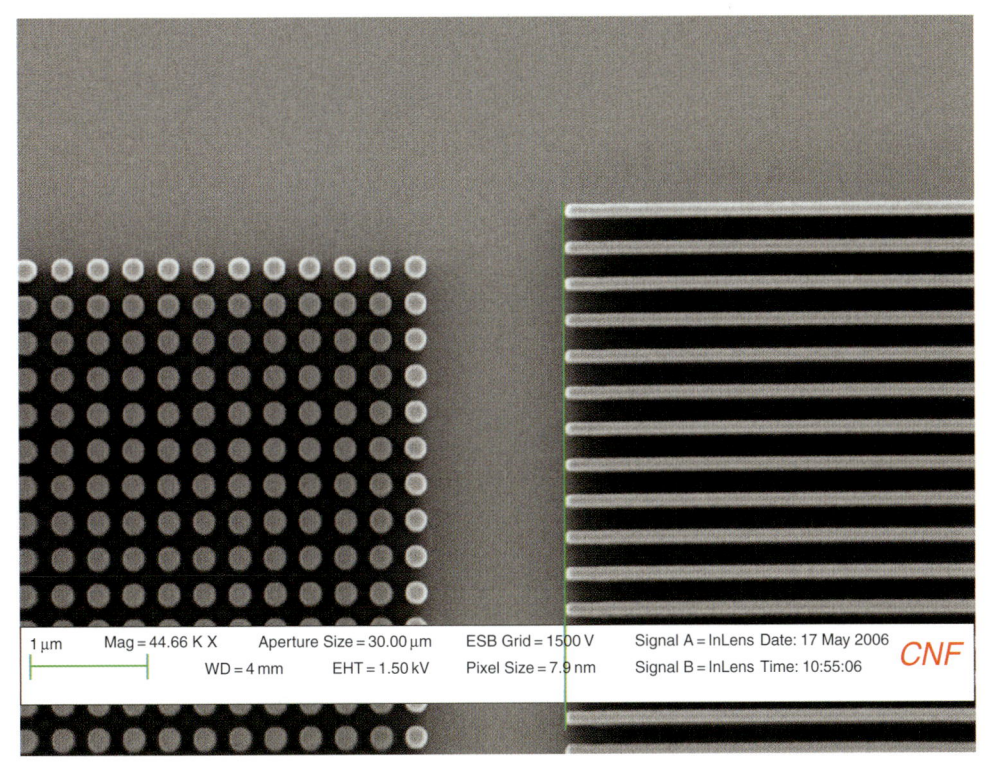

Figure 30 SEM image of corrected line ends.

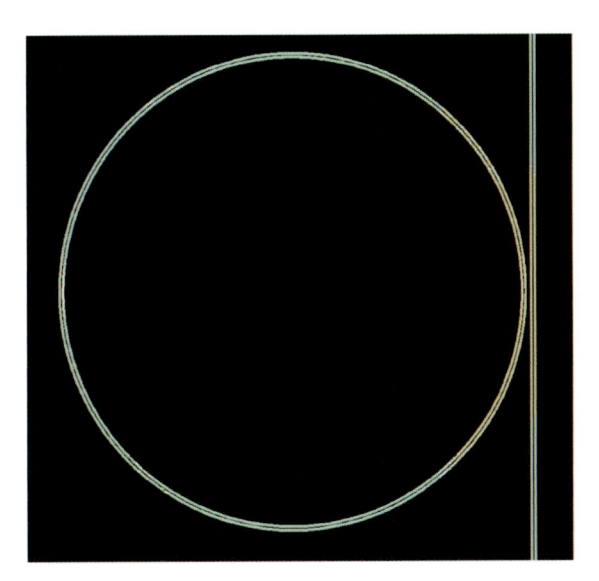

Figure 31 A design for a slotted ring resonator with proximity correction. Courtesy of Kyle Preston, Cornell University.

Acknowledgments

We wish to thank our many colleagues who generously offered images and descriptions of their research accomplishments for inclusion in this chapter. Special thanks to our colleagues at the Cornell Nanoscale Science and Technology Facility, to Rob Ilic for performing the SCELETON and BEAMER simulations featured in **Table 1** and follow on figures, and to Daron Westly for editing and improving the resist chart in **Table 2**.

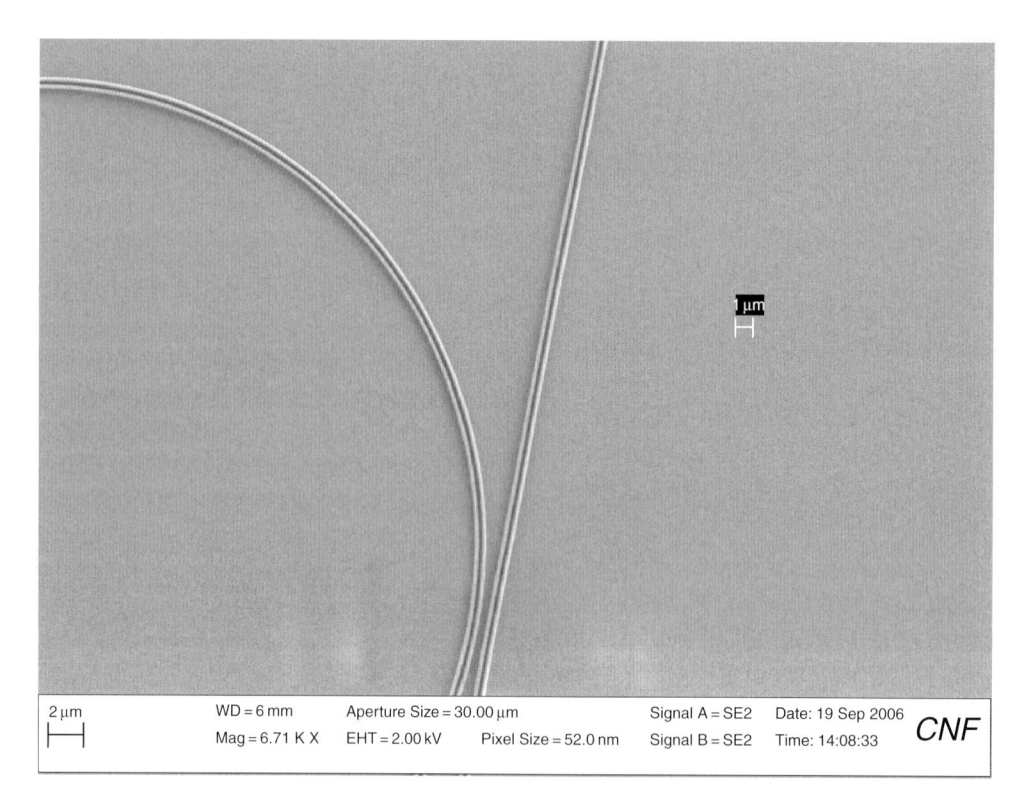

Figure 32 SEM image of slotted ring resonator made with proximity correction. Courtesy of Kyle Preston, Cornell University.

References

1. Eigler DM and Schweizer EK (1990) Positioning single atoms with a scanning tunnelling microscope. *Nature* 344: 524.
2. Timp G (ed.) (1999) *Nanotechnology*, pp. 1–5. New York: AIP Press/Springer.
3. Feynman R (1959) There's plenty of room at the bottom. *Speech Delivered on December 29th 1959 at the Annual Meeting of the American Physical Society*. http://www.zyvex.com/nanotech/feynman.html (accessed June 2009).
4. Newman TH, Williams KE, and Pease RFW (1987) High resolution patterning system with a single bore objective lens. *Journal of Vacuum Science and Technology B: Microelectronics and Nanometer Structures* 5(1): 88–91.
5. Ourmazd A (1988) Private Communication.
6. McCord MA and Rooks MJ (1997) *Handbook of Microlithography, Micromachining, and Microfabrication*, vol. 1, pp. 139–250. Bellingham, WA: SPIE Press.
7. Swanson L and Schwind G (2009) *Handbook of Charged Particle Optics*, Orloff J (ed.), 2nd edn., pp. 1–28. Boca Raton, FL: CRC Press.
7a. Suzuki K (2007) Electron Beam Lithography System. In: *Microlithography: Science and Technology,* Suzuki K and Smith BW (eds.) CRC Press, Boca Raton.
8. Pease Stanford University, private communication (2009). Prof. Dieter Kern at U. Tuebingen also supplemented the account with the earliest reference: G. Mollenstedt u. R. Speidel, *Elektroenoptischer Mikroschreiber unter elektronenmikroskopischer Arbeitskontrolle,* Physikalische Blatter (1960) 192–198.
9. Broers AN (1988) Resolution limits for electron-beam lithography. *IBM Journal of Research and Development* 32(4): 502–513.
10. Mankiewich PM, Jackel LD, and Howard RE (1984) Measurements of electron range and scattering in high voltage e-beam lithography. *Journal of Vacuum Science Technology B: Microelectronics Processing and Phenomena* 3(1): 174–176.
11. Jones G, Blythe S, and Ahmed H (1987) Very high voltage (500 kV) electron beam lithography for thick resists and high resolution. *Journal of Vacuum Science and Technology B* 5(1): 120–123.
12. Yang J and Berggren K (2007) Using high contrast salty development of hydrogen silsequioxane for sub-10-nm half pitch lithography. *Journal of Vacuum Science and Technology B* 25(6): 2025.
13. SCELETON is a Monte Carlo simulator from XLith currently marketed by Synopsis, Inc., Mountainview, CA.
14. BEAMER is an EBL layout and proximity correction software tool from GenISys GmbH, Munich, Germany.
15. Kyser DF (1983) Spatial resolution limits in e-beam nanolithography. *Journal of Vacuum Science and Technology B* 1: 1391.
16. Tennant DM (1999) *Nanotechnology*, Gregory T. (ed.), p. 171. New York: AIP Press/Springer.
17. Slot E, Wieland MJ, de Boer G, *et al.* (2008) MAPPER: High throughput maskless lithography. *Proceedinggs of SPIE* 6921: 69211P.
18. Petric P, Bevis C, Carroll A, *et al.* (2009) REBL: A novel approach to high speed maskless electron beam direct write lithography. *Journal of Vacuum Science and Technology B* 27(1): 161–166.

19. Takeya K, Fuse T, Kinoshita H, and Parker N (2008) Simulation of robustness of a new e-beam column with the 3rd-order imaging technique. In: Schellenberg FM (ed.) *Emerging Lithographic Technologies XII. Proceedings of the SPIE*, 6921:69212J–69212J.
20. Kotsugi T, Fuse T, Kinoshita H, and Parker N (2008) Shaped beam technique using a novel 3rd-order imaging approach. In: Schellenberg FM (ed.) *Emerging Lithographic Technologies XII. Proceedings of SPIE* 6921: 69211V.
21. Hatzakis M (1969) Electron resists for microcircuit and mask production. *Journal of Electrochemical Society* 116: 1033.
22. Howard RE, Hu EL, Jackel LD, Grabbe P, and Tennant DM (1980) 400 angstrom linewidth e-beam lithography on thick silicon substrates. *Applied Physics Letters* 36(7): 592.
23. Hu EL, Howard RE, Jackel LD, Fetter LA, and Tennant DM (1980) Ultrasmall, superconducting tunnel junctions. *IEEE Transactions Electron Devices* ED-27(10): 2030.
24. Mackie S and Beaumont S (1985) Materials and processes for nanometer lithography. *Solid State Technology* 28: 117.
25. Ocola LE, Tennant DM, and Ye PD (2003) Bilayer process for T-gates and G-gates using 100-kV e-beam lithography. *Microelectronic Engineering* 67–68: 104–108.
26. Dobisz E and Peckerar M (2007) *Microlithography: Science and Technology*, Suzuki K and Smith B (eds.), 2nd edn, ch. 15. New York: CRC Press.
27. Fert A (2008) Nobel lecture: Origin, development, and future of spintronics. *Reviews of Modern Physics* 80(4): 1517.
28. Wood R (2009) Future hard disk drive systems. *Journal of Magnetism and Magnetic Materials* 321(6): 555–561.
29. Braganca P, Ozatay O, Garcia A, Lee O, Ralph D, and Buhrman R (2008) Enhancement in spin-torque efficiency by nonuniform spin current generated within a tapered nanopillar spin valve. *Physical Review B: Condensed Matter Material Physics* 77(14): 144423.
30. Cui X, Lee LM, Heng X, Zhong W, Sternberg PW, Psaltis D, and Yang C (2008) Lensless high-resolution on-chip optofluidic microscopes for *Caenorhabditis elegans* and cell imaging. Proceedings of the National Academy of Sciences of the United States of America. arXiv:http://www.pnas.org/content/early/2008/07/25/0804612105.abstract.pdf (accessed June 2009).
31. Zhang Y, Petta JR, Ambily S, Shen Y, Ralph DC, and Malliaras GG (2003) 30 nm channel length pentacene transistors. *Advanced Materials* 15(19): 1632–1635.
32. Lipson M (2004) Overcoming the limitations of microelectronics using Si nanophotonics: Solving the coupling, modulation and switching challenges. *Nanotechnology* 15(10): 622–627.
33. Almeida VR, Barrios CA, Panepucci RR, et al. (2004) All-optical switching on a silicon chip. *Optics Letters* 29(24): 2867–2869.
34. Terris B (2009) Fabrication challenges for patterned recording media. *Journal of Magnetism and Magnetic Materials* 321(6): 512–517.
35. Litvinov D, Parekh V, Chunsheng E, et al. (2008) Nanoscale bit-patterned media for next generation magnetic data storage applications. Nanotechnology 2007. 7th IEEE Conference on. Piscataway, NJ, USA, pp. 395–398.
36. Yang X, Xiao S, Wu W, et al. (2007) Challenges in 1 teradot/in^2 dot patterning using electron beam lithography for bit-patterned media. *Journal of Vacuum Science and Technology* B 25(6): 2202–2209.
37. Cheng JY, Mayes AM, and Ross CA (2004) Nanostructure engineering by templated self-assembly of block copolymers. *Nature Materials* 3(11): 823–828.
38. Lu M, Tennant DM, and Jacobsen CJ (2006) Orientation dependence of linewidth variation in sub-50-nm Gaussian e-beam lithography and its correction. *Journal of Vacuum Science and Technology B* 24(6): 2881–2885.
39. Chao W, Harteneck BD, Liddle JA, Anderson EH, and Attwood DT (2005) Soft X-ray microscopy at a spatial resolution better than 15 nm. *Nature* 435: 1210.
40. Jefimovs K, Vila-Comamala J, Pilvi T, Raabe J, Ritala M, and David C (2007) A zone doubling technique to produce ultra-high resolution X-ray optics. *Physical Review Letters* 99: 264801.
41. Jefimovs K, Bunk O, Pfeiffer F, Grolimund D, van der Veen JF, and David C (2007) Fabrication of Fresnel zone plates for hard X-rays. *Microelectronic Engineering* 84: 1467–1470.
42. Stein A, Evans-Lutterodt K, Bozovic N, and Taylor A (2008) Fabrication of silicon kinoform lenses for hard X-ray focusing by electron beam lithography and deep reactive ion etching. *Journal of Vacuum Science and Technology B* 26(1): 122–127.
43. Evans-Lutterodt K, Stein A, Bozovic N, Taylor A, and Tennant DM (2007) Using compound kinoform hard X-ray lenses to exceed the critical angle limit. *Physical Review Letters* 99: 134801.
44. Tiberio R, Limber J, Galvin G, and Wolf E (1989) Electron beam lithography and resist processing for the fabrication of T-gate structures. *Proceedings SPIE* 1089: pp. 124–131.
45. Ocola LE, Tennant DM, and Ye PD (2003) Bilayer process for T-gates and G-gates using 100-kV e-beam lithography. *Microelectronic Engineering* 67–68: 104–108.
46. Moser J, Panepucci R, Huang Z, et al. (2003) Individual free-standing carbon nanofibers addressable on the 50 nm scale. *Journal of Vacuum Science and Technology B: Microelectronics and Nanometer Structures* 21: 1004–1007.
47. Retterer ST, Melechko A, Hensley DK, Simpson ML, and Doktycz MJ (2008) Positional control of catalyst nanoparticles for the synthesis of high density carbon nanofiber arrays. *Carbon* 46(11): 1378–1383.
48. Kuemmeth F, Ilani S, Ralph D, and McEuen P (2008) Coupling of spin and orbital motion of electrons in carbon nanotubes. *Nature* 452(7186): 448–452.
49. Gunawan O, Sekaric L, Ma jumdar A, et al. (2008) Measurement of carrier mobility in silicon nanowires. *Nano Letters* 8(6): 1566–1571.
50. Kyser D and Viswanathan N (1975) Monte Carlo simulation of spatially distributed beams in electron-beam lithography. *Journal of Vacuum Science and Technology* 12: 1305–1308.
51. Chang THP (1975) Proximity effect in electron-beam lithography. *Journal of Vacuum Science and Technology* 12(6): 1271–1275.
52. Kratschmer E (1981) Verification of a proximity effect correction program in electron-beam lithography. *Journal of Vacuum Science and Technology* 19(4): 1264–1268.
53. Braganca P, Krivorotov I, Ozatay O, et al. (2005) Reducing the critical current for short-pulse spin-transfer switching of nanomagnets. *Applied Physics Letters* 87: 112507–112509.
54. Almeida VR, Panepucci RR, and Lipson M (2003) Nanotaper for compact mode conversion. *Optics Letters* 28(15): 1302–1304.
55. Kern D (1980) A novel approach to proximity effect correction. In: Bakish R (ed.) *Proceedings of the Ninth International Conference on Electron and Ion Beam Science and Technology*, pp. 326–339. St. Louis, MO: Electrochemical Society.
56. Wind SJ, Rosenfield MG, Pepper G, Molzen WW, and Gerber PD (1989) Proximity correction for electron beam lithography using a three-Gaussian model of the electron energy distribution. *International Symposium on Electron, Ion, and Photon Beams* 7(6): 1507–1512.
57. Eisenmann H, Waas T, and Hartmann H (1993) PROXECCO-proximity effect correction by convolution.

Journal of Vacuum Science and Technology B 11: 2741–2745.

58. Peckerar M, Sander D, Srivastava A, Foli A, and Vishkin U (2007) Electron beam and optical proximity effect reduction for nanolithography: New results. *Journal of Vacuum Science and Technology B* 25: 2288–2294.

59. Pati YC, Teolis A, Park D, et al. (1990) An error measure for dose correction in e-beam nanolithography. *Journal of Vacuum Science and Technology B* 8: 1882–1888.

60. Simecek M, Rosenbusch A, Ohta T, and Jinbo H (1998) A new approach of e-beam proximity effect correction for high-resolution applications. *Japanese Journal of Applied Physics* 37(Part 1, No. 12B): 6774–6778.

61. Frye R, Rietman E, and Cummings K (1990) Neural network proximity effect corrections for electron beam lithography, systems, man and cybernetics, 1990. *Conference Proceedings, IEEE International Conference on* 1990: 704–706.

62. Lee S-Y, Jacob JC, Chen C-M, McMillan JA, and MacDonald NC (1991) Proximity effect correction in electron-beam lithography: A hierarchical rule-based scheme-pyramid. *Journal of Vacuum Science and Technology B: Microelectronics and Nanometer Structures* 9(6): 3048–3053.

63. Lee S (2005) A flexible and efficient approach to e-beam proximity effect correction-pyramid. *Surface and Interface Analysis* 37(11): 919–926.

64. Wuest R, Robin F, Hunziker C, Strasser P, Erni D, and Jackel H (2005) Limitations of proximity-effect corrections for electron-beam patterning of planar photonic crystals. *Optical Engineering* 44(4): 043401.

65. Manakli S, Docherty K, Pain L, et al. (2006/08) New electron beam proximity effects correction approach for 45 and 32 nm nodes. *Japanese Journal of Applied Physics. Part 1, Regular Papers, Brief Communications and Review Papers (Japan)* 45(8A): 6462–6467.

66. Hirai Y, Kikuta H, Okano M, Yotsuya T, and Yamamoto K (2000) Automatic dose optimization system for resist cross-sectional profile in a electron beam lithography. *Japanese Journal of Applied Physics* 39(12B): 6831–6835.

67. Murali R, Brown DK, Martin KP, and Meindl JD (2006) Process optimization and proximity effect correction for gray scale e-beam lithography. *Journal of Vacuum Science and Technology B* 24: 2936–2939.

68. Anbumony K and Lee S-Y (2006) True three-dimensional proximity effect correction in electronbeam lithography. *Journal of Vacuum Science and Technology B: Microelectronics and Nanometer Structures* 24: 3115–3120.

69. Ogino K, Hoshino H, and Machida Y (2008) Process variation-aware three-dimensional proximity effect correction for electron beam direct writing at 45 nm node and beyond. *Journal of Vacuum Science and Technology B: Microelectronics and Nanometer Structures* 26: 2032–2038.

70. Rishton SA and Kern DP (1987) Point exposure distribution measurements for proximity correction in electron beam lithography on a sub-100 nm scale. *Journal of Vacuum Science and Technology B: Microelectronics and Nanometer Structures* 5(1): 135–141.

71. Cui Z and Prewett PD (1998) Proximity correction of chemically amplified resists for electron beam lithography. *Microelectronic Engineering* 41–42: 183–186.

72. Rooks M, Belic N, Kratschmer E, and Viswanathan R (2005) Experimental optimization of the electron-beam proximity effect forward scattering parameter. *Journal of Vacuum Science and Technology B: Microelectronics and Nanometer Structures* 23(6): 2769–2774.

73. Hudek P and Beyer D (2006) Exposure optimization in high-resolution e-beam lithography. *Microelectronic Engineering* 83(4–9): 780–783.

5 Status of UV Imprint Lithography for Nanoscale Manufacturing

J Choi, P Schumaker, and F Xu, Molecular Imprints, Inc., Austin, TX, USA

S V Sreenivasan, Molecular Imprints, Inc., Austin, TX, USA, University of Texas at Austin, Austin, TX, USA

© 2010 Elsevier B.V. All rights reserved.

5.1 Introduction

The ability to pattern materials at the nanoscale over large areas is known to be valuable in a variety of applications [1]. In photonics applications, patterning materials with a resolution of about one-tenth of the wavelength of light can lead to photonic crystals for high-brightness (HB) light-emitting diodes (LEDs) for solid-state lighting applications (**Figure 1**) [2]. In magnetic storage, patterning magnetic materials in the sub-20 nm regime (see **Figure 2**) [3] is expected to extend the areal density growth in hard-disk drives to well above a terabit per square inch. In complimentary metal oxide silicon (CMOS) integrated circuit (IC) fabrication, devices are being fabricated today at about 50 nm half-pitch resolution. For CMOS memory and logic devices, the patterning roadmap extends well below 15 nm half-pitch as published in the 2007 International Technology Roadmap for Semiconductors [4]. An example of a sub-40 nm memory gate pattern using ultraviolet (UV) imprint lithography is shown in **Figure 3** [5]. In the area of nanoelectronics, recent literature in nanowire molecular memory indicates that ultra-high-density memory can be fabricated using patterning at the sub-20 nm scale [6]. Finally, novel biomedical application of nanopatterning are being reported, including a recent study that created nanoparticles of controlled size and shape for targeted drug delivery using UV nanoimprint lithography [7]. **Figure 4** shows pillar structures fabricated directly in a biomaterial using UV nanoimprint lithography. These pillar structures are harvested as monodispersed nanoparticles using a lift-off process. All the above applications require cost-effective nanopatterning with long-range order over large areas. A more detailed discussion of their nanomanufacturing requirements is provided in the next section.

5.1.1 Nanoscale Manufacturing Requirements

It is desirable to have a general purpose nanopatterning approach that can manufacture a variety of nanoscale devices such as those discussed above since the research and development of such an approach can benefit from the collective efforts of researchers in the various device areas. While several nanofabrication techniques are reported in the literature, only a small percentage of them have the potential to be viable in volume manufacturing since manufacturing viability requires that the patterning approach have the following attributes:

● Long-range order in nanostructures: this includes control of size of patterns and the location of the patterns with respect to an ideal grid. For example, in applications such as patterned media and CMOS devices, it is desirable that the nanoscale features have size control that is ~10% of the feature size over large areas (10–100 cm^2). Similarly, if the placement of the pattern is distorted significantly from an ideal grid, it can affect read-write head dynamics in patterned media and overlay of multiple lithography layers in CMOS ICs.

● Patterning of arbitrary structures with varying pattern densities: a general-purpose lithography approach should be able to print varying types of features with varying pattern densities. For example, the photonic crystal structure shown in **Figure 1** is very difficult to pattern using photon-based techniques when the wavelength of the source is substantially larger than the minimum feature size. Also, varying pattern densities are, in general, required for applications such as CMOS logic circuits (**Figure 5**) and patterned media (**Figure 6**) [8].

● Suited for fabricating highly integrated (multi-tier) devices: many nanoscale devices, including electronic circuits and thin film heads for hard-disk applications, require fabrication of multilayered structures with nanoscale overlay. The overlay requirement is typically from one-half to one-fourth of the minimum feature size.

● Low process defectivity to achieve high overall product yield: it is important to ensure that the defect requirements of nanoscale devices are well

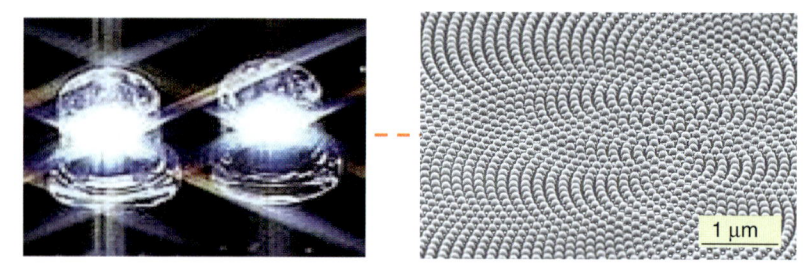

Figure 1 A representative application of nanopatterning in photonics is the use of photonic crystals for improved light extraction and control of directionality in HB LEDs. The figure on the right shows sub-100 nm dense photonic crystals arranged in complex patterns [2].

Figure 2 Sub-20 nm patterning of a hard-disk substrate [2]. Patterned magnetic media is expected to enable extension of areal density in hard disks to one terabit per square inch and beyond.

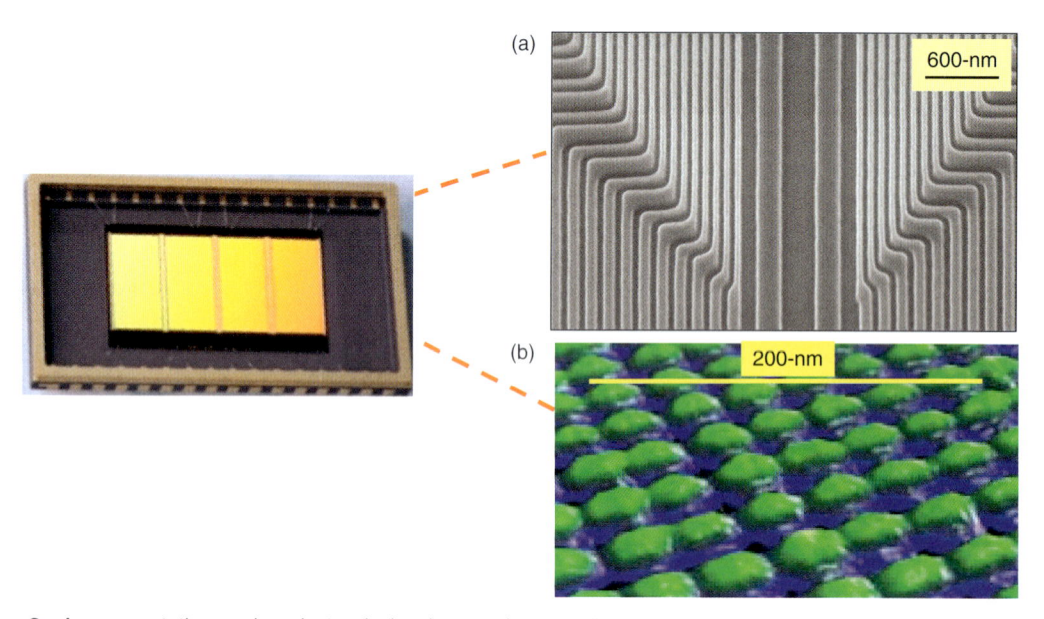

Figure 3 A representative semiconductor device that requires nanolithography is an ultra-high-density memory device. (a) a 38 nm half-pitch gate layer pattern for a flash memory device and (b) a 17 nm half-pitch molecular cross-bar memory device [6].

understood and high device yields are achieved. Acceptable defect requirements vary for various devices with CMOS logic devices being the most demanding.

- Large area, high-throughput processes: this is critical for achieving acceptable data transfer rates in the manufacturing process to achieve cost-effective device manufacturing. The importance of patterned

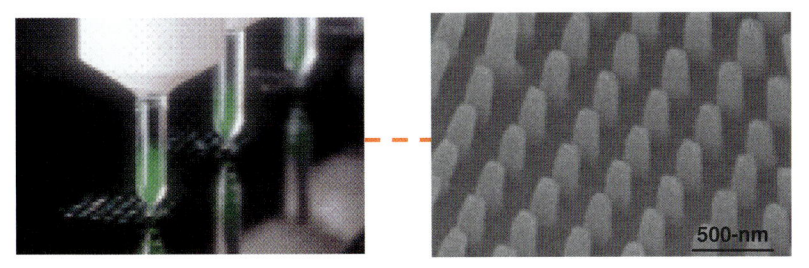

Figure 4 A representative example of biomedical application of nanopatterning involves fabrication of nanoparticles with highly controlled size and shape that has the potential to be used for targeted drug delivery for cancer treatment [7].

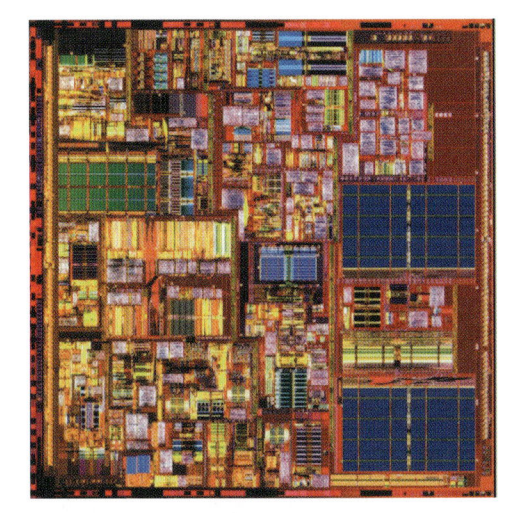

Figure 5 Intel Norwood processor (Pentium 4, 130 nm node process technology) showing pattern density variations.

area and throughput has been addressed in the literature [9]. **Figure 7** is an adaptation from **Figure 1** in Ref. [9] with current information provided for photolithography and imprint lithography. Here, a 7 nm half-pitch (14 nm pitch) is used for imprint lithography based on Ref. [10].

5.1.2 Key Patterning Challenges in Nanoscale Device Applications

Typical patterning requirements for three representative applications are provided in **Table 1**. In addition to basic resolution, which tends to be the primary focus of a good portion of the nanofabrication research literature, other metrics such as pattern uniformity, pattern placement, alignment and overlay, and throughput are also presented. In each row, the most demanding requirement across the three applications is highlighted. While these applications for nanopatterning are not exhaustive, it provides a baseline to establish the critical challenges in the manufacture of nanoscale devices.

From **Table 1**, it can be seen that basic patterning resolution is most demanding for patterned magnetic media, pattern placement requirement in the short range (spatial wavelength in the micron scale) is most aggressive for patterned media, while long-range placement (over tens of millimeters) is most aggressive for silicon ICs. With respect to imprint mask life and defectivity, as well as overlay, silicon ICs are the most demanding application. Under the other requirements, polar pattern layout in patterned media makes this application the most demanding since electron beam tools for mask fabrication are not readily available in polar coordinate format [11, 12]. Patterning over pre-existing topography [13] and printing partial fields on the edge of the wafer [14] are driven by the silicon IC requirements.

The cost metrics include throughput defined in wafers per hour (WPH) and shots per hour (SPH); see definition at the bottom of **Table 1**. The WPH throughput requires high-speed substrate handling and automation, which is well established in the hard-disk industry. The SPH throughput is limited by imprint-specific technical challenges and therefore the silicon IC application is the most demanding application for throughput. Sensitivity to cost of capital and consumables is driven by the hard-disk application. Finally, the substrate nanotopography (see Refs. [15,16] for a definition of nanotopography) is the roughest for III–V substrates used for fabricating LEDs since these compound materials are costly to polish and obtain free of epitaxial defects [17].

The imprint process results will be presented in Section 5.4. For each metric presented in **Table 1**, the most demanding application among all the applications will be used as a reference to understand the status of imprint lithography with respect to each of these metrics.

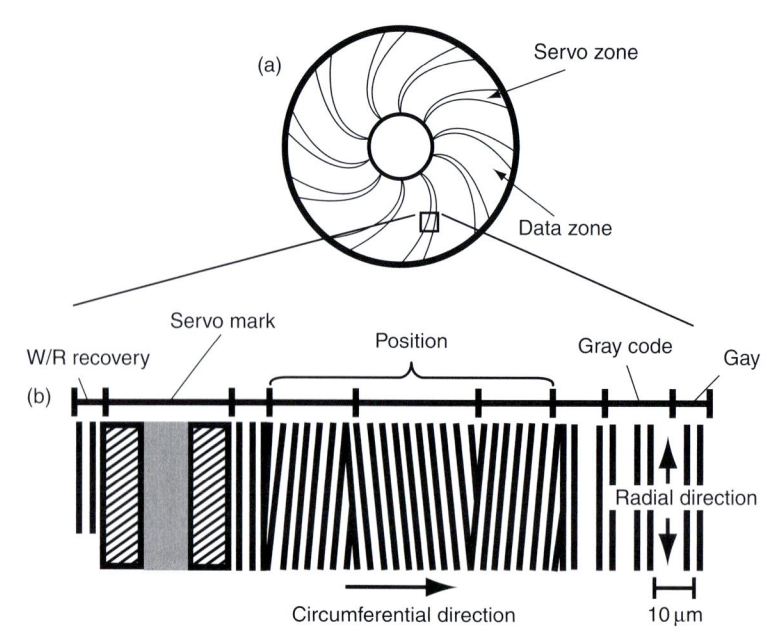

Figure 6 Varying pattern density requirement for patterned media based on data and servo zones in a hard-disk drive described in Ref. [8].

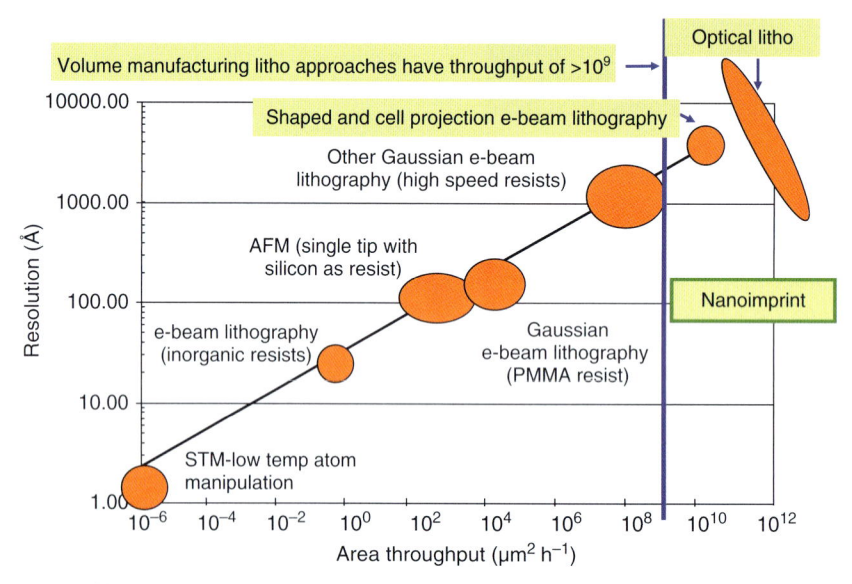

Figure 7 Importance of lithographic throughput for volume manufacturing Adapted with permission from Christie RK, Marrian CRK, and Tennant DM (2003) Nanofabrication. *Journal of Vacuum Science and Technology A* 21(5): S207.

5.2 Top-Down Nanopatterning Options

There are two primary approaches to large-area nanopatterning: bottom-up approach involving self-assembly or directed self-assembly, and top-down approach such as photolithography and imprint lithography. Directed self-assembly processes have historically lacked in their ability to pattern arbitrary structures, obtain long-range order, and achieve adequate throughput for manufacturing applications. In recent years, significant progress has been made by researchers (see, e.g., Ref. [18]) in improving long-range order for

Table 1 Key manufacturing requirements for three representative nano scale devices[a]

Representative applications	Photonic crystals for HB LEDs	Patterned media (>400 GB/sq. in.)	Silicon ICs (sub-32 nm half-pitch)
Performance metrics			
Patterning resolution			
Minimum feature size	<50 nm	<20 nm[c]	<32 nm
Line edge roughness (3σ)	<5 nm	**<2 nm**	<3 nm
Feature variation (3σ)	<5 nm	**<2 nm**	<3 nm
Pattern Placement			
Placement – short range (3σ)	Relaxed	<2 nm	<3 nm
Placement – long range (3σ)	Relaxed	Relaxed	**<3 nm**
Imprint mask life and defectivity			
Mask breakage	Unacceptable	Unacceptable	Unacceptable
Local damage	Relaxed	Somewhat relaxed	<0.01 *defects/cm²*
Alignment and overlay			
Alignment (3σ)	Relaxed	<5 μm	**<3 nm**
Overlay (3σ)	Relaxed	Relaxed	**<7 nm**
Other requirements			
Pattern layout	Cartesian	***Polar***	Cartesian
Patterning over pre-existing features	NA	NA	***Yes***
Edge field printing	NA	NA	***Yes***
Cost metrics			
Throughput[b] (substrate size)	>30 WPH; >30 SPH (50–100 mm)	300–600 ***WPH***; 600–1200 SPH (25–95 mm)	>20 WPH; >2000 ***SPH*** (300 mm)
Sensitivity to capital/consumable cost	Moderate	***High***	Moderate
Nanotopography for acceptable substrate cost (substrate type)	***Rough (III-V substrates)***	Smooth (glass disks)	Smooth (silicon wafers)

[a] These requirements are 'typical' values to motivate the discussion presented in this article; they should not be construed to represent validated industry requirements for these applications.
[b] WPH represents wafers/disks per hour; SPH represents shots per hour. The SPH for patterned media is 2X the WPH since both sides of the hard disks need to be patterned. The SPH for silicon ICs is assumed to be approximately 100X of the WPH since the 300 mm wafers are processed in a step and repeat manner and for typical stepper field size (26 by 32 mm), about 100 full and partial fields can be printed on the wafer.
[c] In each row, the bold and italicized text represents the most demanding requirement across the 3 applications.
NA, not available.

certain types of repeating structures. In this article, bottom-up approaches will not be discussed since it focuses on being able to print arbitrary structures at high throughput.

5.2.1 Photon-Based Top-Down Patterning Techniques

Top-down patterning with photolithography has been the workhorse of the semiconductor industry for over 30 years. In addition to its continued improvement in resolution, the appeal of photolithography has been its long-range order, precision overlay, and its extremely high data transfer efficiency making it attractive in the manufacturing of large-scale integrated devices. The leading edge photolithography solution is 193 nm photolithography (with water immersion of 193 nm-i) and is expected to possess resolution down to about 45 nm half-pitch beyond which double patterning techniques will be required. Double patterning will be expensive and put significant constraints on device designs essentially requiring repeating grating structures of a given pitch across an entire mask design. Critical layers for NAND Flash memory applications are conducive to such constrained designs and therefore, a 193 nm-i along with double patterning is likely to allow sub-45 nm half-pitch patterning for such memory devices. However, logic devices that have more complicated designs are expected to have significant challenges at half-pitch values of 45 nm and below. Further, for applications such as patterned media with its polar coordinate arrangement of sub-20 nm patterns, and LEDs which require complicated sub-50 nm hole and pillar structures (see **Figure 1**) 193 nm-i will simply not possess the resolution. Extreme ultraviolet lithography (EUVL) at 13.2 nm (soft X-ray) wavelengths is being developed by the semiconductor industry to provide patterning capability of arbitrary nanostructures. The EUVL technology has several major challenges that need to be addressed including soft X-ray source power; a high-resolution photoresist with high sensitivity and low line edge roughness, tool complexity and cost, and a reflective mask blank infrastructure [19]. If these challenges are successfully overcome, then EUVL is expected to be viable for sub-30 nm half-pitch lithography for silicon ICs. However, the cost structure and timing of EUVL does not match the requirements of emerging applications such as patterned media and HB LEDs.

In addition to EUVL, other high-throughput next generation lithography silicon IC fabrication techniques that have previously been considered include X-ray lithography (XRL) [20,21], electron projection lithography (EPL) [22], and ion beam projection lithography (IPL) [23]. As of now, excepting EUVL, however, none of these techniques are no longer being pursued by the silicon IC industry in a systematic manner.

Photolithography, EUVL, XRL, IPL, and EPL are all techniques that require a high-cost mask that is typically fabricated using direct-write electron beam tools. These masks are then proposed to be replicated using the high-throughput process and the mask cost becomes acceptable for high-volume device applications such as microprocessors and memory. However, for 'short-run' devices that are not required in high volumes (sometimes also known as application-specific ICs or ASICs), the cost of the mask may no longer be acceptable. During the last decade, other approaches known as mask-less techniques have been considered for such ASICs. These techniques are targeted for medium throughput by using an array of direct write beams or for nanopatterning. In addition, when devices are comprised of specific patterns such as repeating gratings or dots, other techniques such as interferometric lithography (IL) or plasmonic lithography have been developed. Arrayed X-ray with zone plates have been introduced by Prof. Hank Smith's group at MIT [24], and plasmonic patterning was first introduced by Atwater and coworkers from Caltech [25] and later adapted for spinning substrates by Prof. Zhang's group at University of California, Berkeley [26]. IL techniques also have been developed; see Prof. Brueck's review article [27] for IL process and its application. One of the common goals of all these techniques is to eliminate complicated masks.

Among the arrayed lithography techniques, multicolumn electron beam lithography is probably the most developed. To overcome the slow throughput limit, significant research resources have been focused on integrating thousands of arrays of electron beams into a small space. An example of such an effort is the one by 'Mapper' where about 13 000 electron beam columns are being integrated to achieve high-throughput mask-less lithography [28].

Plasmonic lithography utilizes a well-known property of metal surfaces wherein free electrons oscillate when exposed to light. These oscillations, known as evanescent waves, absorb and generate light. The wave length of the evanescent waves is much shorter than that of the original energy source, which enables patterning of fine features. High-throughput plasmonic patterning has been demonstrated by mounting a plasmonic lens on an air-bearing flying at nanoscale proximity to the surface of a substrate, and by modulating the available light to expose the photoresist layer selectively as the substrate moves. An array of plasmonic heads and an equal number of addressable beams are suggested along with a high-speed rotating substrate to satisfy the high-throughput requirement [26].

IL utilizes optical interference between two coherent light beams with opposing incident angles. The optical interference generates sinusoidal intensity variation on a resist-coated substrate to form repeating structures, such as grating patterns, in a photoresist film [27]. Three main control variables to define the interference period are the wavelength of the coherent light, the incident angle, and the refractive index of the photoresist. After exposing the photoresist once, a second interference lithography, oriented normal to the first interference, leads to equally spaced dots. While interferometric techniques can pattern a large area fast and without masks, their application has been limited to only repeating patterns such as gratings or dots. Nonlinear IL for arbitrary two-dimensional patterns has been presented by Bentley [29] but no practical solution was provided to achieve a high-speed patterning capability.

5.2.2 Proximity Mechanical Top-Down Nanopatterning Techniques

Proximity mechanical nanopatterning techniques include atomic force microscope (AFM)–based techniques such as dip-pen lithography, soft lithography, and imprint lithography. As compared to optical schemes, these techniques achieve their nanoscale patterning resolution using mechanical means and have the potential to achieve nanoscale replication at a low cost. Arrayed AFM systems have been used to improve the throughput of AFM-based nanoresolution patterning. These systems are sometimes also known as parallel scanning probe lithography systems. The pioneering work by IBM research laboratory in Zurich [30,31] and that of Quate [32] and his coworkers at Stanford appear to be among the earliest literature in parallel AFM systems for enhanced throughput patterning. Several other researchers, including Mirkin's team at Northwestern [33–35] and researchers in Japan [36,37], have developed various aspects of parallel lithography using AFM tips. While the progress over the last decade in the sense of throughput has been remarkable, these systems are still much slower than photolithography and UV imprint lithography techniques with respect to data transfer rate as defined in Ref. [9].

Figure 8 illustrates soft lithography, thermal, and UV imprinting processes. Whiteside and coworkers

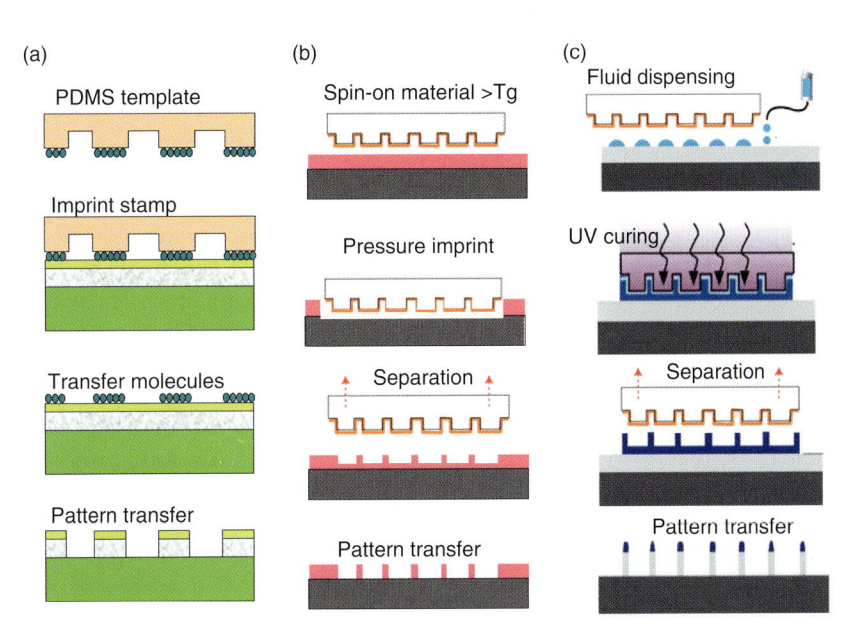

Figure 8 Three nanoimprint lithography schemes developed as (a) soft-lithography, (b) thermal imprinting process, and (c) resist-dispense UV imprint process.

[38] initiated micron- and submicron-size feature patterning using the concept of soft lithography. The mold for soft lithography is generally made with a very flexible material such as polydimethylsiloxane or PDMS [38]. This simple scheme enables patterning without the burden of complicated systems, but also allows the use of flexible low-cost molds instead of rigid molds such as silicon or fused silica used in imprint lithography. Ref. [38] presents a variety of related techniques using PDMS masters including micro-contact printing (μCP), replica molding (REM), micro-transfer molding (μTM), micro-molding in capillaries (MIMIC), and solvent-assisted micro-molding (SAMIM). In μCP, for example, 'ink' adhering to the raised features of a polymer mold is transferred to a substrate using intimate contact of the soft mold and the substrate. This layer acts as an etch mask for subsequent etch processes. This technique has been used very successfully for making micron- and submicron-scale pattern. However, it has some limitations with single-digit nanometer resolution of imprint processes discussed next that essentially mold a liquid to conform to the shape of a rigid imprint mask. The resolution limitations of soft lithography are due to the potential for feature distortions in soft lithography particularly in the presence of nonuniform pattern densities. Further, the use of PDMS type masters does not allow nanoscale control of alignment for fabricating multilayered integrated devices.

The second process shown in **Figure 8** is the hot embossing or thermal nanoimprinting process where the patterned template is in contact with a high-viscosity spin-coated layer. Subsequently, the temperature of the spin-on material is raised above its T_g while applying a high pressure to the stack of the mold and substrate to conform them. Thermal nanoimprint lithography has been known to possess sub-10 nm replication resolution and an attractive cost structure based on the early results from Ref. [39]. Early research related to imprint lithography includes techniques for fabricating optical components such as the one by Napoli and Russell [40]. Compact disks were one of the early applications for a molding technology, similar to nanoimprinting, even though compact disks did not possess sub-100 nm structures. In thermal nanoimprint lithography, both the high temperature and pressure can induce technical difficulty associated with not only its low throughput, due to high-viscous flows and heating-cooling cycle times, but also practical fabrication of multilayer circuits. Thermal processes can induce significant deformations in both the template and substrates if they are not made of thermally matching materials. Due to the high pressure compression between two interfacing surfaces, it is very difficult using this technique to effect nanoscale alignment and overlay of multilayered structures as needed in highly integrated devices. Finally, this method is inherently harsh and can cause damage to the template, thereby impacting process yield. Recently, Schift et al. [41] presented a new thermal imprinting scheme to ramp up the heating and cooling steps so that its overall imprinting cycling can be completed within 10 s. While this addresses the throughput issue to some extent, it still is undesirable for alignment or overlay of multilayered structures and from the perspective of mask protection from damage.

The UV nanoimprint processes uses UV cross-linking materials as imprinting layers. Philips presented a UV micro-molding process using photo polymerization (2P process) where the fluidic layer was cross-linked by UV polymerization process. The work of Philips regarding cross-linked polyacrylates for potential applications such as optical-fiber coatings, aspherical lenses, and information storage systems was presented by Kloosterboer [42]. Recent work in UV nanoimprinting has demonstrated ultra-high-resolution replication including sub-20 nm lithography [3], and feasibility of sub-3 nm lithography [43], (It should be noted that these typical requirements are chosen to motivate the discussion in this article; they do not represent validated industry requirements for these applications.)

The third process in **Figure 8** presents a version of UV nanoimprint process, wherein a low-pressure and room temperature process is achieved due to the use of a low-viscosity UV-curable material that is dispensed as discrete drops of picoliter (pl) volume prior to gap closing between the mask and the substrate to induce the filling of the mask features. This resist–dispense-based UV nanoimprint process has been referred to in the literature as the step and flash imprint lithography (S-FIL) [44]. Alignment error measurement and compensation are carried out when the resist is still in a liquid form, just prior to UV curing. Subsequently, UV curing that solidifies the resist due to cross-linking is carried out, followed by separation of mask from the imprint material left on the substrate. The viscosity of the UV fluid can be tuned low enough so that the fluid filling can be accomplished in the absence of a high compressive pressure, which makes it compatible for low-pressure and room temperature processing. Depending on how the imprinting fluid is applied to the substrate, this group of UV imprint processes can be further divided into drop-on-demand (DoD) process similar to the third process in **Figure 8** [45,46] and spin-on processes [47]. UV imprint process has been implemented for both the step and repeat (S and R) and whole wafer (WW) processes [45–47].

The ability to handle arbitrary pattern densities is critical to a variety of applications including silicon ICs and patterned magnetic media (see **Figures 5 and 6**). Drop dispense UV imprint lithography approach, the third approach in **Figure 8** can handle the varying pattern densities much more effectively than spin-on techniques without compromising on throughput or process control [1,14]. Further, as indicated in the literature, use of spin-on material deposition can lead to nonuniform pressure distribution [48]. This can be further aggravated by the presence of pattern density variations [49]. This process uses low-viscosity UV imprint fluids where an array of small drops of the imprint fluids is dispensed in a DoD manner depending on the pattern density variations of the imprint mask. The uniform low-pressure, room temperature nature of this process, and the transparent imprint templates make it particularly attractive for a high-resolution layer-to-layer alignment [14,50]. Another aspect of UV nanoimprint that assists in the alignment is the presence of a nanoscale layer of low-viscosity liquid between the template and substrate [50]. It has also been demonstrated that this process is inherently very low in defectivity [51,52], and is capable of eliminating soft contaminants from the template during consecutive imprinting process using a self-cleaning process [53,54].

For additional literature related to imprint lithography, the reader can refer to Ref. [55], and to some extent a review article by Guo [56]. Hybrid patterning processes, which use both thermal and UV nanoimprinting, such as nanocasting lithography [57], have also been reported. To obtain very high throughput processes for applications requiring modest imprint resolution, 'roller imprint process' [58,59] and 'sheet imprint process' [60] have been introduced (see **Figure 9**). Recently, a review of processes specifically for patterned media application was documented by Sbiaa *et al.* [61], where various patterning methods and ideas were compared objectively. The patterned media application is important because it represents a high-volume application (similar to semiconductor lithography) and requires very aggressive nanoscale patterning [3, 11, 12, 62]. Patterned magnetic media for hard disk is a big commercial opportunity and imprint lithography appears to be the only approach that provides the resolution and cost structure needed for this application.

Much of the literature discussed here – other than the literature related to UV nanoimprint lithography process of **Figure 8** – presents results addressing only a portion of the manufacturing requirements discussed in Sections 5.1.1 and 5.1.2. Fine-resolution patterning capability of imprint lithography processes for a variety of pattern types has been demonstrated by various research groups. However, to expand beyond such resolution demonstration to manufacturing, it is necessary to address multiple technical issues simultaneously such as large-area fabrication with a good long-range order, high throughput, low defectivity, overlay, etching, and pattern transfer, etc.; and to provide statistically significant data to support the status of each of these issues. This makes the problem very challenging and a systematic interdisciplinary approach that focuses on addressing all the requirements discussed in Sections 5.1.1 and 5.1.2 is needed. The discussion presented in the following sections strives to make

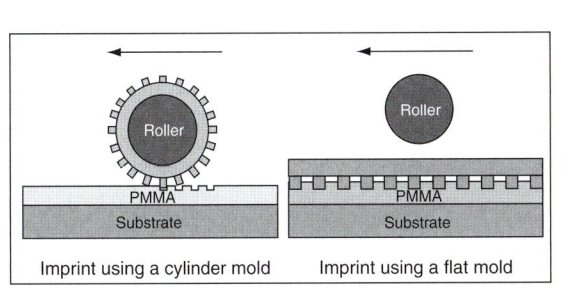

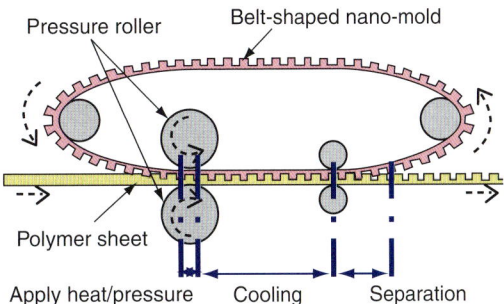

Figure 9 Roller imprint lithography and sheet imprint lithography from Refs. [58,60].

this point by discussing all the trade-offs needed, and the delicate balance in the properties of tools, masks, and materials required to achieve a process that is valuable in manufacturing.

Another important factor that significantly increases the chances of success of such new technologies is leveraging of mature technology infrastructure that has been developed previously for other manufacturing applications. In Section 5.3, it is discussed that fused silica masks and photomask fabrication infrastructure have been used to develop imprint masks; acrylate materials have been chosen as the basis for developing UV imprint materials; and the field size in imprint steppers is chosen to be ≤ 26 mm $\times$ 33 mm to ensure compatibility with the large existing installed base of photolithography tools. These choices are examples of technologies that have previously been developed by other industries for manufacturing applications and are being leveraged here for the development of UV imprint lithography.

Over the years, there have been legitimate concerns about the viability of imprint lithography in manufacturing due to a lack of comprehensive development in areas such as long-range order, overlay, low defectivity, and high throughput. However, in recent years it is becoming evident that imprint lithography is maturing in its manufacturing attributes. It is opening up emerging nanoapplications such as patterned media and HB LEDs due to its versatility in patterning arbitrary nanostructures over large areas on a variety of substrates. It is also demonstrating the potential to become a drop-in replacement to photolithography at sub-30 nm half-pitch lithography, particularly for nonvolatile memory applications where high-density patterning at low cost is highly desirable.

5.2.3 Comparison of Nanoimprint Techniques

In this section, the discussion is confined only to nanoimprint lithography processes, and a systematic comparison of the various aspects of nanoimprinting processes is provided by presenting a one-to-one comparison of competing ideas that have been explored in the literature pertaining to imprint lithography. These one-to-one comparisons have been discussed below.

(1) *Thermal versus UV-curing imprint processes*: a comparison of these two processes is presented in **Table 2**. The comparison is performed with respect to a selected set of relevant manufacturing metrics. It is clear that, in every metric, except for functional material patterning, UV nanoimprint process is seen to be superior to thermal imprinting. Therefore, for sacrificial resist imprint processes, UV nanoimprint is preferred over thermal imprinting for nanoscale manufacturing. (The sub-3 nm results are not strictly dense pitch structures; therefore this result is preliminary. However, it appears to provide compelling evidence of the replication capability. With respect to strict half-pitch results, 7 nm half-pitch (14 nm pitch) appears to be the best result to date using nanoimprint lithography [10].)

(2) *Spin-on versus DoD resist deposition processes*: a comparison of these two processes is presented in

Table 2 Thermal vs. UV imprint processes

Manufacturing metric	Thermal process	UV process
Feature distortion	High in the presence of nonuniform patterns	Low, particularly for resist drop dispense approach
Material viscosity	High under T_g	Typically viscosity material ranging from 1 to 100 cP
	Low above T_g	
Fluid filling	Pressure-driven	Pressure and/or capillary-driven
Material delivery	Spin-on followed by baking	Spin-on of wet film or resist drop dispensing
Process time	Slower due to need to heating and cooling, and due to higher viscosity materials[a]	Faster
Functional patterning	More material options are possible	Material options are more constrained
Overlay capability	Limited to several hundred nanometers	<15 nm is possible [14]

[a] As Chou and others have recently demonstrated, the throughput problem can be overcome for the special case of direct patterning of silicon [106].

Table 3 Spin-on vs. resist DoD UV imprint processes

Manufacturing metric	Spin-on process	DoD process
Pattern density	Best for repeating patterns	Arbitrary patterns are OK
Material viscosity	~50 cP or higher [47]	1 to 10 cP [44,71,1]
Fluid filling	Mainly pressure-driven, assisted by capillary	Mainly capillary-driven
Residual layer control with resist compensation	Not possible	Possible with high level of control (see Section 5.4.1 of Ref. [14])
Resist capital requirement	Automated clean track that can handle wet films	In-built dispense unit with no additional capital cost
Resist material usage	Spin-on of thin films with nanoscale control result in >99% material wastage.	No material wastage, consumable material used is ~0.1% as compared to spin-on films
Step and Repeat (S and R) vs. whole wafer (WW) processing	Best suited for WW. S and R possible but evaporation may lead to process nonuniformity.	S and R and WW are both possible [14,46] and have been deployed commercially

Table 3. The comparison is again performed with respect to a selected set of relevant manufacturing metrics. As observed from the comparison, for several important performances, for reasons of the capital cost and consumable cost, D-o-D resist dispensing leads to a better manufacturing process.

(3) *Step and repeat (S and R) versus whole wafer (WW) imprint processes:* **Table 4** highlights the key compromise between S and R and WW processes – nanoscale overlay versus high throughput. For applications requiring nanoscale alignment for multilayer processing (such as semiconductor ICs, thin film heads for magnetic data storage, and advanced nanowire memory devices), S and R processing is likely needed because of the difficulty in maintaining proper overlay control over large areas. This requires compromised throughput for such applications.

The rest of this article will focus on D-o-D resist dispense UV nanoimprint processing techniques.

5.3 Building Blocks for UV Nanoimprint Lithography

The three primary building blocks that contribute to a UV imprint lithography process are the (1) imprint masks, (2) imprint materials, and (3) imprint tools (**Figure 10**). **Figure 11** provides a side-by-side comparison of a photolithography tool and an imprint tool (A discussion of functional imprint materials versus sacrificial imprint resists is provided in Section 5.3.2. **Figure 10** illustrates an S and R tool, but similar concepts are used in whole substrate imprint tools [46]. The imprint tool figure illustrates how the drop–dispense approach is integrated along with the wafer motion system and the template leveling mechanism in the tool. The imprint material is dispensed onto a field just prior to lowering of the template. Then, the template is lowered to capture the imprint fluid between the template and the wafer. This is followed by UV curing and separation of template prior to moving to pattern on the next

Table 4 S and R vs. WW processing

Comparison metric	S and R	WW
Mask field	≤26 mm × 33 mm[a]	Whole substrate
Nanoscale overlay	<15 nm demonstrated	Sub-100 nm overlay is very difficult
Resist delivery	DoD is best suited to avoid process nonuniformity due to evaporation	Both spin-on and DoD are OK
Throughput	A function of number of imprints per wafer, inherently slower than whole wafer	Superior
Key compromise	Nanoscale overlay at reduced throughput	Modest overlay at higher throughput

[a] This field size is chosen in imprint steppers to allow compatibility with existing photolithography infrastructure [14].

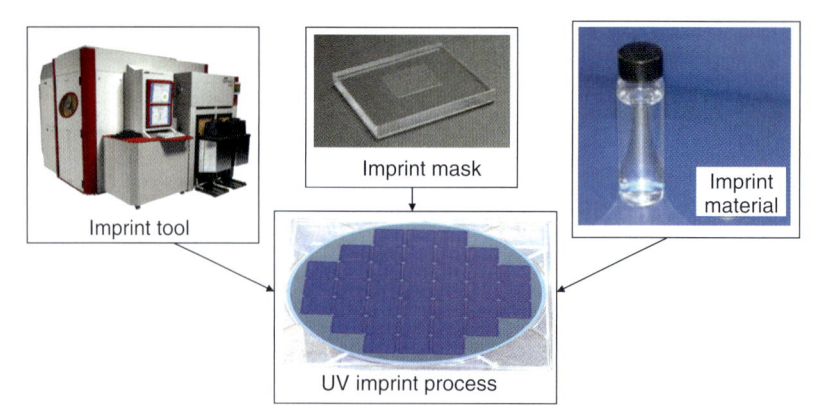

Figure 10 Building blocks of the UV imprint process: tool, mask, and material.

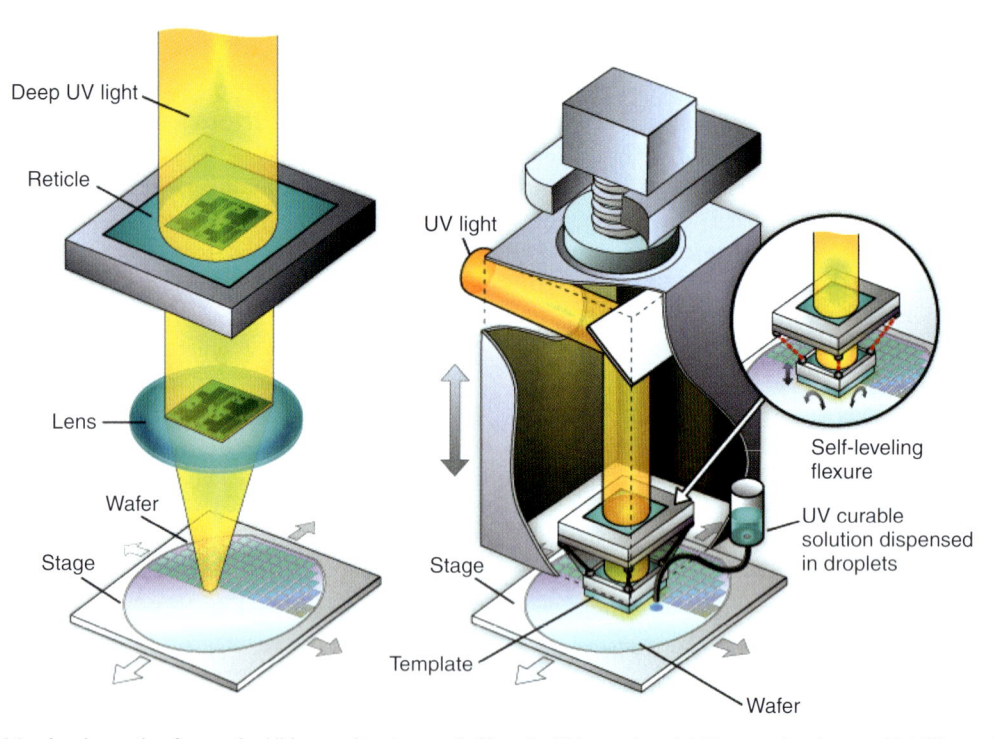

Figure 11 A schematic of an optical lithography stepper (left) and a UV nanoimprint lithography stepper (right) based on the process of

field. The three building blocks are discussed in more detail in the next three subsections.)

5.3.1 UV Imprint Masks

There are essentially two kinds of imprint masks: (1) masks that are made of rigid, fused silica blanks, similar to photomask blanks, that are 6.35 mm thick that are desirable for applications requiring tight overlay such as silicon ICs and thin film heads where the substrate nanotopography is smooth; and (2) imprint masks that are made of thinner substrates that are <1 mm thick for applications that do not require nanoresolution overlay. These thin templates allow patterning over rough nanotopography over large areas, as is the case with III–V substrates (see discussion in Section 5.3). Further, if substrates possess high-spatial-frequency front side defects such as epitaxial processing defects in GaN [17], masks that are made of softer materials can also be used to substantially conform to these high-spatial-frequency defects [17,63].

Imprint masks can be made with commercial photomask materials and processes. This use of existing infrastructure is a significant advantage relative to technologies such as EUVL that requires novel mask substrates with up to 80 reflective films coated on it. However, the 1-1 pattern size requirements between the mask and the wafer patterns (known as 1× masks) do, in principle, test the resolution capability of imprint masks. While photomasks have historically been 4×, the advent of optical proximity correction (OPC) subresolution features, which will soon be no more than 1.3× the minimum feature size on the wafer [4] are accelerating the resolution of mask e-beam writers. In addition, for imprint masks, since the chrome film is only being used as an etch mask (rather than based on optical opacity requirements), it is possible to use thinner chrome and electron beam resist. This has allowed pushing the resolution down to the required 1× dimensions of 32 nm half-pitch by using standard variable shape beam e-beam tools available in mask houses [5,64–66]. Pattern placement is also an issue for 1× masks in overlay-sensitive applications such as silicon ICs and thin film heads for hard disks. The approaching application of double patterning for 193 nm immersion lithography is pressing the existing photomask industry to meet very tight image placement specifications, allowing the progress of pattern placement to sub-3 nm, 3σ, over a 26 by 32 mm field for 1× masks. The use of high-resolution Gaussian beam tools (that are slower with respect to throughput) can even allow sub-20 nm patterning of imprint masks (see resulting pattern in **Figure 12**, Ref. [98]).

Another advantage of imprint masks is that they do not need OPC subresolution features and, therefore, the mask pattern is identical to the patterns eventually required on the wafers; this results in low write times in the electron beam tools [65]. In addition, for high-volume applications such as semiconductor memory and patterned magnetic media, an e-beam master can be used to create many replicas using the imprint process itself. This allows for significant reduction in the impact of the cost of the e-beam master to the cost of ownership of imprint lithography for high-volume device manufacturing [67]. In the semiconductor memory application, mask and replica inspection for defectivity is needed to qualify the imprint masks. It is important to develop such techniques and implement them in such a manner that the cost of the replicas can be maintained low. This is an area of ongoing research and techniques are now being developed [68,69].

In applications such as HB LEDs for solid-state lighting, where the device size is much smaller than mask field size (equal to the size of the wafer), an e-beam master that is only as big as a single device (say a few millimeters on a side) can be fabricated. An imprint lithography-based S and R process can then

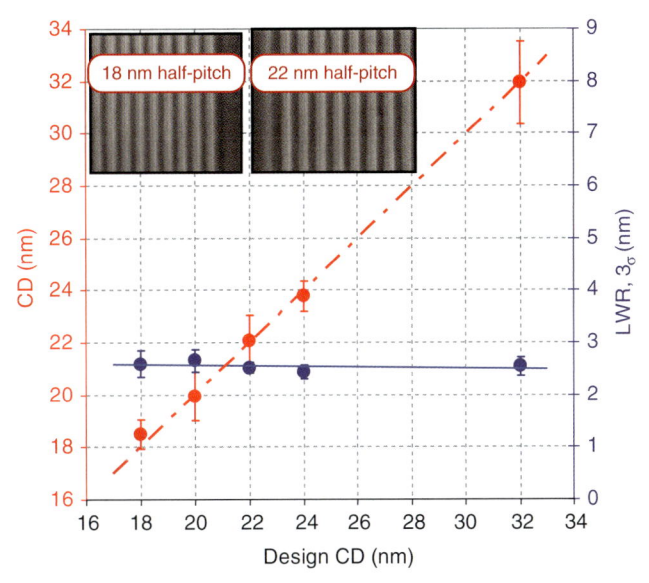

Figure 12 A fixed linewidth roughness of ~2.5 nm (3σ) is observed for patterns varying from 18 nm half-pitch to 32 nm half-pitch [98].

be used to fabricate the large-area replicas for volume production. In this case the cost impact of e-beam master on the eventual device is very small. A detailed discussion of the imprint mask replication process is provided in Ref. [70].

5.3.2 Imprint Materials

Imprint materials that are used to form the pattern on the substrate can be one of two kinds: (1) a sacrificial imprint resist, where the material is used as an etch mask to create patterns in a substrate (similar to photoresists in optical lithography) or (2) a functional imprint material, where the imprinted material directly becomes part of a device. Functional materials are more difficult to design as they have to satisfy both the imprint process constraints for achieving a high-fidelity nanopattern, and satisfy the properties required by the device over the course of the life of the device. In nanoscale imprint lithography, functional materials are not very mature from the perspective of manufacturing (see requirements listed in Section 5.1.1). Only preliminary work has been done in this area, where researchers have directly patterned biomaterials [7] and low-k dielectric materials [104]. However, most of the literature in imprint lithography is directed towards sacrificial imprint resists, and the discussion here is restricted to sacrificial imprint resist materials.

Once the imprint resist is cross-linked using UV exposure, the solid layer formed on the substrate is used as a sacrificial layer for pattern transfer. Pattern transfer can be carried out through either using a regular tone etch process or a reverse tone etch process as illustrated in **Figures 13** and **14**. The standard tone process in **Figure 14** is similar in its form to the trilayer process in traditional 193 nm photolithography. Imprint resist is used as a mask to etch inorganic hard mask (IHM) layer; IHM is, in turn, used to etch organic hard mask (OHM) layer to amplify the aspect ratio. Eventually, OHM serves as a hard mask to etch the substrate to create high-aspect-ratio structures in the substrate. In the reverse tone process (as shown in **Figures 13** and **14**), the etch process results in a pattern tone that is inverse of the tone of the pattern created in the imprint resist.

To achieve a nanoscale manufacturing process that includes good imprint fidelity, low defectivity, high throughput, and precise pattern transfer, every material involved in the imprinting stack has to be optimized carefully. Since the hard mask materials have been studied extensively in the semiconductor fabrication literature, this section will focus on the imprint resist and the adhesion layer as these two materials are specific to imprint lithography.

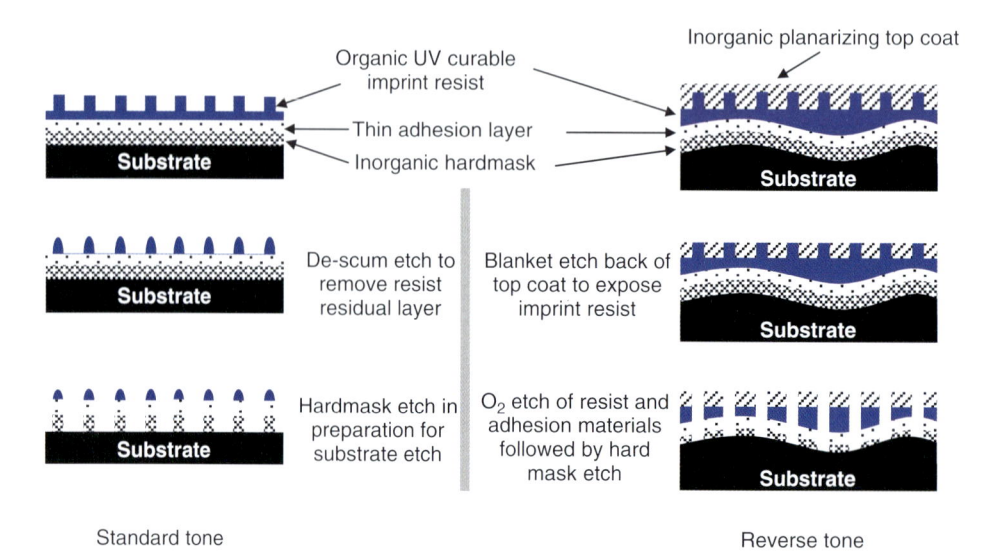

Figure 13 Standard tone and reverse tone etch process for imprint lithography are shown in this figure. Reverse tone process has several advantages [13] including the ability to transfer patterns on substrates with pre-existing topography as shown above. Note the figure is not to scale, for example, the adhesion layer thickness is exaggerated to highlight it.

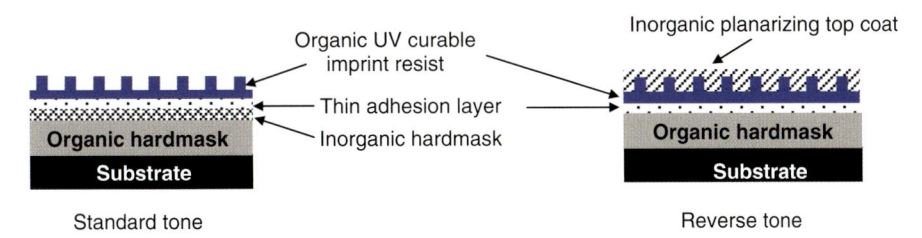

Figure 14 Standard tone and reverse tone film stacks are shown for a bi-layer hard mask. Such bi-layer hard masks are being commonly used in semiconductor fabrication for sub-50 nm half-pitch lithography where high-aspect-ratio (>10:1) nanopatterns are needed in the substrate.

5.3.2.1 Imprint resist

The imprint material requirement can be grouped into fluid (rheological) properties prior to UV curing and solid mechanical properties after UV curing. The fluid properties include viscosity, wettability, evaporation characteristics, and dispensability, whereas the solid mechanical properties include cohesive yield strength, elongation to failure, adhesion to substrate, and ease of release from imprint mask. The optimal imprint materials are formulated with fluid and mechanical considerations discussed above.

Each process step illustrated in part (c) of **Figure 8** possesses a unique requirement for the resist material development. First, fluid should be dispensable, which requires a certain range of viscosity and low evaporation. Once the fluid drops are dispensed on the substrate, it is necessary to enhance the fluid filling motion. Low-viscosity fluids tend to spread faster as compared to viscous fluid. However, low-viscous and high-volatile material can suffer an evaporation problem leading to unpredictable process performance. Therefore, fluid viscosity and evaporation properties need to be carefully balanced. When all features are fully filled, UV cross-linking step follows. Efficient cross-linking materials require short exposure times without causing excessive heating to the template and substrate. Next, during the separation step, it is necessary to ensure that the material adheres to the substrate and not to the template. To reduce the surface energy of the template, a high surface concentration of fluorine is required, but this then restricts the wettability and filling speed, requiring a delicate balance. On the wafer surface, an adhesion promotion film can be used, but this layer needs to be very thin (≤ 2 nm). This can be formulated to ensure adhesion to multiple surface materials and also considering the wettability factor. Adhesion layers will be discussed in Section 5.3.2.2. Each feature of the imprinted pattern needs to remain intact during the separation. This mandates a material with adequate mechanical strength, toughness and Young's Modulus to maximize the aspect ratio that can be used and yet completely prevent the possibility of a feature being left in the template. Adding polar components helps with these properties but excessive amounts increases the surface tension and reduces the fill speed. Not illustrated in part (c) of **Figure 8** is the etching step, where an additional material requirement is to provide a proper etch selectivity (see **Figures 13** and **14** for etch discussions). Finally in defect sensitive applications such as semiconductor IC fabrication, the purity of the resist material must meet the stringent requirements of <10 ppb or better.

Many types of UV-curing chemistries including methacrylate, epoxy, vinyl ether, thiol-ene, and acrylate have been considered for imprint resist. Methacrylate and epoxy have considerably slower polymerization rate; hence not practical for high-throughput volume manufacturing. Most of the UV imprint materials are acrylate-based [71]. Vinyl ether materials that possess much lower viscosities have also been discussed in the literature [72,73]. Vinyl ether relies on cationic polymerization to cure the liquid resist. It is inherently not inhibited by oxygen which scavenges radicals. On the other hand, the cationic initiator is sensitive to the presence of excessive moisture. The main concerns of the vinyl ether are shelf life stability and experimentally observed high separation force against fused silica mask. It is hypothesized that the acidic nature of cationic polymerization may strongly interact with silanol group at fused silica surface to cause release problems. Although one can envision that masks made with different materials will have less interaction with vinyl ether at surface, the use of fused silica masks leverages an existing photolithography infrastructure that has been developed over several decades and is a key factor for success in manufacturing.

Thio-lene provides a promising alternative to acrylate chemistry for UV imprint resists. It has distinct advantages over acrylate system including low oxygen inhibition during cure and low separation force against fused silica mask [74]. Thiol-ene chemistry goes through radical step growth polymerization as shown below:

$$RSH + PI \rightarrow RS \text{ (UV exposure, PI is photoinitator)}$$

$$RS + CH_2CHR' \rightarrow RSCH_2CH\cdot(R')$$

$$RSCH_2CH\cdot(R') + RSH \rightarrow RSCH_2CH_2R' + RS$$

Reduction of oxygen inhibition problem in this case is due to the fact that the thiol group, SH, can recycle the peroxide radical into an active thiyl radical:

$$RSCH_2CH(R')OO + RSH \rightarrow RSCH_2CH_2(R')OOH + RS$$

The disadvantages of thiol-ene system are strong odor of thiol, short shelf life, and limited amount of commercial compounds available. For manufacturing purposes, it is desired that shelf life of resist exceeds several months. Thiol-ene's shelf life is considerably shorter than that although addition of stabilizer will help in prolonging it [75].

In summary, the leading choice of polymerization chemistry for imprint resists is acrylate materials. Acrylate materials possess good curing speed, and are typically one-order magnitude faster than methacrylate chemistry in this respect. It is really advantageous that vast amounts of different acrylates are readily commercially available, which provides large formulation latitude in making imprint resist with different arrays of viscosity, surface tension, and mechanical properties. Acrylates also have excellent shelf life lasting at least 6 months; and can be possibly extended beyond 1 year [74]. The primary limitation of acrylates is that it is susceptible to oxygen inhibition.

Both the liquid and the solidified phase characteristics of an imprint material set need to be optimized as stated earlier. Liquid phase properties to be considered include:

- viscosity,
- drop evaporation rate and multicomponent evaporation uniformity,
- surface tension,
- resist wettability of adhesion layer, and of mask surface covered with release agents,
- UV sensitivity, and
- particle control and trace metal concentrations for electronic applications.

Solid phase properties to be considered include:

- absolute adhesion values,
- selective adhesion ratios,
- mechanical strength, and
- etch resistance

The UV imprint process shown in **Figures 8, 10**, and **11** involves field-to-field (for stepper) or whole substrate dispensing of low-viscosity (<10 cP) UV cross-linkable acrylate imprint resist mixtures. As imprinting mask touches the fluid drops, capillary force pulls mask closer toward imprinting substrate and helps facilitating fluid spreading. The resistant force from viscous dissipation, which is proportional to the viscosity, on the other hand, tends to slow down the fluid spread. It is clear from imprinting throughput perspective that a low-viscosity resist is preferred. A low-viscosity liquid is advantageous in many ways::

- enhanced local spreading leading to insensitivity to pattern density variations, and reduced overall fluid spread time leading to high throughput;
- improved mask life due to a lubricated template-wafer interface avoiding 'hard contact' between mask and wafer; and
- Lubricated (*in situ*) nanoresolution alignment corrections prior to UV exposure.

The limitation of resist viscosity reduction mainly comes from requirements of evaporation control, inkjet dispenser stability, and polymerization speed. The first-generation resist [44] contained large amount of tert-butyl acrylate. This resist operated reasonably well when dispensed at nanoliter volumes. However, practical device fabrication constraints such as need for thin residual layers and ability to handle pattern density variations at acceptable throughput led to the need for picoliter-sized drops. Tert-butyl acrylate has exceptionally low viscosity of 0.9 cP at 20 °C, and it has good mechanical properties; therefore, it is considered as one of the best acrylate components to cut down the viscosity of the imprint resist. Unfortunately, the vapor pressure of tert-butyl acrylate is high at 8.6 Torr at 25 °C and hence effects excessive evaporation at ambient conditions. The fast evaporation problem is exacerbated by picoliter inkjet drop volumes which have high surface-to-volume ratios. The evaporation curve for the material used in Ref. [44] is shown in **Figure 15**. The most volatile component in the original material (namely tert-butyl acrylate) is added at

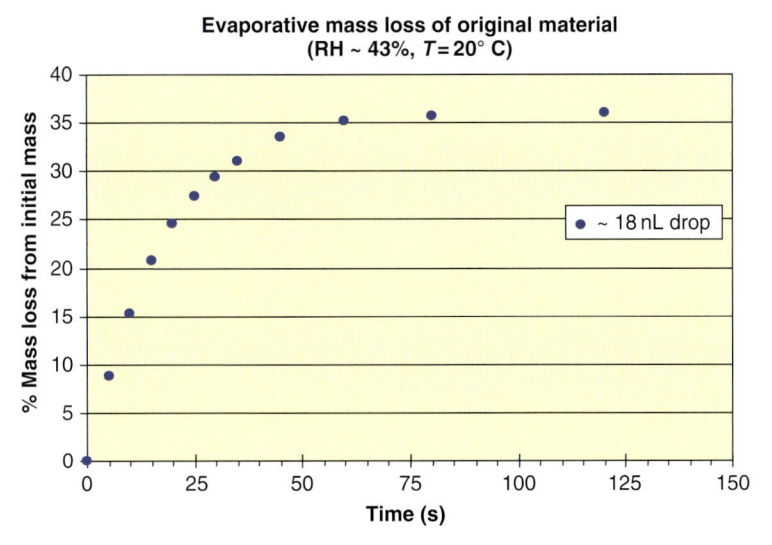

Figure 15 Evaporation characteristics of early acrylate based imprint resists [71].

37 part per hundred by weight. The curve suggests that most of the tert-butyl acrylate is evaporated while the majority of other components are retained after 40 s. This inhomogeneous evaporation results in composition changes during the imprinting process and impacts the mechanical properties of the cured material adversely. This composition change cause a cross-linker concentration variation. The impact of cross-link density on curing speed was studied via real-time Fourier transform infrared spectroscopy (FTIR) technique. **Figure 16** shows the disappearance pattern of acrylate double bond in resist (monomer) as a function of UV exposure time. Increasing cross-link density improves the reaction rate, but excess cross-linking tends to lower the ultimate reaction conversion.

Apart from controlling the absolute evaporation rate, it is important to manage the relative evaporation rates of multiple components in the resist to achieve overall uniform evaporation for an interval of time. In a silicon-containing resist, spatial distribution of silicon content within an imprint field has to be precisely controlled to achieve uniform etch characteristics. Imprinting starts off by depositing a plurality of droplets in a certain pattern to minimize trapping of gases when spreading the imprinting

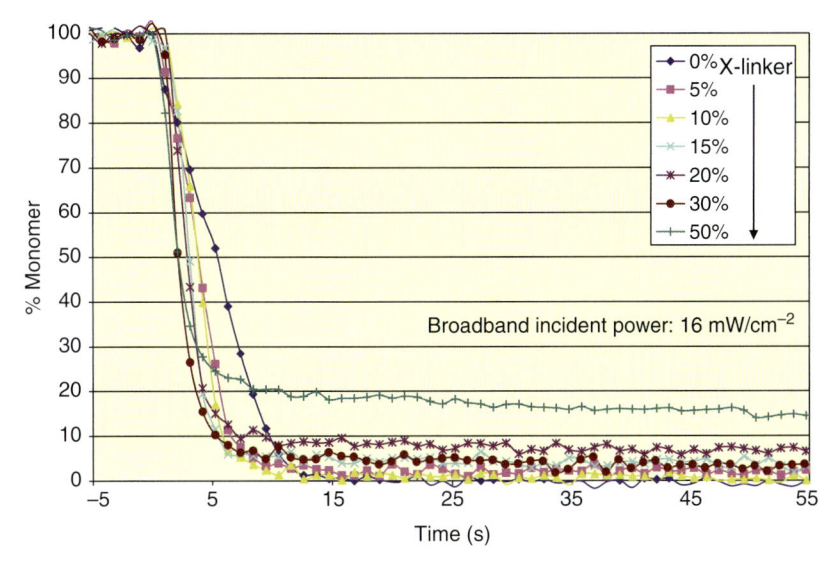

Figure 16 Impact of cross-link density on curing speed.

material over substrate. The drop pattern hence is formed from a plurality of droplets that are deposited sequentially in time. Earlier, it was found that etch characteristics of the solidified imprinting layer differed as a function of the position of the droplet in the drop dispense sequence, referred to as sequential etch differential (SED). It was determined that the SED is attributable to variation, over the area of solidified imprint layer, of the cross-linked and polymerized components, in particular the silicon-containing component, that form solidified imprinting layer. The component variation in solidified imprint layer was observed only when the imprinting material was in a fluid state, due to evaporation. Early silicon-containing resists in which silicon-containing acrylate evaporates substantially faster than the remaining components of the composition were found to present etch nonuniformity. Since drop size is at picoliter level, evaporation of various components is inevitable. However, it was found that by closely matching the evaporation rate of multiple components within an interval of time, the composition change over the solidified imprinting layer was minimized and SED factor was no longer observed. **Figure 17** shows the evaporation rates of various silicon-containing acrylate materials. SED was observed when silicon-containing acrylates A and B were used together with the non-Si-containing acrylate shown in **Figure 4**. On the other hand, Si-containing acrylate C and nonSi-containing acrylate closely match in terms of the evaporation rate. The resulting resist from C and nonSi-containing acrylate showed significantly reduced SED problem. In device fabrication applications, since most practical photolithography and e-beam lithography resists are nonsilicon materials, practical considerations have ensured imprint resists being purely organic with no silicon content. The SED problem discussed here is not an issue for nonsilicon resists. The rest of the discussion will be specific to nonsilicon resists as silicon-containing resists are not being used in imprint lithography applications today.

Imprint resist viscosity, evaporation rate, along with its surface tension will influence inkjet stability [76]. Typical, commercial UV-curable inks are formulated to have 10–20 cP. With software and hardware optimization, inkjettable viscosity range can be expanded to include fluid having low, single-digit cP.

Having a robust release technology that facilitates separation at the interface between the cured resist and the mask is one of the key needs for imprinting material development. A widely adopted release approach is to covalently bond a fluorinated self-assembled monolayer (SAM) material onto template surface [53,77,78]. The SAM layer is susceptible to damages from imprint resist shearing during fluid spread, during mask separation, and UV-induced degradation. Decline of fluorine concentration at mask surface was observed as number of imprints increases which indicates the degradation of the SAM [79]. The relatively short durability of the SAM has adverse effects on defect levels and such short process life is not viable in manufacturing due

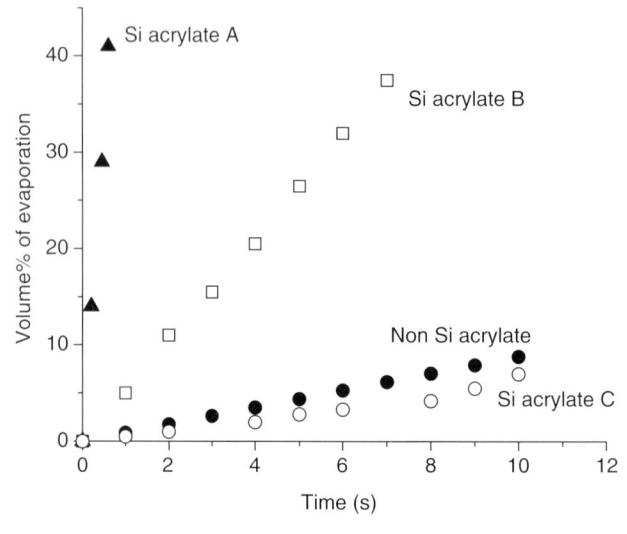

Figure 17 Evaporative rate of different components of imprint resist, 80 pl drop size.

to cost considerations. An *in situ* self-repair mechanism for the release layer during imprinting process could overcome the above-mentioned shortcomings, and is highly desirable [71]. Besides the release performance, wetting property of the release layer situated on the mask surface has to be carefully examined. The choice of release approach is based on multiple factors, some of which can be counterproductive or fundamentally conflicting to each other. High fluorine content is desired for enhancing the release performance of cured resist from mask; on the other end, low fluorine content is desired at the mask surface for improving liquid resist spread and feature filling speeds. Low fluorine content can reduce the hydrophobicity of mask surface and make it more wettable. In the end, a fine balance between the release and wetting performances has to be struck; this, in turn, will influence the choice of the release materials. Recently, IBM team [80] found that metal oxide and nitrides have low release force against certain classes of imprint resists. For example, they work particularly well with POSS methacrylate resists. For the acrylate resists evaluated, the typical fluorinated release agent treatment still outperformed metal oxide and nitrides. The durability and process longevity information of the release layer was not presented in this reference.

A four-point bending fixture [81] was adopted for testing release force of solidified resist from fused silica masks. Two 1"-wide, 3"-long, 1-mm thick glass slides were laid in a cross-direction pattern. Imprint material was cured in-between the slides. Four-point bending compression force was applied to separate the slides. The maximum force/load was taken as the adhesion value. The schematic of such a 4-point bending test is shown in **Figure 18**. Using such tests, the adhesion value of the imprint material to release layer-coated fused silica, and to adhesion layer-coated substrates were measured. The selective adhesion ratio, defined as (imprint material-adhesion layer adhesion force)/(imprint material-mask adhesion force) was measured. When this ratio was measured for the first- and second-generations resists used in Refs. [44] and [71], respectively, it was found to be greater than 5 in each case. Here, the adhesion layer used was a commercial BARC material (DUV30J) supplied by Brewer Sciences [71]. These ratios have been significantly improved using improved designs of the adhesion layers as discussed later.

The imprint resist contact angle against the release agent covered mask surface can significantly impact the quality of feature filling [82]. According to Ref. [82], assuming a typical resist contact angle of 10° against the imprint substrate, feature aspect ratio of 1, and feature width of 1 μm, at 30° contact angle of resist against mask, full feature filling can be achieved. On the other hand, if the contact angle is increased to 45°, the feature can be only partially filled. This illustrates that mask surface cannot be pushed toward unrealistic high degree of hydrophobicity just for the sake of release performance. Mask surface needs to be moderately wettable by imprint resist; otherwise, imprinting quality will suffer.

High mechanical strength will help the cured resist sustain the stress created during mask separation, and maintain the fine feature fidelity. Adding polar components into the imprint resist will increase the mechanical properties; on the other hand, adding too much of the polar components will lead to the increase of liquid resist's surface tension which will result in higher contact angles. **Figure 19** plots the tensile pull mechanical testing results of three generations of imprint resists [44,71]. Both tensile

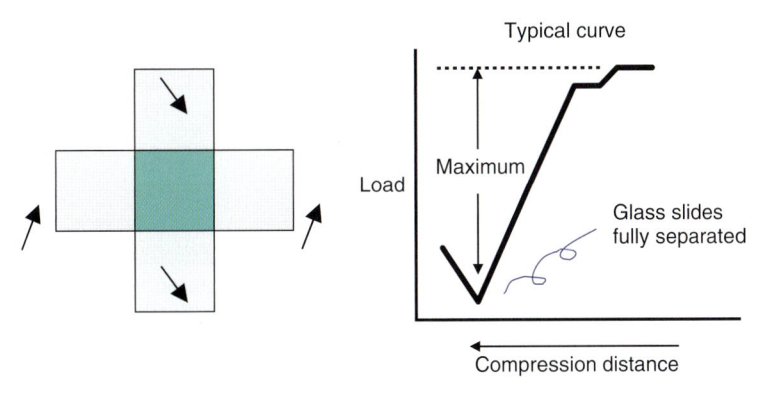

Figure 18 Four-point bending test for adhesion measurement.

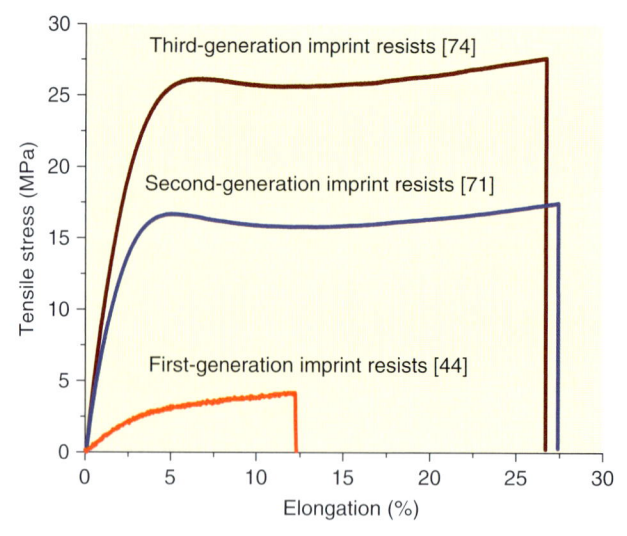

Figure 19 Stress–strain curves of three generation of imprint resists.

modulus and tensile strength have been improved significantly with improved resist formulations. The resist material discussed in Ref. [71] has achieved the desirable characteristics of both strong mechanical properties and high selective adhesion ratio. The imprint process life improvement is shown in **Figure 20**. Top-down SAM image relating the imprint material of Ref. [44] shows significant degradation of 50 nm patterns even at 50 consecutive imprints. The performance of the material used in Ref. [71] shows top-down SAM images of 50 nm lines that show no visible feature degradation after 440 consecutive imprints.

Compared to deep-UV photolithography, imprinting process requires different level of radiation energy. In the case of deep-UV photolithography, chemical amplification of the resist is typically achieved after the light exposure and through postexposure thermal bake; while imprinting process requires the resist's full mechanical properties to be developed through UV exposure and before the mask separation. Curing time can be reduced by introducing more efficient photoinitiators, better matching of photoinitiator absorption peaks, and light source peaks, adding additives to address oxygen inhibition issue, and increase the power of UV lamp.

Etch resistance of the cured resist affects the quality of pattern transfer during reactive ion etching (RIE) processes. In standard tone etch process using organic resists (see, e.g., **Figure 14**), the imprint resist

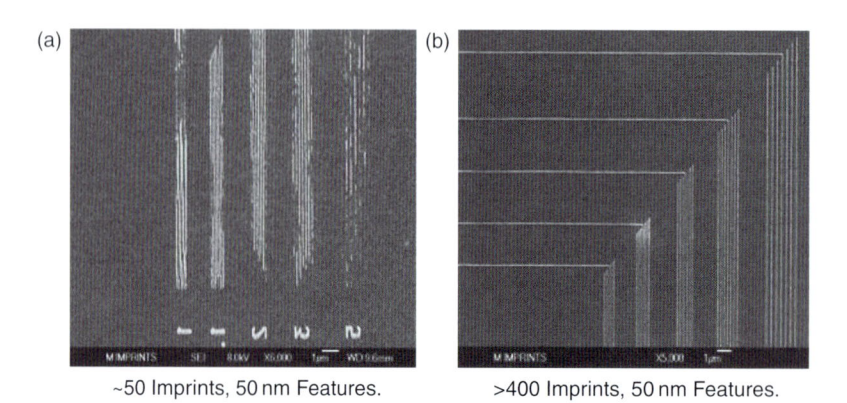

(a) ~50 Imprints, 50 nm Features.

(b) >400 Imprints, 50 nm Features.

Figure 20 Process longevity improvements obtained by addressing inhomogenous evaporation of imprint resist components aggravated by going to small dispensed drop volumes. (a) the imprint material disclosed in Ref. [44], but dispensed at <100 pl volumes; and (b) the imprint material disclosed in Ref. [71]. In each case, the mask and the associated release layer were kept the same during the imprint run.

does not contain any silicon. The resist is imprinted onto an IHM. The organic resist is required to possess a good etch resistance in the IHM etch step. The etch resistance of the organic resist can be adjusted by tuning the ring parameters of the resist components [83].

For electronic applications, including silicon CMOS devices, trace metal content requirement of the resist material is very stringent. Since acrylate resist contains 100% active ingredients, there is no solvent to dilute the trace metal impurity. In certain cases, individual components have to be carefully prepurified before mixing all of them together. After the gross purification step, a final polishing purification step is typically introduced to ensure each individual metal ion stay below 10 ppb level, which matches trace metal content specification of leading photoresists used in the semiconductor industry today.

5.3.2.2 Adhesion layer

As discussed in Section 5.3.2.1, fine feature fidelity of UV imprint lithography is influenced by many solidified phase resist characteristics such as preferential adhesion ratio, release, and mechanical properties. To increase preferential adhesion ratio for a given set of release technology, an adhesion layer or prime layer needs to be applied to the substrate. The adhesion layer needs to be able to be controlled in thickness, for example, be as thin as 1–2 nm, and ideally needs to be deposited onto substrate via spin coating or vapor treatment methods. While spin coating is possible on most substrates, for patterned magnetic media application, vapor treatment of the disks is highly desirable. Hard-disk substrates need patterning on both sides and their format includes a hole in the center of the substrate. This makes it very difficult to use spin coating to put down any sort of films on hard-disk substrates.

The adhesion layer needs to be engineered in such a way that it can work on a variety of different substrates such as Si, Si_xN_y, quartz, glass, GaN, Ta, Al, etc. The adhesion layer also needs to be readily wettable by the imprint resist in its liquid form. Originally, imprint resist was imprinted on transfer layers such as DUV30J BARC available from Brewer Science [44,71]. These BARC materials are designed for photolithography and are not specifically designed with the above-mentioned imprint process requirements of selective adhesion ratio and resist wettability. The recently developed spin coatable adhesion layer (known as TranSpin), and vapor deposited adhesion layer (known as ValMat) have been custom designed for the UV imprint process. The adhesion performance of TranSpin and ValMat against DUV30J on a Ta film deposited on glass disks is shown in **Figure 21**. It is seen that the quality of ValMat deposition and the resulting adhesion performance is influenced by surface cleanliness of substrate. Precleaned substrate provides higher adhesion values. Both TranSpin and ValMat show significant adhesion ratio improvement over DUV30J [74]. **Figure 22** shows the uniformity of the ValMat film on Ta disks via inspection using a Candela tool. Films that are <2 nm thick and have a uniformity of <1 nm 3σ were achieved. By

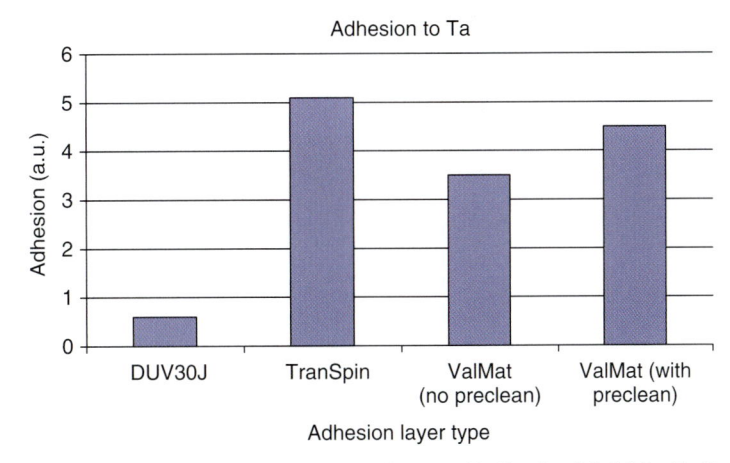

Figure 21 Adhesion comparison of various adhesion layers discussed in Section 5.3.2.2 for Ta films deposited on glass disks.

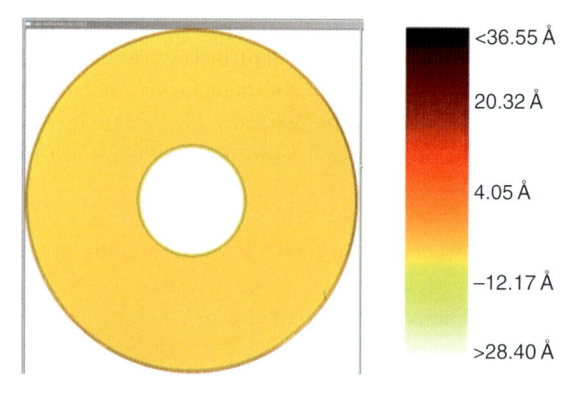

Figure 22 Thin film characterization of ValMat-coated Ta disks using Candela tools by KLA-Tencor shows high uniformity.

combining TranSpin adhesion layer and new imprint resist, process longevity exceeding 10 000 imprints has been recently achieved [51].

5.3.3 UV Imprint Tools

Imprint tools based on DoD UV nanoimprinting are broadly divided into (1) whole substrate tools for applications that do not require nanoresolution overlay (such as patterned media and photonic crystals for LEDs), and (2) steppers for applications requiring nanoresolution overlay and mix-and-match with photolithography (such as silicon ICs and thin film heads for magnetic storage) wherein the field size printing in one patterning step is the same as the industry standard advanced photolithography field size (26 mm by 32 mm).

The basic DoD process shown in part (c) of **Figure 8** can be seamlessly integrated as part of a whole substrate tool or a stepper since in each case the material is dispensed only where needed just prior to the patterning step. This dispense approach not only allows drop tailoring based on mask pattern density variation, it also ensures that a stepper process does not require handling of wet, evaporative films as would be the case with a spin-on UV process. **Figure 11** shows the schematic of a photolithography stepper and a UV imprint stepper based on drop dispense of the imprint material. A large-area tool uses a similar material dispense approach, but simply uses a template that has a patterned region equal to the size of the substrate and is thinner in geometry (<1 mm thick as discussed in Section 5.1) to conform to the short- and long-range nanotopography of substrates such as II–V substrates. The imprint tools include precision self-leveling flexure systems to passively align the template and substrate to be substantially parallel during the imprint process [84]. In addition, by using a drop-dispense approach that can be tailored based on mask pattern variation (**Figure 23**), a highly uniform residual layer can be achieved. Here, the residual layer is the underlying film of imprint material that is always present in the imprint process between the imprint mask and the substrate. This film needs to be thin and uniform for achieving a subsequent etch process that has a high degree of long-range uniformity as discussed in Section 6.

In addition to the dispensing of the imprint fluid and the control of the residual layer, the stepper system has additional precision mechanical systems to achieve nanoresolution alignment and overlay. The alignment subsystem that aligns the imprint mask in x, y, and theta directions with respect to the wafer is based on field-to-field moiré detection alignment scheme developed originally for XRL [85] and subsequently adapted to UV imprint lithography [50]. In addition to the alignment, size (magnification) and shape (orthogonality and trapezoidal corrections) corrections are required to perform

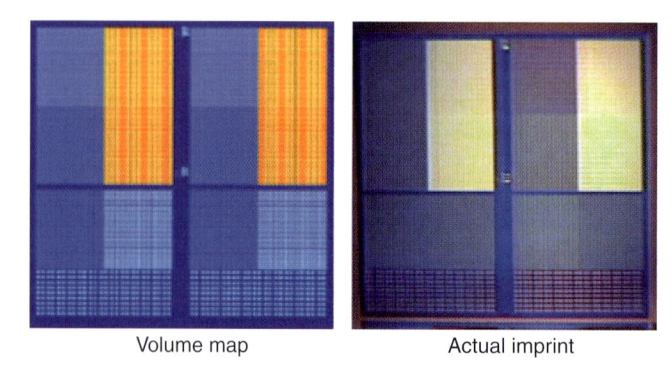

Volume map   Actual imprint

Figure 23 Drop local volume compensation based on imprint mask pattern density.

nanoresolution overlay, particularly in mixing-and-matching to optical lithography. A precision mechanical deformation system that deforms the imprint mask has been developed and implemented as part of the stepper system [86,50]. This allows the stepper to achieve sub-15 nm (mean + 3σ) overlay as discussed in Section 6.

The imprint tools summarized above are now discussed in more detail in the following sections. The requirements of UV nanoimprint tools can be divided into two groups of requirements: (1) basic imprint process requirements that need to be applied across multiple applications, and (2) requirements that are device-specific.

Basic requirements of UV nanoimprint tools include the following factors:

1. handling of the mask and the substrate with ultra-clean processes,
2. imprint material delivery,
3. void-free fluid filling,
4. UV expose, and
5. Separation.

Among the five basic requirements, only one of them (void-free fluid filling) is a concept that is quite sophisticated involving control of nanoscale fluid movement to yield highly uniform residual layers at high throughput, while eliminating any air void at the interface of the master and substrate. **Figure 24** shows air evacuating images as the mask closes the gap on the substrate by starting in a tilted configuration. For the case of whole wafer processes, a similar control of the fluid front has been developed to expel all air in radial direction [46]. Using the combination of fluid local volume compensation (see **Figure 23**) and the fluid front control scheme, even in the presence of wafer topography of more than 2 µm, whole wafer imprints have been developed to yield remarkable residual layer control with less than 5 nm variation in the total indicated range (TIR) [87].

Further, to achieve high-throughput fluid filling needs to be done without moving excessive imprint material around in the presence of pattern density variations in the mask design (again see **Figures 5** and **6** for the importance of this issue). Therefore, it is important for UV imprint tools to be equipped with DoD dispensing capability so that fluid local volume can be matched to that of the pattern density arrangement of the mask (see **Figure 23**). Problems associated with the spin-on imprinting are illustrated in **Figure 25** [88]. Dispensing drop distribution along with local volume can be premapped to compensate not only the local density variation but also tool-related variations such as material evaporation and substrate-master flatness mismatch errors.

Some of the device-specific requirements are discussed next with respect to the three representative application presented in **Table 1**. If a UV imprint tool is developed for integrated device fabrication, S and R patterning is necessary since the substrate is typically significantly larger than the imprinting field and also very tight nanoscale overlay accuracy is required. For data storage substrate patterning, no S and R is required but imprinting over a large area is necessary (up to 95 mm disks). Further, substrate handling is unique for this case since both sides of the substrate are to be imprinted.

A commercially viable imprint stepper solution demands extremely tight tool performance with respect to residual layer control, overlay, throughput, defect management, and master life [14]. Patterning area is required to be matched to that of existing optical lithography processes for the purpose of

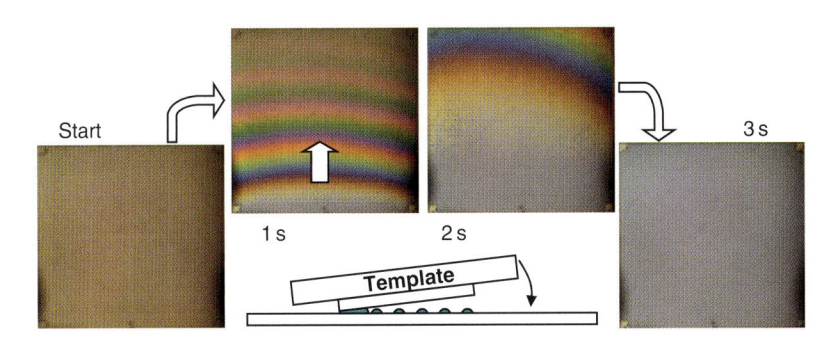

Figure 24 Air evacuation using a tilt imprinting with a mask allows control of the fluid front without causing significant void trapping. Time interval is ~1 s per frame. Mask and wafer are brought into liquid contact from the bottom edge and the fluid front propagates upward to expel the air.

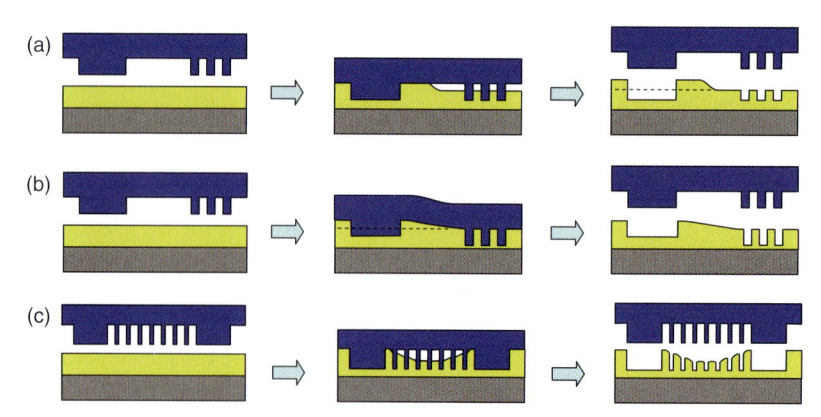

Figure 25 Various defects for the case of spin-on imprinting in the presence of pattern and size variation [88]; shear resistance can cause defects. (a) and (c) nonfilled features; and (b) deformed feature under pressure.

mix-and-match overlay ($\leq$26 mm $\times$ 33 mm). It is also necessary to exclude any excessive force that can induce a distortion to the master and also substrate to maintain their placement error within a desired error budget. Therefore, imprints need to be done with a minimum force between the master and substrate. Such a tight constraint inevitably requires the use of low-viscosity UV imprint fluid and appropriate distribution of resist drops, and if these two factors can be addressed, the process is capillary dominated [89] and does not require external forces that could lead to mask distortions.

Another important aspect of the semiconductor IC application is to have a scheme to control the magnification and also distortion between the master and substrate. XRL processes tried to stretch a membrane mask to control the magnification and also distortion. A similar approach of controlling the master size and displacement can be adapted for UV nanoimprinting. Mask distortion problems in the XRL do not exist for nanoimprinting when the master can be made of a material that is transparent to the UV spectrum and also the thickness of the master can be several orders larger than the pattern height in the mask. Alignment and overlay is discussed separately in a subsection below.

The second application discussed here is pattern magnetic storage media where patterning is needed on both sides of the substrate. This application can adapt a whole wafer process where the entire surface of the substrate is imprinted by a master without S and R. Substrates for data storage media are typically made of either polished Aluminum or fused silica. Their surfaces are tightly controlled to possess a low surface roughness to avoid any collision between low-flying data heads (<10 nm) and media surface. Benefits of DoD dispensing also apply for this application; no edge contamination by materials, material cost saving to lower the overall cost of manufacture, residual layer control, and no additional spin coating capital costs. The overall process flow of the data storage media is well presented in a web site by Hitachi [90]. Separating the mask from imprinted substrates is a challenge due to the large printing areas, and it needs to be done without causing any excessive stresses to prevent a potential catastrophic breakage of the mask. For this application, stage accuracies and stability are not as critical as compared to those of semiconductor IC applications due to micron-scale alignment requirements.

The third application is to patterning features for HB LEDs to increase light extraction efficiency. As compared to the data storage application, substrates for LEDs are not directly suitable for nanoimprinting if the master is made of fused silica or rigid material due to their surface roughness and/or defects as discussed in Section 5.1. Current manufacturing environments for the LED substrates are not as carefully controlled as those of semiconductor IC or hard-disk fabrication facilities, which makes generating clean and smooth substrates for LEDs difficult. Unique requirements for the LED application can be printing on none-flat surface, and high-frequency front-side particles/defects caused by epitaxial growth defects. Therefore, if the surface of the substrate is contaminated with defects, a soft mask can potentially increase the yielding area as discussed in the literature [17,63].

5.3.3.1 Alignment system for UV imprint steppers

Until now, only a handful of publications have addressed the issues of alignment and overlay for nanoimprint processes with statistically significant data and with errors being presented in a manner that is consistent with manufacturing requirements for integrated devices [14]. The most demanding application with respect to overlay is CMOS ICs where overlay budget is significantly less than half of minimum critical dimension (CD) (approximately one-half to one-fourth of the half-pitch). Even though several publications have discussed alignment and overlay for nanoimprinting, their data were limited to a few points of alignment measurement [91,92] until the full advantage of in-proximity and in-liquid alignment capability was demonstrated by researchers from Molecular Imprints, Inc. [50,93,14]. At least one order of magnitude improvement in fully field overlay was demonstrated as discussed in Section 5.4.

A few key requirements for the development of an alignment scheme for high-throughput imprint lithography can be summarized as: (1) absolute real-time alignment error data with a single-point resolution of better than 1 nm, (2) multiple alignment data (>6) capturing simultaneously, (3) no UV blockage from the alignment system, and (4) *in situ* alignment data capturing during gap changes and during *in situ* lubricated alignment corrections just prior to UV exposure. Widely used moiré fringe schemes that capture images vertically from two overlapped gratings can provide sufficient resolution, as it was used by Li and coworkers [92,94]. However, a well-known problem is that 0th order reflecting moiré changes its fringes as the gap changes making it problematic for practical implementation as a robust real-time algorithm. Further, packaging several imaging units within the limited working space above the mask is a very challenging task. Moving microscopes using a stage to capture data from multiple locations can not only negatively affect the throughput, but also generate tool vibrations and particles. Further, prior to UV exposure, the microscope unit needs to be moved out from the UV beam line.

Using a novel alignment scheme originally developed for XRL [95], researchers at Molecular Imprints, Inc developed a multiaxis alignment system that can satisfy all the requirements listed above. Since the moiré fringes are captured as first-order reflecting images, low-NA imaging microscopes can be inclined at a certain angle based on alignment mark periods on the mask and the substrate. The inclined angle is a function of the checkerboard pitch of the substrate layer. Multiple inclined microscopes can be packaged around the mask without blocking cross-linking UV light and therefore do not need to move in and out during the process. Technical discussions about the first-order moiré scheme are available in Refs. [96,97]. For UV imprinting process with low-viscosity material, it is feasible to perform alignment correction continuously from in-proximity to full *in situ* residual films just prior to UV exposure. Course alignment errors are corrected in-proximity and also as the gap between the mask and wafer decreases. Any remaining nanoscale alignment errors are compensated in-liquid as a final step when the mask and wafer are in their final position just prior to UV exposure.

To achieve a sub-25 nm overlay, it is necessary to compensate for orthogonal error and errors in magnification, particularly when mixing-and-matching with photolithography tools. Choi *et al.* [50] presented a mechanically induced high-resolution magnification and orthogonal error compensation scheme. Capabilities of better than 0.1 ppm magnification control and equivalent distortion correction have been demonstrated to achieve better than 20 nm 3σ overlay data as shown in Section 5.4.2.

5.4 UV Nanoimprint Lithography Process Results

This section provides recent process results and discussion for step and flash imprint lithography including (1) resolution, CD control, and line edge roughness; (2) alignment and overlay; (3) defectivity and template life; (4) edge field printing; (5) throughput; and (6) cost considerations.

5.4.1 Resolution, CD Control, and Line Edge Roughness

The resolution of the imprint lithography process is a direct function of the resolution of the imprint mask down to sub-3 nm [43]. The nanopatterns possess good long-range order if the imprint masks are obtained from commercially photomask houses such as DNP and Hoya that have well-established e-beam, etch, and metrology processes. Further, if nonchemically amplified electron beam resists are used, while the imprint mask write times go up somewhat, the line edge roughness of the resulting features are significantly improved [98]. **Figures 26** and **27** show the resolution obtained from imprint masks fabrication using Gaussian beam and variable shaped e-beam tools. In

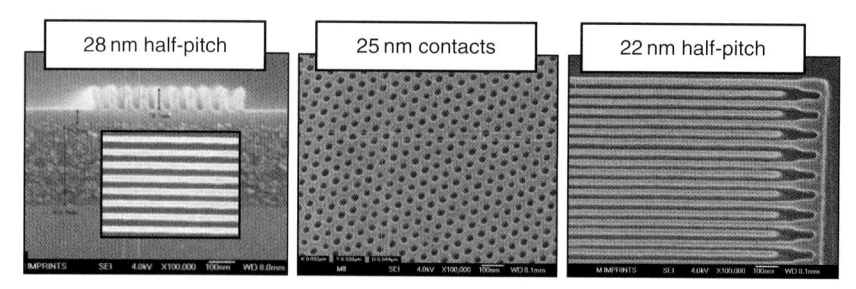

Figure 26 Imprint resolution is directly governed by the imprint mask. Here, sub-30 nm features patterned by using Gaussian e-beam tools are shown [98].

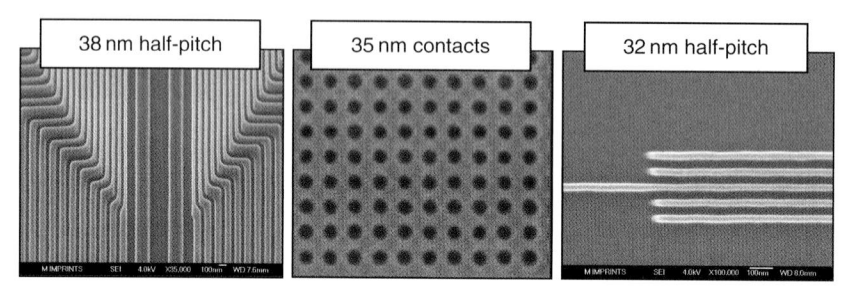

Figure 27 Imprint resolution down to 32 nm half-pitch patterns has been achieved using variable shaped e-beam tools that are industry standard in photomask houses.

addition, a nonchemically amplified e-beam resist known as ZEP520A was used in fabricating these masks. The resulting Gaussian beam resolution and line width roughness of ~2.5 nm, 3σ [Ref. 98] is satisfying the requirements for patterned media presented in **Table 1** and Section 5.3. With respect to CD control, the imprinted features have achieved about 3 nm, 3σ over large areas. Transferring imprinted pattern into an underlying substrate requires a de-scum etch process due to the presence of a residual film underneath the imprint template at the end of the UV-curing step. The thickness and uniformity of this residual film has a direct correlation to the postetch CD and CD control. **Figure 28** provides the quality of residual layer control being achieved in the S-FIL process. This residual film control is affected by several factors in the tool [14] and has been demonstrated to be stable over several months. Details of the de-scum etch process are provided in Ref. [99]. Postetch CD and its control is substantially the same as the imprinted features (correlated to the imprint mask) provided good control of residual layer can be achieved as shown in **Figure 28** [99].

5.4.2 Alignment and Overlay

In this section, alignment is defined as the accuracy with which an imprinted field can register relative to a previously lithographed field at the four corners of the patterned fields. Overlay refers to the accuracy with which an imprinted field can be registered relative to the previous lithographed field at the four corners and several points (typically ~100) over a field. The standard field size chosen here is 26 mm × 32 mm. Also, matched machine overlay (MMO) refers to overlay results obtained by registering an imprint field to a previously imprinted and etched field, where both imprint steps are performed on the same imprint stepper. Further mix-and-match overlay (M and MO) refers to overlay results obtained by registering an imprint field to a previously patterned and etched field that was lithographed using a 193 nm photolithography tool. The results obtained are presented in **Table 5**. The alignment was performed in a field-to-field manner. Since the alignment metrology and corrections can be performed in parallel to fluid filling, use of field-to-field alignment does not limit throughput of the process in its current state [14]. Each data set was obtained from a 200 mm wafer that contained 26 fields. For the alignment results, only four overlay errors at

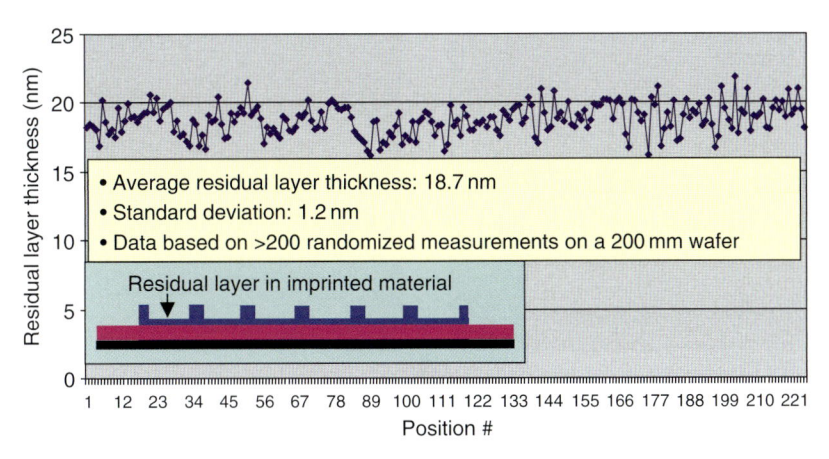

Figure 28 Highly uniform imprint material residual layer of sub-20 nm mean thickness.

Table 5 Alignment and overlay error chart for the experiment described in Section 5.4.2

	X error in nm (mean + 3s)	Y error in nm (mean + 3s)
Alignment error	5	6
MMO	11	15
M and MO	18	24

the four corners were used, while for the overlay results 86 overlay errors were used per field for a total of 2236 overlay error data per wafer. The results in **Table 5** suggest that sub-15 nm full-field overlay for imprint lithography registered to imprint lithography is possible. This is consistent with results independently collected by Toshiba using a Molecular Imprints stepper [100]. Also, the results in **Table 5** suggest that approximately 20 nm full-field overlay for imprint lithography registered to 193 nm photolithography is possible. This result is consistent with the result independently obtained by IBM when they used a Molecular Imprints stepper to pattern a 30 nm critical fin structure as part of a seven-layer storage class memory device [101]. The other layers were fabricated using 193 and 248 nm photolithography tools. The IBM storage class device SAMs are shown in **Figure 29**. To the best of our knowledge, these overlay results are at least 10× better than any comparable results presented in the imprint lithography literature.

5.4.3 Defectivity and Mask Life

UV imprint lithography defects have been studied in detail and have been reported in Refs. [51,52]. The defects studies include: (1) a low-resolution defect studies based on a KLA-Tencor (KT) 2132 inspection tool with a pixel size of 250 nm that is typically used to develop and qualify imprint materials in full wafer imprint processing; and (2) a high-resolution defect study using an electron beam inspection tool (KT-eS32 with a pixel size of 35 nm) to investigate S-FIL defects at sub-50 nm pattern geometries.

Based on the results provided in Refs. [51,52], it has been shown that total S-FIL defectivity is <3 per cm² with some wafers having defectivity of <1 per cm² and many fields having 0 imprint-specific defects, as measured with the KT-2132 (250 nm pixel). Through the use of variable shape electron beam write tools and leading edge mask fabrication processes and facilities, it was possible to obtain imprint masks with zero defects as measured by a KT-576 (90 nm pixel) optical inspection tool. Program defects were used to analyze the sensitivity of imprint mask inspection and the KT-576 was found to have sensitivity below the minimum pixel size of 90 nm, finding defects in contact features as small as 21 nm.

Next, a wafer was imprinted with a mask containing 32–44 nm half-pitch features and the wafer was inspected with a KT-eS32 using a 35 nm inspection pixel. The results were highly encouraging since imprint-specific defectivity was very low (<0.002% of the inspected feature areas). Even some of these defects are believed to be related to false defects due to the fact that the e-beam inspection tool is being used at or near the limits of its capability. These results, coupled with the advent of higher-resolution inspection tools, indicate that there is a path for inspection of 1× imprint masks and wafers with 32 nm half-pitch features.

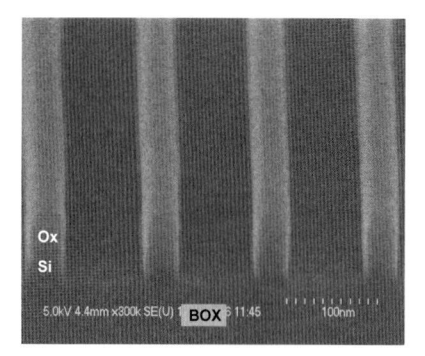

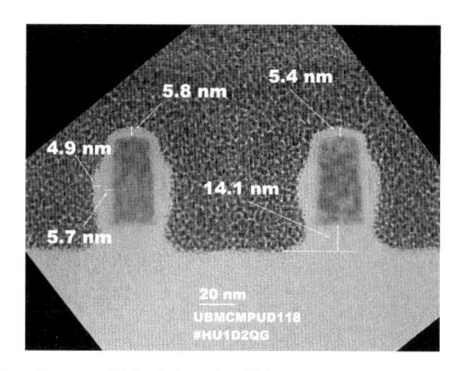

Figure 29 (Left) SEM image of 27nm silicon fins patterned by Step-and-Flash Imprint Lithography at Molecular Imprints, Austin, TX. (Right) TEM image of resulting FinFET structures before implant. Nominally identical structures led to functional 27nm FinFET devices. (Permission to reproduce this image was granted by the IBM Almaden Research Center, San Jose, CA.)

With respect to mask life, there are four kinds of defects that could force the template to fail catastrophically thereby ending its useful life. These are listed below.

1. Gross mask breakage is known to be a problem with high-pressure imprint processes. The UV imprint process using drop dispense of resist is a low imprint pressure process (<1/20th of an atmosphere) and such template breakage is not a problem based on over 10 years of experience with this technology.

2. Fine feature erosion is not a problem with this technology since the imprint monomers do no nanoscale damage to the fused silica based on data collected from >10 000 imprints [74].

3. Soft contamination of the template with the organic imprint material can be caused by sporadic local failure of the release treatment on the template or the adhesion chemistry on the wafer. The UV imprint process recovers in almost all cases and self-cleans the template in subsequent imprints [52–54]. In very rare situations, this self-cleaning may not work and a feature remains plugged. While this is not fully understood, it is believed to be either due to inorganic contaminants in the imprint material or due to poor coverage of the release treatment prior to the very first imprint which prevents the template from self-cleaning.

4. Local microscale template damage due to the template interacting with sporadic, hard inorganic, or metal particles on the wafer. Again, due to the low imprint pressure of the S-FIL process, not all particles cause microscale template damage. In fact, such damage is rare. Further, the damage is very local and has no major impact on applications such as LEDs, patterned media, bio applications, or even highly redundant memory structures. However, such a damage can be a significant problem in silicon-based logic ICs particularly microprocessors that have low levels of redundancy in design. For such highly demanding applications, it is believed that a combination of two approaches will be needed in the future to achieve acceptable process performance. First, a better understanding of all the sources of inorganic particles in the imprint tool and the processes immediately prior to imprint will be required. In addition, *in situ* diagnostic techniques that can identify any remaining rare particle events are required to ensure that the imprint mask can be checked locally for damage and, if necessary, repaired or replaced with a previously replicated imprint mask. *In situ* diagnostic techniques may be realistic since if a particle that has a modest z dimension (say 100 nm) is encountered, it results in a large exclusion zone of size that is a few tens to a few hundreds of microns. This is similar to a large XY hot spot caused by a small back-side particle on a wafer chuck [102]. This detection may be performed in parallel with normal wafer patterning. It is believed that micron resolution imaging processing and machine vision techniques can be applied to analyze low-magnification images of a field postimprint to detect the exclusion zone. Such diagnostic approaches are currently being investigated to make mask life acceptable for the most stringent silicon IC applications.

5.4.4 Throughput

As discussed in Section 5.1 and in **Table 1**, silicon IC application is the most aggressive from the perspective of throughput in SPH. A throughput target of

Table 6 Fluid filling time in seconds for two types of mask patterns

Drop pattern	Fluid front is not controlled — Uniform grid	Fluid front is not controlled — Design-based drop targeting	Fluid front is controlled — Uniform grid	Fluid front is controlled — Design-based drop targeting
High-pattern density variation	>40 s	25 s	30 s	4 s
Low-pattern density variation	30 s	15 s	10 s	3 s

20 WPH which leads to a target of approximately 2000 SPH is based on the fact that imprint tools are likely to be significantly lower in capital cost as compared to 193 nm-i photolithography and EUV tools. A 20 WPH throughput is expected to result in imprint being 1/2 the capital cost of 193-i double patterning in terms of throughput per tool cost. Based on this target, one shot has to be completed in about 1.5 s. UV imprinting requires the dispensing (which can, in principle, be done in parallel to printing), stage moves, fluid filling of the imprint mask, UV exposure, and separation of mask from imprint resist. Of all these steps, only fluid filling is believed to be a fundamental risk as all the other steps are believed to be achievable using known engineering approaches. A fluid filling goal of 1 s for a 26 mm × 33 mm field is assumed here.

Fluid filling in a DoD UV imprint process is affected by various factors such as fluid viscosity, dispense drop resolution, control of fluid front dynamics, and targeting of drops correlated to the mask design (**Table 6**). Also, sharp changes in pattern density in the mask design can cause regions where drop tailoring is suboptimal due to limitations in drop resolution. Most acrylate-based imprint fluids that have been used to date have a viscosity in the range of 4–10 cP [71]. There are vinyl ether materials that have much lower viscosity and if they can be integrated into the S-FIL process, they have the potential to enhance throughput significantly [72]. Also, the current tools use drops that are 6 pl in volume. In future, these drops will approach 1 pl that should assist in improving throughput. **Figure 30** shows a fluid front where the drops are merging to form a contiguous film. It is important to control the geometry of approach between the mask and the wafer to ensure that the drops do not encapsulate large bubbles. This can negatively impact throughput. It is important that the drops merge laterally to form a contiguous film that has the correct volume distribution locally to fill the varying patterns in the mask. In this article, two different mask designs are presented: (1) highly varying pattern density; and (2) Slowly varying pattern density. It can be seen from **Table 3** that if the process is not performed properly, the case of highly varying pattern density can have very slow filling that is more than 10× worse than the optimal case. In summary, by using design-based drop targeting and fluid front control, a fluid filling time that is 4× higher than the targeted value for silicon ICs has been achieved. Continued improvements in material viscosity, drop volume resolution and drop pattern placement optimization is expected to help achieve 1 s fill time for mask patterns with highly varying densities.

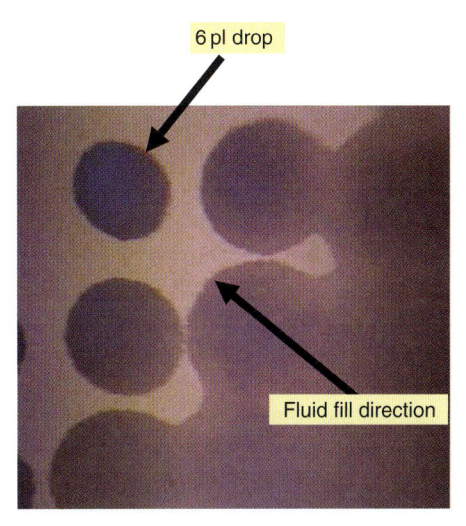

Figure 30 Control of fluid front to avoid trapping of large bubbles between the discrete drops.

5.4.5 Cost Considerations

There are three types of costs that need to be discussed, namely capital costs, consumable costs, and costs due to increased process complexity. The first two kinds of costs will be addressed using patterned media as an example due to its high sensitivity to cost (see **Table 1**). The third kind that arises from increased process complexity is most relevant in silicon IC fabrication due to the large number of steps involved in front-end and back-end processing for interconnects.

A unique advantage of drop dispense UV imprint process with respect to capital cost is that the resulting tool (shown schematically in **Figure 11**) includes a self-contained material dispense module. In spin-on UV or thermal imprint processes, a separate spin coating tool is needed for material deposition. In addition, since hard disks have an inner open region and they need to be patterned on both sides, a high-throughput spin coater (with a throughput of 300–600 WHP) that can address the unique double-sided coating process on disks needs to be developed. In current hard-disk fabrication processes such a coater does not exist. Since patterned media is highly sensitive to cost, this elimination of the coater is significant.

With respect to cost of consumables, the drop dispense approach has virtually no waste. It is estimated that the drop dispense approach will consume about 0.1% of the volume that a spin coating process will consume since spin coating involves significant wastage. This again is significant in a cost-sensitive application like patterned magnetic media.

Finally, with respect to reducing process complexity, a novel patterning approach has been proposed recently that exploits the three-dimensional patterning capability of imprint lithography [103–105]. It has been proposed that a two-tier template can be used to simultaneously pattern via and metal layers for dual damascene fabrication of interconnects (**Figure 31**). Depending on the approach used, this has the potential to significantly change the cost structure of fabricating silicon ICs, particularly logic circuits that tend to have in excess of 10 metal interconnects. Recent processing results have also demonstrated promising electrical yield [105] suggesting that three-dimensional imprinting of interconnects may be an attractive approach to significantly reduce process complexity and cost of silicon IC fabrication.

5.5 Summary and Future Directions

This article provides a review of imprint lithography and its potential use in the manufacture of high-volume nanoscale devices in the areas of terabit density magnetic storage, CMOS IC fabrication, nanowire molecular memory, photonic devices, and biomedical applications. The article focuses on the evolution of drop dispense-based UV nanoimprint lithography over the last few years as it is maturing beyond a research curiosity to become a candidate for large-scale nanomanufacturing in key market segments such as ultra-high-density magnetic media and novel nanoscale memory architectures. In summary, it shows that UV nanoimprint techniques discussed here are promising for manufacturing a broad range of nanoscale devices.

The impact of this technology on mainstream silicon fabrication will require continued progress in imprint mask fabrication, tool overlay, and throughput. It is believed that these problems are of reasonable complexity and can be overcome with appropriate engineering resources. Minimizing defectivity and maximizing imprint mask life is probably a more challenging activity, particularly for logic devices that do not benefit from the kind of architectural redundancy that is present in memory devices. The use of imprint lithography for large-scale logic device fabrication will require significant developments in the area of *in situ* diagnostics for mask defectivity, and cost-effective imprint mask inspection approaches.

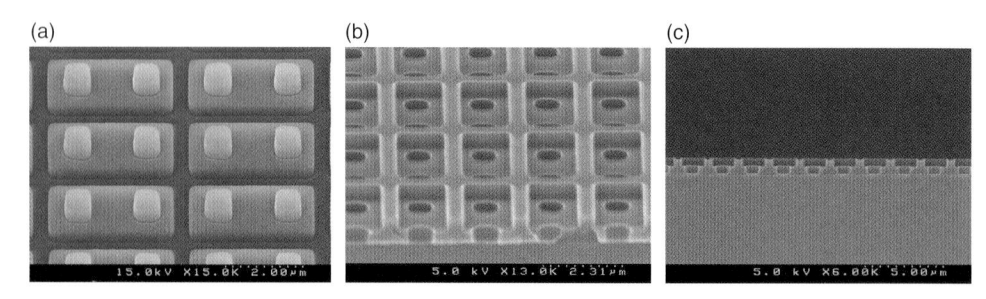

Figure 31 A three-dimensional template is shown in (a) and the corresponding imprinted three-dimensional pattern is shown in (b). (c) A cross-section of the imprinted pattern in (b). These results are from Ref. [103].

Acknowledgments

This work was partially funded by DARPA Contract No. N66001-02-C-8011, NIST Advanced Technology Program Contract No. 70NANB4H3012, and DoD Contract No. N66001-06-C-2003.

References

1. Sreenivasan SV (2008) Nano-scale manufacturing enabled by imprint lithography. *Special Issue: Nanostructured Materials in Information Storage. MRS Bulletin* 33: 854–863.
2. Hershey R, Miller M, Jones C, et al. (2006) 2D photonic crystal patterning for high volume LED manufacturing. *Proceedings of SPIE: Microlithography* 6337: 63370M-1–63370M-10.
3. Schmid G, Doyle G, Miller M, and Resnick D (2007) Fabrication of dense sub-20 nm pillar arrays on fused silica imprint templates. *Presented at the EIPBN Conference*, June 2007. Denver, CO, USA.
4. http://www.itrs.net – International Technology Roadmap for Semiconductors.
5. Yeo J, et al. (2008) Full-field imprinting of sub-40-nm patterns. *Proceedings of SPIE: Advanced Lithography, Emerging Lithographic Technologies XII* 6921: 692107-1.
6. Chen Y, et al. (2003) Nanoscale molecular-switch crossbar circuits. *Nanotechnology* 14: 462.
7. Glangchai LG, et al. (2008) Nanoimprint lithography based fabrication of shape-specific, enzymatically-triggered smart nanoparticles. *Journal of Controlled Release* 125: 263.
8. Ohtsuka Y, et al. (1997) A new magnetic disk with servo pattern embedded under recording layer. *IEEE Transactions on Magnetics* 33(5): 2620.
9. Christie RK, Marrian CRK, and Tennant DM (2003) Nanofabrication. *Journal of Vacuum Science and Technology A* 21(5): S207.
10. Austin MD, et al. (2004) Fabrication of 5 nm line width and 14 nm pitch features by nanoimprint lithography. *Applied Physics Letters* 84: 5299.
11. Bandic ZZ, Dobisz EA, Wu TW, and Albrecht TR (2006) Patterned magnetic media: Impact of nanoscale patterning on hard disk drives. *Solid State Technology Supplement* 49: S7–S19.
12. Yang X and Xioa S (2008) Challenges in electron-beam lithography for high-resolution template fabrication for patterned media. *Invited Presentation at the SPIE Advanced Lithography*, 1 March 2008.
13. Sreenivasan SV, et al. (2005) Using reverse-tone bilayer etch in ultraviolet nanoimprint lithography. *Semiconductor Fabtech*, 25th edition, pp. 107–113, March 2005.
14. Sreenivasan SV, et al. (2008) Critical dimension control, overlay, and throughput budgets in UV nanoimprint stepper technology. *Proceedings of the ASPE 2008 Spring Topical Meeting on "Precision Mechanical Design and Mechatronics for Sub-50 nm Semiconductor Equipment"*, April 2008.
15. Boning D and Lee B (2002) Nanotopography issues in shallow trench isolation CMP. *MRS Bulletin* 27: 761.
16. Muller T, et al. (2001) Techniques for analysing nanotopography on polished silicon wafers. *Microelectronic Engineering* 56(1–2): 123.
17. Singhal S and Sreenivasan SV (2007) Nano-scale mechanics of front-side contamination in photonic device fabrication using imprint lithography. *Proceedings of the ASPE Annual Meeting*, Dallas, TX, October 2007.
18. Gordon J (Keynote Speaker) and Almaden IBM (2007) Directed self assembly – how useful will it be? *IEEE Lithography Workshop*, Rio Grande, Puerto Rico, 10–13 December.
19. Mori I, et al. (2008) Keynote speaker. Selete's EUV program: Progress and challenges. *Proceedings of SPIE: Advanced Lithography, Emerging Lithographic Technologies XII* 6921: 692102-1.
20. Early K, Schattenberg ML, and Smith HI (1990) Absence of resolution degradation in X-ray lithography. *Microelectronic Engineering* 11: 317–321.
21. Carter JD, Pepin A, Schweizer MR, and Smith HI (1997) Direct measurement of the effect of substrate photoelectrons in X-ray nanolithography. *Journal of Vacuum Science and Technology B* 15: 2509–2513.
22. Miura T (2002) Projection electron beam lithography. *Journal of Vacuum Science and Technology B* 20: 2622.
23. Loeschner H, Stengl G, Kaismaier R, and Wolter A (2001). Projection ion beam lithography. *Journal of Vacuum Science and Technology B* 19: 2520.
24. Menon R, et al. (2005) Maskless lithography. *Materials Today* February: 26–33.
25. Kik PG, et al. (2002) Metal nanoparticle arrays for near field optical lithography. *Proceedings of SPIE* 4810: 7–13.
26. Srituravanich W, et al. (2008) Flying plasmonic lens in the near field for high-speed nanolithography. *Nature Nanotechnology* 3: 733.
27. Brueck SRJ (2005) Optical and interferometric lithography – nanotechnology enablers. *Proceedings of the IEEE* 93(10): 1704–1721.
28. Slot E, Wieland MJ, Ten Berge GF, et al. (2008) MAPPER: High throughput maskless electron beam lithography. *Proceedings of the SPIE: Advanced Lithography, Emerging Lithographic Technologies XII* 6921: 69211P–69211P-9.
29. Bentley (2008) Nonlinear interferometric lithography for arbitrary two-dimensional patterns. *Journal of Micro/Nanolithography, MEMS and MOEMS* 7(1): 013004.
30. Lutwyche M, Andreoli C, Binnig G, et al. (1998) Microfabrication and parallel operation of 5×5 2D AFM cantilever arrays for data storage and imaging. *Proceedings of Micro Electro Mechanical Systems, 1998 (MEMS 98), The Eleventh Annual International Workshop*, pp. 8–11. Heidelberg, Germany, 25–29 January.
31. Dürig U, Cross, G, Despont, M, et al. (2000) Millipede – an AFM data storage system at the frontier of nanotribology. *Tribology Letters* 9(1–2): 25–32.
32. Minne C, Adams JD, Yaralioglu G, Manalis SR, Atalar A, and Quate CF (1998) Centimeter scale atomic force microscope imaging and lithography. *Applied Physics Letters* 73: 1742.
33. Piner et al. (1999) Dip pen nanolithography. *Science* 283: 661–663.
34. Hong S and Mirkin CA (2000) A nanoplotter with both parallel and serial writing capabilities. *Science* 288: 1808–1811.
35. Huo F, Zheng Z, Zheng G, Giam LR, Zhang H, and Mirkin CA (2008) Polymer pen lithography. *Science* 321(5896): 1658–1660.
36. Ahn Y, Ono T, and Esashi M (2008) Micromachined Si cantilever arrays for parallel AFM operation. *Journal of Mechanical Science and Technology, The Korean Society of Mechanical Engineers* 22(2) 308–311.

37. Kakushima K, Watanabe M, Shimamoto K, et al. (2003) AFM cantilever array for parallel lithography of quantum devices. *International Microprocesses and Nanotechnology Conference, 2003. Digest of Papers 2003,* pp. 336–337, 29–31 October.
38. Xia Y and Whitesides GM (1998) Soft lithography. *Angewandte Chemie (International ed. in English* 37: 550.
39. Chou SY, Krauss PR, and Renstrom PJ (1996) Nanoimprint lithography. *Journal of Vacuum Science and Technology B* 14(6): 4129.
40. Napoli LS and Russell JP (1988) Process for Forming a Lithographic Mask. US Patent 4,731,155.
41. Schift H, et al. (2007) Visualization of mold filling stages in thermal nanoimprint by using pressure gradients. *Journal of Vacuum Science and Technology* 25(6): 2312–2316.
42. Kloosterboer J (1988) Network formation by chain crosslinking photopolymerization and its applications in electronics. *Advanced Polymer Science* 84: 1.
43. Hua F, Sun Y, Gaur A, et al. (2004) Polymer imprint lithography with molecular-scale resolution. *Nano Letters* 4(12): 2467–2471.
44. Colburn M, et al. (2000) Step and flash imprint lithography for sub-100 nm patterning. *Proceedings of SPIE: Emerging Lithographic Technologies IV* 3997: 453.
45. Resnick D, et al. (2005) Step & flash imprint lithography. *Materials Today* 8(2): 34–42.
46. Resnick D, et al. (2007) High volume full-wafer step and flash imprint lithography. *Solid State Technology* February: 39–43.
47. Otto M, et al. (2004) Reproducibility and homogeneity in step and repeat UV-nanoimprint lithography. *Microelectronic Engineering* 73–74(1): 152–156.
48. Deguchi K, Takeuchi N, and Shimizu A (2002) Evaluation of pressure uniformity using a pressure-sensitive film and calculation of wafer distortions caused by mold press in imprint lithography. *Japanese Journal of Applied Physics* 41(Part 1, 6B): 4178.
49. Sirotkin V, et al. (2007) Coarse-grain method for modeling of stamp and substrate deformation in nanoimprint. *Microelectronic Engineering* 84: 868.
50. Choi BJ, et al. (2005) Distortion and overlay performance of UV step and repeat imprint lithography. *Microelectronic Engineering* 78–79: 633.
51. McMackin I, et al. (2008) Patterned wafer defect density analysis of step and flash imprint lithography. *Journal of Vacuum Science and Technology B* 26(1): 51.
52. McMackin I, et al. (2008) High-resolution defect inspection of step and flash imprint lithography for the 32-nm node and beyond. *Proceedings of SPIE: Advanced Lithography, Emerging Lithographic Technologies XII* 6921: 69211L-1.
53. Bailey T, Choi BJ, Colburn M, et al. (2000) Step and flash imprint lithography: Template surface treatment and defect analysis. *Journal of Vacuum Science and Technology B* 18(6): 3572–3577.
54. Stewart MD, Johnson SC, Sreenivasan SV, Resnick DJ, and Willson CG (2005) Nanofabrication with step and flash imprint lithography. *Journal of Microlithography, Microfabrication, and Microsystem* 4(1): 011002.
55. Sotomayor TS (ed.) (2003) *Alternative Lithography*. Boston, MA: Kluwer.
56. Guo LJ (2004) Recent progress in nanoimprint technology and its applications. *Journal of Physics D: Applied Physics* 37: R123–R141.
57. Sogo K, Nakajima M, Kawata H, and Hirai Y (2006) Low cost and rapid reproduction of fine structures by nano casting lithography. *Micro and Nano Engineering Conference*, 2006.
58. Tan H, et al. (1998) Roller nanoimprint lithography. *Journal of Vacuum Science and Technology B* 16(6): 3926–3928.
59. Ahn, et al. (2007) Roll-to-roll nanoimprint lithography on flexible plastic substrate. *EIPBN*, 2007.
60. http://www.hqrd.hitachi.co.jp/global/news_pdf_e/hrl061110nrde_nanoimprint.pdf.
61. Sbiaa, et al. (2007) Patterned media towards nano-nit magnetic recording: Fabrication and challenges. *Recent Patents on Nanotechnology* 1: 29–40.
62. Rettner CT, Best ME, and Terris BD (2001) Patterning of granular magnetic media with a focused ion beam to produce single-domain islands at >140 Gbit/in2. *Magnetics, IEEE Transactions on* 37(4): 1649–1651.
63. Watts, et al. US Patent No. 7,140,861.
64. Selinidis K, et al. (2008) Full field imprint masks using variable shape beam pattern generators. *Journal of Vacuum Science and Technology B* 26(6): 2410–2415.
65. Yoshitake S and Kamikubo T (2009) Resolution capability of EBM-6000 and EBM-7000 for nanoimprint templates. *European Mask Conference (EMLC) 2009*, 12–15 January, Dresden, Germany.
66. Sasaki, et al. (2009) UV nanoimprint lithograpy template making and imprint evaluation. *European Mask Conference (EMLC) 2009*, 12–15 January, Dresden, Germany.
67. Eynon B (2009) Nanoimprint mask replication reduces lithography CoO. *Semiconductor International*, 1 February 2009. http://www.semiconductor.net/article/CA6636575.html (accessed July 2009).
68. Selinidis K, et al. (2009) Inspection and repair for imprint lithography at 32 nm and below. *Photomask Japan Conference 2009*, 8–10 April 2009.
69. Selinidis K, et al. (2009) Electron beam inspection methods for imprint lithography at 32 nm. *European Mask Conference (EMLC) 2009*, 12–15 January, Dresden, Germany.
70. Miller M, et al. (2008) Step and flash imprint process integration techniques for photonic crystal patterning: Template replication through wafer patterning irrespective of tone. *Proceedings of SPIE: Advanced Fabrication Technologies for Micro/Nano Optics and Photonics* 6883: 68830D.
71. Xu F, et al. (2004) Development of imprint materials for the step and flash imprint lithography process. *SPIE Microlithography: Emerging Lithographic Technologies VIII*, vol. 5374, pp. 232–241, Santa Clara, CA.
72. Kim EK, Stacey NA, Smith BJ, et al. (2004) Vinyl ethers in ultraviolet curable formulations for step and flash imprint lithography. *Journal of Vacuum Science and Technology B* 22(1): 131.
73. Houle FA, et al. (2008) Chemical and mechanical properties of UV-cured nanoimprint resists. *Proceedings of SPIE: Advanced Lithography, Emerging Lithographic Technologies XII* 6921: 69210B-1.
74. Xu F. Unpublished Results from Molecular Imprints Experimental Studies.
75. Davidson RS and Mead CJ (2003) Stabilisation of thiol-ene formulations. *Proceedings of the RadTech Europe Conference*, 2003.
76. Hudd A (2004) UV cure chemistry platforms for industrial applications. *UV Inkjet Session at RadTech's e/5 2004 UV&EB Technology Expo and Conference.*
77. Beck M, Graczyk M, Maximov T, et al. (2002) Improving stamps for 10 nm level wafer scale nanoimprint lithography. *Microeletronic Engineering* 61–62: 441–448.
78. Jung GY, Li Z, Wu W, et al. (2005) Vapor phase self-assembled monolayer for improved mold release in nanoimprint lithography. *Langmuir* 21: 1158–1161.

79. Garidel S, Zelsmann M, Voisn P, Rochat N, and Michallon P (2007) Structure and stability characterization of anti-adhesion self-assembled monolayers formed by vapour deposition for NIL use. *Proceedings of SPIE* 6517: 65172C.
80. Houle FA, Raoux S, Miller DC, Jahnes C, and Rossnagel S (2008) Metal-containing release layers for use with UV-cure nanoimprint lithographic template materials. *Journal of Vacuum Science and Technology B* 26(4): 1301–1304.
81. Taniguchi J, *et al.* (2002) Measurement of adhesive force between mold and photocurable resin in imprint technology. *Japanese Journal of Applied Physics, Part 1* 41: 4194–4197.
82. Kim KD, Kwon HJ, Choi DG, Jeong JH, and Lee ES (2008) Resist flow behavior in ultraviolet nanoimprint lithography as a function of contact angle with stamp and substrate. *Japanese Journal of Applied Physics* 47(11): 8648–8651.
83. Kune RR, Palmateer SC, Forte AR, *et al.* (1996) Limits to etch resistance for 193-nm single-layer resists. *Proceedings of SPIE* 2724: 365–376.
84. Choi BJ, Johnson S, Colburn M, Sreenivasan SV, and Willson CG (2001) Design of orientation stages for step and flash imprint lithography. *Journal of the International Societies for Precision Engineering and Nanotechnology* 25(3): 192–199.
85. Moon EE, Lee J, Everett P, and Smith HI (1998) Application of interferometric broadband imaging alignment on an experimental X-ray stepper. *Journal of Vacuum Science and Technology B: Microelectronics and Nanometer Structures* 16(6): 3631.
86. Cherala A, *et al.* An Apparatus for Varying the Dimensions of a Substrate During Nano-Scale Manufacturing. US Patent No. 7,170,589.
87. Lentz, *et al.* (2007) Whole wafer imprint patterning using step and flash imprint lithography: A manufacturing solution for sub 100 nm patterning. *Proceedings of SPIE: Advanced Lithography, Emerging Lithographic Technologies XI* 6517: 65172F-1.
88. Cheng X and Guo LJ (2004) One-step lithography for various size patterns with a hybrid mask-mold. *Microelectronics Engineering* 71: 288.
89. Colburn M, Choi J, Sreenivasan SV, Bonnecaze R, and Willson CG (2004) Ramifications of lubrication theory on imprint lithography. *Microelectronic Engineering* 75(3): 321–329.
90. http://www.hitachigst.com – Hitachi Global Storage Technologies – Patterned Magnetic Media.
91. Fuchs, *et al.* (2004) Interferometric *in situ* alignment for UV-based nanoimprint. *Journal of Vacuum Science and Technology B* 22(6): 3242–3245.
92. Li *et al.* (2006) Sub-20-nm alignment in nanoimprint lithography using Moiré Fringe. *Nano Letters* 6(11): 2626–2629.
93. Melliar-Smith CM (2007) Lithography beyond 32 nm: A role for imprint? *Proceedings of SPIE: Advanced Lithography, Key Note Presentation* 6517: 21.
94. Muhlberger M, *et al.* A Moire method for high accuracy alignment in nanoimprint lithography. *MNE 06*.
95. Moon, *et al.* (1995) Immunity to signal degradation by overlayers using a novel spatial-phase-matching alignment system. *Journal of Vacuum Science and Technology B* 13(6): 2648–2652.
96. Moon, *et al.* (2003) Interferometric-spatial-phase imaging for six-axis mask control. *Journal of Vacuum Science and Technology B* 21(6): 3112–3115.
97. Moon, *et al.* (1998) Application of interferometric broadband imaging alignment on an experimental x-ray stepper. *Journal of Vacuum Science and Technology B* 16(6): 3631–3636.
98. Schmid G, *et al.* (2008) Minimizing line width roughness for 22 nm node patterning with step and flash imprint lithography. *Proceedings of SPIE: Advanced Lithography, Emerging Lithographic Technologies XII* 6921: 690129-1.
99. Brooks CB, LaBrake DL, and Khusnatdinov N (2008) Etching of 42-nm and 32-nm half-pitch features patterned using step and flash imprint lithography. *Proceedings of SPIE: Advanced Lithography, Emerging Lithographic Technologies XII* 6921: 69211K-1.
100. Yoneda I, *et al.* (2008) Study of nanoimprint lithography for applications toward 22-nm node CMOS devices. *Proceedings of SPIE: Advanced Lithography, Emerging Lithographic Technologies XII* 6921: 692104-1.
101. Hart MW (2007) Step and flash imprint lithography for storage-class memory. *EIPBN Presentation*, May 2007, Denver, CO, USA.
102. Nimmakayala P and Sreenivasan SV (2004) Compliant pin chuck system for minimizing the effect of backside particles on wafer planarity. *American Vacuum Society Journal of Vacuum Science and Technology B* 22(6): 3147–3150.
103. Schmid, G, *et al.* (2006) Implementation of an imprint damascene process for interconnect fabrication. *Journal of Vacuum Science and Technology B* 24(3): 1283–1291.
104. Jen WL, *et al.* (2007) Multi-level step and flash imprint lithography for direct patterning of dielectrics. *Proceedings of SPIE: Advanced Lithography, Emerging Lithographic Technologies XI* 6517: 65170K-1.
105. Chao B, *et al.* (2008) Dual-damascene BEOL processing using multilevel step and flash-imprint lithography. *Proceedings of SPIE: Advanced Lithography, Emerging Lithographic Technologies XII* 6921: 69210C-1.
106. Chou, *et al.* (2002) Ultrafast and direct imprint of nanostructures in silicon. *Nature* 417: 835–837.

6 Picoliter Printing

E Gili, M Caironi, and H Sirringhaus, University of Cambridge, Cambridge, UK

© 2010 Elsevier B.V. All rights reserved.

6.1 Introduction

Since the observations by Lord Rayleigh in 1878 that a liquid stream has the tendency to break up into individual droplets [1], several ink-jet printing technologies have been developed. Although ink-jet printing has been historically extremely successful in the graphics market, several other applications have more recently been developed in fields as diverse as electronics and biology. These novel applications are based on the idea of using this technology to position functional inks on a substrate. These inks can then be transformed, usually by a thermally activated process, in functional materials with structural, electrical, optical, chemical, or biological properties.

This work will review the recent progress on ink-jet printing of functional materials, focusing in particular on applications where it is essential to deliver small amounts of functional inks in drops with picoliter or sub-picoliter volume. The technique of choice for this applications is drop on demand (DOD) ink-jet printing, where a transducer is employed to control the ejection of droplets from a nozzle. This can be achieved using different approaches, which will be reviewed individually. The most employed of these solutions uses a piezoelectric transducer to create a pressure wave that triggers the ejection of the droplets. For this reason, in this work we will focus on piezoelectric DOD ink-jet printing, with particular attention to the applications relevant to the electronic industry.

Two main classes of ink-jet printers have been developed for electronic applications. Single-nozzle printers and printheads are mainly used in research and development because of their flexibility and simplicity. Nevertheless, multi-nozzle systems and dispensers have been developed over the last 10 years to increase the throughput of the printers and to make them suitable for volume manufacturing. In parallel, a considerable effort has been made by the chemical industry to provide functional inks suitable for ink-jet printing. For example, concentrated liquid suspensions of powders have been developed to print three-dimensional (3D) objects [2]. In another application, inks composed of solutions of biomolecules have been used to print protein biochips for clinical diagnostics and drug development [3].

More specifically, functional inks have enabled a wide range of applications in the area of displays and microelectronics. For example, light emitting polymers (LEPs) have been printed to fabricate polymer light emitting diodes (PLEDs) [4]. Bottom-gate organic field effect transistors have also been fabricated depositing all the active materials (semiconductor, gate dielectric, and electrodes) using lithography-free printing techniques [5,6]. Nevertheless, the development of highly conductive, printable inks necessary to fabricate conductor lines with low parasitics has proved challenging. Recently, the introduction of inks containing metal nanoparticles or metal-containing organic complexes has allowed to print lines with high conductivity using low-temperature processes. In the following, we will discuss the different types of metal inks available for display and microelectronic applications.

Among the large range of applications of ink-jet printing of functional materials, in this work we will focus on display applications (organic light emitting diodes (OLEDs), liquid crystal displays (LCDs)) and on electronic applications (active matrix backplanes, sensors, radio frequency identification tags (RFIDs), digital lithography and ink-jet etching). It will be clear that for this range of applications the main fundamental limitation of ink-jet printing is the intrinsic low resolution of the technique. On the other hand, from a manufacturing point of view it has proven difficult to achieve high yield and deposition uniformity to transfer ink-jet printed electronic devices to the production floor. In order to improve the resolution of ink-jet printing, for example, in order to print short-channel field effect transistors (FETs), a novel approach will be reviewed. This is a self-aligned technique that relies on the movement of a droplet on a substrate, induced by forces exerted on the ink by a contrast in surface energy. This can be achieved by pre-patterning high-resolution structures defining areas with different surface energy on a substrate. An alternative, lithography-free approach will also be discussed. This technology is particularly desirable to fabricate high-performance organic transistors without the need of expensive lithographic techniques. Using this technique, organic thin film transistors (OTFTs) with transition frequencies higher than 1 MHz have been demonstrated. This technology has the potential of extending the range of applications of organic electronics to those systems requiring higher-frequency operation, opening up a whole range of market opportunities for ink-jet printing within the electronic industry.

6.2 Drop on Demand Ink-Jet Printing of Functional Materials

6.2.1 Principle of Operation

Ink-jet printers are dispensers that propel droplets of a fluid toward a substrate. This can be achieved in two different ways: continuous and DOD printing [7].

In continuous ink-jet printers (**Figure 1(a)**) a stream of liquid is continuously projected from an orifice. A piezoelectric crystal creates an acoustic wave as it vibrates within the cavity and causes the stream of liquid to break into droplets at regular intervals, due to Rayleigh instability [1]. During their formation, an electric charge is then imparted to the drops by an electrode. Finally, a second pair of electrodes steers the charged droplets by means of an electric field. The droplets can either be directed onto the substrate, or steered toward a catcher, which allows to recirculate the unused liquid.

In DOD ink-jet printers (**Figure 1(b)**), a transducer driven by an electrical signal generates a pressure wave in the cavity of the printhead, which is full of ink. The key difference is that the transducer only generates a signal when a drop needs to be ejected to print a dot on a substrate. The pressure wave has to be optimized to eject only one drop for every electrical pulse applied to the transducer. Once ejected, the droplet is deposited onto the substrate.

Continuous ink-jet printing is the system of choice for high-throughput, low-resolution industrial graphics printing systems. In fact, the high velocity ($\sim$50 m s^{-1}) of the ink droplets allows for a long distance between printhead and substrate and for a high printing speed. Moreover, as the liquid is ejected continuously from the printhead, this system is not susceptible to nozzle clogging. On the other hand, continuous printing has a few disadvantages that hinder its application for printing functional materials. First, the system has poor flexibility, as it is usually designed for only one type of ink. Second, continuous ink-jet usually has stringent requirements on ink viscosity. Third, the liquid of choice needs to be suitable for electrostatic charging.

On the other hand, DOD ink-jet printing is an ideal technique for printing functional materials. First, it is highly flexible, as fluids can be easily interchanged. Second, high-resolution DOD systems are commercially available. Third, the use of

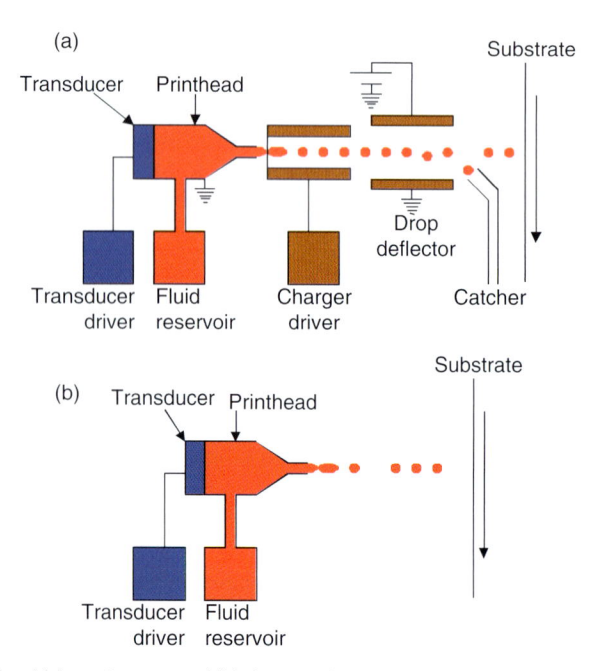

Figure 1 Operating principle of (a) continuous and (b) drop on demand ink-jet printing systems.

mechanical actuators allows to jet a wide range of inks by adapting the shape and intensity of the electrical pulse to their characteristics. Nevertheless, DOD printing has a major drawback due to the intermittent creation of pressure waves in the printhead cavity. For this reason, DOD ink-jet printheads are more susceptible to clogging than continuous printing systems.

Several types of DOD ink-jet printheads have been reported:

- Thermal printheads
- Piezoelectric printheads
- Acoustic printheads
- Electrohydrodynamic printheads

In thermal printheads, the transducer is composed of a heating element that is electrically activated by a pulse of current. When the transducer is heated, the liquid contained in the printhead cavity is locally heated and it evaporates, forming a bubble, which generates a pressure wave in the cavity. Although thermal DOD ink-jet printing of nonaqueous fluids has been reported [8], the use of this technology to print functional materials has proved challenging.

The other three types of DOD printheads have all been used to print inks containing functional materials, and they will be reviewed in detail in the following sections. An important figure of merit of a DOD printhead is the printing resolution, which is dependent on the minimum drop size that can be achieved. **Table 1** compares the drop volume and diameter achievable with the technologies most commonly used to print functional materials. Electrohydrodynamic printheads allow to obtain the smallest drops and the highest resolution reported to date. Nevertheless, only piezoelectric printheads have been scaled up to multi-nozzle dispensing systems that have sufficient throughput for applications in manufacturing.

6.2.2 Piezoelectric DOD Printing

The most common approach for printing inks containing functional materials is to use piezoelectric printheads. In these dispensers, a fluid is ejected through a nozzle due to a pressure pulse generated by a piezoelectric transducer. The minimum volume of the drops that can be generated is mainly limited by the nozzle size. Piezoelectric printheads generating drops with calibrated volume as small as 10 pl [9] (26 µm drop diameter) are used in manufacturing, while 1 pl (12 µm drop diameter) dispensers are used in research and development.

Piezoelectric printheads incorporate a transducer composed of a piezoelectric material attached to the nozzle, connected with a signal generator. When a voltage is applied, the piezoelectric material changes shape or size, thus reducing or increasing the cross-section of the cavity full of ink. This generates a pressure wave in the fluid forcing a droplet of ink from the nozzle. An example of the cycle leading to the ejection of a drop in a piezoelectric printhead is schematically shown in **Figure 2** and can be divided in three phases. For each phase, **Figure 2** shows a schematic picture of the ink cavity and of the nozzle. In the inset, a typical pulse is shown in the $V(t)$ diagram.

When the printhead is inactive, or before the start of the pulse, a small voltage is applied to the piezoelectric element to keep it slightly deflected (**Figure 2(a)**). When the pulse starts (phase 1), a step signal is applied to the piezoelectric element to relax it and expand the ink cavity (**Figure 2(b)**). This generates a negative pressure (expansion) wave that moves both toward the supply and the orifice ends of the cavity. At the supply end of the cavity, the pressure wave is reflected as a positive (compression) wave. During this phase, fluid is pulled into the chamber through the inlet. The voltage then stays constant for the duration of the dwell time.

During phase 2 of the pulse, the piezoelectric element bends inward and compresses the fluid in the cavity (**Figure 2(c)**). This generates a compression wave that, if properly timed, can add to the reflected compression wave from phase 1, thus generating a reinforced pressure wave. When this wave reaches the orifice, the acoustic energy is converted

Table 1 Comparison of the drop volume and diameter achievable with the different drop on demand piezoelectric ink-jet technologies most commonly used to print functional materials

Technology	Minimum drop volume (fl)	Minumum drop diameter (µm)	Multi-nozzle printheads availability
Piezoelectric [17]	1000	12	Yes
Acoustic [11]	65	5	No
Electrohydrodynamic [15]	0.7	1	No

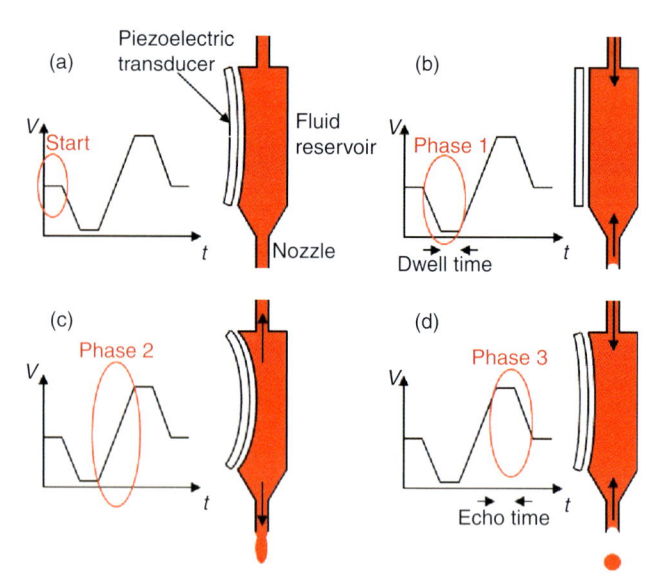

Figure 2 Schematic example of a typical drop ejection cycle in a drop on demand piezoelectric printhead. (a) Initial printhead configuration. (b) Phase 1 (expansion): ink is sucked into the nozzle from the reservoir. (c) Phase 2 (compression): a droplet is ejected. (d) Phase 3 (expansion): additional phase added to eliminate satellite drops. V = voltage applied to the piezoelectric element. t = time.

into kinetic energy, which causes the fluid to start emerging from the orifice at high velocity. When a negative pressure wave, reflected at the supply end of the cavity, reaches the orifice, the fluid is pulled back and a drop is formed.

An additional phase can be added to the pulse to cancel the residual acoustic waves propagating in the tube (**Figure 2(d)**). After the signal is kept constant for the duration of the echo time, it is brought back to its original value, causing a second expansion of the ink cavity. If properly timed, the cavity expansion removes any residual wave propagating in the cavity, avoiding satellite drops to be ejected from the nozzle. Phase 3 is an optional feature of the signal pulse, which can be used to optimize the jetting characteristics of a piezoelectric printhead.

The signal applied to a piezoelectric printhead needs to be adjusted for every fluid that is introduced in the jetting cavity. This is usually done empirically but the suitability of a fluid to be used for a particular printhead should be established before attempting to jet it. The main physical parameters that affect the jetting performance of a fluid are the following:

- Viscosity
- Density
- Surface tension
- Speed of sound in the fluid and, consequently, speed of a pressure wave inside the fluid cavity
- Boiling temperature and, consequently, volatility of the fluid at the nozzle orifice, where evaporation occurs

The design of a printhead and the choice of a particular ink for a piezoelectric printing system involves several factors that should all be taken into account. First, the ink should be formulated within a narrow range of viscosity and surface tension, optimized for the printhead of choice. The general requirements for a fluid to be used in a piezoelectric dispenser are a viscosity of 0.5–40 mPa s (Newtonian fluids) and a surface tension of 20–70 mN m^{-1}. If a fluid containing nanoparticles is used, the diameter of the nanoparticles should be less than 5% of the diameter of the nozzle to avoid jetting instability [10]. Any additive added to modify the physical properties of the ink should be carefully assessed to ensure that it will not affect negatively the performance of the functional material after deposition. Second, no chemical interaction should occur between the ink and the printhead, to prevent structural failure or malfunction. Third, a reliable and stable jet should be achieved before an ink can be used for printing. In particular, the presence of satellite drops may affect the printing resolution. Moreover, nozzles can become clogged due to the presence of air bubbles in the ink, evaporation of ink at the nozzle plate or presence of particles of size similar to the diameter of the nozzle. Fourth, if a multi-nozzle printhead is to be used successfully, the direction and speed of the drops ejected has to be highly uniform in order to ensure an accurate positioning of a large number of drops in well-defined substrate locations.

Piezoelectric ink-jet printers are the most widely used systems for printing functional materials. Although other solutions have been investigated, only piezoelectric printheads have been scaled up to fabricate multi-nozzle dispensers used in industrial applications. For this reason, this is the technique on which we will focus in the following sections.

6.2.3 Acoustic DOD

One of the main drawbacks of piezoelectric printheads is their tendency to become clogged. This is particularly critical when attempting to deposit inks composed of particle suspensions in solvents with low boiling temperature. In this case, nozzles can become clogged due to evaporation of solvent on the nozzle plate. An alternative solution to overcome these limitations is acoustic ink-jet printing [11]. This system has a nozzleless printhead, in which an acoustic beam is focused by a high-frequency acoustic lens on a free liquid surface. The acoustic wave is created by a high-frequency transducer attached to the back of the lens. The lens focuses the acoustic energy, forming a pressure wave that overcomes the surface tension of the liquid standing over the lens and expels a drop from the liquid surface.

An acoustic lens focuses sound in much the same way that an optical lens focuses light. Snell's law describes the refraction of sound as it passes through an interface between two materials of differing sound speed. An acoustic lens provides the appropriate material thicknesses to focus a parallel wavefront of sound to a single focal point. The liquid surface is at the focal plane of the acoustic beam, which is confined to a spot with a diameter of the order of the acoustic wavelength. The acoustic energy transmitted by the beam to the liquid surface causes a mound or liquid to rise and, consequently, a single droplet to be expelled. Stable formation of drops with volume as small as 65 fl (5 µm diameter) has been reported [11].

The nozzleless design of acoustic ink-jet printheads minimizes clogging, improving jetting reliability. OTFTs fabricated using a polymeric semiconducting layer printed by acoustic ink-jet printing have been reported [12]. Nevertheless, the scalability of this technology to multi-ejector printheads, necessary for applications in manufacturing, has not been demonstrated to date.

6.2.4 Electrohydrodynamic-Jet Printing

In electrohydrodynamic ink-jet printers an electric field is used to eject sub-femtoliter drops from very fine glass capillaries [13, 14]. While thermal and piezoelectric printers use acoustic energy to destabilize the fluid contained in a nozzle, electrohydrodynamic-jet (e-jet) printers achieve the same result using an electric field. Park *et al.* [14] describe a setup in which the external surface of very fine glass capillaries, with internal diameter as small as 0.3 µm, is coated with gold. A self-assembled monolayer is subsequently deposited on the gold surface, which becomes hydrophobic, while the internal surface of the capillary remains hydrophilic. A pressure controller is used to control the ink pressure inside the tube. A direct current (DC) voltage is applied between the gold coating on the nozzle and a substrate, which rests on an electrically grounded metal plate. The voltage generates an electric field that causes mobile ions in the ink to accumulate near the surface of the nozzle, creating a flow of ink that, if correctly optimized, can generate a very small drop of ink. The droplet travels from the nozzle to the substrate, where its final location is controlled by a mechanical *x*–*y* stage.

The electrodynamic control of the formation of the droplets, combined with the sub-micrometric diameter of the nozzle, allowed to generate the smallest drops reported in the literature to date. Park *et al.* [14] reported printing lines of polyethyleneglycol with a minimum width of 700 nm. They also reported using e-jet to print a range of inks including poly(3,4-ethylenedioxythiophene) doped with polystyrene sulfonic acid (PEDOT/PSS) and a carbon nanotube ink.

Using a similar experimental setup, Sekitani *et al.* [15] from the University of Tokyo fabricated organic TFTs by printing the source and drain of the devices by e-jet printing. The ink used for this work consisted of silver nanoparticles dispersed in a suitable solvent. Drops with a volume of 0.7 fl and a diameter of 1 µm were obtained using e-jet printing, yielding lines as narrow as 2 µm. This resolution allowed to print TFTs with channel length of 1 µm. Nevertheless, the printed lines were both narrow and very thin. For this reason, 10 passes were necessary to obtain continuous lines with a resistivity low enough for practical applications. Furthermore, to date this system has not been scaled to multi-nozzle operation, which would be necessary for its application in manufacturing high-performance printed circuits.

6.3 Tools and Materials for Piezoelectric DOD Ink-Jet Printing

6.3.1 Ink-Jet Printers

Ink-jet printing of functional materials for electronic applications is more demanding than conventional graphic printing. In particular, ink-jet printers have to fulfill the following requirements:

- Precise control over drop formation, ejection, and directionality.
- Imaging system to assess the stability of the drop formation.
- Precise control on the final location of the drops over the substrate to achieve a typical drop placement accuracy of ± 5 μm.

In order to achieve this objectives, ink-jet printers are composed of several complementary subsystems. The printhead controls drop formation and ejection. The design and performance factors of piezoelectric printheads will be discussed in the next section. Furthermore, the final location of the drops on the substrate is dependent on the position of the printhead at the time of jetting. This can be controlled in two different ways: the stage which supports the substrate can move in both x- and y-direction under a fixed printhead, or both the stage and the printhead can move in perpendicular directions. A heated stage is essential in order to control the spreading of the drops on the substrate and the coffee stain effect [16]. When printing multiple layers, an alignment camera is necessary to pre-align the printhead to pre-existing patterns on the substrate. A separate camera is needed to visualize the droplets during process optimization. This allows the user to optimize the jetting parameters and assess the stability and reliability of the jetting conditions. A signal generator is necessary to drive the piezoelectric transducer in the printhead. A fluid storage and delivery system is needed to transport the ink to the printhead and to control the meniscus pressure. The fluid reservoir must be flexible enough to be quickly filled with solvent to thoroughly clean the delivery system after or during a jetting session. A waste collection system must be provided to collect the ink used during jet optimization and to dispose of any cleaning solvent that is forced through the printhead. Finally, an optional cleaning module can be provided to help maintaining the printhead clean during its operation.

Several ink-jet printer manufacturers are specialized in producing systems tailored for deposition of functional materials. It is outside the scope of this work to provide an exhaustive list of ink-jet printer manufacturers. In the following, a few examples of single- and multi-nozzle systems will be presented to show the evolution of printing systems from research to manufacturing.

The first tools were developed for research and development and only supported single-nozzle printheads. These solutions are still available, because of the simplicity of operation, higher flexibility, and lower cost of single-nozzle printers and printheads. As these tools have only one nozzle, the waveform generator and the positioning system are as simple as possible. Typically, these systems have a fixed printhead and an x–y positioning stage, on which the substrate is held. **Figure 3(a)** shows an example of a custom-built printer based on printheads and droplet generator produced by Microdrop Technologies (Norderstedt, Germany). This printer has the added feature of a transparent substrate holder, which allows to visualize the drops landing on transparent substrates through a camera positioned underneath the x–y stage.

In order to develop ink-jet printing processes suitable for a manufacturing environment, multi-nozzle printers have been developed to address throughput and deposition uniformity issues typical of single-nozzle systems. In order to assess the suitability of a process for a multi-nozzle printing platform, small research and development systems are currently available on the market. A typical example is the Dimatix Materials Printer (DMP), manufactured by Fujifilm Dimatix (**Figure 3(b)**). This is a compact tool designed for prototyping and testing of materials and substrates. It supports a 16-nozzle cartridge, also developed by Fujifilm Dimatix to jet 1 and 10 pl drops [17]. The cartridge is designed to allow users to quickly fill it with test inks in order to test new material formulations.

To ease the transfer of a process to the manufacturing environment, multi-nozzle ink-jet printers with the same printheads used on production systems have been developed. This solution is particularly attractive because, once a laboratory process has been optimized on a development tool, it can be easily transferred to a production system with no

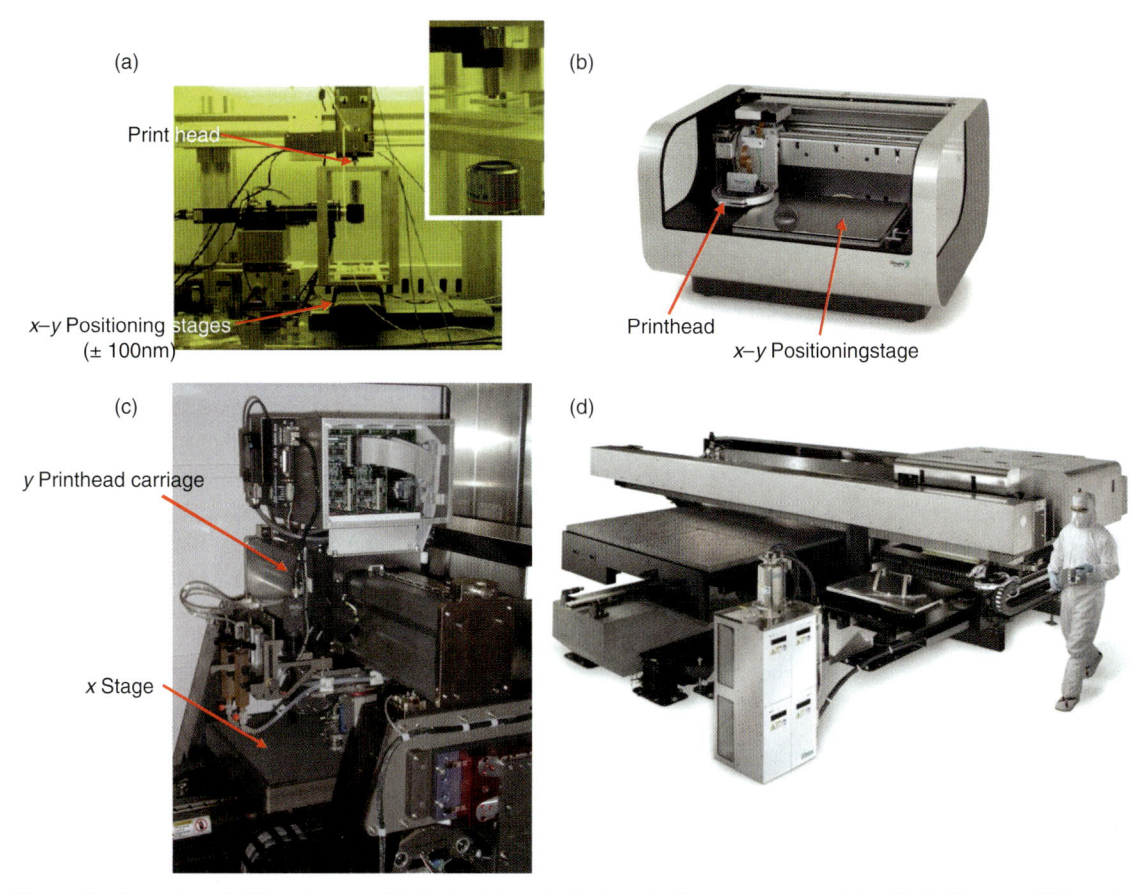

Figure 3 Examples of different types of ink-jet printers. (a) Custom-built single-nozzle printer. (b) Multi-nozzle R&D printer manufactured by Fujifilm Dimatix. (c) Multi-nozzle development printer (Ulvac 120L) manufactured by ULVAC GmbH. (d) Generation 8 multi-nozzle production printer (Ulvac M-Series) manufactured by ULVAC GmbH (b) Reproduced with permission from http://www.dimatix.com, Fujifilm Dimatix. (d) Reproduced with permission from http://www.ulvac.eu, ULVAC GmbH.

need of additional development. An example is the Ulvac 120L ink-jet printer, manufactured by Ulvac GmbH (see **Figure 3(c)**). This is an industrial multi-nozzle ink-jet system designed for process development, based on the Ulvac 140P production system architecture. This tool is ideal to demonstrate the suitability of a technology for manufacturing. It can be used to fabricate prototype devices, to achieve better understanding of the tolerances of new processes, and to test new inks or printheads. The design of the system facilitates quick changing of the ink material and of the solvent. It can also be used to test material formulations that are potentially incompatible with each other. The flexibility of the printhead support module allows different printheads to be installed on the same tool with little hardware change. The Ulvac 120L supports printheads jetting 10 pl drops, and 1 pl printheads are currently being integrated in the system.

Recently, the evolution of the LCD display industry toward ever-larger panel sizes has led to the development of very large ink-jet printing systems to reduce the processing complexity and the material consumption of the fabrication process. An example of this move to ever-larger processing equipment is the Ulvac M-Series Generation 8 ink-jet printer, which is shown in **Figure 3(d)**. This production system can handle a maximum substrate size of 2400 mm × 2400 mm.

6.3.2 Piezoelectric Printheads

The printhead is the most complex and technically demanding part of an ink-jet printer. For this reason, there are specialized companies on the market that fabricate printheads, relying on tool manufacturers to design the positioning and fluid delivery systems. The most obvious specification of a printhead is the number of nozzles and the drop volume, but high-performance printheads for functional materials also have to fulfill other criteria. As the materials are often dissolved in aggressive solvents, it is essential that the printhead is made of materials resistant to a wide range of solvents. Moreover, a particular printhead is optimized for inks within a narrow range of viscosity and surface tension. To extend the range of fluids that can be used with a particular dispenser, a heating element can be added to reduce the viscosity of the ink. To avoid flooding, the nozzle plate or the outside surface of the printhead should be hydrophobic. This is usually achieved by depositing hydrophobic coatings in these areas. Another essential parameter is the maximum jetting frequency at which the printhead can reliably operate. This is a critical factor in devices designed to operate in a manufacturing environment. Often printheads incorporate particle filters and degassing systems, to prevent particles and gas bubbles from reaching (and possibly clogging) the nozzles. The drop directionality also depends on the design and manufacturing tolerances of the printhead. This is an essential factor affecting the drop placement accuracy of a printer. The maximum drop speed that can be achieved also depends on the design of the printhead. Finally, in the case of printheads with multiple nozzles, the crosstalk between nozzles has to be minimized.

It is beyond the purpose of this work to provide an exhaustive review of the different manufacturers of piezoelectric printheads available on the market. In the following, two examples of single- and multi-nozzle piezoelectric printheads will be discussed. **Figure 4**(a) shows an image of an MJ-A style single-nozzle printhead fabricated by MicroFab Technologies. This dispenser is composed of a glass capillary surrounded by a piezoelectric material. A protective case surrounds and protects the assembly. A cable is connected to the piezoelectric transducer to provide the electrical signal. When a voltage is applied, the piezoelectric material squeezes or expands the glass capillary, triggering the ejection of a drop. The MicroFab MJ-A printheads can be used to jet fluids with viscosity lower than 20 mPa s and surface tension of 20–70 mN m^{-1}.

An example of a multi-nozzle printhead is shown in **Figure 4**(b). This is an SX3 printhead manufactured by Fujifilm Dimatix. This printhead has 128 individually addressed nozzles. A main ink chamber is connected to each nozzle, and a piezoelectric strip is located on both sides of the chamber. The nozzles are individually addressed by electrical connections that are fed through flexible fittings. The nozzle plate is composed principally of a rock trap and of a protective layer, on which a hydrophobic coating is deposited. The ink chamber is double ported for ease of flushing. This printhead has 27 μm wide nozzles, which deliver a calibrated drop volume of 10 pl. The nozzles are made of silicon and fabricated using MEMS technology, which allows to obtain very precise tolerances to improve the jetting capability and reproducibility of this printhead. The Dimatix SX3 printhead can be used to jet fluids with viscosity of 8–20 mPa s and surface tension of 24–36 mN m^{-1}. The choice between single- or multi-nozzle printheads depends on the application. For material testing and small prototype fabrication, single-nozzle printheads are the best choice. In fact, they are flexible in their applications and easy to handle. The control electronics is very simple, as it only drives one nozzle. Moreover, single-nozzle dispensers are easy to clean when clogged, because a fluid forced through the

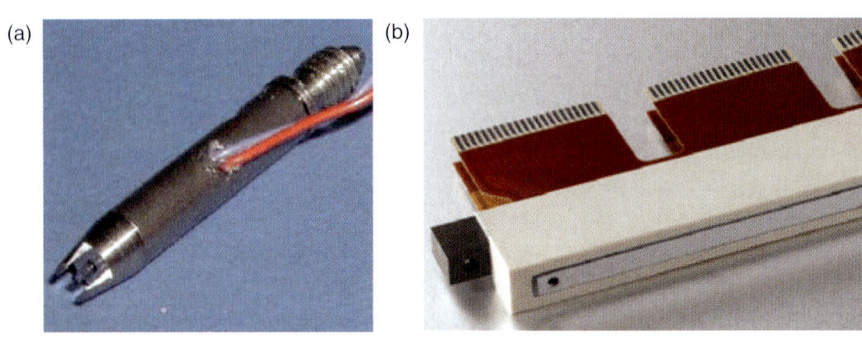

Figure 4 Examples of different types of piezoelectric printheads. (a) A single-nozzle printhead (MJ-A style) fabricated by MicroFab Technologies. (b) A 128-nozzles printhead (Spectra SX3) fabricated by Fujifilm Dimatix. (a) Reproduced with permission from http://www.microfab.com, MicroFab Technologies. (b) Reproduced with permission from http://www.dimatix.com, Fujifilm Dimatix.

capillary will naturally remove any obstruction. The drop speed does not need to be calibrated as far as the jetting is stable. The main disadvantage of single-nozzle systems is the low throughput. Moreover, the necessity of scanning a single nozzle across a substrate introduces a considerable delay between the deposition of the first and last drops. Therefore, different solvent evaporation profiles can appear across the substrate, affecting the yield of the fabricated devices.

On the other hand, multi-nozzle printheads have much higher throughput, as they can print many drops in parallel. Printheads with enough nozzles to print a large pattern in a single pass can be fabricated, minimizing the evaporation profiles across the substrate. Nevertheless, multi-nozzle dispensers are more difficult to handle and to operate than single-nozzle printheads. They are typically difficult to clean because, when the solvent is forced through the nozzles, it avoids the clogged nozzles and flows out of the open ones, making it very difficult to remove particles obstructing the flow of the ink. Moreover, multi-nozzle printheads are usually designed for a small range of fluids. It is also essential that all the nozzles eject droplets with the same speed and in the same direction. If this condition is not fulfilled, the printed pattern can be distorted. Finally, the necessity to control a large number of nozzles requires complicated control electronics.

6.3.3 Printable Metallic Inks for Ink-Jet Printing of Low-Resistivity Conductors

Ink-jet printing is a very attractive technique for printing functional materials for electronics because complete devices can be printed using only one tool, with no need of high-vacuum systems or large clean rooms. As discussed already, many different functional inks have been developed for a wide range of applications. In this section, we will concentrate on the use of metallic inks for ink-jet printing of the high-conductivity lines needed to fabricate high-frequency TFTs. For this application, picoliter and sub-picoliter ink-jet printers have been employed to pattern the source, drain, and gate electrodes with high resolution, in order to minimize the size and the parasitic capacitance of the devices. **Table 2** compares the different approaches compatible with low-temperature processing to print metallic tracks with high electrical conductivity.

An interconnection technology, which is fully compatible with ink-jet printing, is based on organic conductors. The most widely used printable organic conducting material is a water-based ink PEDOT/PSS, with a conductivity range of 30–150 S cm^{-1} (e.g., Clevios conductive polymers, produced by H. C. Starck GmbH) [18]. Given the large interconnection distances and the small cross-section of the conducting tracks needed for many potential printed electronic applications, the use of organic conductors for interconnects results in unacceptably high parasitic resistance and therefore low switching speed.

Recently, a new class of inks containing metallic nanoparticles, suitable for ink-jetting, has received much interest. Several companies now offer inks containing gold or silver nanoparticles. These two materials are preferred for ink-jet applications because they do not grow insulating oxides when deposited in presence of oxygen. This simplifies the processing considerably, as enclosing an ink-jet printer in a oxygen-free atmosphere introduces considerable technical challenges.

The melting temperature of nanometer-sized metallic particles has been found to be dependent on the size of the particles [19]. In particular,

Table 2 Comparison of the different approaches compatible with low-temperature processing to print metallic tracks with high electrical conductivity

Material	Printed line conductivity (S cm^{-1})	Annealling temperature (°C)	Bulk material conductivity (S cm^{-1})
Aqueous dispersion of PEDOT/PSS (Clevios) [18]	30–150	100	N/A
Gold nanoparticle (annealed) [20]	3.3×10^5	150	4.5×10^5
Gold nanoparticles (laser sintered) [21]	1.7×10^5	Laser	4.5×10^5
Organic complex-based silver (TEC-IJ) [25]	4.5×10^5	130	6.2×10^5
Silver seed plus electroless copper plating [27]	5.5×10^5	N/A	5.9×10^5

particles with a diameter lower than 5 nm exhibit a marked drop in melting temperature that, in some cases, makes them suitable for processing on flexible substrates. In order to be diluted in a solvent suitable for ink-jet printing without forming larger aggregates, the metallic nanoparticles need to be encapsulated in organic carbon chains (e.g., alkylthiols), which makes the particles soluble in common nonpolar solvents such as xylene. By optimizing the chain length of the encapsulant carbon chain, Huang et al. [20] demonstrated ink-jet printed gold lines with a conductivity $\sigma = 3.3 \times 10^5$ S cm^{-1} after annealing at 150 °C, compared with a value for bulk gold of 4.5×10^5 S cm^{-1}. An alternative approach to lower the processing temperature even further is laser curing of ink-jet printed metal nanoparticles [5]. In this approach, ink-jet printed gold nanoparticles were sintered by irradiation with a focused laser beam of suitable wavelength to deposit enough thermal energy to induce highly localized melting of the nanoparticles. Ink-jet printed lines with $\sigma = 1.7 \times 10^5$ S cm^{-1} were fabricated using this approach[21]. Local laser heating allowed to reduce the heat-affected zone, lowering the processing temperature considerably. Furthermore, selective laser sintering using a very narrow beam enabled to improve the resolution of the ink-jet printed lines. This was achieved by washing away the unsintered nanoparticles after laser irradiation of the ink-jet printed metal lines.

Although gold and silver nanoparticle inks are currently used in research and in many industrial applications owing to their chemical stability and high conductivity, the growing price of silver may be a concern for its future applications in printed electronics. For this reason, recently several research organizations have tried to develop alternative printable inks based on low-cost materials. With this objective in mind, copper is an excellent candidate due to its low cost and high electrical conductivity. Nevertheless, the development of printable copper inks has proved very difficult, mainly because copper is not stable in air, but has a strong tendency to oxidize. In particular, pure nonnoble metal nanopowders show very high reactivity in oxidizing atmosphere, due to their very high specific area. Moreover, high temperatures are typically required to anneal copper inks, which prevents their use on flexible polymeric substrates. In an attempt to solve these problems, precursor-based inks have been used to print copper conductive lines [22]. Nevertheless, this approach requires annealing temperatures of the order of 200 °C in a reducing atmosphere. This requires sophisticated equipment and therefore is not compatible with low-cost device fabrication.

In a promising approach to develop copper inks compatible with ink-jet printing, a capping polymer has been used to synthesize air-stable copper nanoparticles [23, 24], requiring temperatures of more than 250 °C in order to obtain bulk-like conductivity. In order to reduce processing temperatures, several companies have developed copper inks that can be annealed at temperatures compatible with polymeric substrates. For example, Novacentrix has developed copper nanoparticle inks suitable for ink-jet printing that can be sintered within microseconds using electromagnetic irradiation. This approach allows to produce copper films with a conductivity only 3 times higher than bulk copper and at process temperatures compatible with flexible substrates such as PET.

Printable inks composed of metal-containing organic complexes dissolved in a solvent were developed as an alternative to nanoparticle metallic inks. An example of this class of inks are the TEC-IJ transparent silver inks, produced by InkTec Corporation. In this case, the ink does not contain nanoparticles, but instead is composed of a silver-containing organic complex dissolved in an anisole-based solvent. During annealing, at first the solvent contained in the ink evaporates. In this phase, nucleation and growth of silver nanoparticles is observed. This is followed by a sintering step, when the silver nanoparticles join together to form a continuous metal line. With this approach, ink-jet printed lines with $\sigma = 4.5 \times 10^5$ S cm^{-1} were obtained after annealing at 130 °C, compared with a value for bulk siver of 6.2×10^5 S cm^{-1} [25]. This is the lowest sintering temperature achieved by thermal annealing of metallic inks demonstrated to date, and it is fully compatible with several commercially available flexible polymeric substrate materials, such as polyethylene terephthalate (PET) and polyethylene naphthalate (PEN) [26]. Moreover, the absence of nanoparticles facilitates the jetting of this class of inks. Nanoparticle inks are more likely to cause nozzle clogging than organic complex-based inks, because the nanoparticles can form large aggregates at the nozzle orifice due to evaporation of solvent on the nozzle plate.

An alternative approach to pattern metal tracks with very high conductivity is a printing plus electroless plating process. In this method, a seed layer is ink-jet printed, followed by electroless deposition of a thick layer of the metal required by the application. This technique is very flexible, as different metals can be plated using the same seed layer. For example, copper, silver, gold, nickel, and cobalt can all be deposited using this technique. Moreover, the conductivity of the film is very close to that of bulk material [27]. The possibility of growing thick metal layers without the need of multiple printing passes allows to tune the line resistivity without affecting the throughput. Finally, electroless plating is a low-temperature technique, compatible with flexible substrates. The process temperature only depends on the technique chosen to cure the seed layers. Nevertheless, the resolution that can be achieved with this technique is limited. Therefore, it is used mainly in applications in which low-resolution, high-conductivity metal tracks are needed, for example, to fabricate antennas for radio frequency identification tags (RFIDs).

6.4 Applications of Ink-jet Printing of Functional Materials

6.4.1 Overview of Ink-Jet Printing Applications

In the last 10 years, a growing number of applications has been found to be suitable for ink-jet printing of functional materials. In each of these, inks with a wide range of compositions are used to deposit materials with specific electrical, structural, optical, chemical, or biological properties. The chosen applications benefit from the possibility of depositing small, controlled quantities of materials in precise locations on a substrate and at a low temperature. The low cost of the technique and its extreme flexibility (every run can be tailored to a specific application, as no template is needed) make this technique even more appealing to an extremely varied range of technological sectors.

The two applications that have been more successful have been both developed for the display market: OLEDs and LCDs. Another application which is on the verge of becoming mainstream is the use of ink-jet printing to fabricate OTFTs to be integrated in active matrix backplanes, flexible sensors, and RFIDs. These are all examples of additive processes, but ink-jet printing has also been used to provide more flexibility to conventional subtractive processes that are associated with optical lithography. This has led to the development of digital lithography and ink-jet etching. These applications will be described in more detail in the following sections.

It is beyond the scope of this work to provide an exhaustive analysis of all the techniques that have been developed to exploit the advantages of ink-jet printing of functional materials. Nevertheless, the main fields of application will be outlined in the following.

Antennas and interconnection lines with very high conductivity have been fabricated by a combination of ink-jet printing and electroless plating [28]. In this approach, a printed metallic layer forms a seed layer for the growth of a thick metal film to make high-conductivity conduction lines. This combines the flexibility of ink-jet printing with the need of fabricating high-Q passive components for RFID applications. The possibility of patterning conductive structures on virtually any type of substrate has allowed embedded passive components [29] and solder bumps to be ink-jet printed on printed circuit boards (PCBs). By using this technology, these components can be deposited directly on the PCBs, making expensive lithographic and assembly processes redundant.

The mechanical properties of sintered particle suspensions have been exploited to fabricated 3D objects made, for example, of structural ceramics [2]. This approach can be extremely advantageous particularly in the field of rapid 3D prototyping. Ink-jet printing of structural and/or conductive materials has also been used in the field of micro-electro mechanical systems [30]. In this case, the low cost of ink-jet printing is ideal for low-volume manufacturing and prototyping.

Finally, pioneering applications of ink-jet printing have been explored in chemistry and biology. For example, ink-jet printing has been used to fabricate polymer micro-arrays and libraries [31]. In this case, the possibility of synthesizing in parallel large numbers of chemically different polymers by delivering very small quantities of solutions with different compositions in different points of a substrate has been exploited. A similar application within the realm of biology has been demonstrated by ink-jet printing of protein biochip arrays [3]. The ability to identify thousands of analytes in only a few microliters of biological sample can become an enabling tool in drug development and clinical diagnosis.

6.4.2 Ink-Jet Printing Applications in the Display Industry

The most promising current application of ink-jet printing in volume manufacturing is within the LCD industry. For this application, ink-jet printing has been used to reduce processing costs and material consumption compared to conventional photolithographic and printing techniques. **Figure 5** shows a schematic cross-section of an amorphous silicon TFT bonded to a color filter substrate to make an LCD display. The maximum substrate size handled by the recently introduced Generation 8 equipment is 2400 mm × 2400 mm. At these size, large-area patterning using lithographic equipment becomes extremely challenging, and maskless printing techniques become viable. Finally, the single-pass printing capability of these large printers combined with the advantages of using a contactless technology makes ink-jet extremely attractive for this application. Ink-jet printing has been introduced in LCD technology to deposit the alignment layers for the liquid crystal, the color filters and the spacers to control the cell gap.

Coating of polyimide by ink-jet printing has been utilized to deposit functional materials for aligning the liquid-crystal molecules. Ink-jet printing was found to perform better than conventional flexographic deposition techniques as it yielded lower foreign-particle concentration and better layer thickness uniformity. Moreover, the material cost was found to be reduced by 40% and the waste of cleaning solvents was minimized [32].

The need of depositing different materials on the same layer to fabricate filters with different colors requires several process steps using a conventional subtractive approach. On the other hand, printing the color filters using an ink-jet printer allows to avoid 75% of the process steps and, therefore, reduces the cost by 60%. Furthermore, the additive approach allows to reduce the consumption of the filter materials by 80% [32].

Ink-jet technology has also been used to deposit spacer balls to control the LCD cell gap [33]. For this application, it is critical to ensure that the spacer balls do not encroach in the pixel area.

Within the display industry, ink-jet printing is suitable for the fabrication of OLEDs. This technology allows to fabricate extremely thin displays, as it does not require back-lighting. Moreover, the absence of a filter and the narrow range of wavelengths emitted both contribute to the exceptional color brightness of OLEDs. Finally, unlike LCDs, the viewing quality of OLEDs is not compromised even at very narrow viewing angles. In order to fabricate these devices, the necessity to deliver light emitting polymers to each pixel of the display makes the use of ink-jet printing mandatory. The technology is used to deliver the different semiconductors that define the three primary colors of each pixel. Ink-jet is also used to deliver the hole-injection material, typically PEDOT/PSS. The main requirement for this application is to deposit a controlled amount of polymer within pre-patterned polyimide wells on a substrate [4]. In order to achieve this, a self-aligned process has been developed to compensate for any imprecision in the alignment of the printhead with the substrate and for variations of drop directionality in multi-nozzle printers. In this approach, two successive plasma treatments modify the surface energy of the

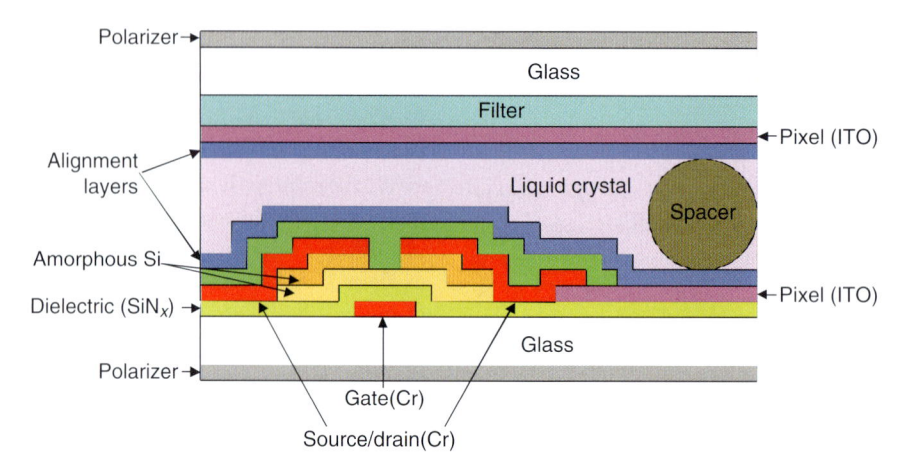

Figure 5 Schematic cross-section of a LCD display with active matrix backplane composed of amorphous silicon TFTs.

substrate, making it hydrophilic, and of the polyimide walls of the wells, which become hydrophobic. When a drop reaches the substrate, if misaligned it could overlap the walls of a well. As far as the misalignment is not excessive, in this case the drop de-wets from the polyimide walls and flows inside the well.

6.4.3 Fabrication of Organic Thin Film Transistors by Ink-Jet Printing

The fabrication of solution processed TFTs has recently become a viable technology, as the mobility of FETs with polymer semiconductors has reached or exceeded that of devices using an amorphous silicon active layer ($\mu = 0.5$–$1 \text{ cm}^2 \text{ V}^{-1} \text{ s}^{-1}$) [34]. In the past, only p-type semiconducting polymers achieved these performances. Nevertheless, recently n-type semiconducting polymers with electron mobilities up to $0.85 \text{ cm}^2 \text{ V}^{-1} \text{ s}^{-1}$ have become available, paving the way to the development of all-printed organic CMOS circuits [35]. A wide range of applications has, therefore, become possible. Manufacturing integrated circuits by ink-jet printing offers several advantages, such as:

- Low process cost due to process simplification; no need to fabricate templates and reduced material consumption due to the additive approach.
- Lower equipment investment if all-printed technologies can be implemented.
- Low-temperature processes that allow the fabrication of circuits on flexible displays.
- A high-throughput technology compatible with roll-to-roll and wide-area processing.
- A more environmental-friendly technology compared with conventional subtractive processes, where large amounts of waste solvents are needed to remove unwanted materials.

The most obvious application of ink-jet printed TFTs is to fabricate active matrix backplanes for flexible displays [36]. The low-frequency operation of this applications does not require high-mobility active layers and, therefore, is compatible with solution processed polymeric semiconductors. Nevertheless, the necessity of defining devices with sub-10 µm channel length is very demanding in terms of the resolution of the systems. Moreover, the necessity of printing different materials, to be used for the active layer, dielectric, electrodes, and interconnects of the circuits requires a printhead compatible with a broad range of different solvents.

A novel approach to overcome the resolution limitations of the ink-jet printing systems will be described in the following sections.

Another area in which there has been a lot of interest in ink-jet processing is the fabrication of all-printed RFIDs [28]. This is a very competitive market where a new technology could only compete with the mainstream silicon CMOS circuits by introducing extreme cost savings. The most likely frequency band in which solution-processed RFID tags could drive the fabrication costs down is the high frequency band at 13.56 MHz. In this frequency range, the coil can be integrated in a planar process and the systems are only moderately susceptible to the interference from metals and liquids. Therefore, 13.56 MHz RFIDs are very promising for use in RF barcodes for item-level tagging.

The requirements and current technology status for a printed RFID tag can be discussed separately for the three main blocks that compose the device: antenna, rectifier and digital section [37]. As mentioned above, the fabrication of the antenna stage has already been demonstrated by a combination of ink-jet printing and electroless plating [28]. On the other hand, the rectifier and circuit protection stages, which filter the high-frequency signal to provide power for the digital block, are currently open to innovations and improvements. An approach that has been demonstrated in the past uses nonprinting processes to fabricate a diode-based rectifier capable of operating at high frequencies. A better, all-printed approach would feature diode-connected transistors. Nevertheless, fabricating transistors capable of operating at 13.56 MHz has proved to be extremely challenging. A novel approach to address this issue will be presented in the following sections. The final stages, the digital and modulation stages, mostly operates at a frequency much lower than the carrier frequency, so its operation is less challenging than that of the rectifier stage. Nevertheless, the divider necessary to generate the clock signal for the digital stage needs to operate at 13.56 MHz. Moreover, the fabrication of a programmable read-only memory to store the barcode is yet another challenging issue in the development of an all-printed RFID tag.

A final application that has recently received a lot of attention in the field of ink-jet printed organic electronics is the fabrication of chemical and mechanical sensors on flexible substrates. A remarkable example of this area of application is the work presented by Someya et al. [38] on flexible pressure sensors for the fabrication of E-skins. In this approach, an active matrix composed of organic

TFTs is fabricated by ink-jet printing of silver nanoparticle inks and polyimide precursors on a flexible substrate. The active matrix circuitry can sense the pressure applied on a pressure-sensitive rubber layer deposited on the whole surface of the sensor. The flexibility of the circuitry could allow for this sensor to be used in a large-area detection system.

6.4.4 Digital Lithography

One of the main intrinsic advantages of ink-jet printing is that no fixed printing master is needed to transfer a pattern on a layer. This could drive down the patterning costs. Moreover, ink-jet printing permits automatic pattern registration if the deposited structures need to be aligned to existing features on a flexible substrate, which could be distorted. Using a subtractive approach, ink-jet printing can be used to deposit materials that can replace photoresist. The printed pattern can then be transferred to the underlying layer by wet or dry etching [39]. This technique has been demonstrated by fabricating both a-Si and organic TFTs using a subtractive process. To find the best compromise between fine printing resolution (requiring high contact angle) and good ink adhesion to the substrate (requiring spreading of the drops on the substrate), phase-change masks have been patterned by ink-jet printing. These are wax-based inks that are maintained slightly above their solidification temperature while jetting. As a consequence, when the droplets come into contact with the substrate, they immediately solidify, allowing high-resolution printing even though they have a good adhesion to the substrate.

6.4.5 Direct Ink-Jet Etching of a Polymeric Substrate

Ink-jet printing can be used as a substractive process if the fluid printed is a suitable solvent that dissolves a material previously deposited on a substrate. This approach was used as a practical way to create vias to fabricate organic circuits. This technique could be an important step toward the integration of all-printed organic circuits. Nevertheless, it is essential that all the layers to be patterned are soluble in the solvent which is ink-jetted on the substrate. Moreover, the polymer to be etched has to be removed faster than the evaporation of the solvent from the substrate. Using this technique combined with p-type organic FETs and ink-jet printing of PEDOT resistors, inverters operating up to the frequency of 250 Hz were fabricated [40].

6.5 Self-Aligned Ink-Jet Printing of Organic Transistors

As in the case of other printing-based manufacturing approaches, standard ink-jet printing shows a clear limitation in terms of minimum achievable feature size. This is given by the minimum volume of ink that can be deposited onto the substrate and its characteristic spreading, and by the droplet placement accuracy. Standard piezoelectric DOD printers can eject droplets with a volume bigger than 1 pl and allow a placement accuracy not better than $\pm 5\,\mu m$ [9]: given these limitations, the minimum channel length (L) achievable in a transistor is 10 µm. In order to develop an ink-jet printing-based technology with micrometer and sub-micrometer feature size, different approaches were proposed to address the intrinsic limitations of the standard printing technique. One approach was to reduce the droplet volume to the sub-femtoliter range with the electrohydrodynamic-jet printers described in Section 6.2.4, leading to the realization of transistors with L as low as 1 µm [15]. Since this is a very low throughput technique, it could be implemented only in the steps where critical features are actually involved, while relying on a standard high-throughput printing technique elsewhere. In order to reach the micrometer and sub-micrometer resolution range, while preserving the throughput, other approaches were adopted, basically relying on the droplet movement on the substrate induced by forces exerted on the ink by a contrast in surface energies. These techniques will be reviewed here.

6.5.1 High-Resolution Ink-Jet Printing Assisted by Surface-Energy Patterns

In order to increase the resolution achievable with standard printing heads, it is mandatory to control the flow and the spreading of the ink on the substrate. This can be achieved by pre-patterning areas with different surface energy on a substrate. In the case of a water-based ink, if a pattern containing hydrophobic and hydrophilic areas is defined, the droplets landing on the substrate will be subjected to a driving force that pushes them away from hydrophobic areas toward hydrophilic areas, leading to a confinement of the ink (see Section 6.4.2). Furthermore, the contact angle of the ink on both areas can be controlled, thus allowing to limit the spreading of the ink. In practice, the process and its resolution are complicated by many factors, like the surface tension of the ink and its viscosity, the receding contact angle (which, if too small, can induce a pinning of the fluid) and the drying rate (which can dynamically influence the other parameters). Notwithstanding the complexity of the mechanism, the viability of surface-energy patterns assisted ink-jet printing in achieving high-resolution fabrication of printed organic TFTs was successfully demonstrated by defining channel lengths as short as 5 µm [41]. In this work, a hydrophobic polyimide pattern was defined by means of standard photolithography on a hydrophilic glass substrate. After that, PEDOT/PSS contacts were printed with high resolution, thanks to the confinement offered by the polyimide structures (see **Figure 6**).

The same concept was then developed on substrates patterned by means of processes more appealing from the point of view of manufacturing, like soft lithographic techniques [36] and digital direct-write laser patterning [42]. These provided a viable tool to develop ink-jet printed backplanes for displays [43], where a micron-scale resolution is sufficient. To address more demanding applications, such as the fabrication of logic circuitry surrounding the display, a higher resolution is needed in order to enhance TFTs speed by scaling down the channel length to the sub-micrometer range. The question regarding the highest resolution achievable with a surface energy-pattern assisted printing was investigated by fabricating patterns with feature sizes down to 200 nm by means of electron-beam lithography. On this basis, de-wetting of the ink from hydrophobic ribs was demonstrated, where droplets were split into two halves with sub-micrometer-scale resolution [44]. Even if in this case the need to pre-define a pattern by lithography increases the process complexity and costs, the technique could prove an exceptional resolution capability achievable with standard ink-jet printing.

6.5.2 Lithography-Free Self-Aligned Printing with Sub-Micrometer Resolution

A technique that allowed to achieve a sub-micrometer resolution using ink-jet printing without the need of pre-patterning the substrate would represent a more viable option in terms of manufacturing. This possibility was first demonstrated by defining channels in the sub-hundred-nanometer range through the self-alignment of PEDOT/PSS electrodes [45]. The fundamental mechanism that allows to achieve such a high resolution with standard ink-jet printing equipment and without the need for any susbtrate prepatterning or precise alignment still relies on the ink droplet motion induced by a contrast in surface energy on the substrate. This self-aligned printing (SAP) technique can be summarized in the following three steps:

1. the first conductive pattern is ink-jet printed on the substrate;
2. the surface energy of the printed pattern is selectively lowered without altering the wettability of the substrate, so that a strong-enough contrast in surface energy is induced between the pattern and the substrate; and
3. a second conductive pattern is ink-jet printed on top of the first one with a small overlap, without requiring fine alignment.

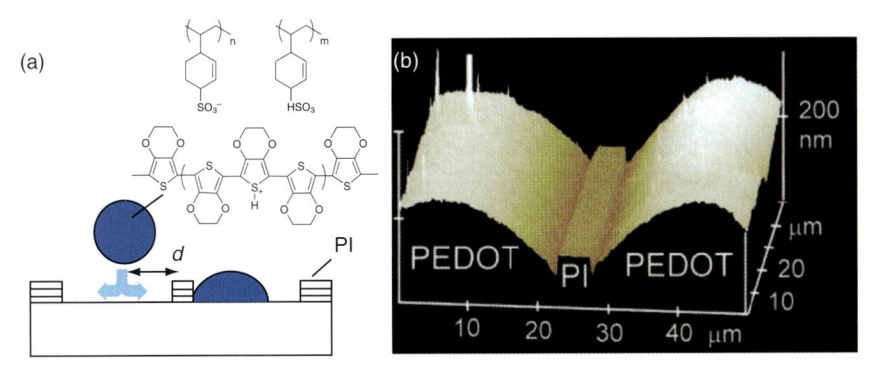

Figure 6 (a) Schematic diagram of high-resolution ink-jet printing onto a pre-patterned substrate. (b) AFM image showing accurate alignment of ink-jet printed PEDOT/PSS source and drain electrodes on hydrophilic glass substrate, separated by a hydrophobic polyimide (PI) line with $L = 5$ µm. With permission from Sirringhaus H, Kawase T, Friend RH, et al. (2000) High-resolution inkjet printing of all-polymer transistor circuits. *Science* 290(5499): 2123–2126.

The droplets deposited in the second pattern experience a driving force that induces them to de-wet from the first pattern. Depending on several parameters, like the surface energy contrast, the ink viscosity and its drying rate, the flow of the second pattern can be controlled to create a very small gap between the two patterns (**Figure 7(a)**). No top-down step is involved in the fabrication of these narrow gaps and the process is compatible with the alignment resolution, which is available in standard printing systems (see Section 6.3). This approach can be extended to inks of very different nature: in the following, we will review the process involving PEDOT/PSS based inks, while, in the next section, the SAP of a metal nanoparticles based ink will be reported.

Different techniques were employed to lower the surface energy of the PEDOT/PSS pattern without altering that of the substrate, like a selective treatment of the surface or a segregation of surfactants.

In the first case, the first PEDOT/PSS pattern, printed on a glass substrate, was subjected to a post-deposition CF_4 plasma treatment. This treatment created a fluorinated surface layer on the PEDOT/PSS pattern, therefore lowering its surface energy, while leaving the glass surface in a high-energy state. This selective modification created the necessary surface energy contrast that allowed the second pattern, printed in partial overlap with the first one, to de-wet and to define a second electrically isolated contact at a sub-micrometer distance (**Figure 7(b)**). Typical gaps obtained with this method were shorter than 100 nm [45]. Among the parameters that had to be controlled in order to obtain a clean and complete de-wetting, the receding contact angle of the ink flowing off the first pattern was found to be particularly critical. This had to be kept close to the advancing contact angle in order to avoid pinning of the contact line due to surface inhomogeneities.

In an alternative approach, suitable surfactants, with polar head groups and nonpolar tail groups, were added to the PEDOT/PSS ink. The surfactants segregated to the surface of the water-based ink and created a low-surface-energy layer with the head polar group in contact with the ink and the nonpolar tail group at the interface with the air [46]. This approach gave results comparable to the plasma treatment described above. Nevertheless, it was found to show a much narrower process window when the second pattern was printed with an overlap, showing a higher sensitivity to droplet positioning. The optimum case was found by printing the second droplets apart from the first pattern so that they could spread against it [43, 45] (**Figure 7(c)**).

By adopting the SAP technique to define the PEDOT/PSS source and drain contacts of a FET on a glass substrate, an all-solution-processed top-gate, bottom-contacts device with a channel length of ≈100 nm was demonstrated [45]. In **Figures 8(a)** and **8(b)** are shown the transfer and output charcateristic curves of a poly(9,9-dioctylfluorene-co-bithiophene) (F8T2) based device (a sketch of the device

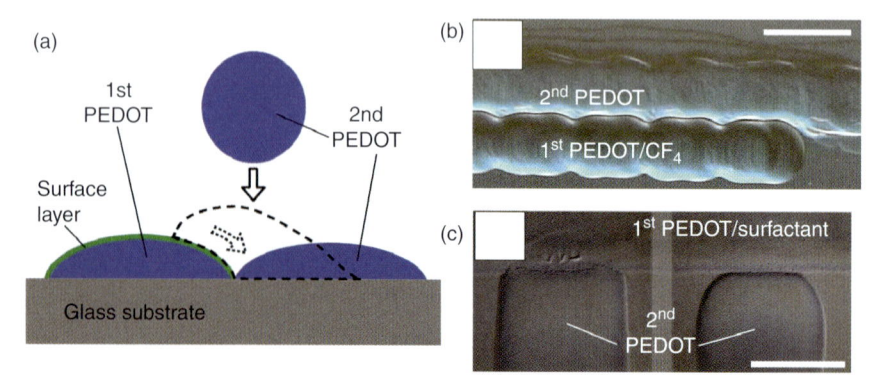

Figure 7 (a) Sketch of the SAP process: a PEDOT/PSS droplet is printed in partial overlap (dashed line) with a previously printed, surface modified hydrophobic pattern; due to surface energy contrast, the droplet experiences de-wetting. (b) Optical micrograph of PEDOT/PSS self-aligned printed contacts; the surface energy of the first pattern has been selectively modified using a CF_4 plasma. (c) Optical micrograph of self-aligned PEDOT/PSS printed contacts; the surface energy or the first pattern has been selectively modified due to benzalkonium chloride surfactant segregation; the image on the left shows incomplete de-wetting without annealing of the first pattern; the image on the right shows complete de-wetting upon annealing of the first pattern. Reproduced with permission from Sele CW, von Werne T, Friend RH, and Sirringhaus H (2005) Lithography-free, self-aligned inkjet printing with sub-hundred nanometer resolution. *Advanced Materials* 17(8): 997–1001.

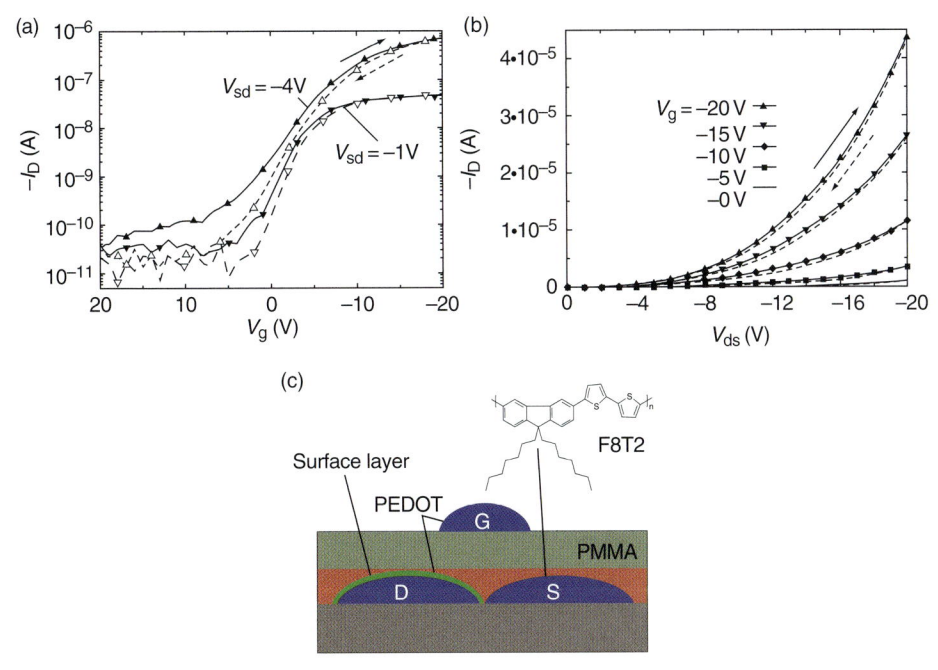

Figure 8 (a) Transfer and (b) output characteristics of an SAP transistor with $W = 350\,\mu m$, whose schematic section is depicted in (c) (D: drain, S: source, G: gate). Solid lines: forward sweep; dashed lines: reverse sweep. Reproduced with permission from Sele CW, von Werne T, Friend RH, and Sirringhaus H (2005) Lithography-free, self-aligned inkjet printing with sub-hundred nanometer resolution. *Advanced Materials* 17(8): 997–1001.

architecture is depicted in **Figure 8(c)**). As it is evident from the output curves, the FET shows short-channel effects, due to a nonideal scaling of the dielectric thickness (d). In this case, a standard spin-coated, 120–130 nm thick PMMA layer was used as the gate dielectric. However, the device showed clear field-effect behavior and, therefore, showed SAP as a high-resolution bottom-up promising process for organic FET fabrication. For this technique to be viable, an ideal d/L ratio has to be obtained, as it will be described in Section 6.6.

6.5.3 Self-Aligned Printing of Metal Nanoparticles Inks

The low conductivity of polymer conductors like PEDOT/PSS poses a severe limit to the currents obtainable in short-channel transistors. Therefore highly conducting printable metal inks compatible with SAP were needed to extend the process applicability and to demonstrate the generality of the conceptual approach. Metal nanoparticles inks (see Section 6.3.3), previously adopted in defining electrodes in low-resolution printing processes [47, 48], represented an obvious option. The first successful demonstration of highly conductive electrodes defined by SAP technique was achieved with gold nanoparticles inks, where xylene and/or cyclohexylbenzene (CHB) were used as solvents, adopting a selective surface treatment approach [49]. After printing and sintering of the first gold pattern on glass or silicon/SiO$_2$, the substrate is subjected to a vapor deposition of a self-assembled monolayer (SAM) of 1H,1H,2H,2H-perfluoro-1-decanethiol (PFDT). The thiol, having a strong affinity with gold, grows selectively on the first printed pattern and lowers its surface energy. On the other hand, PFDT does not chemically bond to the substrate and the physically deposited molecules can be easily washed away by solvent rinsing. The result is that the substrate is unaffected by the SAM treatment and, therefore, a strong surface energy contrast is achieved. Then, when a second pattern is printed in partial overlap with the first pattern, the ink de-wets from the first pattern and defines a narrow self-aligned channel. The channel length is found to depend on fluid dynamical conditions, which were investigated by varying several parameters of the ink, such as the drying rate, the viscosity, and the surface tension, in a controlled experiment where the first gold pattern was defined by lithography. In particular, it is found that the channel length can be controlled from ≈50 nm to a few micrometers by adding an increasing amount of CHB to a xylene solution of the gold nanoparticles. CHB has higher surface tension and boiling point with respect to xylene; therefore, a formulation with more CHB shows a higher repulsion from the hydrophobic SAM-treated gold and a longer drying time, which allows the ink to de-wet further before its contact line is pinned to the substrate because of solute precipitation due to solvent evaporation (**Figure 9(a)**).

The enhancement of the currents in an F8T2 transistor obtained thanks to the highly conducting printed metal electrodes (conductivity in the 10^5 S cm^{-1} range) with respect to the PEDOT/PSS electrodes (conductivity lower than 1 S cm^{-1} with the adopted formulations) can be appreciated in **Figure 9(b)**. The compared devices were realized with the same top-gate architecture (**Figure 8(c)**), with identical channel width (W) and comparable L, the only difference being represented by the ink used to pattern source–drain contacts. The increase in the on currents is higher than a factor of 5, especially at low driving voltages.

The SAP technique promises to have technological relevance since it shows to work, after a careful optimization of the parameters, in many different conditions. Self-aligned metallic contacts were defined both by printing the second pattern on a previously lithographically defined one (see **Figures 10(a)** and **10(b)**) and, more interestingly, in an all-printed process, without any need for pre-patterning (see **Figures 10(c)** and **10 (d)**). The SAP technique could be implemented both with a custom-made gold nanoparticles ink, and by adopting a commercially available gold nanopaste, suitably diluted [50]. Moreover, SAP patterns were printed with various single-nozzle piezo-driven commercial heads, like MicroDrop and Microfab (see Section 6.3.2), and both with low (4 Hz) and high (up to 1 kHz) droplet ejection frequencies.

The channel width could be nearly reduced to the dimension of one droplet by printing the contacts in a T-shape (see **Figure 10(c)**). This allows to easily monitor and to limit defects in the fabrication process resulting in a further reduction of leakage currents. In 20–50 μm wide contacts, with a gap of about 200 nm (**Figure 10(d)**), this can be lower than 1 pA at a bias voltage as high as 40 V. This very low leakage is beneficial for the device operation, especially if we consider not only transistors but also detecting applications, and is made possible by a reliable and high-yield printing process (details will be reported elsewhere).

The reliability and versatility of the process are promising features for its development and transfer to high-throughput multi-nozzle printing systems.

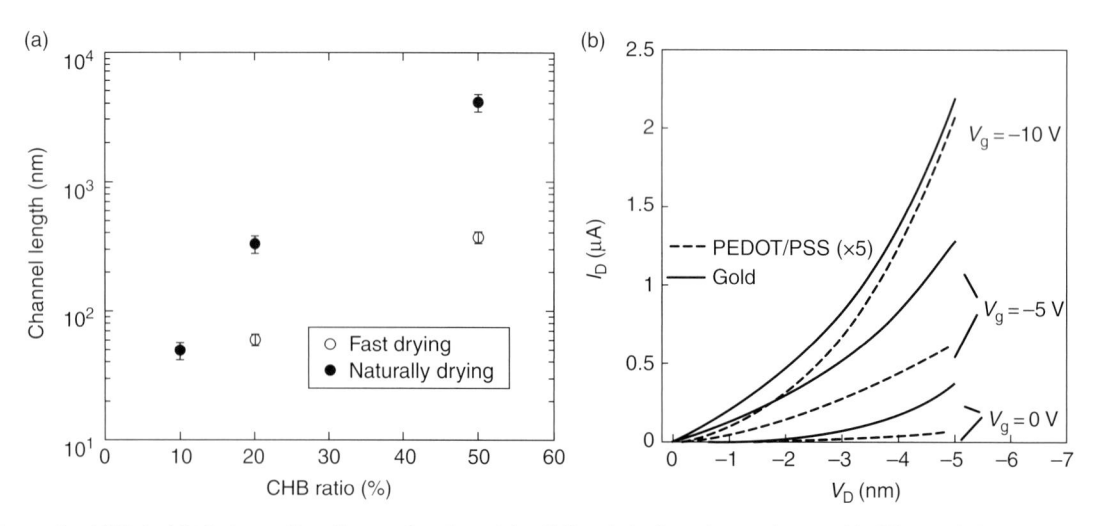

Figure 9 (a) Plot of SAP channel lengths as a function of the CHB ratio in the solvent mixture with different drying conditions. Fast drying: the substrates were heated to ensure a uniform drying time of the ink of around 2–3 s. Naturally drying: the ink was allowed to dry at room temperature. (b) Comparison of the output characteristics of SAP transistors with $W = 350$ μm and $L = 60$ nm, with gold (solid lines) and PEDOT/PSS (dashed lines) source and drain electrodes. The output current of the PEDOT/PSS device is multiplied by a factor of 5 in order to be visible on the same graph. With permission from Zhao N, Chisea M, Sirringhaus H, Li Y, Wu Y, and Ong B (2007) Self-aligned inkjet printing of highly conducting gold electrodes with submicron resolution. *Journal of Applied Physics* 101(6): 064513.

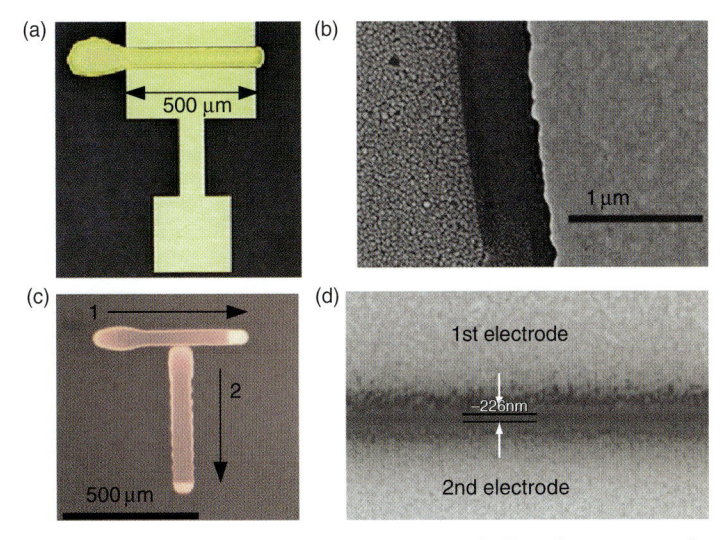

Figure 10 (a) Optical micrograph of an ink-jet printed gold electrode self-aligned to evaporated gold contacts and (b) an SEM picture showing the sub-micrometer channel. (c) Optical micrograph of T-shaped, all-printed self-aligned gold contacts, and (d) the relative SEM image showing a channel with sub-micron channel length. (b) Reproduced with permission from Noh YY, Zhao N, Caironi M, and Sirringhaus H (2007) Downscaling of self-aligned, all-printed polymer thin-film transistors. *Nature Nanotechnology* 2(12): 784–789.

6.6 Fully Downscaled, Self-Aligned Printed Polymer Thin-Film Transistors

6.6.1 Downscaling Requirements

In order to become a fabrication tool enabling the embedding of electronic functionality in flexible substrates, SAP has to be integrated in a printing process capable to fully downscale the device while retaining the requirements for low-cost, large-area, flexible electronics manufacturing.

SAP has shown to be a promising tool for nanopatterning of source and drain contacts for organic transistors in a bottom-up approach capable to overcome the poor resolution achievable with standard mass-printing tools. However, the downscaling of the channel length is not enough to achieve high performing, fast-switching FETs. To fulfill basic downscaling requirements, all the critical sizes governing the transistor functionality have to be correctly scaled. In order to fully control the channel through the gate contact, it is necessary that the dielectric thickness is correspondingly reduced, at least resulting in $d < L/4$. This has to be done by keeping the gate leakage current low and by limiting the overlap (x_{OV}) between the gate and the source and drain contacts (see **Figure 11**), which has a strong influence on the transition frequency (f_T) of the transistor. Moreover, a compatible solution-processable semiconductor with a relatively high mobility ($\mu > 0.1\ cm^2\ V^{-1}\ s^{-1}$) is needed to obtain enough drain current. The transistor architecture also has to be carefully chosen in order to preserve good injection from the contacts even with sub-micrometer channels. To complete the device architecture, a printing process capable of defining the gate metal level needs to be implemented.

6.6.2 Self-Aligned Printed Polymer TFTs with Thin Gate Dielectrics

In previous studies, where clear field effect in submicrometer transistors fabricated with SAP contacts was clearly demonstrated (see Section 6.5.2), it was shown that a nonideal scaling of the gate dielectric thickness with channel length led to severe degradation of the performance of SAP transistors. Strong short-channel effects with the loss of current saturation were observed, negatively affecting the driving capability of the transistors and preventing their use as building blocks of logic gates.

A suitable scaling of the dielectric, compatible with solution process and top-gate architecture, is therefore needed. The approach based on crosslinkable polymers, demonstrated in bottom-gate architectures [50], could not be applied, given the high crosslinking temperatures involved, which are typically higher than 150 °C. These temperatures

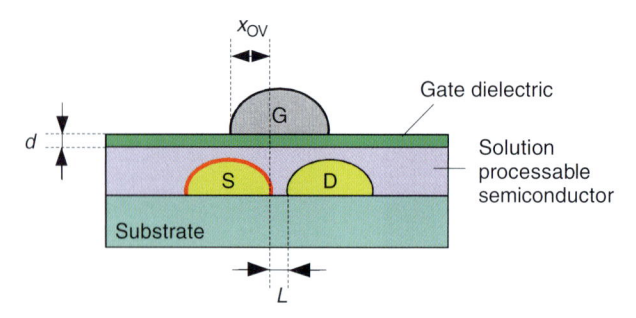

Figure 11 Schematic cross-section of an all-printed, bottom-contact, top-gate transistor showing its main features. The printed electrodes are represented in yellow (S; source, D: drain), defining the channel length *L*. The printed gate contact is shown in grey (G: gate). *d* = dielectric thickness. x_{OV} = gate overlap with the source and drain electrodes, which is the main source of gate parasitic capacitance.

would cause, in most cases, unacceptable damage to the underlying semiconductor layer. Moreover, a common additional problem is the difficulty in finding suitable orthogonal solvents. Therefore, it was necessary to choose a different approach, which was successfully identified in the selection of a suitable blend of a polymer dielectric and a crosslinking agent, essentially a trichlorosilyl-alkane, both soluble in the same orthogonal solvent [51]. The great advantage of this approach is the spontaneous cross-linking that can develop at room temperature in the presence of moisture and oxygen. Several dielectric systems were studied and their performance assessed and reported in detail [52]. The best performing blend (C-PMMA) consisted in poly(methylmethacrylate) (PMMA) as the base polymer dielectric, 1,6-*bis*(trichlorosilyl)hexane as the crosslinker and n-butyl acetate as the orthogonal solvent (**Figure 12(a)**). When the blend is exposed to air, a physical crosslinking takes place, where the chlorosilane reacts with water to form a siloxane network through the PMMA, therefore reducing the free volume of the composite crosslinked material [50]. C-PMMA was shown to have performances comparable to those obtainable with thermal SiO_2 grown on Si [53], with a ≈30–50 nm thick C-PMMA layer exhibiting a breakdown field higher than $3\,\text{MV cm}^{-1}$ and a leakage current as low as $10–100\,\text{nA cm}^{-1}$ at a field of $2\,\text{MV cm}^{-1}$ (**Figure 12(b)**).

The scalability and robustness of the C-PMMA dielectric allowed to fabricate correctly scaled sub-micrometer channel SAP devices, where clean saturation in the output could be achieved at low applied voltages. In a study where F8T2 was adopted as the semiconductor, it was shown that, when $d < L/4$, the transistor shows long-channel behavior, while for $d \geq L/3$ the outputs increasingly deviate from saturation, showing nearly linear characteristics at

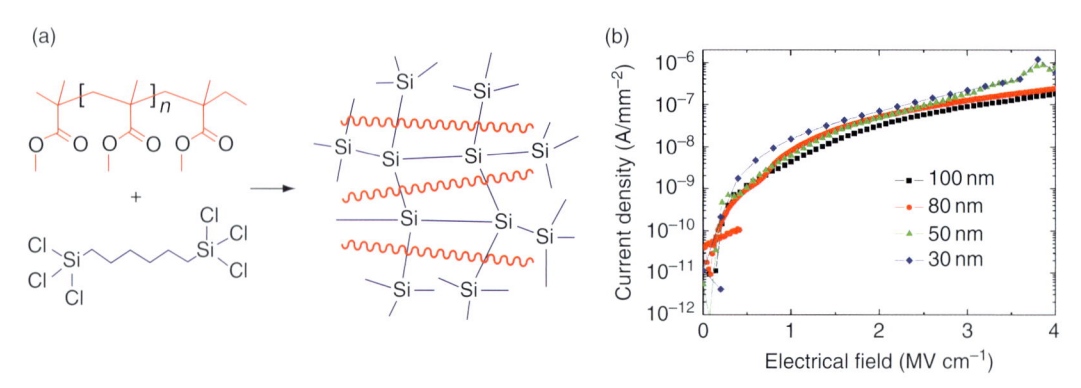

Figure 12 (a) Schematic representation of the cross-linked PMMA system (C-PMMA): 1,6-*bis*(trichlorosilyl)hexane (blue, left) is mixed with a PMMA solution and upon reacting with water it forms a siloxane network (right) throughout the polymer chains. (b) Leakage current density versus applied voltage of C-PMMA polymer films sandwiched between gold electrodes. Reproduced with permission from Noh YY, Zhao N, Caironi M, and Sirringhaus H (2007) Downscaling of self-aligned, all-printed polymer thin-film transistor. *Nature Nanotechnology* 2(12): 784–789.

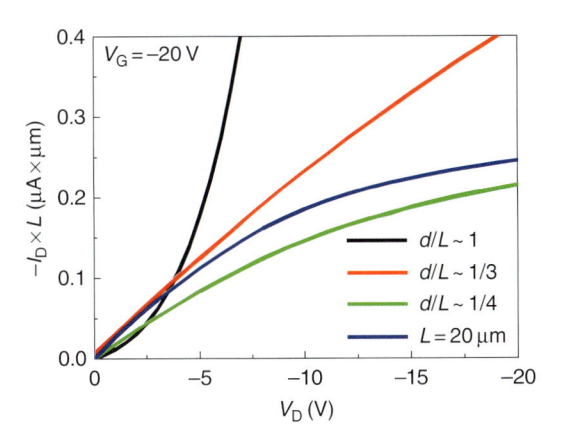

Figure 13 Output characteristics of SAP F8T2 transistors with a ≈100 nm thick C-PMMA gate dielectric. Curves measured on devices with different channel lengths are plotted: $L \approx 100$ nm (black line), $L \approx 300$ nm (red line), $L \approx 400$ nm (green line) and $L = 20$ μm (blue line). Reproduced with permission from Noh YY, Zhao N, Caironi M, and Sirringhaus H (2007) Downscaling of self-aligned, all-printed polymer thin-film transistor. *Nature Nanotechnology* 2(12): 784–789.

$d = L/3$ and strongly pronounced superlinear behavior for $d = L/1$ (**Figure 13**).

It is important to point out that, for a successful downscaling of the device, it is mandatory to retain a good charge injection from the contacts to the semiconductor also in the sub-micrometer channel, where the contact resistance becomes a critical parameter. Especially when environmentally stable polymers such as F8T2 are used, showing high HOMO levels, high injection barriers can develop at the contacts. In order to observe a correct scaling behavior as in the previous study, it is therefore necessary to modify the source–drain contacts with an SAM capable of shifting the metal work functions and allowing a better matching of the energy levels [54]. For the same reason, the chosen top-gate architecture that allows to distribute carrier injection, thanks to the current crowding effect [55], is found to be beneficial in limiting the contact resistance effects.

An example of an SAP FET showing clean saturation, completely processed from solution, without the use of any mask, is depicted in **Figure 14**(a). The SAP structure was fabricated on a glass substrate by ink-jet printing of a gold nanoparticles ink (**Figure 10**(c)) to realize a device with $L \approx 200$ nm and $W \approx 35–40$ μm. F8T2 was used as the semiconducting layer and it was spun from solution on previously PFDT treated electrodes. After spincasting of a ≈50 nm thick C-PMMA dielectric layer, the gate contact was then realized by ink-jet printing of a highly conducting silver complex-based ink produced by InkTec Corporation. This was possible thanks to the low sintering temperature of the ink (<150 °C), compatible with a top-gate architecture fabrication process. The characteristic curves of the device (**Figures 14**(b) and **14**(c)), reflecting a mobility $\mu > 10^{-3}$ cm^2 V^{-1} s^{-1}, demonstrate the possibility to achieve long-channel behavior in an SAP FET fabricated with a mask-free and lithography-free process.

6.6.3 Self-Aligned Gate Printing

The reduction of the channel length allows to extract more current from a transistor but is not sufficient to obtain the maximum device operation speed if the parasitic gate capacitance is not minimized. A figure of merit of the highest operation frequency achievable in an FET is the transition frequency f_T (or unity-gain frequency), corresponding to the

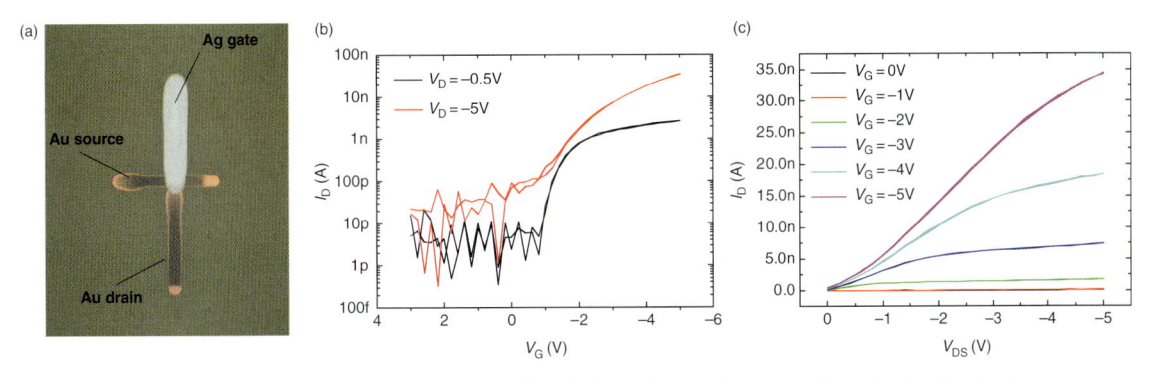

Figure 14 (a) Optical micrograph of an all-printed F8T2 self-aligned transistor with gold electrodes ($L \approx 200$ nm, $W \approx 40$ μm), a ≈50 nm thick C-PMMA gate dielectric and a silver gate contact. In (b) and (c), the transfer and output curves of the device are plotted.

crossover point at which an alternating current (AC) modulated channel current in response to a gate-voltage modulation becomes equal to the parasitic current flowing through the capacitance between the gate and source–drain [50]. **Equation 1** provides an expression for f_T based on the gradual channel approximation [56, 57]:

$$f_T = \frac{g_m}{2\pi(C_{gd} + C_{gs})} \quad (1)$$

where g_m is the transistor transconductance, C_{gd} is the gate–drain capacitance and C_{gs} is the gate–source capacitance. Since $g_m \propto L^{-1}$, equation shows that f_T can be increased proportionally to L^{-1} if C_{gd} and C_{gs} are constant. However, if also the gate- and source–drain capacitances could be scaled, limiting the total gate capacitance area to the geometrical channel area $L \times W$, f_T would scale with L^{-2}. Therefore, it is evident that, in order to boost the operational frequency of an SAP FET, the control of the gate overlap capacitance is crucial.

As an example, **Figure 15** shows that f_T for a top-gate, bottom-contacts, micrometer-scale F8T2 transistor is around 1 kHz ($L = 5\,\mu$m). When the channel length is reduced of an order of magnitude in an F8T2 SAP device, the drain current correspondingly increases, but if no attempt is made in order to control the overlap x_{OV}, this may result in an even larger parasitic gate current than in the micrometer-scale case. In the reported example, this is due to the fact that the micrometer-scale device shows an interdigitated contact pattern where the ratio of the channel area to the total area of the pattern is higher than in the case of the SAP device, in which the width of the metal contacts is comparable but the channel is much shorter. This limits the final f_T to 1 kHz, thus completely loosing the advantage of a scaled L. Even if a more controlled printing of the gate line was performed, the overlap between the gate and the source–drain would, however, be associated with the printed gate linewidth, which is of the order of 50–100 μm. This would create an overlapping area ≈ 2 orders of magnitude wider than the actual channel area, introducing an unacceptable parasitic capacitance.

This limitation extends to all the state-of-the-art volume printing techniques, not capable of dispensing small liquid volumes in order to significantly reduce the linewidths. Even if a considerable linewidth reduction could be achieved, for example, using a sub-femtoliter printing system [15], this would have to be compatible with large-area manufacturing and, more importantly, the issue of fine gate-channel alignment would still be problematic. For these reasons it is important that the gate is also defined with a self-aligned process.

To achieve this, different solutions were proposed in the literature. One approach used a self-aligned surface energy pattern, defined thanks to a topographic contrast used during patterning of a lower layer [58] or to the self-aligned photopatterning of a SAM [59]. A more general approach, fully compatible with large-area manufacturing of top-gate SAP FETs, was proposed, in which the source and drain contacts were used as a photomask to develop a trench in a thick photoresist film deposited on top of the dielectric layer [50].

The process comprises the following steps:

1. a thick (1–2 μm) UV photosensitive dielectric layer is deposited on top of the thin dielectric layer;
2. the substrate is irradiated with light through the back, selectively exposing the channel region, thanks to the masking action provided by SAP contacts (**Figure 16(a)**);

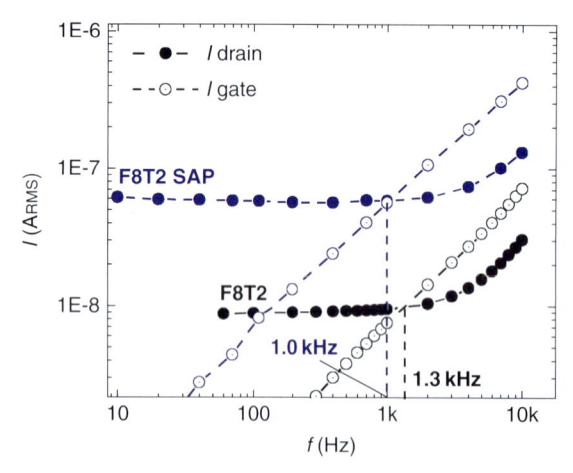

Figure 15 Root-mean-square (RMS) values of drain (filled circles) and gate (open circles) currents versus frequency of an F8T2 transistor (black: $L = 5\,\mu$m, $W = 500\,\mu$m) and of an SAP F8T2 transistor with sub-micron channel length (blue: $L \approx 200$ nm, $W = 500\,\mu$m). The devices are all biased in the trans-diode regime ($V_G = V_D$) and the AC signal V_{gs} has an amplitude in the 120–250 mV range. From the interception (dashed lines) between the drain and gate current curves the frequency of transition (f_T) can be extracted: although the drain current is increased thanks to the self-aligned sub-micrometer channel, the f_T for the SAP FET is strongly limited by an increased gate current due to a noncontrolled gate parasitic capacitance.

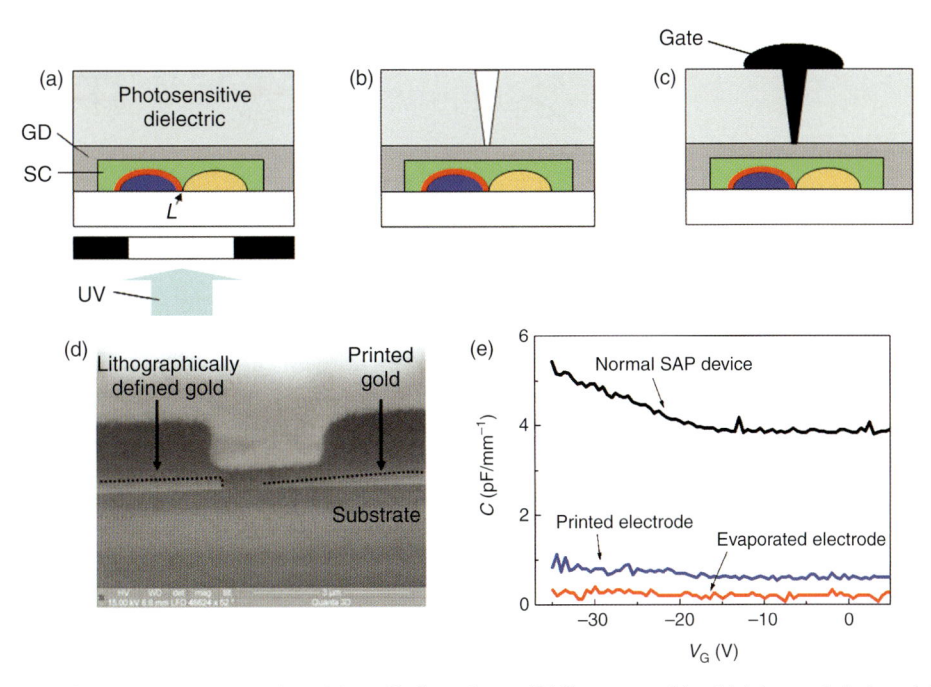

Figure 16 (a–c) Schematic representation of the self-aligned gate (SAG) process: (a) a thick layer of photoresist is spun on top of the thin dielectric layer (L: channel length, SC: semiconductor, GD: gate dielectric); (b) the sample is exposed from the bottom to UV light, using the SAP channel as a photomask; a trench, self-aligned to the channel, is developed; and (c) a gate contact is printed in the trench. (d) SEM cross-section of an SAP-SAG FET. (e) Capacitance-voltage characteristics of SAP FETs (semiconductor: F8T2; gate dielectric: C-PMMA) with self-aligned printed source–drain electrodes. In these devices, the first contact was defined by evaporation and photolithography (evaporated electrode) and the second was defined by SAP (printed electrode). The gate contact was made of ink-jet printed PEDOT/PSS. The black line shows the overlap capacitance between the gate and the source/drain electrodes of a SAP device without SAG. The blue line shows the overlap capacitance between the gate and the printed electrode in an SAG-SAP device. The red line shows the overlap capacitance between the gate and the evaporated electrode in an SAG-SAP device. Reproduced with permission from Noh YY, Zhao N, Caironi M, and Sirringhaus H (2007) Downscaling of self-aligned, all printed polymer thin-film transistors. *Nature Nanotechnology* 2(12): 784–789.

3. the photoresist is developed, obtaining a trench structure self-aligned to the edges of the source and drain contacts (**Figure 16(b)**);
4. a wide gate electrode is ink-jet printed with no need for fine alignment (**Figure 16(c)**).

Thanks to this self-aligned gate (SAG) process, the gate dielectric is thin only relative to the channel, where this is required for correct gating of the device (**Figure 16(d)**). Everywhere else the photoresist provides a thick spacer that strongly reduces the overlap capacitance, which is found to be as low as 1 pF mm^{-1}, up to an order of magnitude lower than in reference devices fabricated without the SAG structure (**Figure 16(e)**). Interestingly, the narrow trench region, where the dielectric is thin, contributes only for ≈10% of the overall overlap capacitance, which is still dominated by the surrounding regions where the thick second dielectric is present. Therefore, a further reduction of the overlap capacitance is possible by employing a thicker resist layer.

The overall self-aligned fabrication process of the transistor, comprising the SAP and SAG processes, is compatible with well-known polymer semiconductors, as F8T2 and the high-mobility poly(2,5-*bis*(3-alkylthiophen-2-yl)thieno[3,2-*b*]thiophene (pBTTT) : no degradation was observed during UV exposure or standard development process. The adoption of pBTTT allowed the fabrication of high-performance transistors with field-effect mobility in the range of 0.1 cm^2 V^{-1} s^{-1}, operating at 5–8 V (**Figure 17(a)**). The SAG process enabled full downscaling of the devices, which reached f_T values of around 1.6 MHz, with a strong improvement compared to a pBTTT transistor incorporating only SAP electrodes, which showed f_T limited to 40 kHz (**Figure 17(b)**).

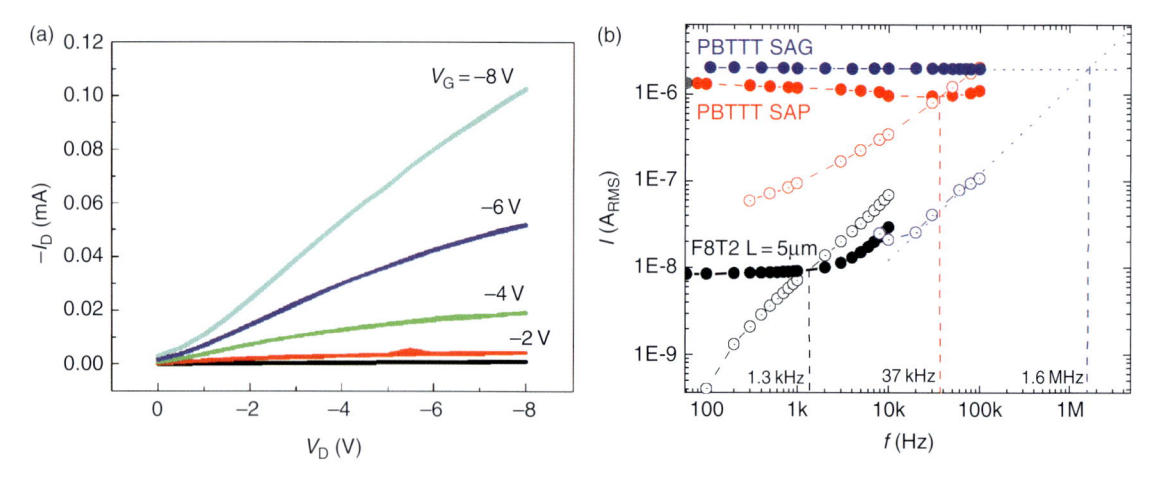

Figure 17 (a) Output characteristics of an SAG-SAP PBTTT-based FET ($L \approx 200\,\mu m$, $W = 500\,\mu m$, $d \approx 50\,nm$). (b) f_T measurements for an SAG-SAP PBTTT FET (blue lines), a normal SAP PBTTT FET without SAG (red lines) and a reference long-channel F8T2 device (black lines). Reproduced with permission from Noh YY, Zhao N, Caironi M, and Sirringhaus H (2007) Downscaling of self-aligned, all-printed polymer thin-film transistors. *Nature Nanotechnology* 2(12): 784–789.

The integration of SAP and SAG techniques offers a viable process for the fabrication of high-performance organic transistors, paving the way to the realization of fast-switching organic logic compatible with the manufacturing requirements of low-cost, large-area flexible circuits.

Although specifically developed in the field of organic transistors, this can be regarded as an example of how the control of surface energy and of wetting properties of the inks, as well as the development of novel architectures, can be effectively used to overcome stringent limitations of ink-jet printing and make this technique suitable for high-frequency electronic applications.

6.7 Conclusions and Outlook

In this article we have reviewed the current status of ink-jet printing of functional materials, focusing on the application of organic electronics and on the fabrication of organic TFTs. With the aim to provide an overview of this emerging field, besides reporting the fundamentals of ink-jet technology, we described the basic tools and materials that enable it.

Recently, the viability of ink-jet printing as a deposition technique was demonstrated for a vast number of functional materials and some basic processes were successfully developed, especially for electronic applications. This fostered the further development of specialized printing equipment and inks. In the former case, tools ranging from single-nozzle systems suitable for basic research, to advanced, multi-nozzle, high-throughput prototyping and production machines have been introduced on the market. In the latter case, the engineering of functional inks and their commercialization is providing both researchers and engineers with more and more solutions for the development of ink-jet printing based processes.

One of the most challenging aspects that are now being addressed is represented by the reliability of this technology: disregarding the specific applications, process uniformity and yield represent key issues for the future spreading of ink-jet printing as a low-cost tool suitable for volume manufacturing in many different areas. The emerging use of ink-jet printing in the volume manufacturing of LCDs demonstrates that these challenges can be overcome. However, in each specific application, this requires an intense interdisciplinary effort of equipment engineering, ink, and process development to provide tailored inks and print systems that are able to meet the very challenging yield and uniformity requirements of large-area electronics applications. At the research level, there are ample opportunities for using ink-jet printing as a tool for better understanding of the structure–processing–property relationships of functional materials and to develop novel device architectures that make use of the unique ability of ink-jet printing to deliver small quantities of liquid to defined positions on an (almost) arbitrary substrate.

Acknowledgments

The Engineering and Physical Sciences Research Council (EPSRC), the Cambridge Integrated Knowledge Center (CIKC), and Plastic Logic Ltd. are gratefully acknowledged for financial support and help in the development of the printing equipment.

References

1. Rayleigh FRS (1878) On the instability of jets. *Proceedings of the London Mathematical Society* S1–10(1): 4–13.
2. Derby B and Reis N (2003) Inkjet printing of highly loaded particulate suspensions. *MRS Bulletin* 28(11): 815–818.
3. Zaugg FG and Wagner P (2003) Drop-on-demand printing of protein biochip arrays. *MRS Bulletin* 8(11): 837–842.
4. Shimoda T, Morii K, Seki S, and Kiguchi H (2003) Inkjet printing of light-emitting polymer displays. *MRS Bulletin* 28(11): 821–827.
5. Ko SH, Pan H, Grigoropoulos CP, Luscombe CK, Fréchet JMJ, and Poulikakos D (2007) All-inkjet-printed flexible electronics fabrication on a polymer substrate by low-temperature high-resolution selective laser sintering of metal nanoparticles. *Nanotechnology* 18(34): 1–8.
6. Cho JH, Lee J, Xia Y, Kim B, He Y, Renn MJ, Lodge TP, and Frisbie CD (2008) Printable ion-gel gate dielectrics for low-voltage polymer thin-film transistors on plastic. *Nature Materials* 7(11): 900–906.
7. Le HP (1998) Progress and trends in ink-jet printing technology. *The Journal of Imaging Science and Technology* 42(1): 49–62.
8. Lindner TJ (2005) Printed electronics – hp's technology beyond ink on paper. *Presentation at Printed Electronics USA*.
9. Creagh LT and Mcdonald M (2003) Design and performance of inkjet print heads for non-graphic-arts applications. *MRS Bulletin* 28(11): 807–811.
10. MicroFab Technologies, Inc., Technote 99-02: Fluid Properties Effects. http://www.microfab.com (accessed July 2009).
11. Elrod SA, Hadimioglu B, Yakub KBT, et al. (1989) Nozzleless droplet formation with focused acoustic beams. *Journal of Applied Physics* 65(9): 3441–3447.
12. Paul KE, Wong WS, Ready SE, and Street RA (2003) Additive jet printing of polymer thin-film transistors. *Applied Physics Letters* 83(10): 2070–2072.
13. Murata K, Matsumoto J, Tezuka A, Matsuba Y, and Yokoyama H (2005) Super-fine ink-jet printing: Toward the minimal manufacturing system. *Microsystem Technologies* 12(1): 2–7.
14. Park JU, Hardy M, Kang SJ, et al. (2007) High-resolution electrohydrodynamic jet printing. *Nature Materials* 6(10): 782–789.
15. Sekitani T, Noguchi Y, Zschieschang U, Klauk H, and Someya T (2008) Organic transistors manufactured using inkjet technology with subfemtoliter accuracy. *Proceedings of the National Academy of Sciences* 105(13): 4976–4980.
16. Soltman D and Subramanian V (2008) Inkjet-printed line morphologies and temperature control of the coffee ring effect. *Langmuir* 24(5): 2224–2231.
17. Fujifilm Dimatix (2008) Materials printer & cartridges dmp-2831 & dmc-11601/11610 – datasheet. http://www.dimatix.com (accessed July 2008).
18. Starck H. C. Gmbh, Clevios conducting polymers for organic thin film transistors. http://www.clevios.com (accessed July 2008).
19. Buffat P and Borel JP (1976) Size effect on the melting temperature of gold particles. *Physical Review A* 13(6): 2287.
20. Huang D, Liao F, Molesa S, Redinger D, and Subramanian V (2003) Plastic-compatible low resistance printable gold nanoparticle conductors for flexible electronics. *Journal of the Electrochemical Society* 150(7): G412–G417.
21. Chung J, Ko S, Bieri NR, Grigoropoulos CP, and Poulikakos D (2004) Conductor microstructures by laser curing of printed gold nanoparticle ink. *Applied Physics Letters* 84(5): 801–803.
22. Hong CM and Wagner S (2000) Inkjet printed copper source/drain metallization for amorphous silicon thin-film transistors. *Electron Device Letters, IEEE* 21(8): 384–386.
23. Park B, Kim D, Jeong S, Moon J, and Kim J (2007) Direct writing of copper conductive patterns by ink-jet printing. *Thin Solid Films* 515(19): 7706–7711.
24. Luechinger NA, Athanassiou EK, and Stark WJ (2008) Graphene-stabilized copper nanoparticles as an air-stable substitute for silver and gold in low-cost ink-jet printable electronics. *Nanotechnology* 19(44): 445201–445206.
25. Inktec Corporation, Transparent silver ink-jet inks. http://www.inktec.com (accessed July 2008).
26. Macdonald WA (2004) Engineered films for display technologies. *Journal of Materials Chemistry* 14(1): 4–10.
27. Kao CY and Chou KS (2007) Electroless copper plating onto printed lines of nanosized silver seeds. *Electrochemical and Solid-State Letters* 10(3): D32–D34.
28. Subramanian V, Frechet JMJ, Chang PC, Huang DC, Lee JB, Molesa SE, Murphy AR, Redinger DR, and Volkman SK (2005) Progress toward development of all-printed rfid tags: Materials, processes, and devices. *Proceedings of the IEEE* 93(7): 1330–1338.
29. Moderegger E, Leising G, Plank H, et al. (2004) Inkjet printing of embedded passive components. *Materials Research Conference Proceedings – Spring Meeting*.
30. Fuller SB, Wilhelm EJ, and Jacobson JM (2002) Ink-jet printed nanoparticle microelectromechanical systems. *Microelectromechanical Systems, Journal of* 11(1): 54–60.
31. Gans B-J and Schubert US (2003) Inkjet printing of polymer micro-arrays and libraries: Instrumentation, requirements, and perspectives. *Macromolecular Rapid Communications* 24(11): 659–666.
32. Baker R (2008) Ink jet's role as a manufacturing tool. *Presentation at IMI 17th Annual Ink Jet Printing Conference*.
33. Albertalli D (2005) Gen 7 fpd inkjet equipment – development status. *Society for Information Display (SID)* 36(1): 1200–1203.
34. Sirringhaus H (2005) Device physics of solution-processed organic field-effect transistors. *Advanced Materials* 17(20): 2411–2425.
35. Yan H, Chen Z, Zheng Y, et al. (2009) A high-mobility electron-transporting polymer for printed transistors. *Nature* 457: 679–686.
36. Sirringhaus H, Bürgi L, Kawase T, and Friend RH (2003) Polymer transistor circuits fabricated by solution processing and direct printing. In: Kagan CR and Andry P (eds.) *Thin Film Transistors*, pp. 427–474. New York: Marcel Dekker.
37. Subramanian V (2007) Radio frequency identification tags. In: Bao Z and Locklin J (eds.) *Organic Field-Effect*

Transistors vol. 128, pp. 489–506. Boca Raton, FL: CRC Press.
38. Someya T, Sakurai T, Sekitani T, and Noguchi Y (2007) Printed organic transistors for large-area electronics. *IEEE Polytronic Conference Proceedings*, pp. 6–11.
39. Wong WS, Daniel JH, Chabinyc M, Arias AC, Ready SE, and Lujan RA (2006) Thin-film transistor fabrication by digital lithography. In: Klauk H (ed.) *Organic Electronics: Materials, Manufacturing and Applications*, pp. 271–293. Weinheim: Wiley VCH.
40. Kawase T, Sirringhaus H, Friend RH, and Shimoda T (2001) Inkjet printed via-hole interconnections and resistors for all-polymer transistor circuits. *Advanced Materials* 13(21): 1601–1605.
41. Sirringhaus H, Kawase T, Friend RH, et al. (2000) High-resolution inkjet printing of all-polymer transistor circuits. *Science* 290(5499): 2123–2126.
42. Burns SE, Cain P, Mills J, Wang J, and Sirringhaus H (2003) Inkjet printing of polymer thin-film transistor circuits. *MRS Bulletin* 28(11): 829–833.
43. Sirringhaus H, Sele CW, Ramsdale C, and von Werne T (2006) Manufacturing of organic transistor circuits bysolution-based printing. In: Klauk H (ed.) *Organic Electronics, Materials, Manufacturing and Applications*, pp. 294–322. Weinheim: Wiley VCH.
44. Wang JZ, Zheng ZH, Li HW, Huck WTS, and Sirringhaus H (2004) Dewetting of conducting polymer inkjet droplets on patterned surfaces. *Nature Materials* 3: 171–176.
45. Sele CW, Friend RH, von Werne T, and Sirringhaus H (2005) Lithography-free, self-aligned inkjet printing with sub-hundred-nanometer resolution. *Advanced Materials* 17(8): 997–1001.
46. Porter MR (1994) *Handbook of Surfactants*. New York: Blackie Academic and Professional.
47. Li Y, Wu Y, and Ong BS (2005) Facile synthesis of silver nanoparticles useful for fabrication of high-conductivity elements for printed electronics. *Journal of the American Chemical Society* 127(10): 3266–3267.
48. Szczech JB, Megaridis CM, Gamota DR, and Zhang J (2002) Fine-line conductor manufacturing using drop-on demand pzt printing technology. *IEEE Transaction on Electronics Packaging Manufacturing* 25: 26–33.
49. Zhao N, Chiesa M, Sirringhaus H, Li Y, Wu Y, and Ong B (2007) Self-aligned inkjet printing of highly conducting gold electrodes with submicron resolution. *Journal of Applied Physics* 101(6): 064513.
50. Noh YY, Zhao N, Caironi M, and Sirringhaus H (2007) Downscaling of self-aligned, all-printed polymer thin-film transistors. *Nature Nanotechnology* 2(12): 784–789.
51. Yoon MH, Yan H, Facchetti A, and Marks TJ (2005) Low-voltage organic field-effect transistors and inverters enabled by ultrathin cross-linked polymers as gate dielectrics. *Journal of the American Chemical Society* 127(29): 10388–10395.
52. Noh Y and Sirringhaus H (2009) Ultra-thin polymer gate dielectrics for top-gate polymer field-effect transistors. *Organic Electronics* 10(1): 174–180.
53. Frank DJ, Dennard RH, Nowak E, Solomon PM, Taur Y, and Wong H-SP (2001) Device scaling limits of si mosfets and their application dependencies. *Proceedings of the IEEE* 89(3): 259–288.
54. Campbell IH, Kress JD, Martin RL, Smith DL, Barashkov NN, and Ferraris JP (1997) Controlling charge injection in organic electronic devices using self-assembled monolayers. *Applied Physics Letters* 71(24): 3528–3530.
55. Maitra K and Bhat N (2004) Impact of gate-to-source/drain overlap length on 80-nm cmos circuit performance. *Electron Devices, IEEE Transactions on* 51(3): 409–414.
56. Sze SM (1981) *Physics of Semiconductor Devices*. New York: Wiley.
57. Sedra AS and Smith AS (1998) *Microelectronic Circuits*, 4th edn., New York: Oxford University Press.
58. Stutzmann N, Friend RH, and Sirringhaus H (2003) Self-aligned, vertical-channel, polymer field-effect transistors. *Science* 299(5614): 1881–1884.
59. Ando M, Kawasaki M, Imazeki S, Sasaki H, and Kamata T (2004) Self-aligned self-assembly process for fabricating organic thin-film transistors. *Applied Physics Letters* 85(10): 1849–1851.
60. Mcculloch I, Heeney M, Bailey C, et al. (2006) Liquid-crystalline semiconducting polymers with high charge-carrier mobility. *Nature Materials* 5: 328–333.

7 Molecular Printboards: From Supramolecular Chemistry to Nanofabrication

R Salvio, J Huskens, and D N Reinhoudt, University of Twente, Enschede, The Netherlands

© 2010 Elsevier B.V. All rights reserved.

7.1 Introduction

Nanotechnology deals with structures and objects that typically range from 1 to 100 nm. The properties of such small objects can be dramatically different compared to the properties of these in bulk. One requirement of nanotechnology is to precisely position molecules and/or nanoparticles (NPs) on surfaces, so that they may be addressed and manipulated for bottom–up construction of nanoscale devices. The interface chemistry of the nanostructures and the substrates plays a very important role in the positioning and immobilization of these structures. It is attractive to separate the interface design from the nanostructure fabrication in order to have control over the interfacial properties. This requires the development of versatile and generally applicable interface chemistry to manipulate the nanostructures and control their position.

Parameters such as stoichiometry, binding strength and dynamics, packing density and order, and reversibility should be controllable. Covalent immobilization does not always fulfill these criteria. Physisorption or chemisorption offers reversibility and the possibility of error correction. Nevertheless, the predictability of binding stoichiometry and thermodynamic binding parameters is insufficient and thus control is limited. Conversely, designed host–guest or receptor–ligand supramolecular interactions can provide control over these parameters. Self-assembled monolayers (SAMs) on solid surfaces can be used to covalently attach these receptors. The fixation to a substrate leads to a multivalent display of supramolecular interaction sites. The density of these sites is an important parameter in the binding of complementary multivalent guests. Multivalency [1, 2], which describes the interaction between multiple interacting sites on one entity with multiple interacting sites on another, is therefore the fundamental principle governing the stabilities and dynamics of such systems and offers the means to control the binding properties of an entity binding to a substrate. This control can be exerted by systematic variation and optimization of the number of interacting sites, the intrinsic binding strength of an individual interacting pair, and the geometry of the multivalent building blocks.

Apart from the implications for nanofabrication described in this article, multivalency has a profound impact on biology [1]. Contacts between cells and viruses or bacteria are initiated by multivalent protein–carbohydrate interactions. The monovalent individual interactions can be fairly weak; however, the combined multivalent display at such biological interfaces makes the interactions strong. Qualitatively, this pathway is quite well understood, but quantitative details often lack for such systems. The supramolecular interface systems can be regarded as model systems for biological interfaces and their study can lead to a more quantitative understanding of multivalent binding at biologically relevant systems [3].

In this article, we discuss the concept and the applications of the molecular printboard. In general, a molecular printboard is defined as a monolayer of host molecules on a solid substrate onto which guest molecules can be attached with control over position, binding strength, and binding dynamics. More specifically, we focus on the work with cyclodextrin (CD) molecules immobilized in monomolecular layers on gold, silicon wafers, and on glass. Guest molecules (e.g., adamantane and ferrocene derivatives) bind to these host surfaces through supramolecular hydrophobic interaction. Multivalent interactions are exploited to tune the binding strength and dynamics of the interaction of guest molecules with the printboard. Molecules can be positioned onto the molecular printboard using supramolecular microcontact printing (μCP) and supramolecular dip-pen nanolithography (DPN) owing to the specific interaction between the ink and the substrate. In this way, nanoscale patterns can be written and erased and two-dimensional (2D) and 3D assemblies can be fabricated with a very high level of control. We show that the printboard is a very versatile tool for positioning and assembly of various molecules and materials on surfaces and that it is possible to exploit this system to study properties at the single-molecule level.

7.2 The Concept of Multivalency

Multivalency comprises the multiple interactions that take place between a multivalent host and a multivalent guest [1, 2, 4]. In the most simple case, a divalent guest and a divalent host interact to form a 1:1 complex. Multivalent systems are characterized by an intermolecular binding event followed by (an) intracomplex (intramolecular) assembly step(s). This feature makes such systems distinctly different, both thermodynamically and kinetically, from monovalent and multiple monovalent (between one multivalent and multiple monovalent entities) systems that lack these intramolecular steps (**Figure 1**).

A quantitative comparison between inter- and intramolecular binding events in a multivalent interaction can be accomplished by adopting the effective concentration or effective molarity (EM) terminology [5]. The effective concentration symbolizes a physically real concentration of one of the interacting functionalities as experienced by its complementary counterpart. The concept of effective concentration originates from the study of cyclization reactions [5–7]. Effective concentration is conceptually very similar to the more general concept of EM. Whereas effective concentration is based on concentrations calculated or estimated from physical geometries of complexes, EM uses the ratio of intra- and intermolecular rates or association constants [5]. In other words, the EM is used as an empirical quantity relating the overall stability constant of the multivalent system to the one of the monovalent parent systems. Conversely, the effective concentration is the theoretical prediction, for example, from molecular modeling incorporating linker lengths, flexibilities, etc., of that quantity, thus providing a theoretical estimate of what EM should be when only statistical, entropic, and multivalency factors are taken into account. Therefore, the comparison between the two values provides a handle to evaluate whether additional cooperative or anticooperative effects occur. When $EM = C_{eff}$, the data can be explained by assuming independent, noncooperative interactions only, while when it is observed that $EM \neq C_{eff}$, this may indicate the existence of positive or negative cooperative effects. The analysis to dissect possible multivalency and cooperativity effects has been used to describe the thermodynamics of binding of the divalent complex between a bis-adamantyl calix[4]arene guest (**2**) and a bis-CD host (**1**) (**Figure 2**) [8]. The comparison between the overall 1:1 binding constant of this divalent complex and the intrinsic monovalent binding constant yielded an EM of approximately 3 mM, while a C_{eff} of 2 mM was estimated from the linker lengths of the guest and host in the monovalently linked intermediate. This excellent agreement, together with the fact that the binding enthalpy of the divalent complex was twice the value for the monovalent complex, led to the conclusion that this divalent system could be well described with multivalency effects only, thus without cooperativity.

7.3 Multivalency at Interfaces and the Molecular Printboard

Molecular recognition by synthetic receptors in solution has reached a high degree of sophistication [9, 10]. However, for possible applications at the device level, such receptors should be confined in

(a) Monovalent interaction

(b) Multiple monovalent interactions

(c) Multivalent interaction

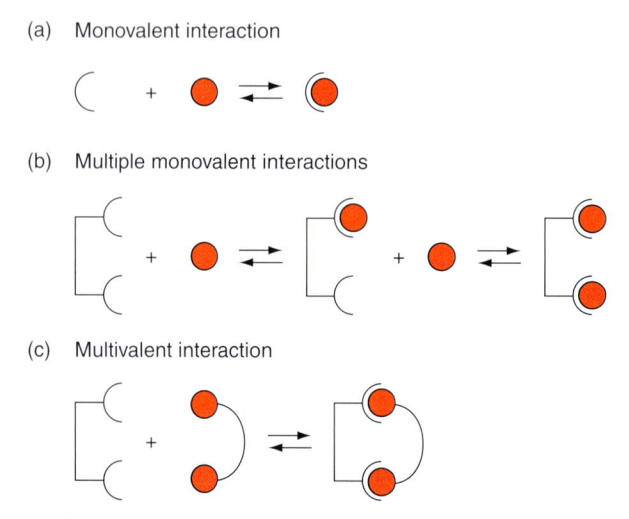

Figure 1 Schematic representation of monovalent, multiple monovalent, and multivalent binding events.

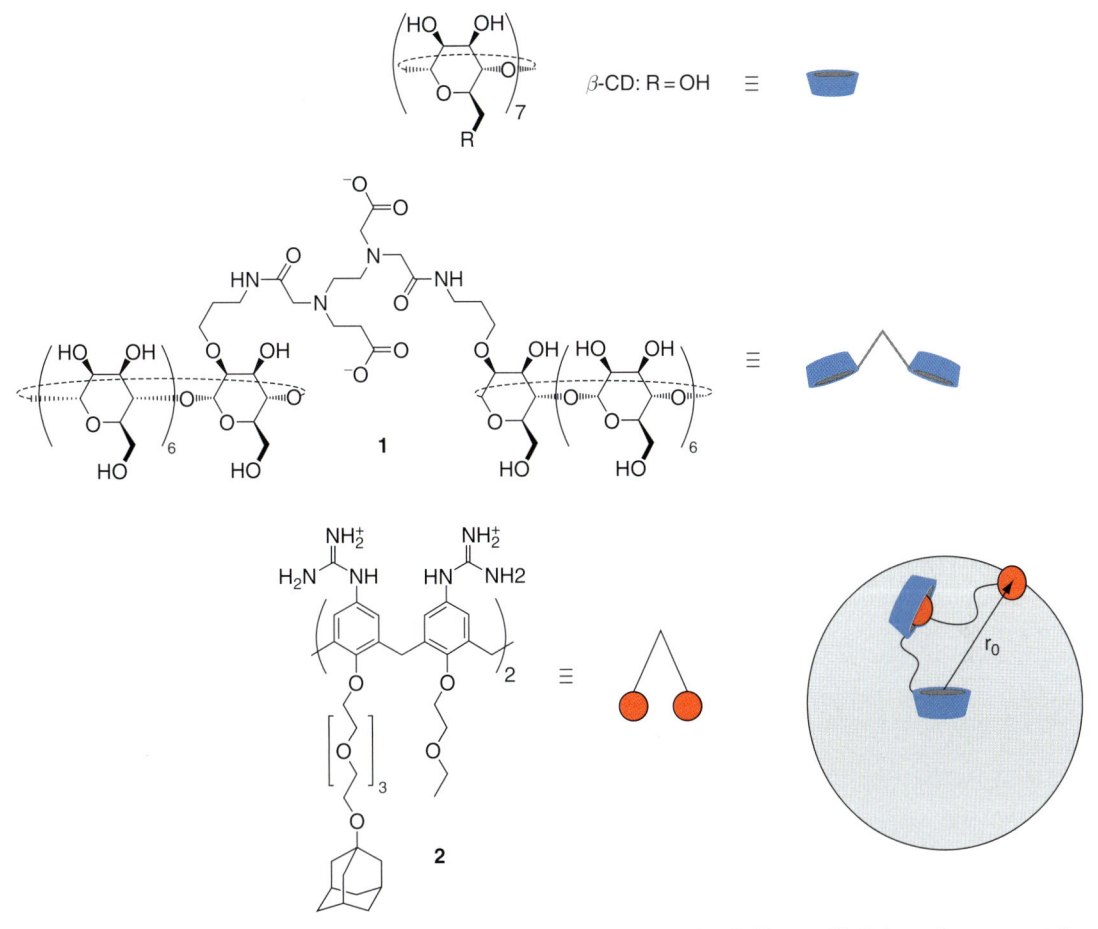

Figure 2 β-cyclodextrin (CD), EDTA-linked CD dimer (**1**) and a bis(adamantyl)-calix[4]arene (**2**). Schematic representation of the concept of effective concentration for the interaction of the two bifunctional molecules (bottom and right).

space. SAMs of molecules on gold have the advantage of a high degree of molecular organization [11, 12]. The most-studied monolayers are based on thiols as anchoring groups, although stable monolayers of dialkyl sulfides have also been reported [13–16]. To obtain devices for the transduction of molecular recognition into macroscopic properties, the self-assembly of receptor molecules, such as resorcin[4]arene [17–20], calix[4]arene [21, 22], and carceplex derivatives [21] (see **Figure 3**), on gold has been reported. These were the first attempts to functionalize a surface with synthetic receptors able to perform a multivalent recognition. In these molecules, four dialkyl sulfide tails have been attached, for coordination, to the gold surface. In these studies, it appeared that multiple points of attachment can be advantageous for the quality and stability of SAMs. Four dialkyl sulfide units are necessary to fill the space underneath the cavity head group of, for example, a resorcin[4]arene (**Figure 3(a)**) in order to obtain dense, well-packed monolayers. Binding studies carried out with a quartz crystal microbalance and surface plasmon resonance proved that these monolayers on gold are able to complex guest molecules [23, 24].

The next step in the fabrication of a surface provided with molecular receptors was the use of β-CD (**Figure 2**). CDs are attractive host molecules for sensing purposes, as they can accommodate a variety of organic guest molecules [25]. The binding constants in water of hydrophobic guests that geometrically fit in the β-CD cavity typically range from 10^4 to 10^5 M^{-1} (see **Figure 4**) [26, 27].

Sulfur-modified CD derivatives have been used for the preparation of monolayers on gold [28–35]. Kaifer and coworkers used monolayers of per-6-deoxy-(6-thio)-β-CD with seven thiol attachment units directly linked to the CD, for the complexation

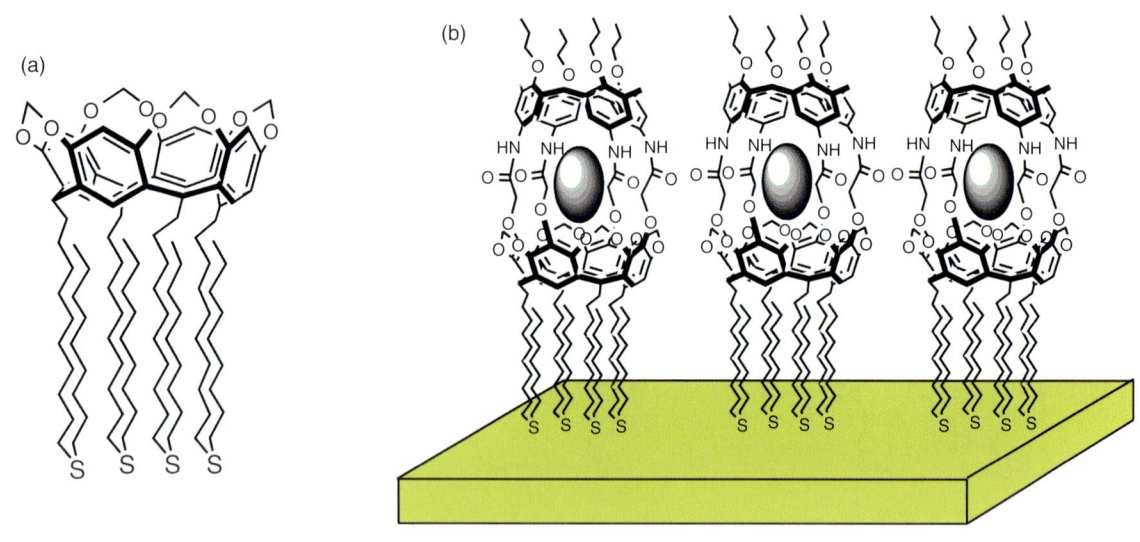

Figure 3 (a) Example of a cavitand adsorbate based on a resorcin[4]arene scaffold. (b) Schematic representation of a monolayer of carceplex molecules trapping a guest in their cavity.

	K (M^{-1})	ΔH (kcal mol^{-1})	$T\Delta S$ (kcal mol^{-1})
Ferrocene methanol	$1.0 \cdot 10^4$	−6.1	−0.7
tert-butylphenyl acetamide	$3.0 \cdot 10^4$	−5.2	0.9
adamantyl acetamide	$6.8 \cdot 10^4$	−5.9	0.7

Figure 4 Binding constants and thermodynamic parameters for the binding of hydrophobic guests inside the β-CD cavity in water at 25 °C.

of ferrocene [28]. Wenz and coworkers have thoroughly studied monolayers of β-CD substituted with 1 or with a mixture of 2–4 thiol units [29–31]. Moreover, several groups focused more on possible applications rather than characterization and used β-CDs with only one thiol group for attachment to gold surfaces [32, 34, 35]. However, molecular dynamic calculations indicated that CD monolayers with only one attachment point are assembled into a random, quasi-two-layer system [36]. This imperfect structure renders only half of the β-CDs available as hosts. It has been proven that multiple points of attachment will result in robust layers in which the space underneath the β-CD head group is completely filled [37, 38]. This could prevent intercalation in the monolayer and render all the cavities available for molecular recognition. Furthermore, it has been proven that the use of a long chain length results in a better-ordered and better-packed monolayer [37]. Besides, a long spacer can improve the performance and the binding properties in terms of both kinetic and thermodynamic stability [38].

The SAM of the β-CD heptathioether derivative **3** (**Figure 5**) has been extensively characterized with different analytical techniques, including electrochemistry, wettability measurements, X-ray photoelectron spectroscopy (XPS), time-of-flight–secondary ion mass spectrometry (TOF–SIMS), and atomic force microscopy (AFM) (**Figure 6**).

The main conclusions were that the molecules form a monolayer with the secondary sides of the CD ring exposed to the solution, that the SAMs were

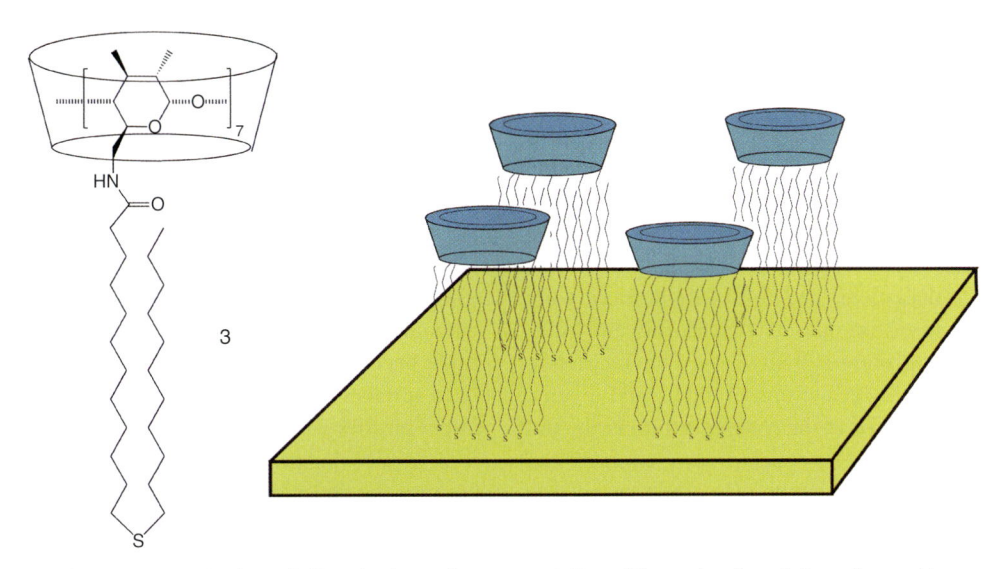

Figure 5 β-Cyclodextrin adsorbate (left) and schematic representation of the molecular printboard on gold.

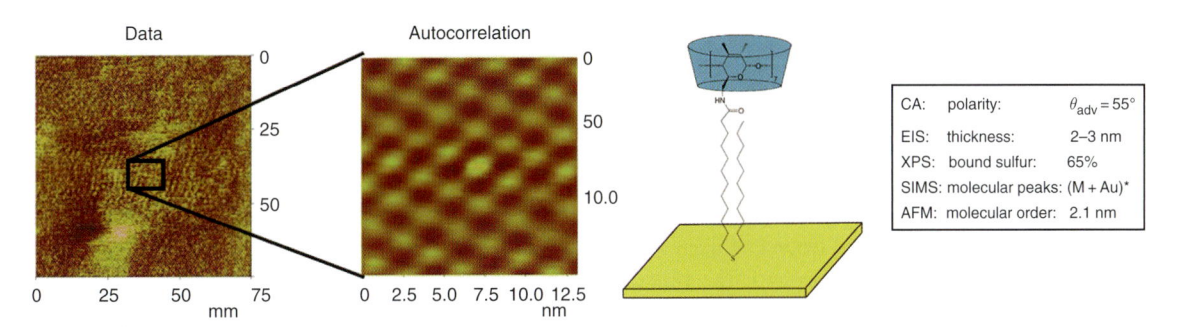

Figure 6 An AFM image of a β-cyclodextrin adsorbate monolayer in tapping mode and corresponding autocorrelation-filtered image of the same monolayer taken from an ordered area (left). Characterization data of the monolayer (right).

comparatively well ordered, and that they were densely packed in the alkyl regions of the adsorbate, leading to a CD cavity lattice periodicity of approximately 2 nm, which was confirmed by AFM [38]. Host–guest studies, performed with small monovalent guests, showed that the molecular recognition properties of the β-CD cavities are unaltered by the surface immobilization as was shown (1) by the identical stability constant obtained for these guests in binding to the β-CD SAMs and to native β-CD in solution [26], and (2) by AFM pull-off experiments with a variety of guests immobilized on an AFM tip [39]. The association and dissociation are fast on the experimental timescale, as is to be expected for such monovalent systems, providing rapid reversibility to the system which was thought to be beneficial when sensor systems were envisaged.

Only in case of bulky steroidal guests, a difference in terms of binding constants was observed in the case of the β-CD adsorbate [26]. The lower binding constants are probably due to the fact that, in case of binding to native β-CD, the bulky guests normally protrude from the cavity at both sides of the β-CD, a possibility which is precluded when one side of the β-CD is functionalized with long alkyl chains.

The binding of guest molecules onto a β-CD monolayer was studied also *in silico* using molecular dynamics (MD) free-energy simulations to describe the specificity of guest:β-CD association. Good agreement with experimental thermodynamic measurements was found for differences of binding enthalpy between three phenyl guests: benzene, toluene, and *t*-butylbenzene [40]. Partial and full methylation of the secondary rim of β-CD decreases host rigidity and significantly impairs binding of the guests. According to the simulations, the β-CD cavity is also very intolerant of guest charging, penalizing the oxidized state of ferrocene by at least 7 kcal mol^{-1} [40]. The modeling was also extended to charged CD hosts. Shifting to high pH or alternatively grafting a charged sidearm onto β-CD created three distinct types of anionic cavities. Electronic structure calculations and MD simulations were used to measure host–guest charge transfer and binding strengths in these systems [41]. The simulation showed that steric recognition of uncharged organic molecules is retained at the charged printboards, and that improved guest–host electrostatic contacts can strengthen binding of larger inks while penalizing small inks, enhancing the level of discrimination. A prudent choice of complementary host–guest shape and charge states thus provides a means of tuning both ink binding strength and specificity at molecular printboards.

7.4 Immobilization of Multivalent Guests on the Molecular Printboard

After the study of the binding of simple monovalent guests, such as adamantane and ferrocene derivatives [26], the next step was to investigate the binding behavior of more complex molecules with multiple guest units.

Adamantyl-functionalized poly(propylene imine) (PPI) dendrimers (**Figure 7**) were deposited on the β-CD monolayer [42]. Slower dissociation kinetics, and thus a shift from reversible to irreversible binding were observed upon increase of the dendrimer generation, thus the number of interactions with the β-CD SAM. For the larger dendrimer generations, individual dendrimer molecules could be visualized using AFM showing that they were attached strongly enough to withstand the forces exerted by the AFM tip. It became apparent that such receptor-functionalized surfaces could be used as assembly platforms for larger entities with considerable lifetimes. Hence the term 'molecular printboards' was coined [8, 42, 43] because the deposition of the material can be, under specific conditions, permanent and irreversible.

It is clear that the key to control the binding thermodynamics and kinetics and to stimulus-dependent control arises from multivalency. By tuning the type of monovalent interaction, the number of these interactions, and their geometry on both multivalent guest and host platforms, one can vary the association and dissociation rates practically at will so that the whole range from labile to stable complexes can be accessed. It turns out that the number of interactions needed to obtain kinetically stable assemblies can be rather low, even for interaction motifs with moderately weak intrinsic interaction strengths.

The binding of the divalent calix[4]arene **2** onto the molecular printboard was studied (**Figure 8**) and the data were compared with its behavior in solution

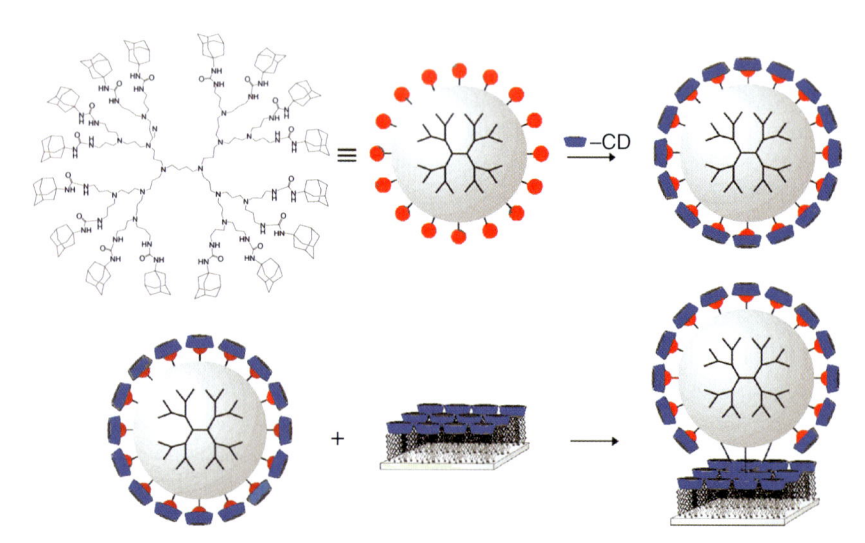

Figure 7 Generation-3 adamantyl-functionalized poly(propylene imine) dendrimer and the formation of assemblies with β-CDs and onto a β-CD SAM on gold.

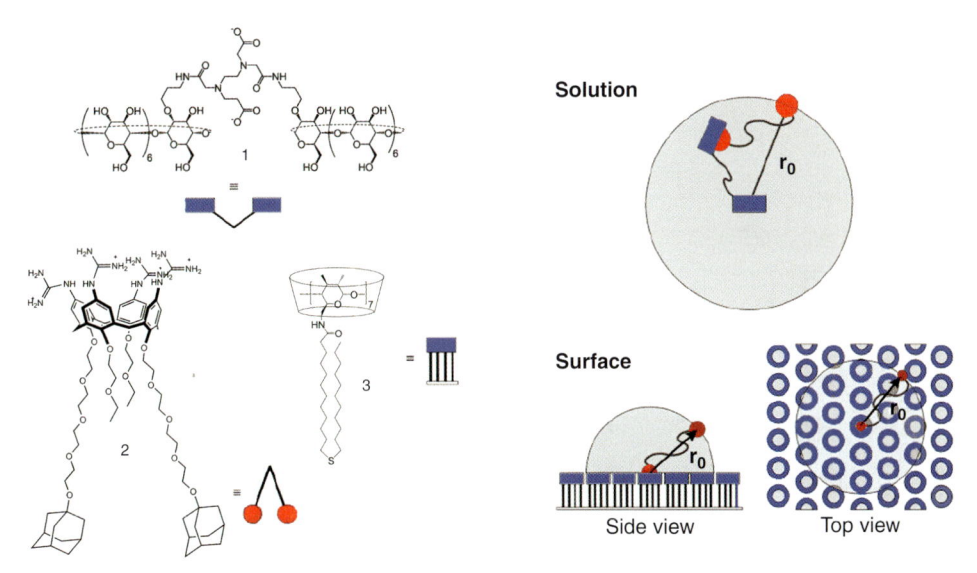

Figure 8 Schematic comparison of the binding between the bis-adamantyl-functionalized calix[4]arene **2** and the βcyclodextrin dimer **1** in solution and between **2** and the β-CD SAM.

[8]. A clear distinction with the situation in solution was the fact that the overall binding constant at the β-CD SAM was 2–3 orders of magnitude larger than in solution. This effect is due to multivalency. The stability constant increases from about $10^7 \, M^{-1}$ for binding to the CD dimer to 10^9–$10^{10} \, M^{-1}$ for binding to the CD SAM due to a higher effective concentration (∼0.2 M) at the CD SAM.

In the literature, other examples of multivalent specific recognition on surfaces have also been reported. An example of relevant biological interest was reported by Whitesides and coworkers. They investigated the binding of vancomycin (Van) monomer and its corresponding dimer to D-Ala-D-Ala and D-Ala-D-Lac [44, 45]. They developed mixed SAMs on gold that consist of an adsorbate with Na-Ac-L-Lys-D-Ala-D-Ala (L*) and with carboxylic acid groups, the mole fractions of which were both about 0.5. Surface plasmon resonance (SPR) spectroscopy experiments on this monolayer showed that the adsorption of Van is comparable to the adsorption of a divalent Van derivative. These experiments also indicated that the binding of Van to L* at a SAM is comparable to binding in solution. They also established that binding of the divalent Van derivative to the SAM was much stronger than the binding of the monomeric Van derivative. These examples clearly illustrate that the effect of multivalency on the surfaces can be remarkably higher than in solution, if the monolayer is densely packed.

In the studies cited above, the binding stoichiometry of the multivalent guests are easily predictable from their structure. For dendrimeric guests, it is harder to know the binding stoichiometry to the β-CD SAMs because it cannot be estimated from basic molecular modeling.

In this respect, the study of the binding behavior of the electroactive ferrocenyl-functionalized PPI dendrimers (**Figure 9**) [46, 47] is interesting. These dendrimers are decorated with ferrocene units that are known to be easily oxidized both chemically and electrochemically. The binding constant of the oxidized ferrocene to the cavity of the β-CD is extremely poor compared to that of the nonoxidized species. This property allows desorption of these dendrimers from the β-CD monolayer upon oxidation (**Figure 9**).

A quantitative understanding of the binding of multivalent dendrimers comes from electrochemistry on these electroactive systems, which provides an independent experimental measure of the number of interactions. The overall binding constants, determined by SPR, of the thermodynamically reversible dendrimer generations 1–3 was evaluated in terms of multivalency. A model for describing the multivalency in a quantitative manner for the binding of such multivalent molecules at the β-CD interfaces was developed, incorporating the effective concentration concept as well as possible competition with monovalent hosts in solution [48]. All data on the dendrimer binding led to the conclusion that the

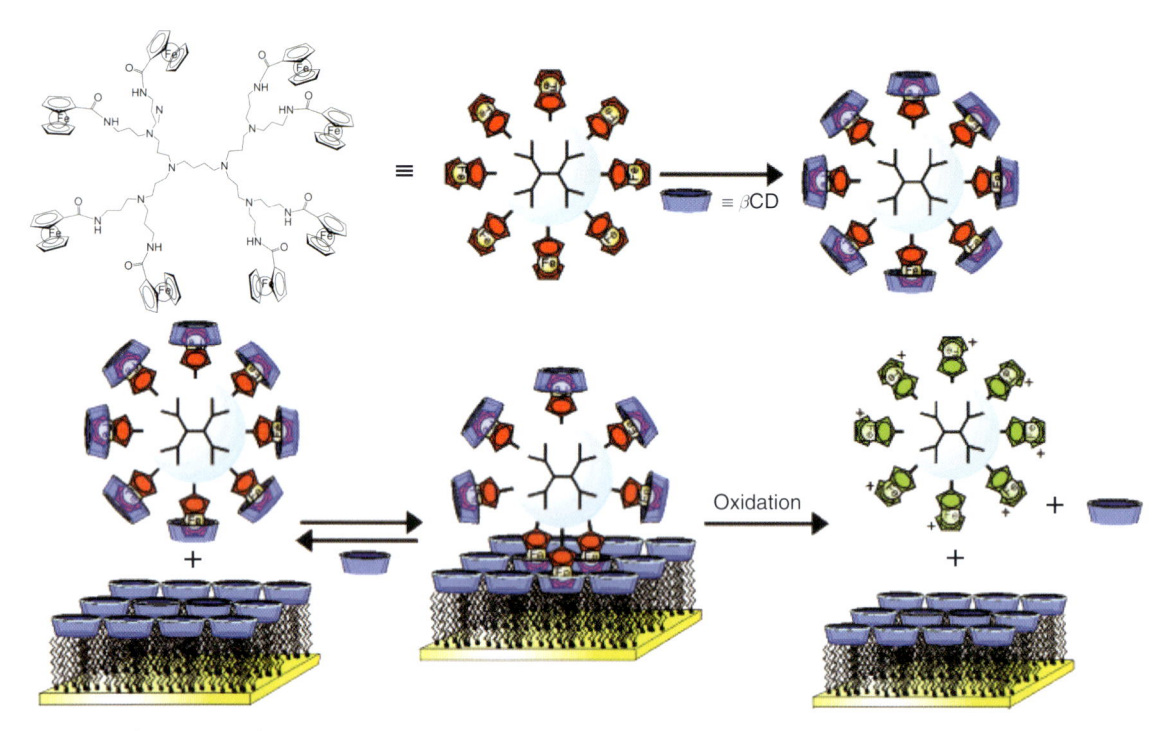

Figure 9 Generation-2 ferrocenyl-functionalized PPI dendrimer and its formation of water soluble assemblies with β-CD molecules (top); the adsorption/desorption of the dendrimer onto/from the printboard.

binding enhancement stems solely from the multivalency effect. The surface coverage of electroactive ferrocenyl groups, in a full monolayer of these dendrimers on a β-CD SAM, as determined by cyclic voltammetry (CV), was compared to the known surface coverage of the CD host molecules. This provided ratios of bound versus unbound ferrocenyl groups (see **Figure 10**). This electrochemical determination of the stoichiometry also worked for the higher generations 4 and 5, which showed dissociation rates that were too slow to allow stability constant determinations by SPR [46].

Straightforward extension of the thermodynamic model provides reliable K values for such systems. Comparison of the experimentally observed binding stoichiometries for these dendrimers to molecular models revealed a geometric rule for the binding stoichiometry. As long as a dendritic branch can stretch without violating common bond lengths and angles to reach a neighboring free β-CD-binding site at the β-CD SAM, it will bind, thus contributing to the overall stoichiometry. This was confirmed by modifying the dendrimer skeleton and the spacer length between the dendrimer amino end groups of

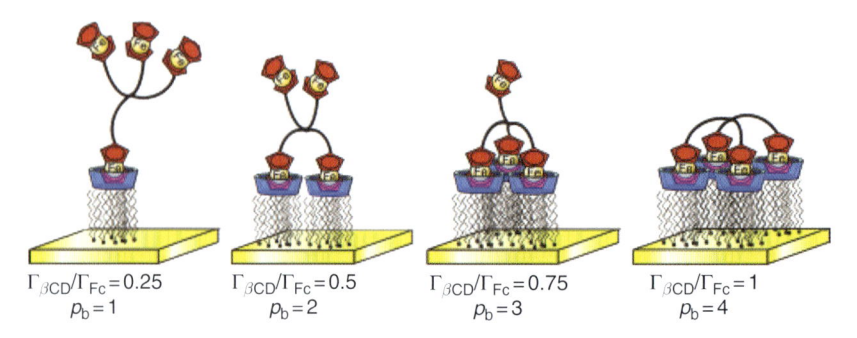

Figure 10 Schematic representation of the four possible binding stoichiometries of generation-1 ferrocenyl dendrimer to the β-CD SAM. The numbers of bound sites (p_b) and the predicted coverage ratios (Γ_{CD}/Γ_{Fc}) are reported below.

the parent dendrimers and the ferrocenyl groups attached to them [47]. These modifications led to different binding stoichiometries but always followed this geometric rule. Since replacing the ferrocenyl groups with adamantyls does not change the geometry of the dendrimers significantly, the stoichiometry data for the ferrocenyl dendrimers could be directly applied to the adamantyl dendrimers.

Extension of the multivalency of the model guest systems, together with the well-defined properties of the β-CD SAMs, gave quantitative thermodynamic data for multivalent binding events occurring at these interfaces. The data and model lead to the conclusions that (i1) binding events at these β-CD SAMs can be explained by multivalency only, without the need for assuming cooperativity, and that (2) crude molecular models (e.g., CPK) suffice to estimate whether an unused binding site of the guest can reach a neighboring free host site and thus provide easy estimates of the (maximal) binding stoichiometries even when these are not experimentally accessible. The mathematical model also provides a clear-cut way to estimate dissociation rate constants [48].

In addition, computational chemistry studies helped to understand the binding behavior of dendrimers [49–51].

MD simulations, with explicit treatment of solvent molecules, were performed to probe the conformational space available to dendrimer in both free and bound environments. It was shown that accurate treatment of both pH effects and binding conformations gives calculated binding modes in line with known binding multivalencies. The steric frustration causing small, low-generation dendrimer inks to bind to the printboard using only a subset of the available anchor groups was identified and quantified. Furthermore, it was shown that the enhanced binding energy of multisite attachments offsets the steric strain [51]. MD simulations were also carried out to measure the effective local concentration of unbound ink anchor groups at the printboard for a variety of binding modes. These simulations allowed us to describe the conformational space occupied by partially bound inks and to estimate the likelihood of an additional binding interaction. By simulating the shift from divalent to monovalent binding mode, it was shown that the released anchor quickly moves to the periphery of the dendrimer binding hemisphere; however, re-approaches the printboard and remains in the vicinity of an alternative binding site [50]. Secondary electrostatic interactions between the protonated dendrimer core and hydroxyl groups at the entrance to the β-CD cavities give flattened dendrimer binding orientations and may aid dendrimer diffusion on the printboard, allowing the dendrimer to walk along the printboard by switching between different partially bound states and minimizing complete unbinding to bulk solution [50].

7.5 Writing Patterns of Molecules on the Molecular Printboard

As emphasized previously, bottom–up nanotechnology has to start with the precise positioning of molecules and the molecular printboard is an excellent and versatile tool for this purpose. It is possible to print or write on the β-CD SAMs by μCP and DPN as shown in **Figure 11**. μCP has been developed by Xia and Whitesides [52] for the preparation of patterns of thiols onto gold substrates. Such transfer was extended by Mirkin and co-workers to writing with molecules on such surfaces using DPN [53]. Various types of molecules were deposited onto different substrates by DPN which led to arrays of, for example, DNA [54], proteins [55], and NPs [56]. Registry capabilities have been demonstrated as well,[57] and a multipen nanoplotter, able to produce parallel patterns with different ink molecules, has been developed [58].

Supramolecular μCP was applied for the first time on the molecular printboard with the divalent calix[4]arene **2** (**Figure 8**) [43]. Patterns on the β-CD SAM were visible directly after printing and remained clearly visible when rinsed with substantial amounts of water or aqueous NaCl (see **Figure 12**).

Even prolonged washing with an aqueous solution of β-CD, to promote competition between the host–guest binding in solution and the printboard, did not completely remove the patterns. Moreover, the SIMS image confirmed the presence of the calixarene on the surface (**Figure 12**).

A DPN procedure was also applied to write patterns of the same molecule on the printboard [43]. Silicon nitride AFM tips were dipped into an aqueous solution of the calixarene and scanned across β-CD- as well as OH-modified SAMs. The transfer of ink was observed for both SAMs, but stable patterns after rinsing were obtained only when the printboard was used in the experiments (**Figure 13**). The resolution that can be attained by this technique is well below 100 nm.

In addition, the fourth and fifth generations of adamantyl-functionalized PPI dendrimer guests

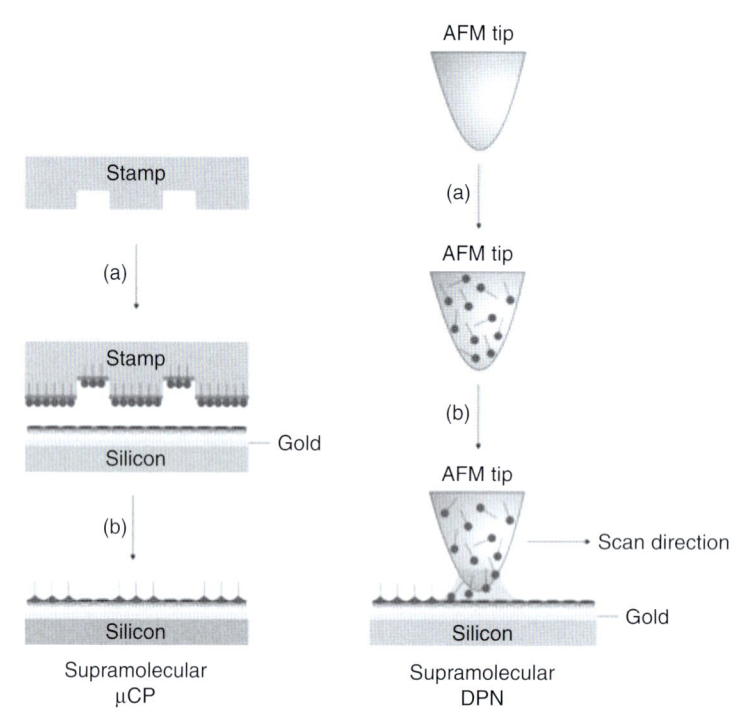

Figure 11 Schematic representation of supramolecular μCP and supramolecular DPN of printboard-compatible guests on the β-cyclodextrin printboard on gold.

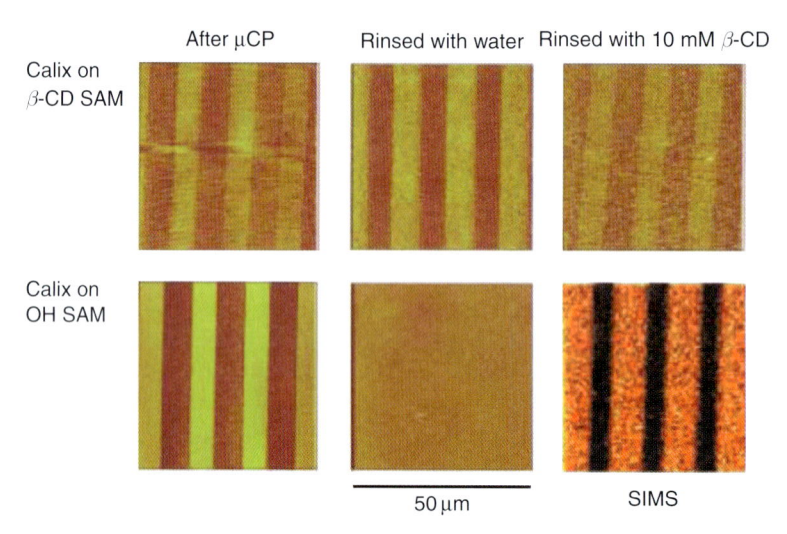

Figure 12 AFM friction force images of patterns, obtained by μCP, of the calix[4]arene **2** (brighter areas) on β-CD (top) and OH-terminated SAMs (bottom): before rinsing, after rinsing with water, or after rinsing with 10 mM aqueous β-CD. TOF–SIMS image of a β-CD-terminated SAM after printing: the bright areas indicate the presence of the molecular ion peak of the calixarene **2**. Reproduced with permission from Auletta T, Dordi B, Mulder A, et al. (2004) Writing patterns on molecular printboards. *Angewandte Chemie (International ed. in English)* 116: 373–377.

(**Figures 14 and 15**) were transferred by μCP and DPN to form assemblies that are extremely stable upon rinsing in the case of the fifth-generation dendrimer [59].

Furthermore, polymeric materials such as poly(-isobutene-*alt*-maleic acid)s modified with p-*tert*-butylphenyl or adamantyl groups were immobilized onto β-CD SAMs (see **Figure 16**) [60].

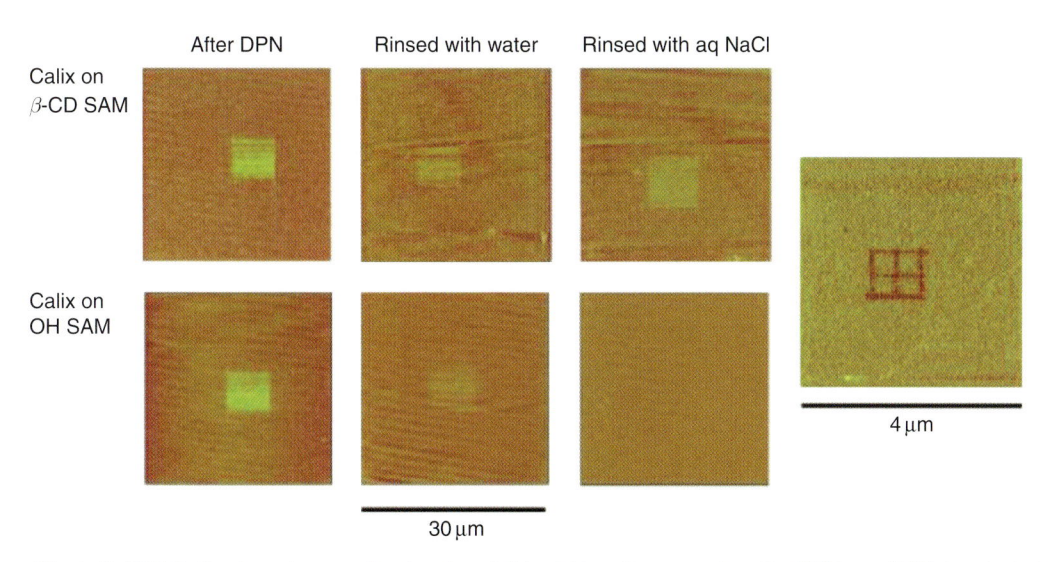

Figure 13 Left: AFM friction force images showing, from left to right, patterns produced by DPN on a β-CD-terminated SAM (top) and an -OH terminated SAM (bottom) using the calix[4]arene **2** as ink before rinsing, after rinsing with water, and with aqueous NaCl. Right: AFM friction force images in air of arrays of lines with mean widths of 60 nm produced by DPN. Reproduced with permission from Auletta T, Dordi B, Mulder A, et al. (2004) Writing patterns on molecular printboards. *Angewandte Chemie (International ed. in English)* 116: 373–377.

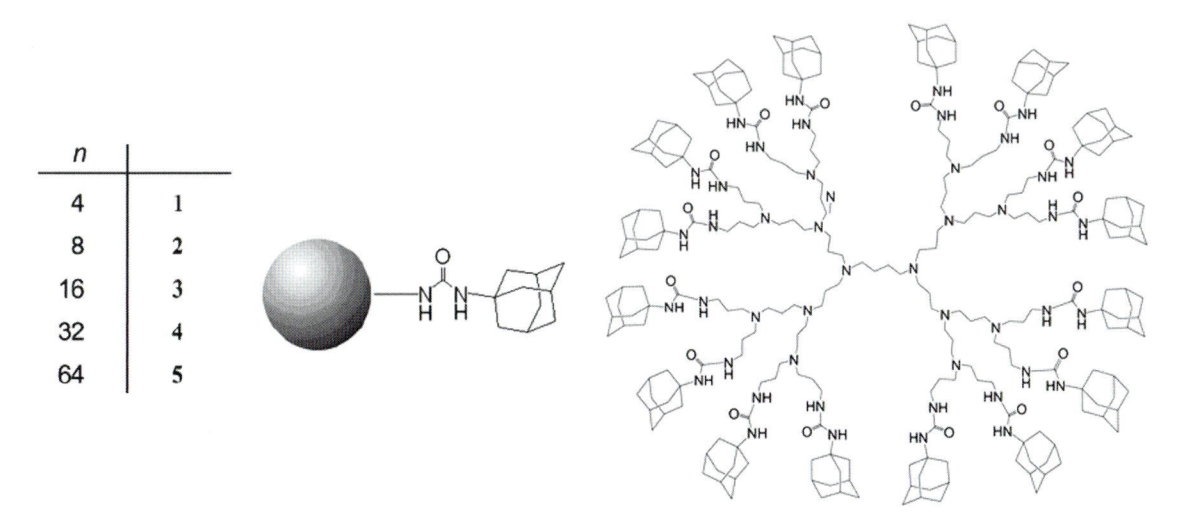

n	
4	1
8	2
16	3
32	4
64	5

Figure 14 Right: Structure of the generation 3 adamantyl-functionalized poly(propylene) imine dendrimer guest (G3-PPI-(Ad)$_{16}$). Left: table with the number of the adamantyl groups depending on the dendrimer generation.

The adsorption of this polymeric material was strong, specific, and irreversible. A drastic change of conformation of the polymers from an average spherical shape in solution to a completely flattened thin layer when adsorbed on the β-CD SAMs was observed. According to SPR spectroscopy and AFM, the adsorbed polymer layer was only 1 nm thick. Apparently, most of the guest units are employed in the binding (**Figure 17**) as is supported by the absence of specific binding of β-CD-modified gold NPs to the polymer surface assemblies.

A monolayer of β-CD was fabricated not only on a gold substrate but also on a SiO$_2$ or glass surface. The preparation of the monolayer requires, in this case, a multistep synthetic procedure starting from a cyanoterminated monolayer (**Figure 18**) [61].

The use of the molecular printboard on silicon oxide widens the scope of the supramolecular patterning [43, 62]. μCP of the calix[4]arene **2** onto the printboard on glass/SiO$_2$ gave results similar to the β-CD-terminated SAM on gold with regard to ink transfer and pattern stability. The printing of a

Figure 15 An array of lines, made by DPN, 3 μm long with average widths of approximately 60 nm of the multivalent dendrimer guest G5-PPI-(Ad)$_{64}$ on the molecular printboard. Reproduced with permission from Bruinink CM, Nijhuis CA, Peter M, et al. (2005) Supramolecular microcontact printing and dip-pen nanolithography on molecular printboards. *Chemistry – A European Journal* 11: 3988–3996.

Figure 16 Chemical structure of poly(isobutene-*alt*-maleic acid)s. R = p-*tert*-butylphenyl or adamantyl.

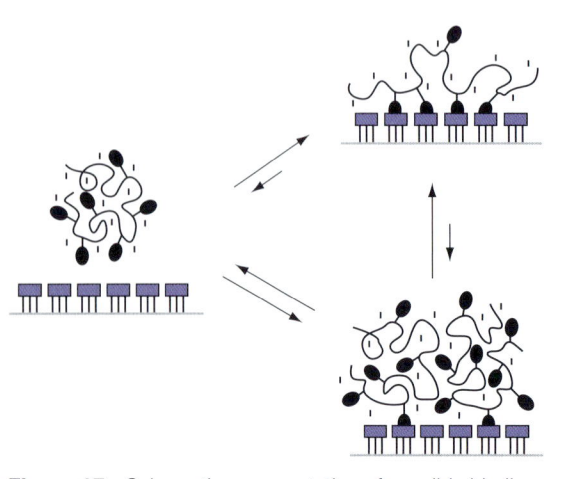

Figure 17 Schematic representation of possible binding modes of guest polymers onto β-CD SAMs.

fluorescent dendritic wedge (**Figure 19**) allowed pattern detection by confocal microscopy [43].

A systematic investigation of the immobilization of such molecules was also carried out with fluorescent divalent guests [61, 62]. Titration curves by fluorescence detection gave stability constants that were in line with the data obtained for the gold substrates [61]. The patterns of these molecules on a β-CD SAM were compared with patterns fabricated with the same procedure on a poly(ethylene glycol) (PEG) reference monolayer. In the latter case, the patterns could be instantly removed by rinsing with water. The patterns on the β-CD monolayer displayed long-term stability when stored under nitrogen, whereas patterns at PEG monolayers faded within a few weeks due to the diffusion of

Figure 18 Synthesis scheme for the preparation of β-CD SAM on SiO$_2$/glass surfaces. (i) Red Al, toluene 40 °C; (ii) DITC, toluene, 50 °C; (iii) per-6-amino β-CD, H$_2$O, 50 °C.

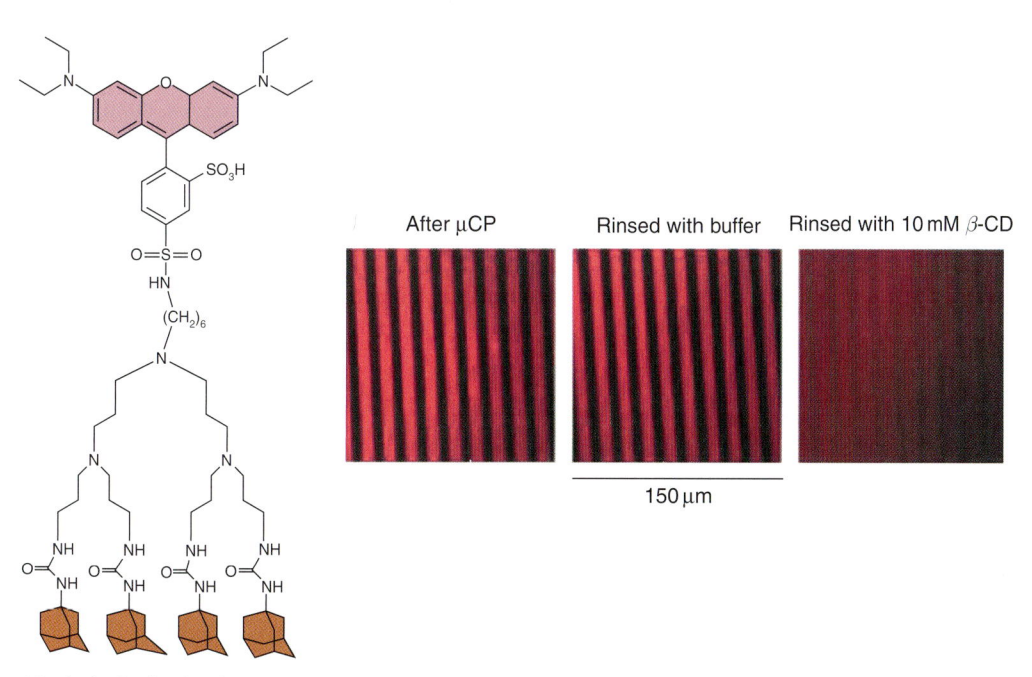

Figure 19 Left: rhodamine-functionalized dendritic wedge. Right: confocal microscopy images after μCP of this fluorescent guest on the molecular printboard on SiO$_2$ before and after rinsing.

fluorescent molecules across the surface. Moreover, the application of two dyes, one in a printing step and the second one in a subsequent solution assembly step, showed that alternating patterns of dyes could be obtained [62]. The second dye was found almost exclusively in the areas left vacant after the preceding printing step, which showed that the first dye was bound in a stable fashion and that exchange of dyes in the subsequent solution assembly step did not occur to a noticeable extent.

More high-resolution patterning, down to linewidths of approximately 200 nm, was achieved by DPN, using the calix[4]arene, an adamantyl dendrimer, or the fluorescent dye guest molecules as the ink [62].

One of the divalent fluorescent guest molecules was also used in the binding to CD vesicles of about 100 nm [63]. Binding constants are similar to values obtained for flat β-CD SAMs. Vesicles consisting of both α- and β-CD of varying ratios were employed to test the hypothesis of receptor clustering in these mobile layered architectures [64]. Indeed, binding of the divalent dyes showed that binding to vesicles with a fraction of β-CD yielded always significantly higher binding constants than expected when assuming random mixing of these receptors. Whether this clustering stems from demixing of the receptor molecules in the vesicles before guest binding or from active clustering upon guest binding is an unsolved issue.

7.6 Stepwise Assembly of Complex Structures and Stimulus-Dependent Desorption on/from the Molecular Printboard

Weak supramolecular interactions, when used in a multivalent fashion, can provide thermodynamically and kinetically stable assemblies, both in solution and at interfaces. To exploit this stability for nanofabrication, it is important to make sure that: (1) when the assembly occurs at an interface, the complex stays at the position where it was originally formed, (2) directed assembly can therefore be applied to obtain patterns of such supramolecular complexes, and (3) additional building blocks with other or identical binding motifs can be employed in subsequent assembly steps to extend the supramolecular structure, thus leading to materials of increasing complexity. It is possible to obtain stable assemblies and stimulus-dependent reversal of various types of multivalent supramolecular entities, from molecules to polymers and biomolecules, onto the molecular printboards. The use of basic motifs allows extending the assemblies to larger systems.

Here, we discuss how the level of complexity can be increased. This goal can be achieved by exploiting interactions between the bound guests and the new guests to immobilize them on the surface.

Negatively charged fluorescent dyes were bound to fifth-generation PPP adamantyl dendrimers by μCP [65]. The guests act like a kind of molecular boxes. This two-step procedure realizes an architecture where the dendrimers are bound by multivalent host–guest interactions, whereas the dyes are immobilized inside the dendrimer cores by electrostatic interactions. A more complicated architecture was obtained by printing lines of dendrimers in one direction, one dye in an orthogonal direction, and finally assembling a second dye from solution (see **Figure 20**). This yielded a dicolored block pattern showing good selectivity and directionality even for electrostatically bound dyes [65].

If the surfaces, after the immobilization of the dendrimers and the encapsulation of the fluorescent

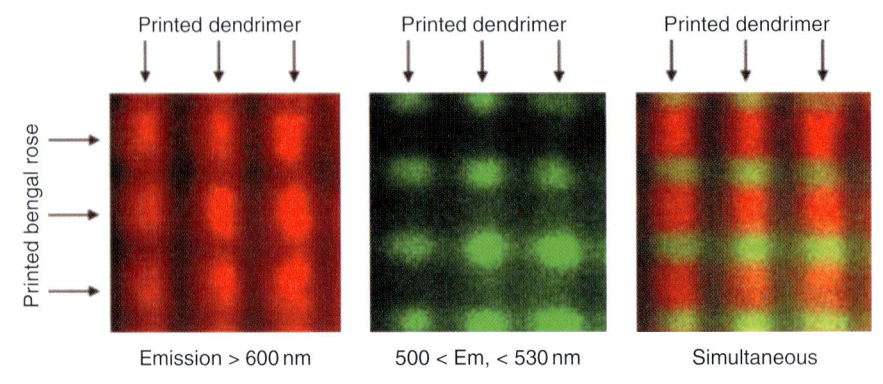

Figure 20 Confocal microscopy images (50 × 50 μm^2) of a β-CD SAM on glass after μCP of the generation-5 adamantyl dendrimer, followed by cross-printing of Bengal Rose, and subsequent filling with fluorescein. The substrates were simultaneously exited at 488 nm and 543 nm and images were recorded by measuring the emission above 600 nm (left), between 500 and 530 nm (center), and simultaneous (right). Reproduced with permission from Onclin S, Huskens J, Ravoo BJ, and Reinhoudt DN (2005) Molecular boxes on a molecular printboard: Encapsulation of anionic dyes in immobilized dendrimers. *Small* 1(8–9): 852–857.

dye, were rinsed with an aqueous 100-mM phosphate buffer solution at pH 9, an evident decrease in fluorescence intensity was observed. This basic solution led to deprotonation of the dendritic molecule and/or exchange of the fluorescein molecules for non-fluorescent phosphate anions. If the substrate was rinsed with aqueous HCl solution to protonate the tertiary amines in the interior of the dendrimer, and dipped in an aqueous Bengal Rose solution, fluorescence was observed again. These experiments indicate that the immobilized dendrimer boxes allow consecutive handling and rinsing steps without degradation of the μCP pattern [65].

Another example of increased complexity level is the stepwise buildup of supramolecular capsules on a molecular printboard [66]. The capsule used in the nanofabrication is composed of two calix[4]arenes that are held together by complementary charged groups on each building block. One of the calix[4]-arenes was functionalized with four adamantyl groups which allowed attachment to the β-CD SAMs (**Figure 21**). The capsule was assembled in a stepwise fashion: adsorption of the adamantyl-functionalized half of the capsule on the β-CD SAM followed by binding of the second half to complete the capsule from solution (**Figure 21**). The architecture could also be disassembled in a stepwise fashion by first rinsing with a competitive calix[4]arene complementary to the top half of the immobilized capsule, followed by rinsing with a polar organic solvent to weaken the β-CD host–guest interactions (the tetravalently bound calix[4]arene was attached too strongly to be removed by rinsing with 10 mM β-CD). In most cases, the binding of adamantyl derivatives to β-CD SAMs can be reversed by competition with a host in solution [43]. As pointed out previously, this becomes progressively more difficult with increasing numbers of interactions of the multivalent complexes [42, 60].

Here, we discuss the molecular printboard as an excellent platform to carry out stepwise assemblies onto SAMs on surfaces. We also discuss an example of stepwise assembly of biologic material on surfaces by Whitesides and coworkers. Bifunctional polymers consisting of Van and fluorescein were adsorbed to SAMs consisting of D-Ala-D-Ala and tri(ethylene glycol) groups (**Figure 22**) [67] SPR studies revealed that the adsorbed polymer desorbed only very slowly from the surface ($k_{off} = 10^{-6}\,s^{-1}$). When soluble ligands were added, the dissociation rate increased by a factor of about 50. This very strong interaction with the surface was attributed to multivalent interactions between the multiple Van groups at the polymer and multiple D-Ala-D-Ala groups present on the SAM. The fluorescein groups present in the polymer directed

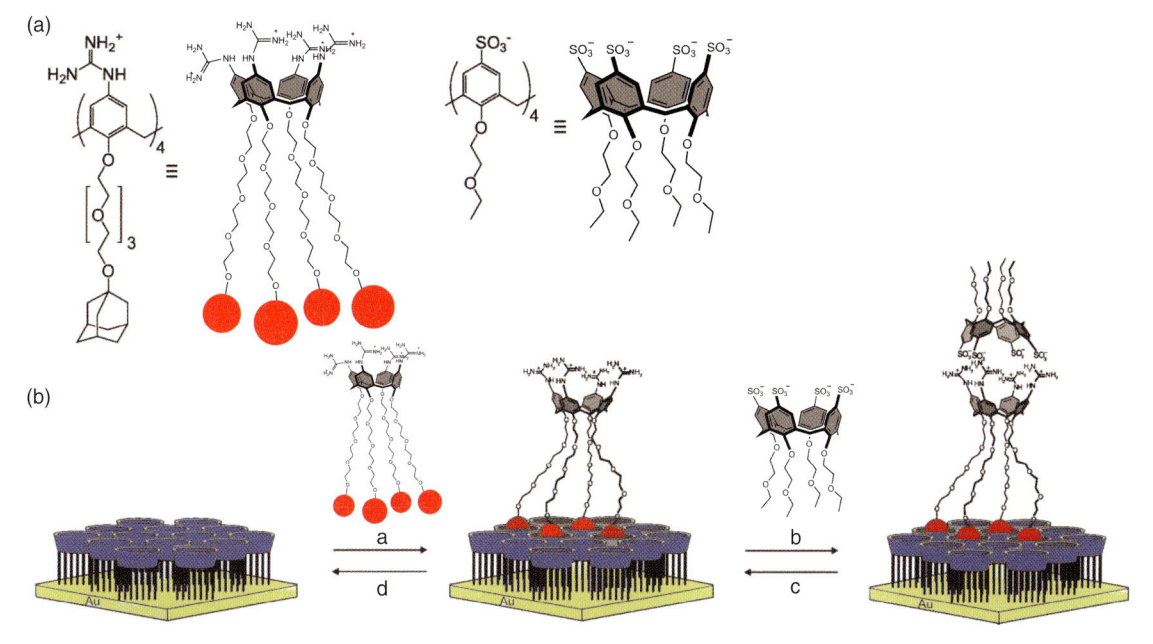

Figure 21 (a) Tetra-(adamantyl)-calix[4]arene and tetrasulfonate-calix[4]arene, the building blocks of the capsule. (b) Schematic representation of the stepwise build-up of the capsule at the molecular printboard.

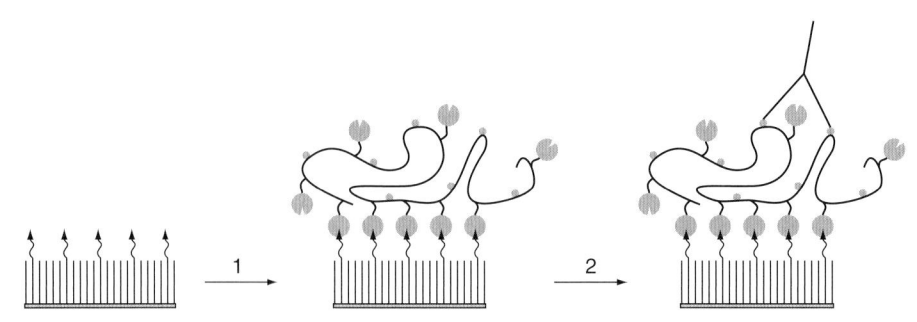

Figure 22 The absorption of a bifunctional polymer with vancomycin and fluorescein groups to SAMs consisting of D-Ala-D-Ala groups and tri(ethylene glycol) groups (1): the adsorption of an antifluorescein antibody to such SAMs to which the bifunctional vancomycin–fluorescein polymer was adsorbed (2) [67].

the assembly of antifluorescein antibodies (ABs) toward the polymer (**Figure 22**). The affinity of the AB for binding to the SAM is enhanced by a factor of 570 due to divalency. Thus, the bifunctional polymer acts as a bridge between the SAM and the immunoglobulin through two independent interactions, the polyvalent interaction between the SAM and the polymer, and the divalent interaction between the immunoglobulin and the polymer.

Our group investigated the desorption of the electroactive ferrocenyl PPI dendrimers by an external stimulus (**Figure 9**) [46–48, 68, 69]. Oxidation of the ferrocenyl groups to ferrocenium cations leads to a dramatic decrease of the affinity for the β-CD cavity. CV provided proof of the assembly and disassembly scheme (**Figure 9**) [46]. Oxidation of the ferrocenyl groups of a full monolayer of fragment crystallizable (Fc) dendrimers on a β-CD SAM, which occurred for all ferrocenyl groups at the same potential, led to complete desorption of the dendrimers, as indicated by a combination of CV experiments with SPR [47]. Subsequent reduction showed that only part of the oxidized dendrimers were reduced back and readsorbed, thus leading to lower charge densities for subsequent CV scans. When these dendrimers were added to the electrolyte solution in contact with the Fc dendrimer monolayer, oxidation led to complete desorption; however, upon reduction, the dendrimer monolayer was fully reconstituted. This procedure was fully reversible at various scan rates.

Another example of the molecular printboard as a platform to create supramolecular assemblies is given in **Figure 23** [70].

In these experiments, the binding of a supramolecular complex at a multivalent host surface was achieved by combining the orthogonal β-CD host–guest and metal ion-ethylenediamine coordination motifs (**Figure 23**). The binding to a β-CD SAM of the complex between an adamantyl-functionalized ethylenediamine derivative and Cu^{II} or Ni^{II} was studied as a function of pH by means of SPR spectroscopy. A heterotropic, multivalent binding model at interfaces describes the multivalent enrichment at the surface. The Cu(II) complex showed divalent binding to the CD surface with an enhancement factor higher than 100 compared to the corresponding divalent complex in solution. Similar behavior was observed for the Ni(II) systems. The remarkable result is that at pH 6, at the multivalent surface the formation of the divalent CuL_2 complex is favored, whereas the monovalent CuL is the majority species in solution. This behavior can be attributed to the high C_{eff} of CD sites present at the surface and the close to optimal linker lengths between the two adamantyl groups relative to the periodicity of the β-CD lattice (about 2 nm) [71].

Further increase of the complexity of the assembly at the surface was achieved in the immobilization of biological materials by means of step-by-step assembly. Streptavidin (SAv) was attached to β-CD SAMs through orthogonal host–guest and SAv–biotin interactions [72, 73].

The orthogonal linkers consist of a biotin functionality for binding to SAv and adamantyl functionalities for host–guest interactions at β-CD SAMs (see **Figure 24**). SAv, complexed to an excess of monovalent linker **3** in solution and then attached to a β-CD SAM, could be removed from the surface by rinsing with a 10-mM β-CD solution (**Figure 25**).

When SAv was attached to the β-CD SAM through the divalent linker **4**, it was impossible to remove SAv from the surface with this procedure. This is due to the SAv binding pockets oriented toward the β-CD SAM, resulting in (labile) divalent and (stable) tetravalent β-CD–adamantyl interactions for the mono- and divalent

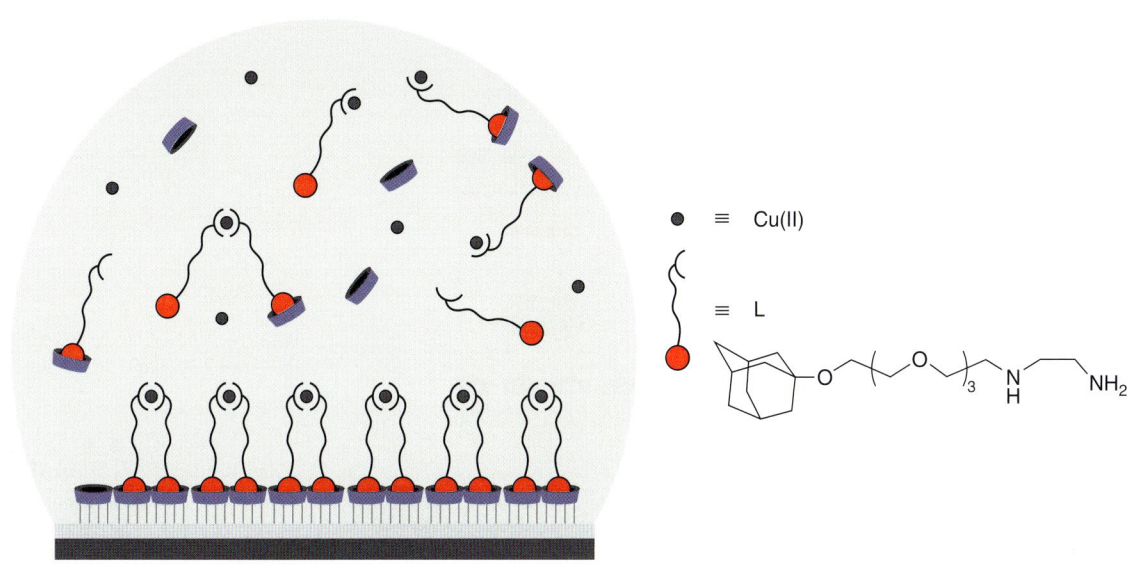

Figure 23 Schematic illustration of the assembly of the Cu(II) with two molecules of adamantyl-functionalized ethylenediamine derivative on the molecular printboard.

Figure 24 Compounds used to realize the assembly scheme in **Figure 25**.

linkers, respectively. This was confirmed by experiments with different relative concentrations. When the [linker]/[SAv] ratio is reduced, in case of the divalent linker, a clear trend was observed. The less linker was used, the more protein could be removed from the surface. It was proven that the orthogonality of the binding motifs and the stability of the divalent linker at the β-CD SAM allow the stepwise assembly of the complex on the printboard by first adsorbing the linker and then the SAv.

The attachment of proteins to surfaces is a key step in many biotechnological applications [74, 75].

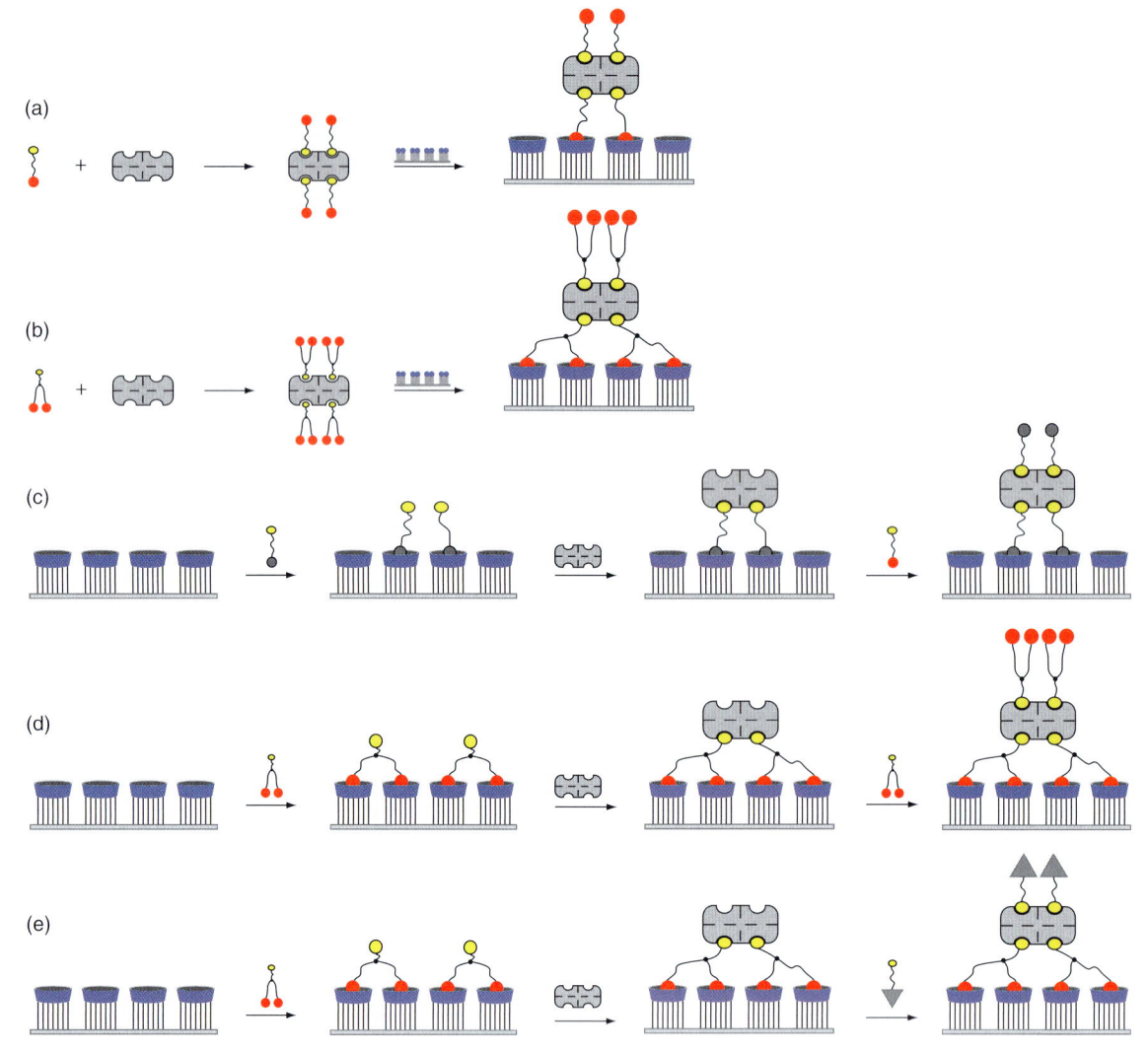

Figure 25 Adsorption schemes for the assemblies of SAv through monovalent and divalent linkers.

For many of these, one needs control over the adsorption strength and reversibility, protein orientation, and retention of biological function. Such requirements can only be met when the binding of the protein to the surface is specific. In other words, it is important to avoid the nonspecific adsorption of the protein on the surface. There are different options to prevent the nonspecific adsorption of proteins onto surfaces, such as adding surfactants or bovine serum albumin (BSA) to protein solutions [76, 77]. Another well-known method is the use of SAMs that are protein-resistant, such as oligo(ethylene glycol) (OEG) monolayers [78–80]. The prevention of nonspecific interactions by using SAMs with OEG chains is attributed to loose packing and the well-hydrated nature of these SAMs [79].

To avoid nonspecific binding, the hexa(ethylene glycol) mono(adamantyl ether) **6** was used [81], a compound with a single adamantyl group for a predictable, specific, and reversible interaction with the β-CD SAM, and a hexa(ethylene glycol) chain for preventing nonspecific protein adsorption. In the absence of this compound, the SPR experiments showed significant nonspecific adsorption for SAv, maltose-binding protein (MBP), and BSA proteins, but even a low concentration of **6** was sufficient to suppress nonspecific interactions. The binding of **6** temporarily blocks the β-CD cavities and exposes the hexa(ethylene glycol) tails to the solution. The adsorption of SAv by the divalent linker **4** was studied on a β-CD SAM covered with **6**. This guest was reversibly bound to the surface. As expected, the multivalency effect made the binding of the SAv-(**4**)₄ complex much stronger than the binding of the monovalent linker **6**, even if the latter was in excess. Removal of the protein–linker assemblies was not observed after rinsing with 10 mM of β-CD, proving the strong interaction with the molecular printboard.

This method of avoiding nonspecific interactions not only applies to SAv, but also to the histidine-tagged MBP (His₆-MBP) that is a representative of the class of bioengineered His-tagged proteins [82, 83]. The binding of this protein to the printboard was achieved using the complex between linker **7** (**Figure 24**) and Ni^{II}. A solution containing a mixture of **7**, Ni^{II}, and the protein was flowed over the β-CD SAM already covered with the monovalent compound **6** (**Figure 26**) [81, 84]. SPR experiments show, also in this case, adsorption of the protein.

There is a considerable interest in building bionanostructures at surfaces for sensing purposes. As pointed out previously, the main factors are the control of the orientation, functionality, and specificity of protein adsorption [85–91]. ABs are often present in sensors because they can be used as medical diagnostic tools [92]. Control over orientation when immobilizing ABs to surfaces for sensor purposes is of utmost importance because this constitutes, for a large part, the effectiveness of the ABs to detect antigens [93–96]. One way to achieve this is the use of Fc receptors, such as protein A (PA), protein G (PG), or protein A/G (PA/G) [97]. An AB binds with its Fc fragment to PA or PG, and therefore, the fragment antigen binding (Fab) fragments of the AB are directed toward the solution, and as such, are capable of binding antigens from solution. Miniaturization is important for biological assays because it allows faster diagnostics with small amounts of sample. Currently, there are numerous applications for microchips [98–100]. Protein functionality and the inhibition of nonspecific adsorption are key issues in this field. Immunoassays, with immobilized ABs in microchannels, are important biological assays because small quantities of antigens can be detected [101, 102].

The formation of AB assemblies on the molecular printboard based on host–guest, protein–ligand, and protein–protein interactions was investigated [103]. In this study, the buildup of a structure consisting of the divalent bis(adamantyl)-biotin linker **4**, SAv, biotinylated protein A (btPA), and an Fc fragment of a human immunoglobulin G (IgG-Fc) was achieved (see **Figure 27**).

Patterns of such a structure were obtained through μCP of the divalent linker at the molecular printboard, followed by the subsequent attachment of the proteins (**Figure 28**).

Fluorescence microscopy showed that the assembly of these bionanostructures on the β-CD SAM is highly specific. On the basis of these results, bionanostructures were made in which whole ABs were used instead of the IgG-Fc. These ABs were bound to the SAv layer either through biotinylated protein G (bt-PG) or through a biotinylated AB (**Figure 29**).

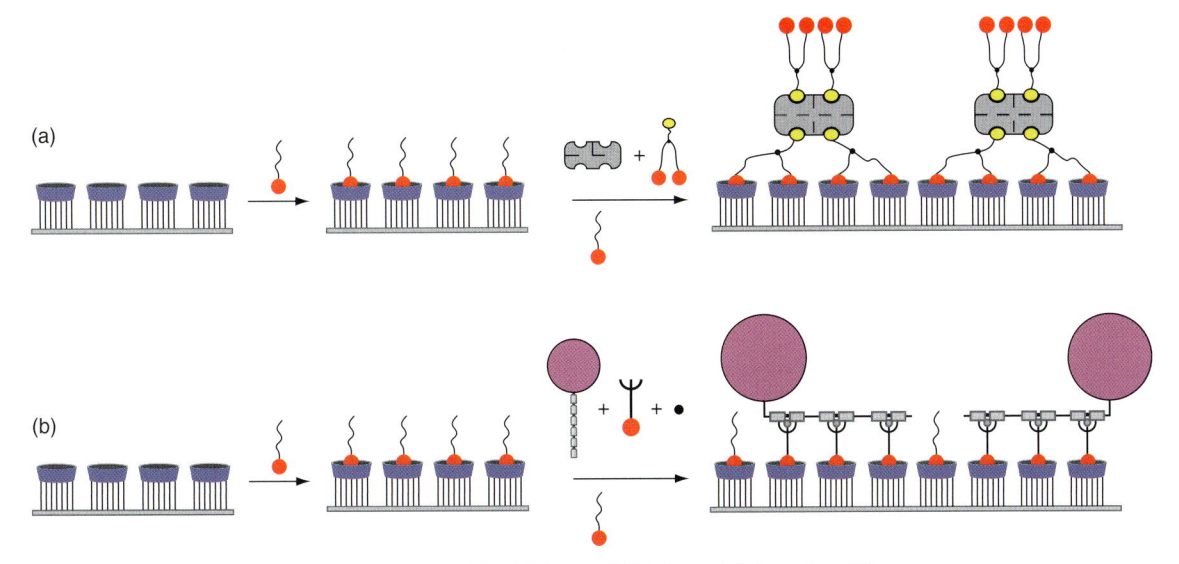

Figure 26 Adsorption schemes for the assembly of SAv and MBP through linkers **4** and **7**.

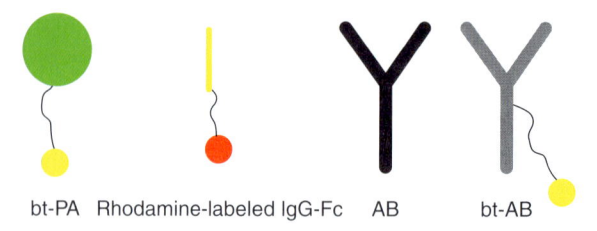

Figure 27 Biotinylated protein A (btPA), rhodamine-labeled IgG-Fc and the general structure of biotinylated antibodies.

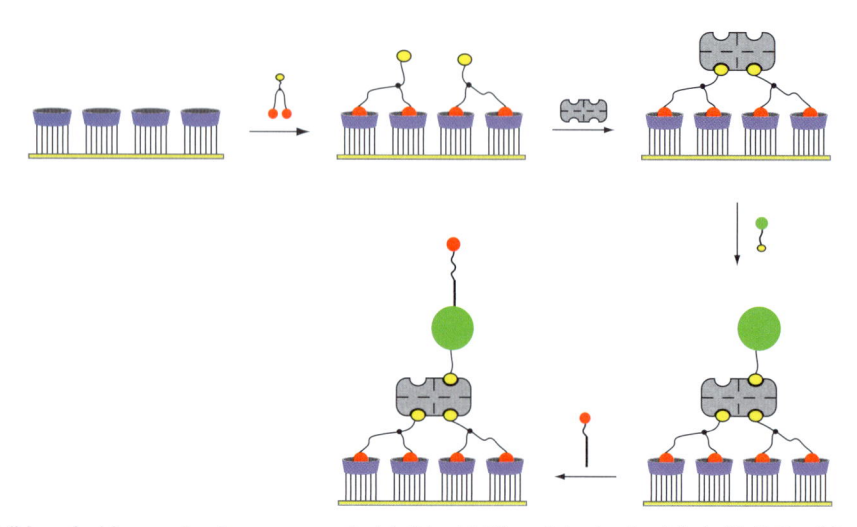

Figure 28 Buildup of a bionanostructure composed of 4, SAv, bt-PA and rhodamine-labeled IgG-Fc at the β-CD SAM.

The immobilization of ABs to the molecular printboard was used to create a platform for lymphocyte cell count purposes [103]. Monoclonal antibodies (MABs) were attached to the SAv layer using bt-PG or through nonspecific adsorption. The binding specificity of the immobilized cells was the highest on bt-PA, which is attributed to an optimized orientation of the ABs. An approximately linear relationship between the numbers of seeded cells and counted cells demonstrated that the platform is potentially suitable for lymphocyte cell counting.

For the immobilization of the β-CD monolayers and the subsequent specific attachment of proteins inside channels, a microchip was fabricated with one large channel that splits into four smaller channels (**Figure 30**) [104].

The β-CD monolayer was attached to the microchip as the reagents were flowed through the channel. Static contact-angle measurements followed the same trend as the advancing contact angles measured on planar substrate [61]. To test whether or not host–guest interactions of β-CD monolayers inside microchannels are comparable to those on planar substrates, an adamantyl-terminated fluorescent dendritic wedge was immobilized in the microchannel by rinsing a solution of it in 10-mM β-CD in water. The fluorescence image, recorded after subsequent rinsing for 10 min with water, clearly showed that the molecule is present in the channel. After flowing 10 mM of β-CD in water, the intensity decreased significantly but the compound was not completely removed. There was no trace of fluorescence only after rinsing with methanol. These results are comparable with β-CD monolayers on glass and gold [43, 62].

For the immobilization of SAv on the inner walls of the microchannel, linker **4** was washed over the surface followed by a wash with SAv [72, 103]. This assembly process leaves two biotin binding pockets available for further functionalization which can be used to attach the fluorescently labeled biotin moieties **8** and **9** (**Figures 31 and 32(a)**), or biotinylated PA or PG. In turn, these can bind to fluorescently labeled human-IgG (**5**) and goat-IgG (**6**), respectively (**Figures 32(b) and 32(c)**). To show that the channels can be addressed individually, SAv was assembled in all channels through divalent linker **4**. The latter was adsorbed from inlet A (see **Figure 30**) and, from the same side, SAv was subsequently

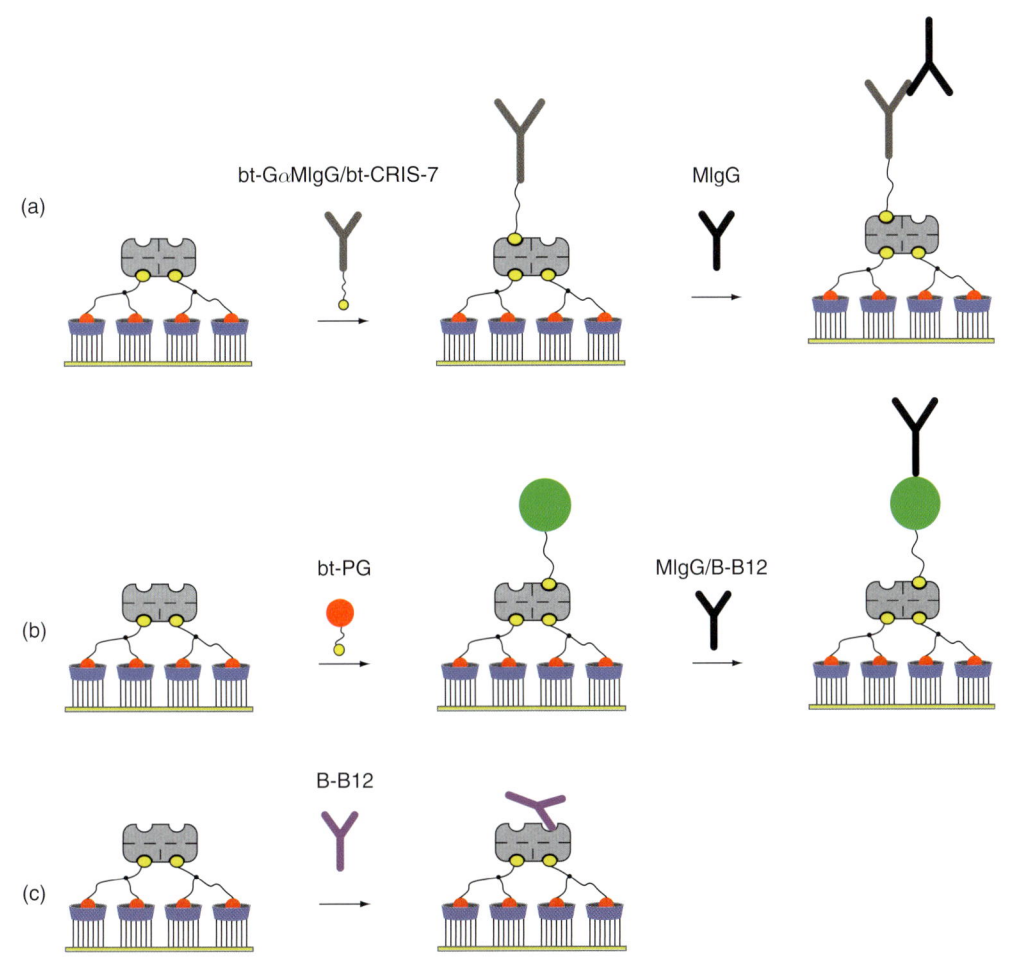

Figure 29 AB bionanostructure to β-CD SAM covered with **4** and SAv. Adsorption of bt-GαMIgG or bt-CRIS-7 and the subsequent attachment of MIgG to bt-GαMIgG (a); adsorption of bt-PG and the subsequent attachment of MIgG or B-B12 (b), and the (nonspecific) adsorption of B-B12 onto SAv. Schemes in (a) (bt-GRMigG + MigG) and (b) (bt-PG + MIgG) apply to the study on the assembly of Abs, while the attachments of bt-CRIS-7 (a) and B-B12 (b, c) apply to the cell adhesion studies.

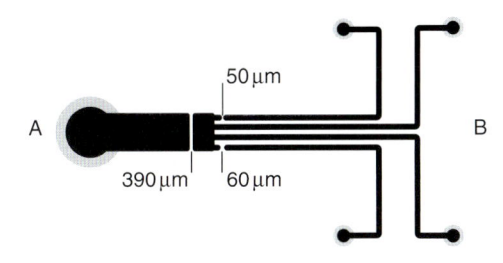

Figure 30 Design of the chip used for the study of the molecular printboard in microchannels.

flowed through the channel. Two different fluorescently labeled biotin derivatives (**3** and **4**) were introduced from the small inlets at side B through alternating channels to create assemblies according to the procedure shown in **Figure 32(a)**.

Imaging with green excitation light showed that **8** was immobilized in two channels (**Figure 33**). Imaging the channels with blue excitation light showed that **9** was immobilized in the other two channels. The combined image shows the four channels, with alternating **8** and **9**, which proved the possibility of individual channel functionalization by using intrinsically reversible supramolecular interactions.

Similarly, **10**, labeled with a fluorescein isothiocyanate (FITC) label, and **11**, labeled with Alexa fluor-568 (**Figure 31**), were assembled through bt-PA and bt-PG, respectively (**Figures 32(b) and 32(c)**). This was achieved by introducing bt-PA and bt-PG into the channels simultaneously from side B on the SAv-**4**-coated β-CD monolayer, in the alternating fashion shown above for **8** and **9**. Subsequently, compounds **10** and **11** were

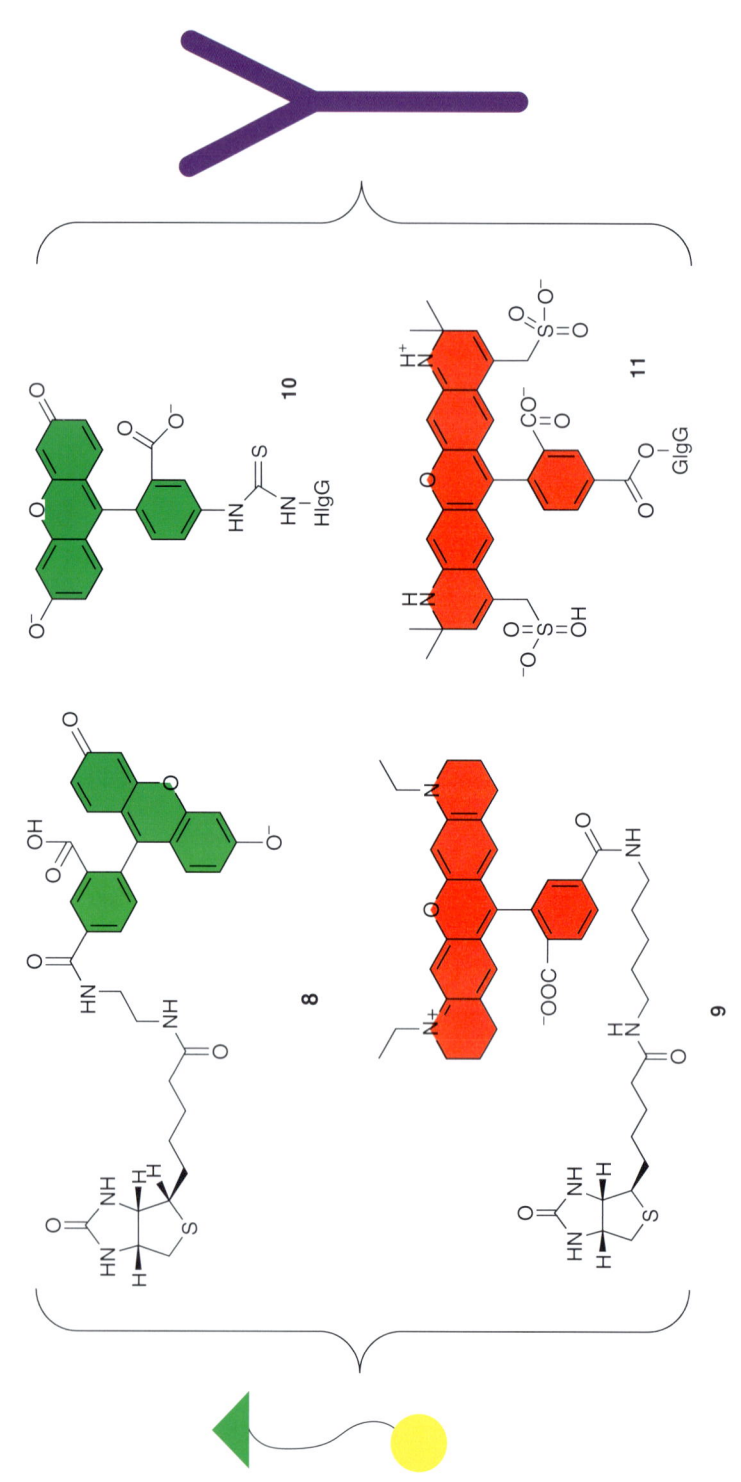

Figure 31 Compounds used in the microchip studies: biotin-4-fluorescein (**8**), Atto 565-biotin (**9**), human IgG-fluorescein (**10**), goat Alexafluor-568-IgG (**11**).

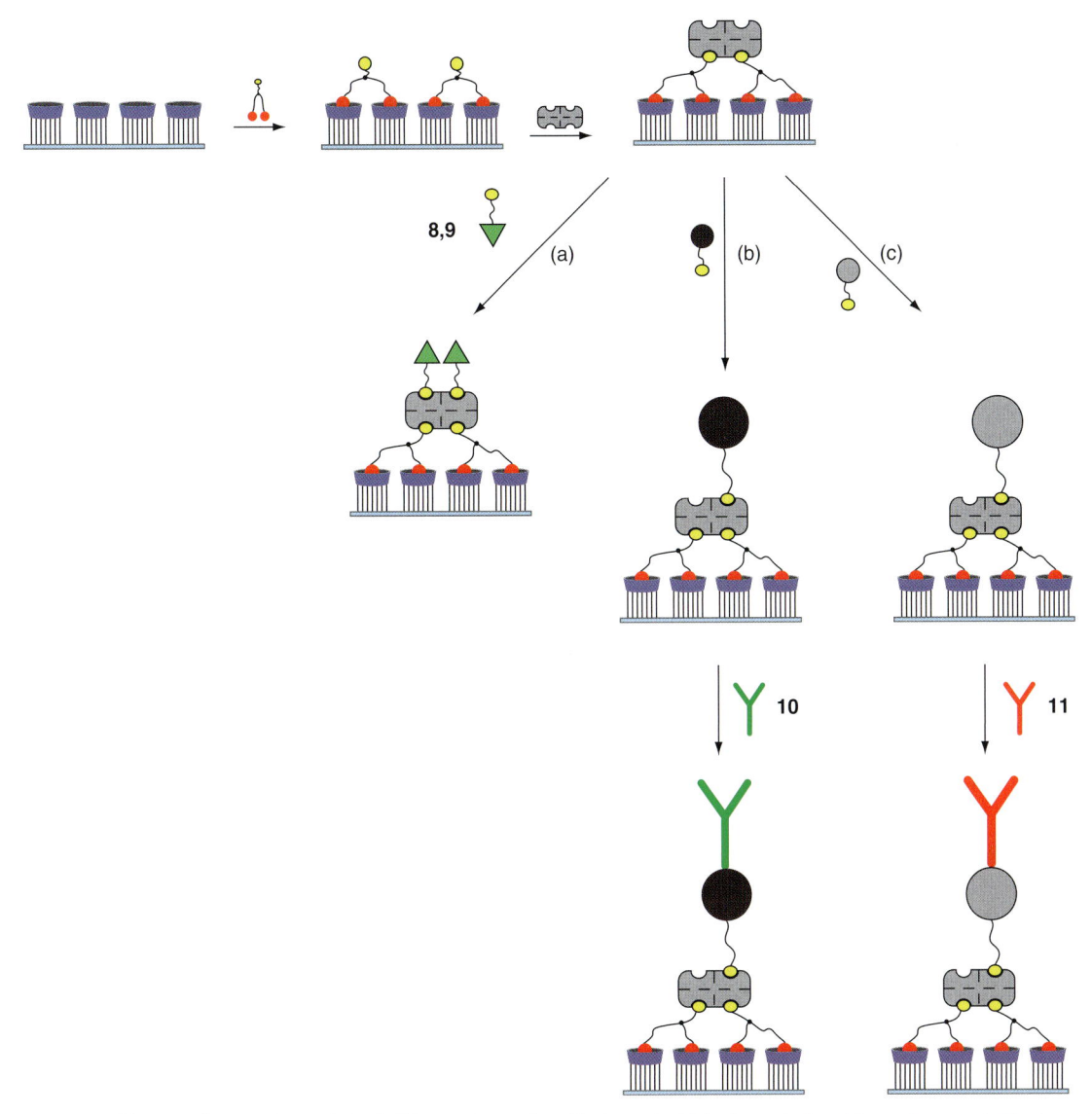

Figure 32 The attachment of protein to β-CD monolayer: the attachment of SAv through **4** and the subsequent attachment to SAv of a (a) fluorescently labeled biotin derivative (**8,9**) or (b) IgG **10**, or IgG **11** through biotinylated PA or PG.

simultaneously passed through the channels in an alternating fashion from side B.

Protein assays demand the detection of only specific ABs. To prove that this is possible in this system, divalent linker **4** and SAv were immobilized in the β-CD-covered channels from side A, followed by the immobilization of bt-PA also from this side, which resulted in all channels being covered with bt-PA (**Figure 32(b)**). Subsequently, compounds **10** and **11** were simultaneously introduced through alternating channels of the chip in the reverse direction (from side B) for 30 min. After rinsing with water for 20 min only two channels had been modified with a fluorescent IgG (**10**). Compound **11** was not immobilized, as expected, because **11** does not bind to PA. These experiments show that the channels can be addressed separately and that they can be modified in such a manner that the immobilization of the ABs is specific.

Another fascinating example of ordered assemblies of biological material through this biotin–SAv interaction is the immobilization of cytochrome c (cyt c) on the molecular printboard [105]. A biotinylated cyt c was obtained according to a literature procedure with biotin-LC-NHS, which has a spacer arm of 2.24 nm, and reacts to free amino positions at

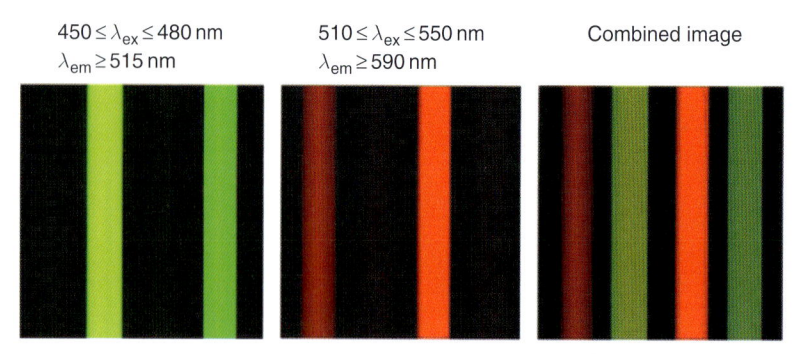

Figure 33 Fluorescence microscopy images recorded with blue (left) and green (center) excitation light of the chip functionalized with biotin derivatives (**8** and **9**) in alternating channels. The combined image is also shown. Reproduced from Ludden MJW, Ling XY, Gang T, et al. (2008) Multivalent binding of small guest molecules and proteins to molecular printboards inside microchannels. *Chemistry – A European Journal* 14: 136–142.

the surface of the protein [106]. With this method it is possible to obtain a cyt *c* modified with multiple biotin moieties (bt-cyt *c*). Control experiments with SPR proved that when cyt *c* is flowed onto a SAv monolayer there is no specific adsorption because the protein can be easily removed after rinsing. In contrast, bt-cyt *c* was strongly attached to the surface and was stable upon rinsing (**Figure 34**). Cyt *c* displays a Soret band in the UV-Vis at $\lambda = 408$ nm which shifts to lower wavelengths upon denaturation [107]. The spectra of the assembly of cyt *c* on the surface clearly show the presence of the Soret band, proving that no denaturation occurred.

Scanning electrochemical microscopy (SECM) was used to determine the surface coverage of cyt *c* (**Figure 35**) [105].

The redox reactions that occur at the ultra-microelectrode and at the surface are shown in **Figure 35**. The SECM experiments on β-CD SAM were a modification of the method developed before for the ferrocene-terminated dendrimers, because a monolayer of cyt *c* offers much less redox equivalents [68]. The UME was positioned at a certain distance from the surface and a potential pulse was applied in order to reduce the mediator $[Ru(NH_3)_6]^{3+}$. Different experiments were performed that included changing

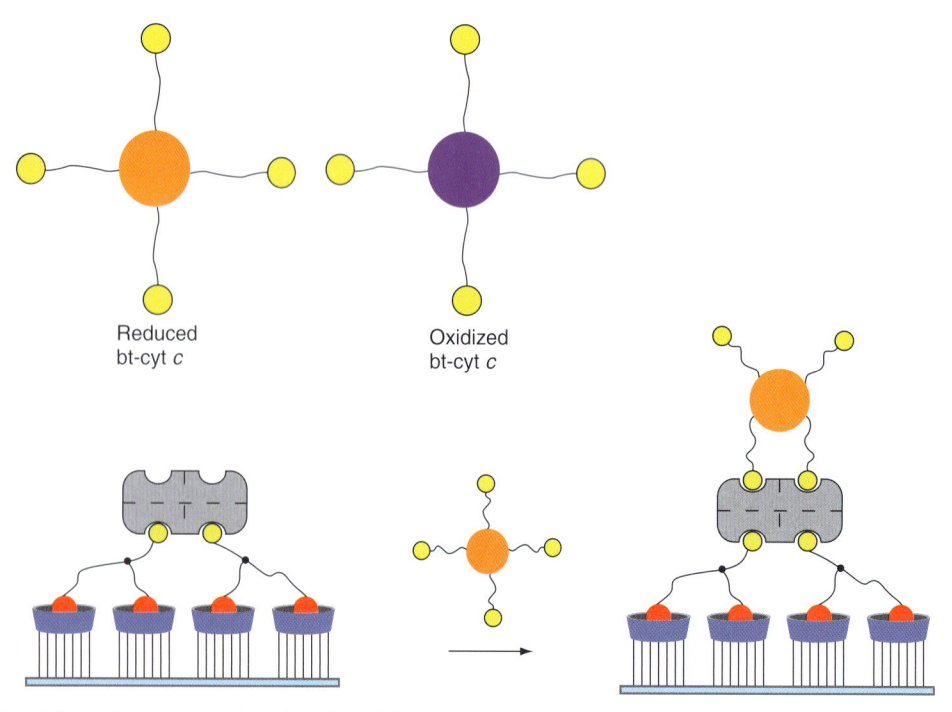

Figure 34 Schematic representation of the immobilization of the cytocrome *c* on the molecular printboard through biotin-SAv interactions.

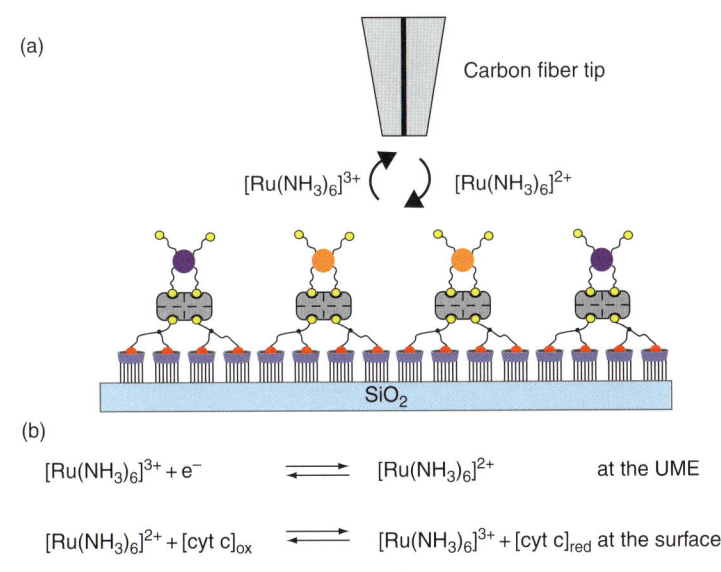

Figure 35 (a) Schematics of the SECM experiment. [Ru(NH$_3$)$_6$]$^{3+}$ is reduced at the tip and diffuses back to the UME, which results in a negative feedback current. (b) Redox reactions taking place at the UME and at the surface.

the position and the distance of the UME on the surface covered with cyt *c*. Control experiments with an inert surface were also performed. These experiments led to the conclusion that [Ru(NH$_3$)$_6$]$^{3+}$ acts as a mediator in the redox process as illustrated in **Figure 35** and therefore cyt *c* immobilized on the surface maintains its redox properties. It is important to note that the coverage of cyt *c* is in full agreement with the expected binding stoichiometry of the resulting bionanostructure, which shows that it can be in principle controlled through the design and the use of linker molecules with other valencies [72].

7.7 Assemblies of NPs and 2D and 3D Nanofabrication on the Molecular Printboard

Functionalized NPs can act as model systems in between solution and surface chemistry and as a tool for nanofabrication. In this section, we discuss the functionalization of NPs with host sites that allow: (1) the controlled aggregation with guest-functionalized dendrimers in solution, (2) the specific adsorption onto patterned substrates, and (3) the fabrication of larger architectures using the layer-by-layer (LBL) methodology. The molecular printboard turns out to be a very powerful tool for this.

The functionalization of the β-CD with thiol end groups allows attachment to the gold NP surface. The gold NPs were formed in the presence of CD heptathiol to yield CD-functionalized gold NPs in one step (**Figure 36**) [108]. The NPs prepared this way were well dispersed and showed an average size of about 3 nm.

The aggregation of the CD gold NPs in solution was studied in the presence of multivalent adamantyl-functionalized PPI dendrimers [108]. UV/Vis experiments, monitoring the λ_{max} and the intensity of the gold NP plasmon adsorption band, showed that addition of dendrimers (generations 1–3) leads to pronounced aggregation and eventually even to irreversible precipitation. In contrast, the addition of a monovalent guest did not lead to any change.

Figure 36 Preparation of cyclodextrin-functionalized gold nanoparticles.

β-CD-modified silica NPs were prepared by reacting carboxylic active ester-terminated silica NPs with β-CD heptamine [109]. Silica NPs functionalized with glucosamine, having similar surface properties but lacking host–guest recognition properties, were used as a reference. The β-CD-modified NPs exhibited pH-dependent aggregation, due to the presence of free amino and carboxylic acid groups on the particle surface, which was corroborated by zeta potential measurements. The functionalization with β-CD was further confirmed by host–guest studies in solution. Moreover, in this case, the addition of the adamantyl-terminated dendrimer led to aggregation of the particles. The NPs strongly adsorbed onto a β-CD monolayer on silicon onto which adamantyl-terminated dendrimers were pre-adsorbed (see **Figure 37**) [109].

The multivalent, supramolecular aggregation of β-CD-functionalized gold NPs was applied in the LBL [110] assembly scheme depicted in **Figure 38** [111]. UV–Vis, surface plasmon resonance, ellipsometry, and AFM data confirmed the LBL assembly and showed a linear dependence of the layer thickness with the number of bilayer adsorption steps with a thickness increase of approximately 2–3 nm per bilayer [111].

The integration of top–down and bottom–up nanofabrication schemes is a key issue in nanotechnology. Combination of LBL assembly with nanotransfer printing works with assembly schemes shown above for achieving micrometer structures [112]. Nanoimprint lithography (NIL) allows a sub-10-nm resolution in pattern replication and has recently been put forward by the semiconductor industry as the most promising technique to be implemented in future chip-fabrication processes. The full integration of LBL assembly and NIL therefore leads to the fabrication of 3D nano objects of arbitrary shapes on substrates, where the x,y dimensions are determined by NIL and the z dimension by the LBL assembly. A nice example of combination of these two techniques was achieved using the molecular printboard (**Figure 39**) [113, 114]. In this procedure, NIL was used as a tool to pattern SAMs on silicon substrates because of its ability to pattern in the μm and nm ranges [113]. The polymethylmethacrylate (PMMA) template behaves as a physical barrier, preventing the formation of a SAM in the covered areas of the substrate. After polymer removal, SAM patterns were obtained [114]. Integration of the LBL assembly and NIL was achieved by adopting process conditions such that none of the processing steps interfered with or damaged preceding fabrication steps. The PMMA patterns function as a mask for the LBL assembly and can be removed afterwards. The bare silicon oxide areas between the PMMA structures were functionalized consecutively with aminosilane, diisothiocyanate, and CD heptamine, finally resulting in CD monolayers on silicon oxide in the empty areas [61]. In order to keep the PMMA structures, the fine tuning of the chemistry is essential. Therefore, the first silane attachment was performed from the gas phase, while the second and third steps were performed in ethanol and water, respectively, in order not to damage, swell, or dissolve the PMMA. Ellipsometry on full substrates (without PMMA structures) confirmed the expected layer-thickness increases for all steps. AFM imaging confirmed the expected layer thickness for a patterned sample, which was subjected to acetone and ultrasonication in order to remove the PMMA structures.

In this LBL assembly method, it is essential that nonspecific adsorption of these components onto the PMMA structures does not pose a problem because the polymer is dissolved in the final liftoff step together with the material deposited onto it.

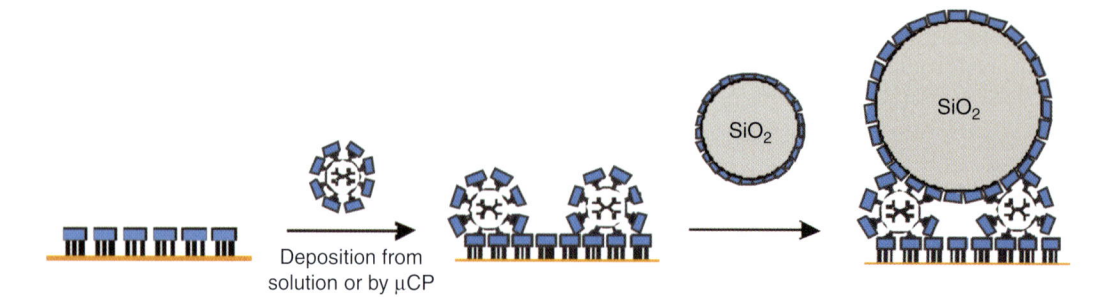

Figure 37 Schematic representation of the assembly of β-CD-functionalized silica nanoparticles on β-CD molecular printboards on silicon oxide through multivalent host–guest interactions employing adamantyl-terminated PPI dendrimers as supramolecular glue.

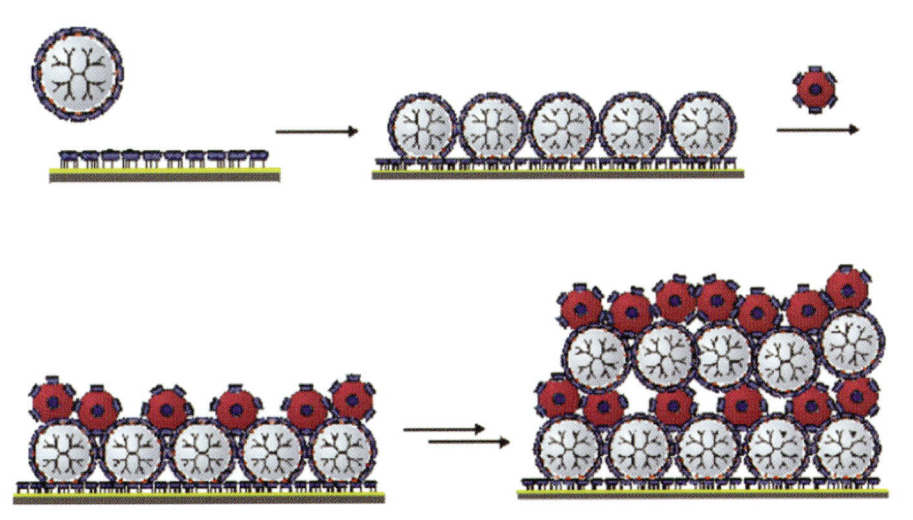

Figure 38 Layer-by-layer assembly scheme for the alternating adsorption of adamantyl-modified dendrimers and β-CD-functionalized gold nanoparticles onto the molecular printboard.

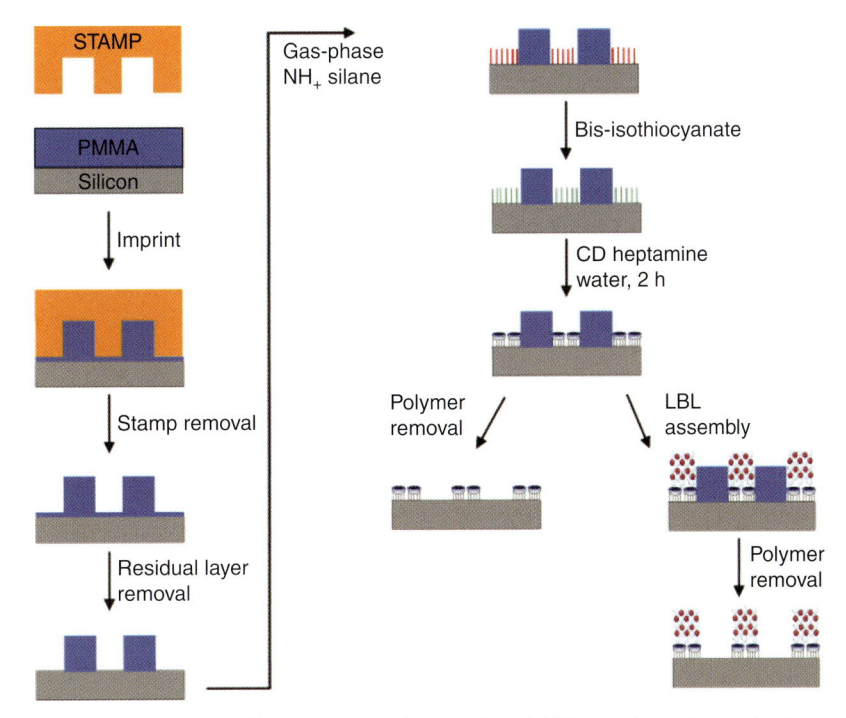

Figure 39 Integrated nanofabrication scheme incorporating nanoimprint lithography and layer-by-layer assembly.

The application of this LBL assembly scheme to the NIL-patterned β-CD substrates led to the expected layer growth of approximately 2 nm per layer in the micrometer-sized β-CD areas between PMMA structures. This was visualized by AFM (**Figure 40**) after removal of the polymer in acetone, using sonication. This shows that the growth of the multilayer structures is not different from the growth on unpatterned substrates and that the liftoff step does not lead to removal or damaging of the multilayer structures. The latter was confirmed by exposing multilayer substrates to the same liftoff procedure. This yielded comparable layer thicknesses and qualities before and after the acetone treatment.

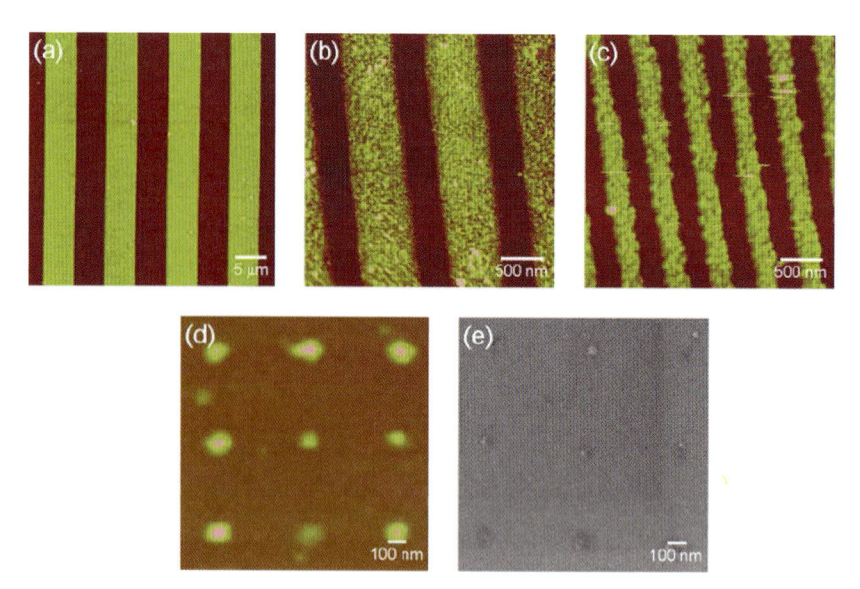

Figure 40 Patterned assemblies of generation-5 adamantyl-functionalized dendrimers and β-CD-functionalized Au nanoparticles prepared by LBL assembly on an NIL-patterned β-CD SAM, followed by PMMA removal (a–e). Contact mode AFM (a–d) and SEM (e) show micrometer structures of multilayers assembled along stripes of different sizes (a–c) and dots (d, e).

Building up ordered nanostructures from particles has attracted a lot of interest due to the need for miniaturization [115]. The interesting chemical, electronic, and surface properties that arise from such individual or organized nanometer-sized objects make them suitable for electronic, optical, and biological applications [116]. In general, there are two approaches to assemble nanostructured materials, namely, physical assembly and supramolecular assembly. The physical assembly processes include convective or capillary assembly [117, 118], spin coating [119], and colloidal epitaxy [120]. In particular, convective assembly has been frequently employed to arrange nonfunctionalized particles into hexagonally ordered and close-packed single or multilayered particle lattices [121]. The major driving force for the assembly is the evaporation of water from the particle suspension [122]. Mobile particles in a thin liquid film are convectively assembled as a result of the hydrodynamic forces induced by the influx of water close to the drying edge. The assembly process starts when the thickness of the solvent layer becomes smaller than the particle diameter [123]. This technique provides limited control over the structure of the particle lattices as well as the dimensions and complexity of the final assembly.

Supramolecular assembly utilizes coupling chemistry to precisely direct and control the deposition of functionalized particles onto the substrate. Typical chemical approaches can be electrostatic interactions [124], host–guest interactions [111], and thiol-based self-assembly [125]. In this assembly, the organization of the particle lattice will no longer depend solely on surface tension and long-range attractive forces that act in a lateral direction, but rather on the competition between lateral attractive forces and vertical supramolecular interactions between functionalized particles and the surface. In this respect, the molecular printboard provides a useful tool for the supramolecular assembly of NPs.

The formation of particle monolayers by convective assembly was studied *in situ* with three different kinds of particle–surface interactions [126], adsorption onto native surfaces, with additional electrostatic interactions, and with supramolecular host–guest interaction on the molecular printboard (see **Figure 41**). In this study, carboxylate- and β-CD-functionalized poly(styrene-*co*-acrylic acid) particles (PS-COOH and PS-CD) were used [126].

The convective assembly was carried out on a horizontal deposition setup with the method illustrated in **Figure 42**. A droplet of particle suspension was introduced into the gap between a mobile substrate and a fixed glass slide while the temperature was controlled between 4 and 20 °C. The substrate was shifted to the left at a constant velocity. Particles assemble on the surface in the assembly zone (**Figure 42(a)**) as a result of the convective flow of

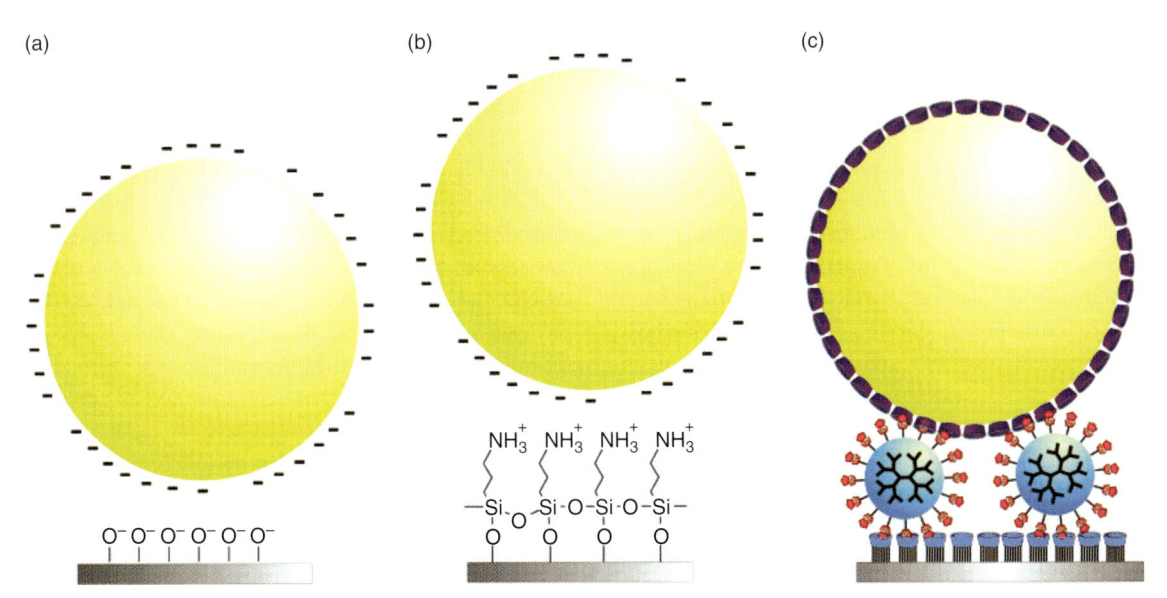

Figure 41 Schematic illustration of three different assembly schemes of particles. (a) Convective assembly of PS-COOH particles on native SiO$_2$ surface on silicon. (b) Convective assembly with additional electrostatic interactions of PS-COOH particles on NH$_3^+$ (SAM). (c) Convective assembly with supramolecular interactions of PS-CD with complementary pre-adsorbed ferrocenyl-functionalized dendrimers on a β-CD SAM.

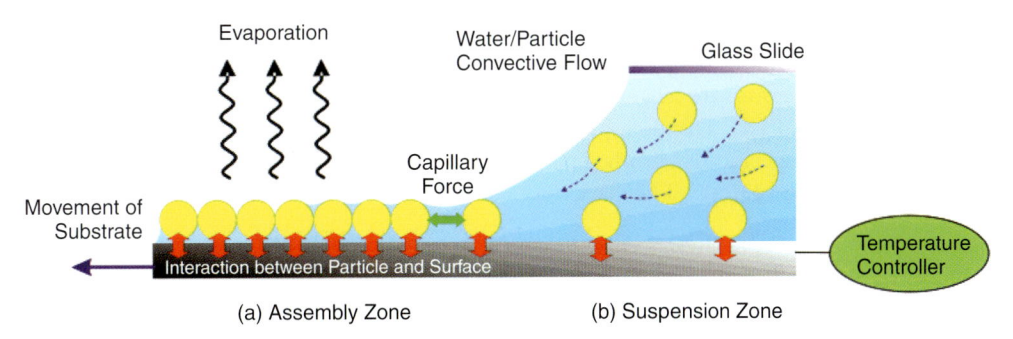

Figure 42 Schematic illustration of the adsorption and assembly of particles on a substrate. Reproduced with permission from Ling XY, Malaquin L, Reinhoudt DN, Wolf H, and Huskens J (2007) An *in situ* study of the adsorption behavior of functionalized particles on self-assembled monolayers via different chemical interactions. *Langmuir* 23: 9990–9999.

particles induced by the evaporation solvent at the assembly zone. In the suspension zone (**Figure 42(b)**), the particle suspension resembles the bulk. Depending on the particle–substrate interaction, adsorption from the suspension onto the substrate may also occur in the suspension zone when specific interactions occur.

In one case, the PS-COOH particles were assembled onto native silicon oxide surfaces. The same particles were also assembled onto a protonated amino-functionalized (NH$_3^+$) SAM. The assembly of β-CD-functionalized polystyrene (PS-CD) particles was studied onto a β-CD SAM with pre-adsorbed ferrocenyl-functionalized dendrimers.

The deposition behavior of particles from surfaces was evaluated by reducing the temperature below the dew point, thus initiating water condensation. Particle lattices on native oxide surfaces formed the best hexagonal close-packed (hcp) order and could be easily desorbed by reducing the temperature below the dew point. The electrostatically modified assembly resulted in densely packed, but disordered particle lattices. The specificity and selectivity of the supramolecular assembly process were optimized by the use of ferrocenyl-functionalized dendrimers of low generation and by the introduction of competitive interactions by native β-CD molecules during the assembly. The supramolecular particle lattices were nearly hcp. Both electrostatically and supramolecularly formed lattices of particles were strongly attached to the surfaces and could not be removed by condensation.

As an extension of the procedure for the fabrication of nanostructures, substrates patterned by NIL were employed to assemble the particles into micrometer lines [126]. **Figure 43** shows the patterned particle lattices formed on nanoimprinted patterns assembled through convective assembly, with electrostatic interactions and with supramolecular host–guest interactions. For the convective assembly, the particles were physically confined by the PMMA polymer barriers into the silicon oxide areas, which resulted in the formation of highly hcp-ordered particle lines. The selective particle assembly is driven by the large chemical contrast between PMMA and silicon oxide. In the beginning of the assembly with additional electrostatic interactions, the particles bind preferentially to the NH_3^+ SAM. The particle lines are disordered and multilayered as a result of the strong electrostatic interactions of the particles with the substrate. As the wetting contrast between the NH_3^+ SAM and the PMMA is relatively low, after a few seconds, the particles started to assemble on the polymer lines as well. In the end, nearly all the polymer features are covered with particles. In the absence of β-CD, a random particle lattice was formed with nonspecifically adsorbed particles everywhere (**Figure 43(c)**). The lack of specificity is attributed to the adsorption of dendrimers onto the PMMA structure as a result of hydrophobic interactions [112].

The nonspecific adsorption of dendrimers on PMMA is reduced as a result of competition by β-CD in solution. Hence, a well-controlled, stable, and single layer of particles is formed in the assembly process with a better degree of order than without β-CD. Potentially, this method may also yield multilayered nanostructures by repeating the adsorption of the ferrocenyl dendrimers and the β-CD functionalized particles as shown before in the LBL process [111]. In conclusion, the use of supramolecular interactions on the printboard can be exploited for fine-tuning the particle assembly behavior that cannot be achieved by physical assembly. The combination of top–down and bottom–up methods gives well-defined 3D nanostructures.

Regeneration of surfaces and reversible attachment is an important issue in nanotechnology [127]. Ferrocenyl-functionalized dendrimers can be deposited on the molecular printboard and then easily desorbed by oxidation [46, 68, 128, 129]. β-CD-functionalized NPs can be reversibly adsorbed and desorbed onto/from the molecular printboard (**Figure 44**) [130].

In this section, the attention was focused on: (1) the electrochemical addressability of the reversible glue, when sandwiched between the SAM and the NPs interfaces, (2) optimization of the conditions for

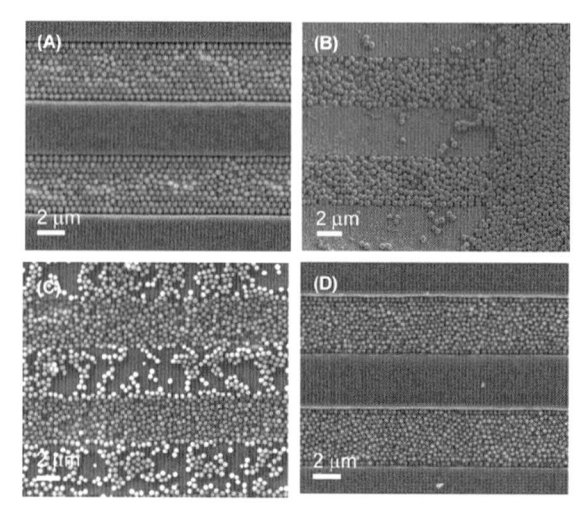

Figure 43 SEM micrographs of NIL-patterned particle lattices formed by adsorption of PS-COOH on native SiO_2 on Si (a) PS-COOH on an NH_3+ SAM (b). PS-CD on a β-CD SAM with pre-adsorbed ferrocenyl-functionalized dendrimers in the absence (c) and presence (d) of 5 mM native β-CD. Reproduced with permission from Ling XY, Malaquin L, Reinhoudt DN, Wolf H, and Huskens J (2007) An in situ study of the adsorption behavior of functionalized particles on self-assembled monolayers via different chemical interactions. *Langmuir* 23: 9990–9999.

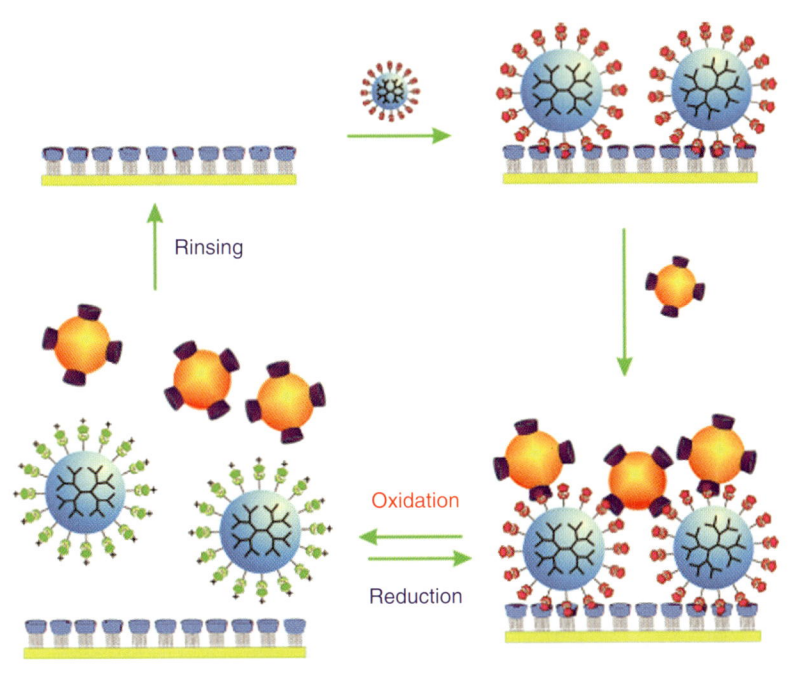

Figure 44 Illustration of the adsorption and desorption of β-CD-functionalized nanoparticles onto and from β-CD SAM with Fc dendrimers as a reversible supramolecular glue.

disassembly of the reversible glue and the NPs off the interface, and (3) possible size effects of the NPs on the reversibility. The repeated adsorption and desorption of β-CD-functionalized gold NPs onto and from the surface were monitored *in situ* by a combined SPR spectroscopy and electrochemistry setup [47]. These experiments proved that the adsorption and desorption of the material deposited on the printboard follow the scheme depicted in **Figure 44**. The layer was repeatedly reduced and oxidized during 50 cycles in a CV experiment, proving that the system is highly reversible. It was also possible to remove the assemblies locally. As a proof of principle, an area defined by an O-ring was exposed to the electrolyte with the counter electrode, whereas the area outside the ring remained dry. The system underwent oxidation only in this restricted area.

7.8 Probing Single-Molecule Interaction by AFM

A fascinating application of the molecular printboard is the probing of individual host–guest interactions by dynamic single-molecule spectroscopy [38, 39, 71]. A silicon nitride AFM tip, covered with 50–70-nm layer of gold, was functionalized with guest molecules provided with a thiol chain, such as the 6-ferrocenyl-hexanethiol. If the functionalization of the tip was carried out in high dilution, the number of guest molecules actually bound to the tip was very low. When the guest-functionalized AFM tip starts approaching the CD monolayer, no force is experienced (see **Figure 45**). As the tip gets closer, the guest molecule(s) bind(s) to the printboard (snap-on) and afterwards the force increases linearly. In the retracting phase, the force goes back to the background value and thereafter it becomes negative and decreases with a linear trend until the pull-off event occurs.

A histogram of the pull-off forces revealed quantization and the values corresponding to the different peaks were multiples of a single force value, which is attributed to the dissociation of a single host–guest pair. This result can be explained by the binding of different numbers of guest molecules at the tip to the β-CD monolayer (**Figure 46**).

The use of different guests such as the adamantane and *t*-butyl-phenyl derivatives (**Figure 47**) shows that the binding forces on the molecular printboard follow the same trend as the association constants measured for model guest compounds with the β-CD in aqueous solution [26]. A model

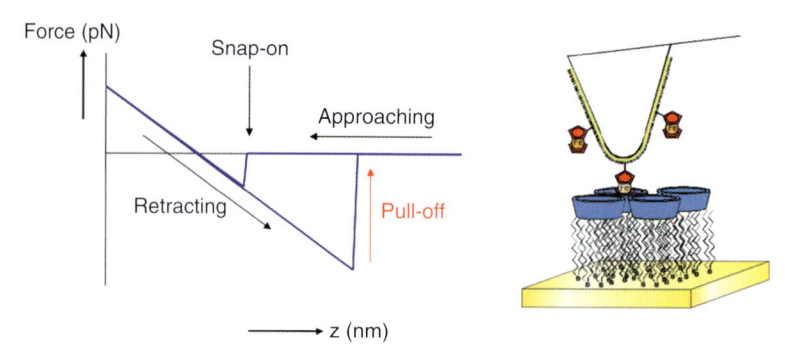

Figure 45 Plot of the force vs. the tip distance in the approaching and retracting stages (left). Schematic representation of an AFM tip functionalized with ferrocene units interacting with the β-CD monolayer (right).

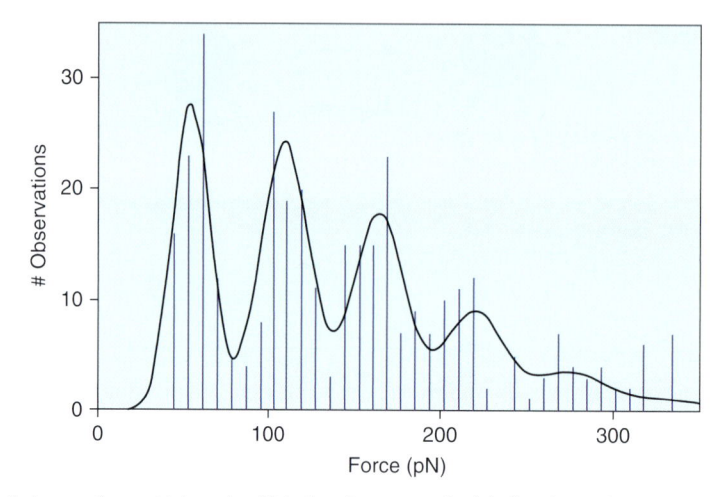

Figure 46 Number of observations of intervals of binding forces vs. the binding force for dynamic single-molecule force spectroscopy experiments carried out with an AFM tip, functionalized with ferrocene units, on the molecular printboard.

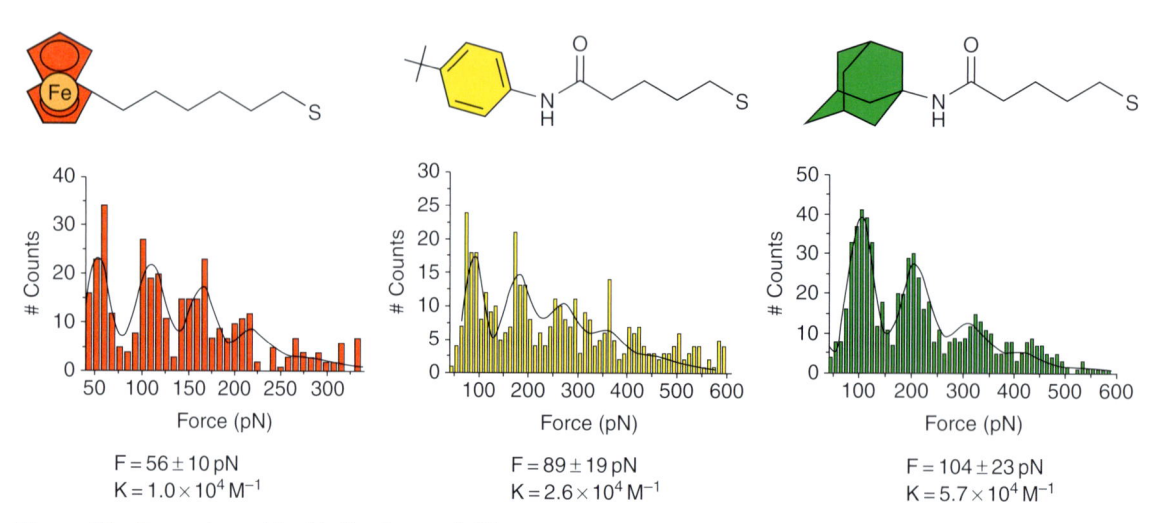

$F = 56 \pm 10$ pN
$K = 1.0 \times 10^4$ M^{-1}

$F = 89 \pm 19$ pN
$K = 2.6 \times 10^4$ M^{-1}

$F = 104 \pm 23$ pN
$K = 5.7 \times 10^4$ M^{-1}

Figure 47 Comparison of the binding forces of different guests and correspondent values of the binding constants with β-cyclodextrin in water.

was developed to quantitatively relate the binding forces to the association constants [39].

7.9 Conclusions

All the different systems discussed in this article show that multivalency is an important tool for the understanding of the assembly of nonbiological nanostructures at surfaces. Molecular printboards are a tool that allows multivalency on surfaces exploiting supramolecular host–guest interactions.

The β-CD SAM has several important advantages: (1) it can be prepared in a relatively easy way; (2) the monolayer is ordered and densely packed and this allows the multivalent guests to interact with it using all the binding units (adamantyl, ferrocenyl, etc.) that are accessible; (3) it is sufficiently stable to carry out fabrication schemes on it and is resistant to several chemical, physical, and electrochemical treatments; and (4) it is water compatible and this feature opens up the possibility to use the printboard for the immobilization of proteins, ABs, and peptides.

At the beginning of the article, we discussed how molecules can be immobilized on the β-CD SAM. It is possible to fine-tune the strength of the interactions changing the number of binding moieties on the guests. μCP and DPN offer the possibility to immobilize guest molecules on very small areas. The use of electroactive dendrimers functionalized with ferrocenyl units provides the possibility of controlled reversible binding and helps to understand the binding stoichiometry of the dendrimers on the molecular printboard.

By increasing the level of complexity, it is possible to create complex assemblies on the printboard using step-by-step procedures. In the case of the immobilization of cyt c [105], it is shown that they can keep their biological activity when immobilized on the surface.

Furthermore, using dendrimers and functionalized (nano)particles, assemblies were fabricated with an LBL procedure. The combination of NIL, for 2D control, and LBL assembly, for height control, is a powerful method for the fabrication of 3D nanostructures. Dynamic single-molecule force spectroscopy on the molecular printboard is a valuable method to measure the interaction strength at a single-molecule level.

References

1. Mammen M, Choi SK, and Whitesides GM (1998) Polyvalent interactions in biological systems: Implications for design and use of multivalent ligands and inhibitors. *Angewandte Chemie (International ed. in English)* 37: 2755–2794.
2. Mulder A, Huskens J, and Reinhoudt DN (2004) Multivalency in supramolecular chemistry and nanofabrication. *Organic and Biomolecular Chemistry* 2: 3409–3424.
3. Huskens J (2006) Multivalent interactions at interfaces. *Current Opinion in Chemical Biology* 10: 537–543.
4. Badjic JD, Nelson A, Cantrill SJ, Turnbull WB, and Stoddart JF (2005) Multivalency and cooperativity in supramolecular chemistry. *Accounts of Chemical Research* 38: 723–732.
5. Mandolini L (1986) Intramolecular reactions of chain molecules. *Advances in Physical Organic Chemistry* 22: 1–111.
6. Galli C and Mandolini L (2000) The role of ring strain on the ease of ring closure of bifunctional chain molecules. *European Journal of Organic Chemistry* 2000(18): 3117–3125.
7. Cacciapaglia R, Di Stefano S, and Mandolini L (2004) Effective molarities in supramolecular catalysis of two-substrate reactions. *Accounts of Chemical Research* 37: 113–122.
8. Mulder A, Auletta T, Sartori A, et al. (2004) Divalent binding of a bis(adamantyl)-functionalized calix[4]arene to beta-cyclodextrin-based hosts: An experimental and theoretical study on multivalent binding in solution and at self-assembled monolayers. *Journal of the American Chemical Society* 126: 6627–6636.
9. Lehn JM (1988) Supramolecular chemistry – scope and perspectives molecules, supermolecules, and molecular devices. *Angewandte Chemie (International ed. in English)* 27: 89–112.
10. Rudkevich DM, Brzozka Z, Palys M, Visser HC, Verboom W, and Reinhoudt DN (1994) A difunctional receptor for the simultaneous complexation of anions and cations – recognition of Kh2po4. *Angewandte Chemie (International ed. in English)* 33: 467–468.
11. Ulman A (1991) *An Introduction to Ultrathin Organic Films*. San Diego, CA: Academic Press.
12. Bard AJ and Rubinstein I (1996) *Electroanalytical Chemistry*. New York: Marcel Dekker.
13. Troughton EB, Bain CD, Whitesides GM, Nuzzo RG, Allara DL, and Porter MD (1988) Monolayer films prepared by the spontaneous self-assembly of symmetrical and unsymmetrical dialkyl sulfides from solution onto gold substrates – structure, properties, and reactivity of constituent functional-groups. *Langmuir* 4: 365–385.
14. Steinberg S and Rubinstein I (1992) Ion-selective monolayer membranes based upon self-assembling tetradentate ligand monolayers on gold electrodes. 3. Application as selective ion sensors. *Langmuir* 8: 1183–1187.
15. Hagenhoff B, Benninghoven A, Spinke J, Liley M, and Knoll W (1993) Time-of-flight secondary-ion mass-spectrometry investigations of self-assembled monolayers of organic thiols, sulfides, and disulfides on gold surfaces. *Langmuir* 9: 1622–1624.
16. Zhang MH and Anderson MR (1994) Investigation of the charge-transfer properties of electrodes modified by the spontaneous adsorption of unsymmetrical dialkyl sulfides. *Langmuir* 10: 2807–2813.

17. Thoden van Velzen EU, Engbersen JFJ, and Reinhoudt DN (1994) Self-assembled monolayers of receptor adsorbates on gold – preparation and characterization. *Journal of the American Chemical Society* 116: 3597–3598.
18. Broos J, Engbersen JFJ, Verboom W, and Reinhoudt DN (1995) Inversion of enantioselectivity of serine proteases. *Recueil des Travaux Chimiques des Pays-Bas* 114: 255–257.
19. Schonherr H, Vancso GJ, Huisman BH, vanVeggel FCJM, and Reinhoudt DN (1997) An atomic force microscopy study of self-assembled monolayers of calix[4]resorcinarene adsorbates on Au(111). *Langmuir* 13: 1567–1570.
20. Thoden van Velzen EU, Engbersen JFJ, and Reinhoudt DN (1995) Synthesis of self-assembling resorcin[4]arene tetrasulfide adsorbates. *Synthesis* 1995(8): 989–997.
21. Huisman BH, Thoden van Velzen EU, Vanveggel FCJM, Engbersen JFJ, and Reinhoudt DN (1995) Self-assembled monolayers of calix[4]arene derivatives on gold. *Tetrahedron Letters* 36: 3273–3276.
22. Huisman BH, Rudkevich DM, vanVeggel FCJM, and Reinhoudt DN (1996) Self-assembled monolayers of carceplexes on gold. *Journal of the American Chemical Society* 118: 3523–3524.
23. Schierbaum KD, Weiss T, Thoden van Velzen EU, Engbersen JFJ, Reinhoudt DN, and Gopel W (1994) Molecular recognition by self-assembled monolayers of cavitand receptors. *Science* 265: 1413–1415.
24. Huisman BH, Kooyman RPH, vanVeggel FCJM, and Reinhoudt DN (1996) Molecular recognition by self-assembled monolayers detected with surface plasmon resonance. *Advanced Matererials* 8: 561–564.
25. Szejtli J and Osa T (1996) *Comprehensive Supramolecular Chemistry, Vol. 3: Cyclodextrins*. Oxford: Elsevier Science.
26. de Jong MR, Huskens J, and Reinhoudt DN (2001) Influencing the binding selectivity of self-assembled cyclodextrin monolayers on gold through their architecture. *Chemistry – A European Journal* 7: 4164–4170.
27. Rekharsky MV and Inoue Y (1998) Complexation thermodynamics of cyclodextrins. *Chemical Reviews* 98: 1875–1917.
28. Rojas MT, Koniger R, Stoddart JF, and Kaifer AE (1995) Supported monolayers containing preformed binding-sites – synthesis and interfacial binding-properties of a thiolated beta-cyclodextrin derivative. *Journal of the American Chemical Society* 117: 336–343.
29. Weisser M, Nelles G, Wohlfart P, Wenz G, and MittlerNeher S (1996) Immobilization kinetics of cyclodextrins at gold surfaces. *Journal of Physical Chemistry* 100: 17893–17900.
30. Nelles G, Weisser M, Back R, Wohlfart P, Wenz G, and MittlerNeher S (1996) Controlled orientation of cyclodextrin derivatives immobilized on gold surfaces. *Journal of the American Chemical Society* 118: 5039–5046.
31. Weisser M, Nelles G, Wenz G, and MittlerNeher S (1997) Guest–host interactions with immobilized cyclodextrins. *Sensors and Actuators B* 38: 58–67.
32. Henke C, Steinem C, Janshoff A, et al. (1996) Self-assembled monolayers of monofunctionalized cyclodextrins onto gold: A mass spectrometric characterization and impedance analysis of host–guest interaction. *Analytical Chemistry* 68: 3158–3165.
33. He PG, Ye YN, Fang YZ, Suzuki I, and Osa T (1997) Organized self-assembled lipoyl-beta-cyclodextrin derivative monolayer on a gold electrode. *Electroanalysis* 9: 68–73.
34. He PA, Ye JN, Fang YH, Suzuki I, and Osa T (1997) Voltammetric responsive sensors for organic compounds based on organized self-assembled lipoyl-beta-cyclodextrin derivative monolayer on a gold electrode. *Analytica Chimica Acta* 337: 217–223.
35. Lahav M, Ranjit KT, Katz E, and Willner I (1997) A beta-amino-cyclodextrin monolayer-modified Au electrode: A command surface for the amperometric and microgravimetric transduction of optical signals recorded by a photoisomerizable bipyridinium-azobenzene diad. *Chemical Communications* 1997(3): 259–260.
36. Qian X, Hentschke R, and Knoll W (1997) Superstructures of cyclodextrin derivatives on Au(111): A combined random planting-molecular dynamics approach. *Langmuir* 13: 7092–7098.
37. Beulen MWJ, Bugler J, Lammerink B, et al. (1998) Self-assembled monolayers of heptapodant beta-cyclodextrins on gold. *Langmuir* 14: 6424–6429.
38. Beulen MWJ, Bugler J, de Jong MR, et al. (2000) Host–guest interactions at self-assembled monolayers of cyclodextrins on gold. *Chemistry – A European Journal* 6: 1176–1183.
39. Auletta T, de Jong MR, Mulder A, et al. (2004) Beta-cyclodextrin host–guest complexes probed under thermodynamic equilibrium: Thermodynamics and AFM force spectroscopy. *Journal of the American Chemical Society* 126: 1577–1584.
40. Thompson D and Larsson JA (2006) Modeling competitive guest binding to beta-cyclodextrin molecular printboards. *Journal of Physical Chemistry B* 110: 16640–16645.
41. Thompson D (2007) In silico engineering of tailored ink-binding ability at molecular printboards. *ChemPhysChem* 8: 1684–1693.
42. Huskens J, Deij MA, and Reinhoudt DN (2002) Attachment of molecules at a molecular printboard by multiple host–guest interactions. *Angewandte Chemie (International ed. in English)* 41: 4467–4471.
43. Auletta T, Dordi B, Mulder A, et al. (2004) Writing patterns of molecules on molecular printboards. *Angewandte Chemie (International ed. in English)* 116: 373–377.
44. Rao JH, Yan L, Lahiri J, Whitesides GM, Weis RM, and Warren HS (1999) Binding of a dimeric derivative of vancomycin to L-Lys-D-Ala-D-lactate in solution and at a surface. *Chemistry and Biology* 6: 353–359.
45. Yan L, Marzolin C, Terfort A, and Whitesides GM (1997) Formation and reaction of interchain carboxylic anhydride groups on self-assembled monolayers on gold. *Langmuir* 13: 6704–6712.
46. Nijhuis CA, Huskens J, and Reinhoudt DN (2004) Binding control and stoichiometry of ferrocenyl dendrimers at a molecular printboard. *Journal of the American Chemical Society* 126: 12266–12267.
47. Nijhuis CA, Yu F, Knoll W, Huskens J, and Reinhoudt DN (2005) Multivalent dendrimers at molecular printboards: Influence of dendrimer structure on binding strength and stoichiometry and their electrochemically induced desorption. *Langmuir* 21: 7866–7876.
48. Huskens J, Mulder A, Auletta T, Nijhuis CA, Ludden MJ, and Reinhoudt DN (2004) A model for describing the thermodynamics of multivalent host–guest interactions at interfaces. *Journal of the American Chemical Society* 126: 6784–6797.
49. Thompson D (2008) The effective concentration of unbound ink anchors at the molecular printboard. *Journal of Physical Chemistry B* 112: 4994–4999.
50. Cieplak M and Thompson D (2008) Coarse-grained molecular dynamics simulations of nanopatterning with multivalent inks. *Journal of Chemical Physics* 128: 234906-1.

51. Thompson D (2007) Free energy balance predicates dendrimer binding multivalency at molecular printboards. *Langmuir* 23: 8441–8451.
52. Xia YN and Whitesides GM (1998) Soft lithography. *Angewandte Chemie (International ed. in English)* 37: 551–575.
53. Piner RD, Zhu J, Xu F, Hong SH, and Mirkin CA (1999) "Dip-pen" nanolithography. *Science* 283: 661–663.
54. Demers LM, Ginger DS, Park SJ, Li Z, Chung SW, and Mirkin CA (2002) Direct patterning of modified oligonucleotides on metals and insulators by dip-pen nanolithography. *Science* 296: 1836–1838.
55. Lee KB, Park SJ, Mirkin CA, Smith JC, and Mrksich M (2002) Protein nanoarrays generated by dip-pen nanolithography. *Science* 295: 1702–1705.
56. Liu XG, Fu L, Hong SH, Dravid VP, and Mirkin CA (2002) Arrays of magnetic nanoparticles patterned via "dip-pen" nanolithography. *Advanced Materials* 14: 231–234.
57. Mirkin CA, Holliday BJ, Eisenberg AH, et al. (2002) Three dimensional assemblies formed via the weak-link approach. *Abstracts of Papers of the American Chemical Society* 223: A12.
58. Hong SH and Mirkin CA (2000) A nanoplotter with both parallel and serial writing capabilities. *Science* 288: 1808–1811.
59. Bruinink CM, Nijhuis CA, Peter M, et al. (2005) Supramolecular microcontact printing and dip-pen nanolithography on molecular printboards. *Chemistry – A European Journal* 11: 3988–3996.
60. Crespo-Biel O, Peter M, Bruinink CM, Ravoo BJ, Reinhoudt DN, and Huskens J (2005) Multivalent host–guest interactions between beta-cyclodextrin self-assembled monolayers and poly(isobutene-alt-maleic acid)s modified with hydrophobic guest moieties. *Chemistry – A European Journal* 11: 2426–2432.
61. Onclin S, Mulder A, Huskens J, Ravoo BJ, and Reinhoudt DN (2004) Molecular printboards: Monolayers of beta-cyclodextrins on silicon oxide surfaces. *Langmuir* 20: 5460–5466.
62. Mulder A, Onclin S, Peter M, et al. (2005) Molecular printboards on silicon oxide: Lithographic patterning of cyclodextrin monolayers with multivalent, fluorescent guest molecules. *Small* 1: 242–253.
63. Falvey P, Lim CW, Darcy R, et al. (2005) Bilayer vesicles of amphiphilic cyclodextrins: Host membranes that recognize guest molecules. *Chemistry – A European Journal* 11: 1171–1180.
64. Lim CW, Ravoo BJ, and Reinhoudt DN (2005) Dynamic multivalent recognition of cyclodextrin vesicles. *Chemical Communications* 2005: 5627–5629.
65. Onclin S, Huskens J, Ravoo BJ, and Reinhoudt DN (2005) Molecular boxes on a molecular printboard: Encapsulation of anionic dyes in immobilized dendrimers. *Small* 1: 852–857.
66. Corbellini F, Mulder A, Sartori A, et al. (2004) Assembly of a supramolecular capsule on a molecular printboard. *Journal of the American Chemical Society* 126: 17050–17058.
67. Metallo SJ, Kane RS, Holmlin RE, and Whitesides GM (2003) Using bifunctional polymers presenting vancomycin and fluorescein groups to direct anti-fluorescein antibodies to self-assembled monolayers presenting D-alanine-D-alanine groups. *Journal of the American Chemical Society* 125: 4534–4540.
68. Nijhuis CA, Sinha JK, Wittstock G, Huskens J, Ravoo BJ, and Reinhoudt DN (2006) Controlling the supramolecular assembly of redox-active dendrimers at molecular printboards by scanning electrochemical microscopy. *Langmuir* 22: 9770–9775.
69. Nijhuis CA, Boukamp BA, Ravoo BJ, Huskens J, and Reinhoudt DN (2007) Electrochemistry of ferrocenyl dendrimer – beta-cyclodextrin assemblies at the interface of an aqueous solution and a molecular printboard. *Journal of Physical Chemistry C* 111: 9799–9810.
70. Crespo-Biel O, Lim CW, Ravoo BJ, Reinhoudt DN, and Huskens J (2006) Expression of a supramolecular complex at a multivalent interface. *Journal of the American Chemical Society* 128: 17024–17032.
71. Schonherr H, Beulen MWJ, Bugler J, et al. (2000) Individual supramolecular host–guest interactions studied by dynamic single molecule force spectroscopy. *Journal of the American Chemical Society* 122: 4963–4967.
72. Ludden MJW, Peter M, Reinhoudt DN, and Huskens J (2006) Attachment of streptavidin to beta-cyclodextrin molecular printboards via orthogonal host–guest and protein–ligand interactions. *Small* 2: 1192–1202.
73. Ludden MJW and Huskens J (2007) Attachment of proteins to molecular printboards through orthogonal multivalent linkers. *Biochemical Society Transactions* 35: 492–494.
74. Zhu H, Klemic JF, Chang S, et al. (2000) Analysis of yeast protein kinases using protein chips. *Nature Genetics* 26: 283–289.
75. MacBeath G and Schreiber SL (2000) Printing proteins as microarrays for high-throughput function determination. *Science* 289: 1760–1763.
76. Quinn CP, Semenova VA, Elie CM, et al. (2002) Specific, sensitive, and quantitative enzyme-linked immunosorbent assay for human immunoglobulin G antibodies to anthrax toxin protective antigen. *Emerging Infectious Diseases* 8: 1103–1110.
77. Kyo M, Yamamoto T, Motohashi H, et al. (2004) Evaluation of MafG interaction with Maf recognition element arrays by surface plasmon resonance imaging technique. *Genes Cells* 9: 153–164.
78. Herrwerth S, Eck W, Reinhardt S, and Grunze M (2003) Factors that determine the protein resistance of oligoether self-assembled monolayers – internal hydrophilicity, terminal hydrophilicity, and lateral packing density. *Journal of the American Chemical Society* 125: 9359–9366.
79. Prime KL and Whitesides GM (1993) Adsorption of proteins onto surfaces containing end-attached oligo(ethylene oxide) – a model system using self-assembled monolayers. *Journal of the American Chemical Society* 115: 10714–10721.
80. Prime KL and Whitesides GM (1991) Self-assembled organic monolayers – model systems for studying adsorption of proteins at surfaces. *Science* 252: 1164–1167.
81. Ludden MJW, Mulder A, Tampe R, Reinhoudt DN, and Huskens J (2007) Molecular printboards as a general platform for protein immobilization: A supramolecular solution to nonspecific adsorption. *Angewandte Chemie (International ed. in English)* 46: 4104–4107.
82. Lata S and Piehler J (2005) Stable and functional immobilization of histidine-tagged proteins via multivalent chelator headgroups on a molecular poly(ethylene glycol) brush. *Analytical Chemistry* 77: 1096–1105.
83. Tinazli A, Tang JL, Valiokas R, et al. (2005) High-affinity chelator thiols for switchable and oriented immobilization of histidine-tagged proteins: A generic platform for protein chip technologies. *Chemistry – A European Journal* 11: 5249–5259.
84. Ludden MLW, Mulder A, Schulze K, Subramaniam V, Tampe R, and Huskens J (2008) Anchoring of histidine-tagged proteins to molecular printboards: Self-assembly,

thermodynamic modeling, and patterning. *Chemistry – A European Journal* 14: 2044–2051.
85. Turkova J (1999) Oriented immobilization of biologically active proteins as a tool for revealing protein interactions and function. *Journal of Chromatography B* 722: 11–31.
86. Biebricher A, Paul A, Tinnefeld P, Golzhauser A, and Sauer M (2004) Controlled three-dimensional immobilization of biomolecules on chemically patterned surfaces. *Journal of Biotechnology* 112: 97–107.
87. Zhang KC, Diehl MR, and Tirrell DA (2005) Artificial polypeptide scaffold for protein immobilization. *Journal of the American Chemical Society* 127: 10136–10137.
88. Zhu H and Snyder M (2003) Protein chip technology. *Current Opinion in Chemical Biology* 7: 55–63.
89. Rosi NL and Mirkin CA (2005) Nanostructures in biodiagnostics. *Chemical Reviews* 105: 1547–1562.
90. Niemeyer CM (2001) Nanoparticles, proteins, and nucleic acids: Biotechnology meets materials science. *Angewandte Chemie (International ed. in English)* 40: 4128–4158.
91. Tiefenauer L and Ros R (2002) Biointerface analysis on a molecular level – new tools for biosensor research. *Colloids and Surfaces B* 23: 95–114.
92. Su CC, Wu TZ, Chen LK, Yang HH, and Tai DF (2003) Development of immunochips for the detection of dengue viral antigens. *Analytica Chimica Acta* 479: 117–123.
93. Oshannessy DJ, Dobersen MJ, and Quarles RH (1984) A novel procedure for labeling immunoglobulins by conjugation to oligosaccharide moieties. *Immunology Letters* 8: 273–277.
94. Chang IN and Herron JN (1995) Orientation of acid-pretreated antibodies on hydrophobic dichlorodimethylsilane-treated silica surfaces. *Langmuir* 11: 2083–2089.
95. Chang IN, Lin JN, Andrade JD, and Herron JN (1995) Adsorption mechanism of acid pretreated antibodies on dichlorodimethylsilane-treated silica surfaces. *Journal of Colloid and Interface Science* 174: 10–23.
96. Buijs J, White DD, and Norde W (1997) The effect of adsorption on the antigen binding by IgG and its F(ab')(2) fragments. *Colloids and Surfaces B* 8: 239–249.
97. Akerstrom B, Brodin T, Reis K, and Bjorck L (1985) Protein-G – a powerful tool for binding and detection of monoclonal and polyclonal antibodies. *Journal of Immunology* 135: 2589–2592.
98. Juncker D, Schmid H, and Delamarche E (2005) Multipurpose microfluidic probe. *Nature Matererials* 4: 622–628.
99. Hultschig C, Kreutzberger J, Seitz H, Konthur Z, Bussow K, and Lehrach H (2006) Recent advances of protein microarrays. *Current Opinion in Chemical Biology* 10: 4–10.
100. Breslauer DN, Lee PJ, and Lee LP (2006) Microfluidics-based systems biology. *Molecular Biosystems* 2: 97–112.
101. Herr AE, Throckmorton DJ, Davenport AA, and Singh AK (2005) On-chip native gel electrophoresis-based immunoassays for tetanus antibody and toxin. *Analytical Chemistry* 77: 585–590.
102. Ymeti A, Greve J, Lambeck PV, et al. (2007) Fast, ultrasensitive virus detection using a young interferometer sensor. *Nano Letters* 7: 394–397.
103. Ludden MJW, Li X, Greve J, et al. (2008) Assembly of bionanostructures onto beta-cyclodextrin molecular printboards for antibody recognition and lymphocyte cell counting. *Journal of the American Chemical Society* 130: 6964–6973.
104. Ludden MJW, Ling XY, Gang T, et al. (2008) Multivalent binding of small guest molecules and proteins to molecular printboards inside microchannels. *Chemistry – A European Journal* 14: 136–142.
105. Ludden MJW, Sinha JK, Wittstock G, Reinhoudt DN, and Huskens J (2008) Control over binding stoichiometry and specificity in the supramolecular immobilization of cytochrome c on a molecular printboard. *Organic and Biomolecular Chemistry* 6: 1553–1557.
106. Brinkley M (1992) A brief survey of methods for preparing protein conjugates with dyes, haptens, and cross-linking reagents. *Bioconjugate Chemistry* 3: 2–13.
107. Latypov RF, Cheng H, Roder NA, Zhang J, and Roder H (2006) Structural characterization of an equilibrium unfolding intermediate in cytochrome c. *Journal of Molecular Biology* 357: 1009–1025.
108. Crespo-Biel O, Jukovic A, Karlsson M, Reinhoudt DN, and Huskens J (2005) Multivalent aggregation of cyclodextrin gold nanoparticles and adamantyl-terminated guest molecules. *Israel Journal of Chemistry* 45: 353–362.
109. Mahalingam V, Onclin S, Peter M, Ravoo BJ, Huskens J, and Reinhoudt DN (2004) Directed self-assembly of functionalized silica nanoparticles on molecular printboards through multivalent supramolecular interactions. *Langmuir* 20: 11756–11762.
110. Decher G (1997) Fuzzy nanoassemblies: Toward layered polymeric multicomposites. *Science* 277: 1232–1237.
111. Crespo-Biel O, Dordi B, Reinhoudt DN, and Huskens J (2005) Supramolecular layer-by-layer assembly: Alternating adsorptions of guest- and host-functionalized molecules and particles using multivalent supramolecular interactions. *Journal of the American Chemical Society* 127: 7594–7600.
112. Crespo-Biel O, Dordi B, Maury P, Peter M, Reinhoudt DN, and Huskens J (2006) Patterned, hybrid, multilayer nanostructures based on multivalent supramolecular interactions. *Chemistry of Materials* 18: 2545–2551.
113. Maury P, Peter M, Mahalingam V, Reinhoudt DN, and Huskens J (2005) Patterned self-assembled monolayers on silicon oxide prepared by nanoimprint lithography and their applications in nanofabrication. *Advanced Functional Materials* 15: 451–457.
114. Maury P, Escalante M, Reinhoudt DN, and Huskens J (2005) Directed assembly of nanoparticles onto polymer-imprinted or chemically patterned templates fabricated by nanoimprint lithography. *Advanced Materials* 17: 2718–2723.
115. Arsenault A, Fournier-Bidoz SB, Hatton B, et al. (2004) Towards the synthetic all-optical computer: Science fiction or reality? *Journal of Materials Chemistry* 14: 781–794.
116. Kharitonov AB, Shipway AN, Katz E, and Willner I (1999) Gold-nanoparticle/bis-bipyridinium cyclophane-functionalized ion-sensitive field-effect transistors: Novel assemblies for the sensing of neurotransmitters. *Reviews in Analytical Chemistry* 18: 255–260.
117. Murray CB, Kagan CR, and Bawendi MG (1995) Self-organization of Cdse nanocrystallites into 3-dimensional quantum-dot superlattices. *Science* 270: 1335–1338.
118. Gates B, Qin D, and Xia YN (1999) Assembly of nanoparticles into opaline structures over large areas. *Advanced Materials* 11: 466–469.
119. Ozin GA and Yang SM (2001) The race for the photonic chip: Colloidal crystal assembly in silicon wafers. *Advanced Functional Materials* 11: 95–104.
120. vanBlaaderen A, Ruel R, and Wiltzius P (1997) Template-directed colloidal crystallization. *Nature* 385: 321–324.
121. Zhang J, Alsayed A, Lin KH, et al. (2002) Template-directed convective assembly of three-dimensional face-centered-cubic colloidal crystals. *Applied Physics Letters* 81: 3176–3178.

122. Denkov ND, Velev OD, Kralchevsky PA, Ivanov IB, Yoshimura H, and Nagayama K (1993) 2-Dimensional crystallization. *Nature* 361: 26.
123. Paunov VN, Kralchevsky PA, Denkov ND, and Nagayama K (1993) Lateral capillary forces between floating submillimeter particles. *Journal of Colloid and Interface Science* 157: 100–112.
124. Decher G, Hong JD, and Schmitt J (1992) Buildup of ultrathin multilayer films by a self-assembly process. 3. Consecutively alternating adsorption of anionic and cationic polyelectrolytes on charged surfaces. *Thin Solid Films* 210: 831–835.
125. Kiely CJ, Fink J, Brust M, Bethell D, and Schiffrin DJ (1998) Spontaneous ordering of bimodal ensembles of nanoscopic gold clusters. *Nature* 396: 444–446.
126. Ling XY, Malaquin L, Reinhoudt DN, Wolf H, and Huskens J (2007) An *in situ* study of the adsorption behavior of functionalized particles on self-assembled monolayers via different chemical interactions. *Langmuir* 23: 9990–9999.
127. Tang CS, Schmutz P, Petronis S, Textor M, Keller B, and Voros J (2005) Locally addressable electrochemical patterning technique (LAEPT) applied to poly (L-lysine)-graft-poly(ethylene glycol) adlayers on titanium and silicon oxide surfaces. *Biotechnology and Bioengineering* 91: 285–295.
128. Nijhuis CA, Yu F, Knoll W, Huskens J, and Reinhoudt DN (2005) Multivalent dendrimers at molecular printboards: Influence of dendrimer structure on binding strength and stoichiometry and their electrochemically induced desorption. *Langmuir* 21: 7866–7876.
129. Nijhuis CA, Dolatowska KA, Ravoo BJ, Huskens J, and Reinhoudt DN (2007) Redox-controlled interaction of biferrocenyl-terminated dendrimers with beta-cyclodextrin molecular printboards. *Chemistry – A European Journal* 13: 69–80.
130. Ling XY, Reinhoudt DN, and Huskens J (2008) Reversible attachment of nanostructures at molecular printboards through supramolecular glue. *Chemistry of Materials* 20: 3574–3578.

8 Molecular Machines and Motors

A Credi, Università di Bologna, Bologna, Italy

© 2010 Elsevier B.V. All rights reserved.

8.1 Introduction

Movement is one of life's central attributes. Nature provides living systems with complex molecules called 'motor proteins', which work inside a cell much the same as ordinary machines built for everyday needs [1,2]. The development of civilization has always been strictly related to the design and construction of devices, from wheel to jet engine, capable of facilitating human movement and traveling. Nowadays, the miniaturization race leads scientists to investigate the possibility of designing and constructing motors and machines at the nanometer scale, that is, at the molecular level. Chemists, by the nature of their discipline, are able to manipulate atoms and molecules and are therefore in the ideal position to develop bottom–up strategies for the construction of nanoscale devices.

Natural molecular motors are extremely complex systems [3]; their structures and detailed working mechanisms have been elucidated only in a few cases and any attempt to construct systems of such a complexity by using the bottom–up molecular approach would be prohibitive. What can be done, at present, in the field of artificial molecular motors [4,5] is to construct simple prototypes consisting of a few molecular components, capable of moving in a controllable way, and to investigate the challenging problems posed by interfacing artificial molecular devices with the macroscopic world, particularly as far as energy supply and information exchange are concerned [6,7]. The study of motion at the molecular level is, undoubtedly, a fascinating topic from the viewpoint of basic research and a promising field for novel applications.

The article starts with an introduction to the concepts of molecular motors and machines, and a description of the bottom–up (i.e., supramolecular) approach to their construction. The characteristics of molecular machines are then discussed with reference to those of macroscopic ones. In the next section, the working principles of some natural molecular motors are illustrated. The subsequent section deals with artificial molecular motors and machines. For space reasons, only selected examples, often taken from our own research, are described. However, other relevant contributions in this field are mentioned in the text or listed in the references. The next section illustrates a few hybrid systems, that is, obtained either by suitable engineering of motor proteins or by integrating natural and artificial molecular devices. In the penultimate section, a few remarkable research achievements toward the utilization of nanomachines in practical devices are presented. In the final section, perspectives and limitations of this kind of systems are discussed.

8.1.1 The Bottom–Up (Supramolecular) Approach to Nanodevices

A device can be very big or very small, depending on the purpose of its use. In the last 50 years, progressive miniaturization of the components employed for the construction of devices and machines has resulted in outstanding technological achievements, particularly in the field of information processing. A common prediction is that further progress in miniaturization will not only decrease the size and increase the power of computers, but could also open the way to new technologies in the fields of medicine, environment, energy, and materials.

Until now, miniaturization has been pursued by a large-downward (top-down) approach, which is reaching practical and fundamental limits [8]. Miniaturization, however, can be pushed further on, as Richard Feynman [9] stated in his famous talk to the American Physical Society in 1959, "there is plenty of room at the bottom". The key sentence of Feynman's talk was the following: "The principle of physics do not speak against the possibility of maneuvering things atom by atom." The idea of the atom-by-atom bottom–up approach to the construction [10] of nanoscale devices and machines, however, did not convince chemists who are well aware of the high reactivity of most atomic species and of the subtle aspects of chemical bond. Chemists know [11] that atoms are not simple spheres that can be moved from one place to another at will. Atoms do not stay isolated; they bond strongly to their neighbors and it is difficult to imagine that the atoms can be taken from a starting material and transferred to another material.

In the late 1970s a new branch of chemistry, called 'supramolecular chemistry', emerged and expanded very rapidly. In the frame of research on supramolecular chemistry, the idea began to arise in a few laboratories [12–14] that molecules are much more convenient building blocks than atoms to construct nanoscale devices and machines. The main reasons at the basis of this idea are: (1) molecules are stable species, whereas atoms are difficult to handle; (2) nature starts from molecules, not from atoms, to construct the great number and variety of nanodevices and nanomachines that sustain life; (3) most of the laboratory chemical processes deal with molecules, not with atoms; (4) molecules are objects that exhibit distinct shapes and carry device-related properties (e.g., properties that can be manipulated by photochemical and electrochemical inputs); and (5) molecules can self-assemble or can be connected to make larger structures. In the same period, research on molecular electronic devices began to flourish [15].

In the following years, supramolecular chemistry grew very rapidly [16,17] and it became clear that the bottom–up approach based on molecules opens virtually unlimited possibilities concerning design and construction of artificial molecular devices and machines. Recently, the concept of molecules as nanoscale objects exhibiting their own shape, size, and properties has been confirmed by new, very powerful techniques, such as single-molecule fluorescence spectroscopy and the various types of probe microscopies [18], capable of visualizing and manipulating single molecules, and even to investigate bimolecular chemical reactions at the single-molecule level.

Much of the inspiration to construct molecular devices and machines comes from the outstanding progress of molecular biology that has begun to reveal the secrets of the natural nanodevices, which constitute the material base of life [1–3]. Obviously, chemists have started to construct much simpler systems, without mimicking the complexity of the biological structures. In the last few years, synthetic talent, that has always been the most distinctive feature of chemists, combined with a device-driven ingenuity, evolved from chemists' attention to functions and reactivity, has led to outstanding achievements in this field. Among the systems reported are molecular tweezers [19], propellers [20], rotors [21], turnstiles [22], gyroscopes [23,24], gears[25], brakes [26], scissors [27], pedals [28], drive trains [29], ratchets [30], rotary motors [31], shuttles [32], elevators [33], muscles [34], valves [35], artificial processive enzymes [36], walkers [37], vehicles [38], and catalytic self-propelled micro- and nano-objects [39,40]. Several excellent reviews, [5–7, 41–48] thematic issues, [49–53], and a monograph [4] dealing with artificial molecular machines and motors are available.

8.1.2 Basic Concepts on Molecular Machines

8.1.2.1 Terms and definitions

In the macroscopic world, devices and machines are assemblies of components designed to achieve a specific function. Each component of the assembly performs a simple act, while the entire assembly performs a more complex, useful function, characteristic of that particular device or machine. In principle, the macroscopic concepts of a device and a machine can be extended to the molecular level [4]. A 'molecular device' can be defined as an assembly of a discrete number of molecular components designed to achieve a specific function. Each molecular component performs a single act, while the entire supramolecular assembly performs a more complex function, which results from the cooperation of the various components.

Nature shows, however, that nanoscale devices and machines can hardly be considered as shrunk versions of macroscopic counterparts because several intrinsic properties of molecular-level entities are quite different from those of macroscopic objects [1]. In fact, the design and construction of artificial molecular machines can take greater benefit from the knowledge of the working principles of natural ones rather than from sheer attempts to apply at the nanoscale macroscopic engineering principles [54]. Biomolecular machines are made of nanometer-size floppy molecules that operate at constant temperature in the soft and chaotic environment produced by the weak intermolecular forces and the ceaseless and random molecular movements. Gravity and inertia motions, which we are familiar with in our everyday experience, are fully negligible at the molecular scale; viscous forces resulting from intermolecular interactions (including those with solvent water molecules) largely prevail and it is difficult to obtain directed motion. This means that while one can describe the bottom–up construction of a nanoscale device as an assembly of suitable (molecular) components by analogy with what happens in the macroscopic world, it should be emphasized that the design principles and the operating mechanisms at the molecular level are different.

Owing to the above reasons, it is not easy to define the functions related to artificial molecular motions. A simple and immediate categorization is usually based on an iconic comparison with motions taking place in macroscopic systems (e.g., braking, locking, shuttling, and rotating). Such a comparison not only presents the advantage of an easy representation of molecular devices by cartoons that clearly explain their mechanical functions, but it also implies the danger of overlooking the substantial differences between the macroscopic and molecular worlds. The following minimal set of terms and definitions may be suggested:

- *Mechanical device.* A particular type of device designed to perform mechanical movements.
- *Machine.* A particular type of mechanical device designed to perform a specific mechanical movement under the action of a defined energy input.
- *Motor.* A machine capable of using an energy input to produce useful work.

Clearly, there is a hierarchy: a motor is also a machine, and a machine, in turn, is also a mechanical device, but a mechanical device might not be a machine or a motor and a machine might not be a motor. It should also be noted that mechanical movements at the molecular level result from nuclear motions caused by chemical reactions. Any kind of chemical reaction involves, of course, some nuclear displacement; however, in the present context, only large-amplitude, nontrivial motions leading to real translocation of some component parts of the system should be considered.

It is also useful to discuss briefly the relation between molecular switches, and molecular machines and motors. A switch is a multistate system whose properties and effects on the environment are a function of its state [55]. Most often, the interconversion between two given states of a molecular switch can take place by the same pathway that is traveled in opposite directions (**Figure 1(a)**). In this case, any mechanical effect exerted on an external system is canceled out when the switch returns to its original state. Switches exist, however, in which the forward and backward transitions between a pair of states follow different pathways. A typical example is provided by a rotary device undergoing a 360° unidirectional rotation through two directionally correlated half-rotations [31,56] (**Figure 1(b)**). Switches of this kind can influence a system as a function of their switching trajectory, and a physical task performed in a cycle is not inherently undone. This is a fundamental requirement if a molecular motor has to be constructed. Therefore, generally speaking, molecular machines are also switches, whose states differ from one another for the relative positioning of the various molecular components. However, to behave as motors, the above-described additional feature is required. Noteworthy, the vast majority of artificial molecular machines reported so far does not exhibit such a behavior and are therefore switches, but not motors. A more thorough discussion on this topic can be found in Ref. [5].

8.1.2.2 Energy supply and monitoring signals

As it happens in the macroscopic world, molecular-level devices and machines need energy to operate and signals to communicate with the operator [57]. The most obvious way to supply energy to a chemical system is through an exergonic chemical reaction. Not surprisingly, the majority of the molecular motors of the biological world are powered by chemical reactions (e.g., adenosine triphosphate (ATP)

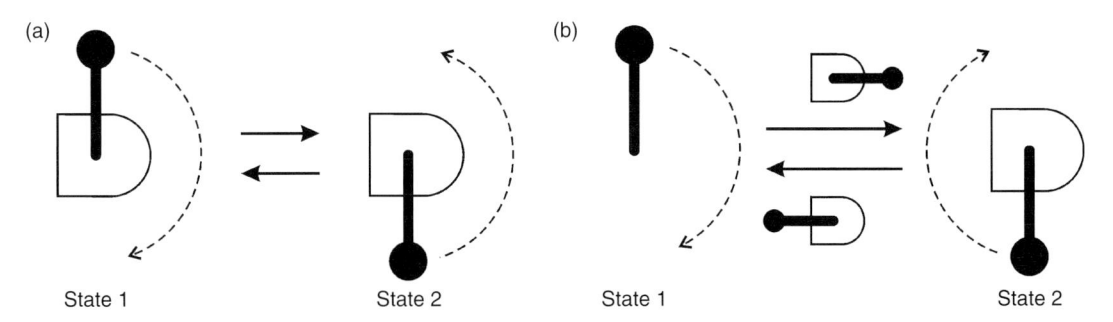

Figure 1 (a) Rotary switch: the interconversion between two states can take place by the same pathway traveled in opposite directions; (b) rotary motor: the forward and backward transitions between two states follow different pathways.

hydrolysis) [1–3]. Richard Feynman [9] observed that "an internal combustion engine of molecular size is impossible. Other chemical reactions, liberating energy when cold, can be used instead." This is exactly what happens in our body, where the chemical energy supplied by food is used in long series of slightly exergonic reactions to power the biological machinery that sustains life.

If an artificial molecular machine has to work by inputs of chemical energy, it will need addition of fresh reactants (fuel) at any step of its working cycle, with the concomitant formation of waste products. Accumulation of waste products, however, will compromise the operation of the device unless they are removed from the system, as it happens in our body as well as in macroscopic internal combustion engines. The need to remove waste products introduces noticeable limitations in the design and construction of artificial molecular machines based on chemical fuel inputs.

Chemists have since long known that photochemical and electrochemical energy inputs can cause the occurrence of 'endergonic' and 'reversible reactions'. In the last few years, the outstanding progress made by supramolecular photochemistry [58] and electrochemistry [59] has thus led to the design and construction of molecular machines powered by light or electrical energy that work without formation of waste products.

In the context of artificial nanomachines, light-energy stimulation possesses a number of advantages compared to chemical or electrochemical stimulation. First of all, the amount of energy conferred to a chemical system by using photons can be carefully controlled by the wavelength and intensity of the exciting light, in relation to the absorption spectrum of the targeted species. Such an energy can be transmitted to molecules without physically connecting them to the source (no wiring is necessary), the only requirement being the transparency of the matrix at the excitation wavelength. Other properties of light, such as polarization, can also be utilized. Lasers provide the opportunity of working in very small spaces and extremely short time domains; and near-field techniques allow excitation with nanometer resolution. Conversely, the irradiation of large areas and volumes can be conveniently carried out, thereby allowing the parallel (or even synchronous) addressing of a very high number of individual nanodevices.

Because molecules are extremely small, the observation of motions at the molecular level, which is crucial for monitoring the operation of a molecular machine, is not trivial. In general, the motion of the component parts should cause readable changes in some chemical or physical properties of the system. Photochemical methods are also useful in this regard. As a matter of fact, photons can play with respect to chemical systems the dual role of 'writing' (i.e., causing a change in the system) and 'reading' (i.e., reporting the state of the system) [58]. This is primarily true in nature, where sunlight photons are employed both as energy quanta in photosynthetic processes, and as information elements in vision and other light-triggered processes. For example, luminescence spectroscopy is a valuable method because it is easily accessible and offers good sensitivity and selectivity, along with the possibility of time- and space-resolved studies [60]. In particular, flash spectroscopic techniques with laser excitation allow the study of extremely fast processes.

The use of light to power nanoscale devices is relevant for another important reason. If and when a nanotechnology-based industry will be developed, its products will have to be powered by renewable energy sources, because it has become clear that the problem of energy supply is a crucial one for human civilization for the years ahead [61]. In this frame, the construction of nanodevices, including natural–artificial hybrids that harness solar energy in the form of visible or near-ultraviolet (UV) light, is indeed an important possibility.

8.1.2.3 Other features

In addition to the kind of energy input supplied to make them work and the way of monitoring their operation, molecular machines are characterized by other features such as (1) the type of motion – for example, translation, rotation, and oscillation – performed by their components; (2) the possibility to repeat the operation in cycles; (3) the timescale needed to complete a cycle; and (4) the function performed.

An important property of molecular machines, related to energy supply and cyclic operation, is their capability to exhibit an autonomous behavior; that is, to keep operating, in a constant environment and without the intervention of an external operator, as long as the energy source is available. Natural motors are autonomous, but most of the artificial systems reported so far are not autonomous because, after the mechanical movement induced by a given input, they need another, opposite input to reset. Obviously, the operation of a molecular machine is accompanied by partial degradation of free energy into heat, regardless of the chemical, photochemical, and electrochemical nature of the energy input.

Finally, the functions that can be performed by exploiting the movements of the component parts in molecular machines are various and, to a large extent, still unpredictable. In natural systems, the molecular motions are always aimed at obtaining specific functions, for example, catalysis, transport, and gating. It is worth noting that the changes in the physicochemical properties related to the mechanical movements in molecular machines usually obey binary logic, and can thus be taken as a basis for information processing at the molecular level.

8.2 Natural Systems

In the last few years, much progress has been made in elucidation of the moving mechanisms of motor biomolecules, owing to the fact that – in addition to the established physiological and biochemical methods – novel *in vitro* techniques have been developed, which combine optical and mechanical methods to observe the behavior of a single protein.

The most important and best-known natural molecular motors are myosin, kinesin, and ATP synthase [3]. The enzymes of the myosin and the kinesin families are linear motors that move along polymer substrates (actin filaments for myosin and microtubules for kinesin), converting the energy of ATP hydrolysis into mechanical work [62]. ATP synthase is the ubiquitous enzyme that manufactures ATP and is a rotary motor [63,64].

Several other biological processes are based on motions, including protein folding or unfolding. Another example is RNA polymerase [2], which moves along DNA while carrying out transcriptions, thus acting as a molecular motor. The structure and operation of myosin, kinesin, and ATP synthase are briefly described below.

8.2.1 Myosin and Kinesin

Enzymes such as myosin and kinesin and their relatives are linear motors that move along polymer substrates converting the energy of ATP hydrolysis into mechanical work; myosin moves along actin filaments in muscle and other cells, and kinesin along microtubules [3,62]. Motion is derived from a mechanochemical cycle during which the motor protein binds to successive sites along the substrate in such a way as to move forward on the average.

In the last few years, much progress has been made in elucidation of the moving mechanisms of natural linear motors, particularly of myosin [65–67] and kinesin [68,69]. This progress has been made because, in addition to the established physiological and biochemical methods, novel *in vitro* techniques have been developed which combine optical and mechanical methods to observe the behavior of a single protein. Particularly useful has been the introduction of fluorescence imaging in which the position of an organic dye attached to a motor head is established with single-nanometer precision by determining the center of the emission pattern.

Structurally, myosin and kinesin are dimeric, having two motor heads, two legs, and a common stalk (**Figure 2**). The head regions (7 nm × 4 nm × 4 nm in size for kinesin) bind to actin or microtubule filaments. The motor cannot lift up both feet at the same time or it will fall off the filament and diffuse away. This means that there must be something in the step of a foot that controls the step of the other foot. Another central question is how the two heads are coupled so that the motor can processively move along its track. Two different mechanisms have been proposed [70]. In the hand-over-hand model (**Figure 2**(a)), ATP binding and hydrolysis cause a conformational change in the forward head (h1) that pulls the rear head (h2) forward, while h1 stays attached to the track. In the next step, h2 stays fixed and pulls h1 forward. In the inchworm model (**Figure 2**(b)), only the forward head catalyzes ATP and leads, while the other head follows. The results obtained for kinesin [68,69], miosyn V, [66,67], and myosin VI [65] have shown that these motors move by walking hand over hand in a fashion that resembles the human gait: every swing of a leg traverses twice the net distance traveled by the motor's center of mass. For example, in the case of kinesin, as the stalk took 8-nm steps, the head was observed to take alternating 16-nm and 0-nm steps. Surprisingly, recent results suggest that the ATP gate that controls the stepwise movement of kinesin along a microtubule track operates independently of the microtubule lattice [71].

Motor proteins generate force and transport cargo unidirectionally. The movement of single-motor molecules on their respective tracks under variable ATP concentrations and loads has been studied [3]. Reconstituting motor proteins in their active state *ex vivo* was initially pursued to study the mechanism by which they generate force. More recently, biomolecular motors have been extensively used to power nanodevices in hybrid systems (Section 8.4).

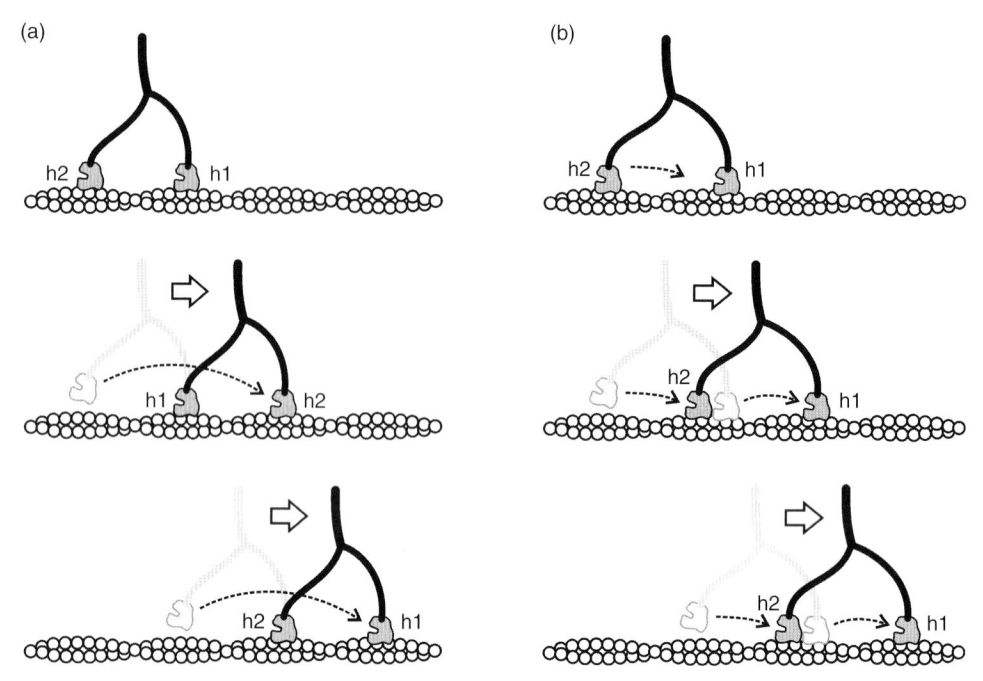

Figure 2 (a) Schematic representation of the hand over hand; (b) inchworm walking mechanisms for natural linear motors [70]. Reproduced with permission from Balzani V, Credi A, and Venturi M (2008) *Molecular Devices and Machines – Concepts and Perspectives for the Nanoworld*, p. 402. Weinheim: Wiley-VCH.

8.2.2 ATP Synthase

This enzyme consists of two rotary molecular motors attached to a common shaft, each attempting to rotate in the opposite direction. The F_1 motor uses the free energy of ATP hydrolysis to rotate in one direction, while the F_0 motor uses the energy stored in a transmembrane electrochemical gradient to turn in the opposite direction. Which motor wins (i.e., develops more torque) depends on cellular conditions. When F_0 takes over, which is the normal situation, it drives the F_1 motor in reverse whereupon it synthesizes ATP from its constituents, adenosine diphosphate (ADP) and inorganic phosphate, P_i. When F_1 dominates, it hydrolyzes ATP and drives the F_0 motor in reverse, turning it into an ion pump that moves ions across the membrane against the electrochemical gradient. The mechanochemistry of ATP synthase has been studied in great detail [72] and new structural information continues to appear [73].

This enzyme consists of two principal domains (**Figure 3**). The asymmetric membrane-spanning F_0 portion contains a proton channel, and the soluble F_1 portion contains three catalytic sites which cooperate in synthetic reactions. The catalytic region is made up of nine protein subunits with the stoichiometry $3\alpha:3\beta:1\gamma:1\delta:1\varepsilon$, approximating to a flattened sphere, 10 nm across and 8 nm high. The flow of protons through F_0 generates a torque which is transmitted to F_1 by an asymmetric shaft, the γ-subunit. This subunit acts as a rotating cam within F_1, sequentially releasing ATPs from the three active sites. The free energy difference across the inner membrane of mitochondria and bacteria is sufficient to produce 3 ATPs per 12 protons passing through the motor.

As mentioned above, the F_0F_1-ATP synthase is reversible, that is, the full enzyme can synthesize or hydrolyze ATP; F_1 in isolation, however, can only hydrolyze it. The spinning of F_1-ATP synthase, that is, the rotary motor nature of this enzyme, was first proposed more than a decade ago [63,64] and directly observed [74] by attaching a fluorescent actin filament to the γ-subunit as a marker. Further data on motor rotation were obtained from other experiments carried out on single molecules of ATP synthase [73,75].

8.3 Artificial Systems

Artificial molecular machines are necessarily built on something much simpler than motor proteins; in principle, they can be designed starting from several

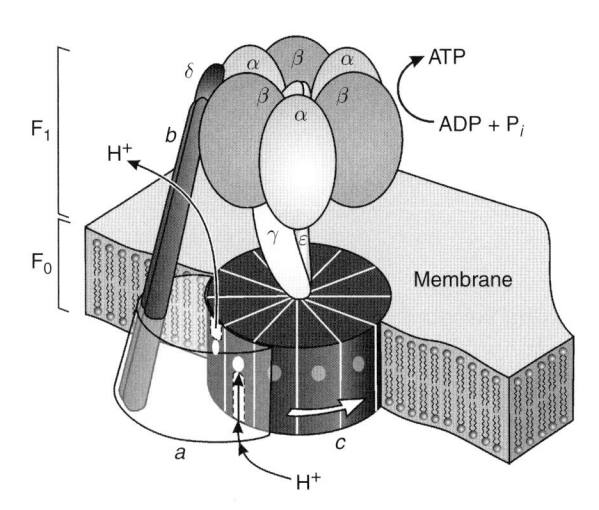

Figure 3 Schematic representation of the structure of F_0F_1-ATP synthase [63, 64]. Reproduced with permission from Balzani V, Credi A, and Venturi M (2008) *Molecular Devices and Machines – Concepts and Perspectives for the Nanoworld*, p. 454. Weinheim: Wiley-VCH.

kinds of molecular and supramolecular species, including DNA [4–7, 19–53]. However, for the reasons described below, most of the systems constructed so far are based on interlocked molecular species, such as rotaxanes, catenanes, and related species [76]. The names of these compounds derive from the Latin words 'rota' and 'axis' for wheel and axle, respectively, and 'catena' for chain. Rotaxanes are minimally composed (**Figures 4(a)** and **4(b)**) of an axle-shaped molecule surrounded by a macrocyclic compound (the 'ring') and terminated by bulky groups ('stoppers') that prevent disassembly. Catenanes are minimally composed of two interlocked macrocyclic compounds (**Figure 4(c)**). Important features of rotaxanes and catenanes derive from noncovalent interactions between the components that contain complementary recognition sites. Such interactions, which are responsible for the efficient template-directed syntheses of these systems [76], include: electron donor–acceptor ability, hydrogen bonding, hydrophobic–hydrophilic character, π–π stacking, electrostatic forces, and, on the side of the strong interaction limit, metal–ligand bonding.

Rotaxanes are appealing systems for the construction of molecular machines because (1) the mechanical bond enables a large variety of mutual arrangements of the molecular components while conferring stability to the system; (2) the interlocked architecture limits the amplitude of the intercomponent motion in the three directions; (3) the stability of a specific arrangement (co-conformation [77]) is determined by the strength of the intercomponent interactions; and (4) such interactions can be modulated by external stimulation. Two interesting molecular motions can be envisaged in rotaxanes, namely translation, that is, shuttling of the ring along the axle (**Figure 4(a)**), and rotation of the ring around the axle (**Figure 4(b)**). Hence, rotaxanes are good prototypes for the construction of both linear and rotary molecular machines. The systems of the first type are termed 'molecular shuttles' [78] and constitute indeed the most common implementation of the molecular machine concept with rotaxanes. In catenanes, the rotation of one molecular ring with respect to the other can be considered (**Figure 4(c)**).

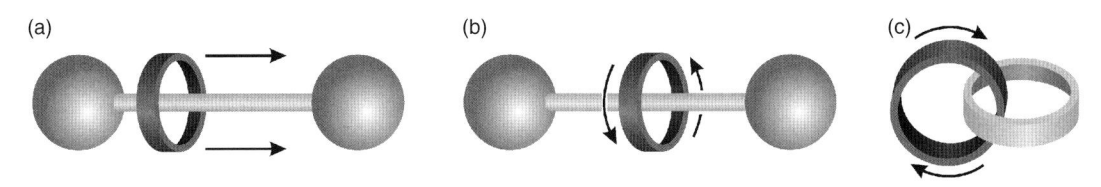

Figure 4 Schematic representation of the intercomponent motions that can be obtained with simple interlocked molecular architectures: (a) ring shuttling in rotaxanes; (b) ring rotation in rotaxanes; (c) catenanes.

Selected examples of molecular machines based on rotaxanes and catenanes are described in the next two sections. Section 8.3.3 deals with rotary motors, that is, molecular machines exhibiting unidirectional rotation. A few examples of nanomachines based on DNA are illustrated in Section 8.3.4.

8.3.1 Molecular Shuttles and Related Species

If, during the template-directed synthesis of a rotaxane, the location of two identical recognition sites ('stations') within its axle component can be arranged (**Figure 5**), a degenerate, co-conformational equilibrium state is obtained in which the macrocyclic component spontaneously shuttles back and forth between the two stations. Compound 1^{4+} (**Figure 6(a)**) is an example of rotaxane that behaves as degenerate molecular shuttle [78].

When the two recognition sites in the axle component are different (**Figure 7**), a rotaxane can exist as two different equilibrating co-conformations, the populations of which reflect their relative free energies as determined primarily by the strengths of the two different sets of noncovalent bonding interactions. In the schematic representation shown in **Figure 7**, it has been assumed that the molecular ring resides preferentially around station A (state 0), until a stimulus is applied that switches off this recognition site (A → A′). As a consequence, the system is brought into a nonequilibrium state that subsequently relaxes (equilibrates) according to the new potential energy landscape. This process implies the motion of the molecular ring to the second recognition site (station B) until a new equilibrium is reached (state 1). If station A is switched on again by an opposite stimulus, the original potential energy landscape is restored, and another co-conformational equilibration occurs through the shuttling of the ring back to station A. In appropriately designed rotaxanes, the switching process can be controlled by reversible chemical reactions (protonation–deprotonation, reduction–oxidation, and isomerization) caused by chemical, electrochemical, or photochemical stimulation. After the first-reported example (compound 2^{4+}, **Figure 6(b)**) [32], a remarkable number of controllable molecular shuttles have been described in the literature [4,5,7,41,42,46–51]. Owing to their bistability, controllable molecular shuttles are also interesting for processing and storing binary information.

It should be noted that a molecular shuttle similar to that shown in **Figure 6(b)** could not perform a net mechanical work in a full cycle of operation because – as for any reversible molecular shuttle – the work done in the forward stroke would be canceled by the backward stroke [5]. As it will be discussed below, more advanced molecular machines and/or a better engineering of their operating environment (e.g., a surface or a membrane) are required to reach this goal.

The energy barrier associated with the shuttling motion can also be modified by modulating the steric hindrance experienced by the ring in the region of the axle comprised between the two stations. This result can be achieved by covalent attachment–detachment of a bulky group [79], metal–ion coordination [80], and photoisomerization [81]. In these systems, the thermodynamic stimulus may not be

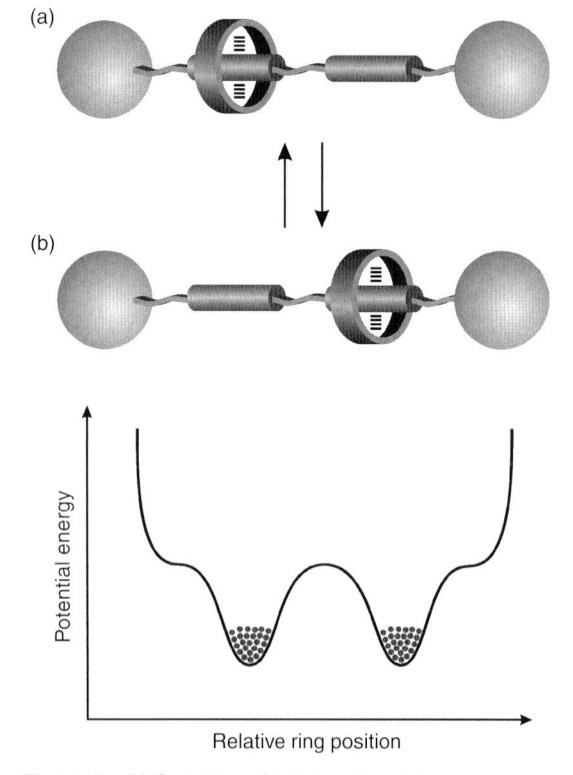

Figure 5 (a) Operation of a two-station rotaxane as a degenerate molecular shuttle; (b) idealized representation of the potential energy of the system as a function of the position of the ring relative to the axle. The number of circles in each potential well reflects the relative population of the corresponding co-conformation in a statistically significant ensemble. Reproduced with permission from Silvi S, Venturi M, and Credi A (2009) *Journal of Materials Chemistry* 19: 2281.

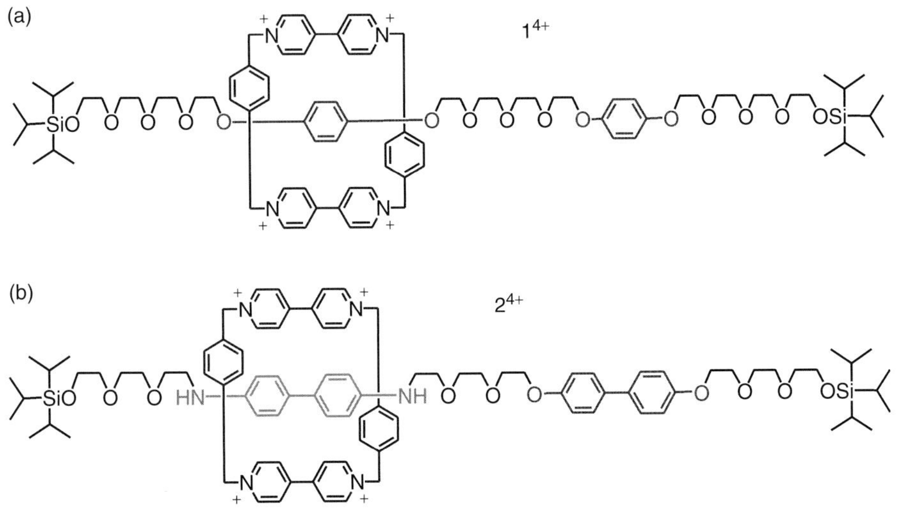

Figure 6 (a) Structure formulas of two-station rotaxanes that behave as a degenerate molecular shuttle; [78] (b) a nondegenerate controllable molecular shuttle [32].

followed by relaxation, leaving the rotaxane molecules into nonequilibrium states (**Figures 7(b) and 7(d)**). This strategy was used to construct rotaxanes in which the axle performs the task of directionally changing the net position of the macrocycle [79]. Such energy-ratcheting effects are crucial for the design of unidirectional molecular transport systems akin to biomolecular motors. Driving the ring distribution away from equilibrium by an information ratchet mechanism was recently reported [81]. For a discussion on ratcheting effects in the context of molecular machines, see Section 8.3.3.3 and Refs. 5, 56, 79, and 82.

8.3.1.1 Chemically driven shuttles, muscles, and elevators

Many examples of rotaxanes whose ring component can be displaced along the molecular axle as a consequence of chemical reactions have been reported [4,5,7,41,42,46–51]. Rotaxane 3^{3+} (**Figure 8(a)**), for example, incorporates a dialkylammonium and a bipyridinium recognition site in its axle component [83]. In acetone or acetonitrile at room temperature, the macrocycle, namely a dibenzo[24]crown-8 (DB24C8), resides exclusively around the ammonium station (**Figure 8(a)**, state 0), as demonstrated by a variety of techniques including ^{1}H nuclear magnetic resonance (NMR) spectroscopy and voltammetry. The preference of the macrocycle for the ammonium site is a result of a combination of [$^+$N—H···O] and [C—H···O] interactions between the $-CH_2NH_2{}^+-$ hydrogen atoms of the axle and the oxygen atoms of the macrocycle. On addition of excess of base (i-Pr$_2$NEt or n-Bu$_3$N) to a solution of 3^{3+}, deprotonation of the ammonium center occurs. As a result, the intercomponent hydrogen bonds are destroyed and the DB24C8 ring shuttles to the bipyridinium station (**Figure 8(a)**, state 1). The original co-conformation is restored by addition of trifluoroacetic acid, because protonation of the ammonium center is followed by the shuttling of the macrocycle back to encircle the ammonium station. The shuttling process in this rotaxane is quantitative and can be followed by ^{1}H NMR spectroscopy and electrochemical techniques. A kinetic study of the shuttling revealed that the base-induced forward step is significantly slower than the acid-induced backward motion of the ring, despite a lower enthalpy of activation for the forward step. This result has been ascribed to entropic factors arising from the rearrangement of counterions in the transition state [84].

A recent exciting development of the chemistry described above is the synthesis of the rotaxane dimer 4^{6+} (**Figure 8(b)**), designed to perform contraction and extension movements under acid–base control [85]. Systems of this kind are referred to as 'molecular muscles' [34,86] because the executed mechanical motion reminds the shortening and lengthening of the functional elements (sarcomeres) present in muscle fibers. An NMR spectroscopic analysis indicates that the initially stable co-conformation for 4^{6+} in acetonitrile at 298 K is the extended one (**Figure 8(b)**, state 0), in which the DB24C8 rings surrounds the ammonium

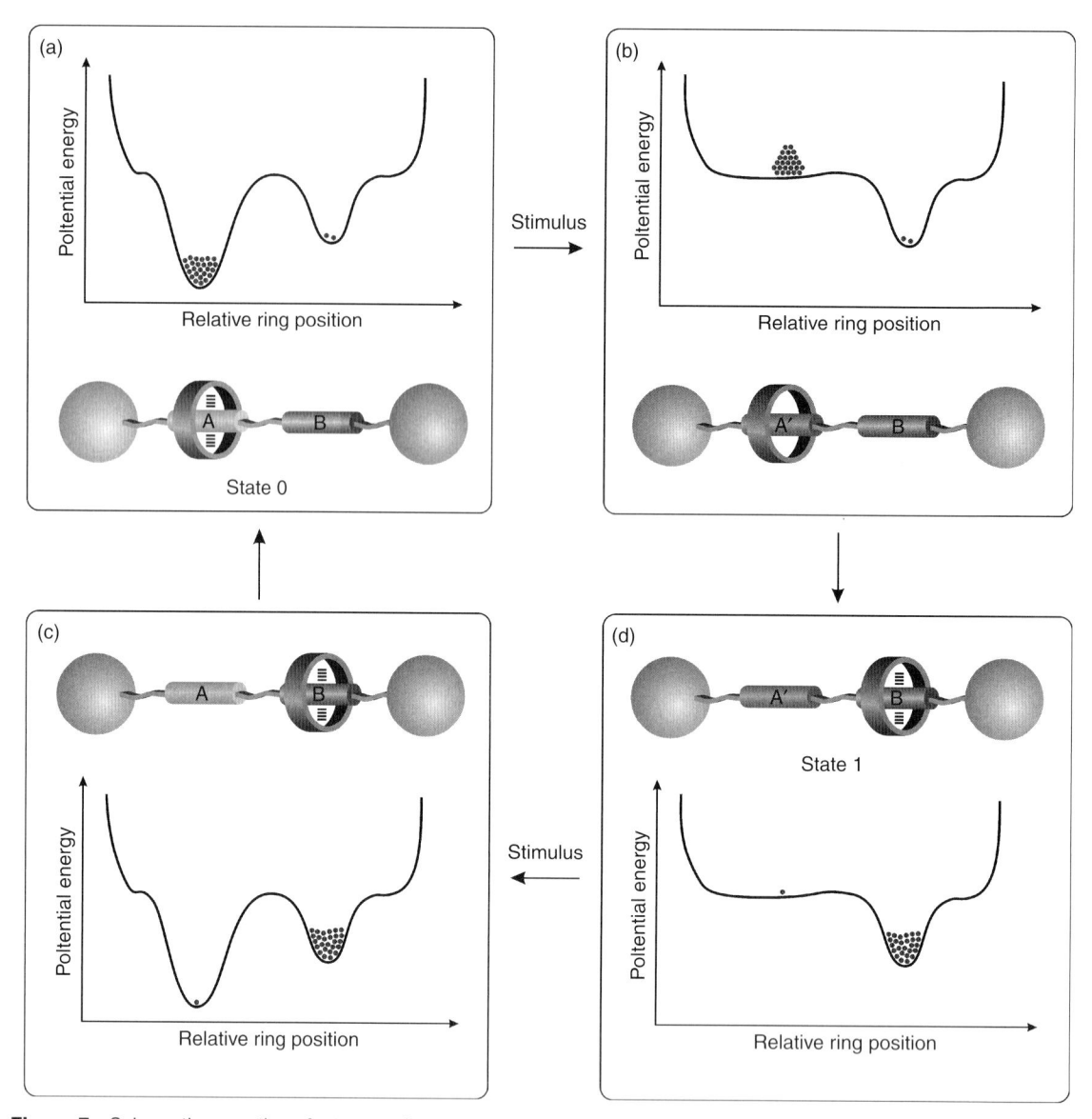

Figure 7 Schematic operation of a two-station rotaxane as a controllable molecular shuttle, and idealized representation of the potential energy of the system as a function of the position of the ring relative to the axle upon switching off and on station A. The number of dots in each position reflects the relative population of the corresponding co-conformation in a statistically significant ensemble. Structures (a) and (c) correspond to equilibrium states, whereas (b) and (d) are metastable states. An alternative approach would be to modify station B through an external stimulus to make it a stronger recognition site compared to station A. Reproduced with permission from Silvi S, Venturi M, and Credi A (2009) *Journal of Materials Chemistry* 19: 2282.

units. Molecular models show that the length of this species is approximately 3.1 nm. The addition of a stoichiometric amount of a strong, non-nucleophilic phosphazene base causes the deprotonation of the two ammonium centers of the rotaxane dimer. As a consequence, the two interlocked filaments glide along one another through the terminal DB24C8 rings until the latter reach the bipyridinium stations (**Figure 8(b)**, state 1). Reprotonation of the dialkylamine units, following the addition of trifluoroacetic acid, results in the return of the DB24C8 rings to the ammonium stations. Because the so-obtained contracted form is about 2.2 nm in length, deprotonation induces a 29% shortening (0.9 nm) of the longitudinal molecular size with respect to the original extended length.

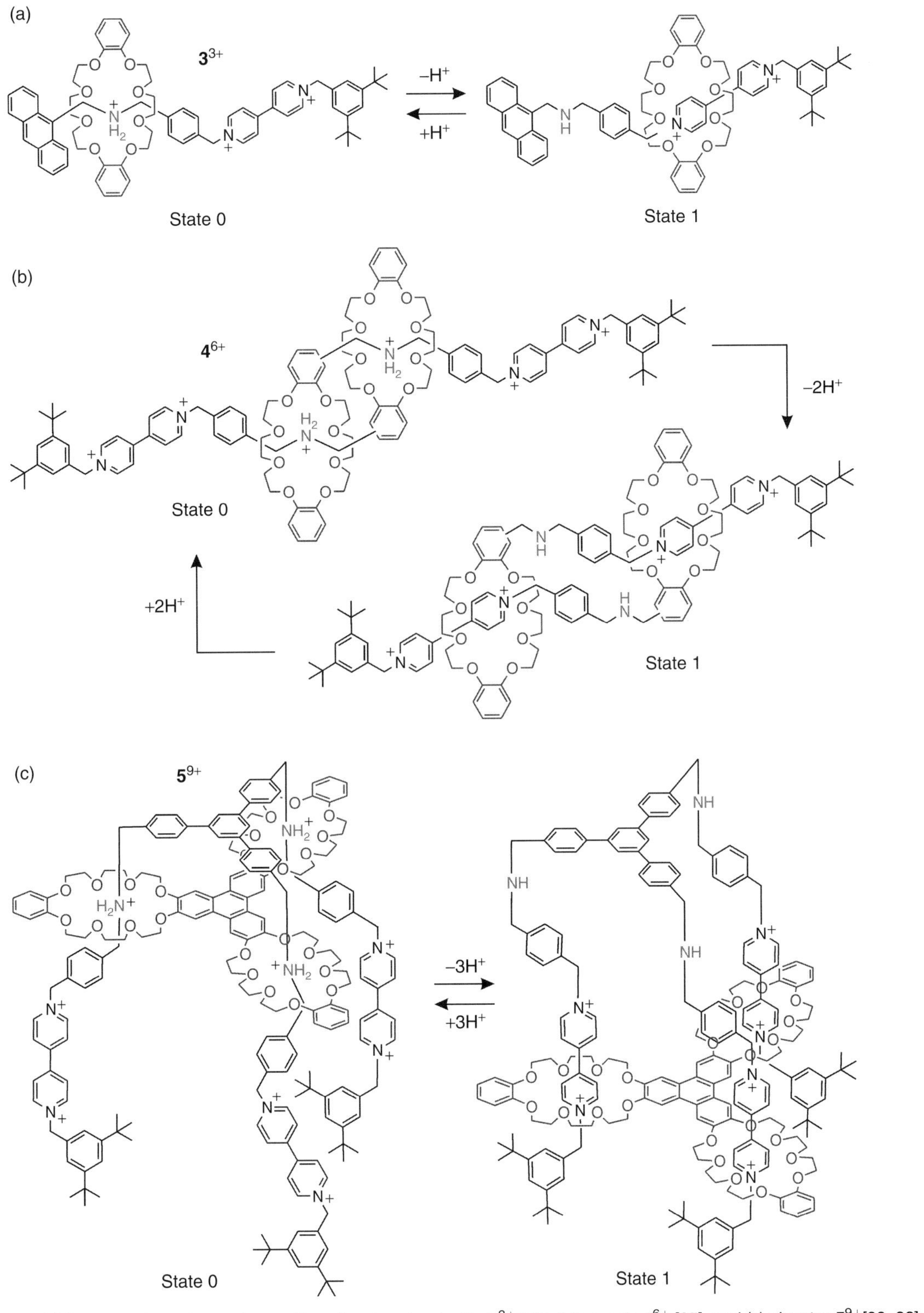

Figure 8 (a) Structure formulas and operation of molecular shuttle 3^{3+} [83], (b) muscle 4^{6+} [85], and (c) elevator 5^{9+} [33, 88] by acid–base inputs. Reproduced with permission from Silvi S, Venturi M, and Credi A (2009) *Journal of Materials Chemistry* 19: 2283.

By using an incrementally staged strategy, the architectural features of the switchable rotaxane 3^{3+} (**Figure 8(a)**) [83] were incorporated into those of a previously investigated [87] triply threaded two-component supramolecular bundle. The result was the construction of two-component molecular devices such as 5^{9+} (**Figure 8(c)**) that behave like nanometer-scale elevators [33,88]. Compound 5^{9+}, which measures about 2.5 nm in height and 3.5 nm in diameter, consists of a tripod component containing two different notches – one ammonium center and one bipyridinium unit – at different levels in each of its three legs. Such legs are interlocked by a tritopic host, which plays the role of a platform that can be stopped at the two different levels.

Initially, the platform resides exclusively on the upper level, that is, with the three rings surrounding the ammonium centers (**Figure 8(c)**, state 0). On addition of a phosphazene base to an acetonitrile solution of 5^{9+}, deprotonation of the ammonium centers occurs and, as a result, the platform moves to the lower level, that is, with the three crown ether rings surrounding the bipyridinium units (**Figure 8(c)**, state 1). The distance traveled by the platform is ~0.7 nm and it was estimated that a force of up to 200 pN, that is, more than 1 order of magnitude larger than that generated by natural linear motors [3], could be obtained. Subsequent addition of acid restores the ammonium centers, and the platform moves back to the upper level. The up-and-down elevator-like motion, which corresponds to a quantitative switching and can be repeated many times, was monitored by ^{1}H NMR spectroscopy, electrochemistry, and absorption and fluorescence spectroscopy [33,88]. Detailed spectroscopic investigations showed that the platform operates by taking three distinct steps associated with each of the three deprotonation processes. In this regard, the molecular elevator is more reminiscent of a legged animal than it is of a passenger or freight elevator.

The base–acid controlled mechanical motion in 5^{9+} is associated with interesting structural modifications, such as the opening and closing of a large cavity (1.5×0.8 nm) and the control of the positions and properties of the bipyridinium legs. This behavior can, in principle, be used to control the uptake and release of a guest molecule, a function of interest for the development of drug delivery systems.

8.3.1.2 Electrochemically driven shuttles

An interesting case of a molecular shuttle controlled by electrochemically induced redox reactions is shown in **Figure 9**. Rotaxane 6^+ has a phenanthroline

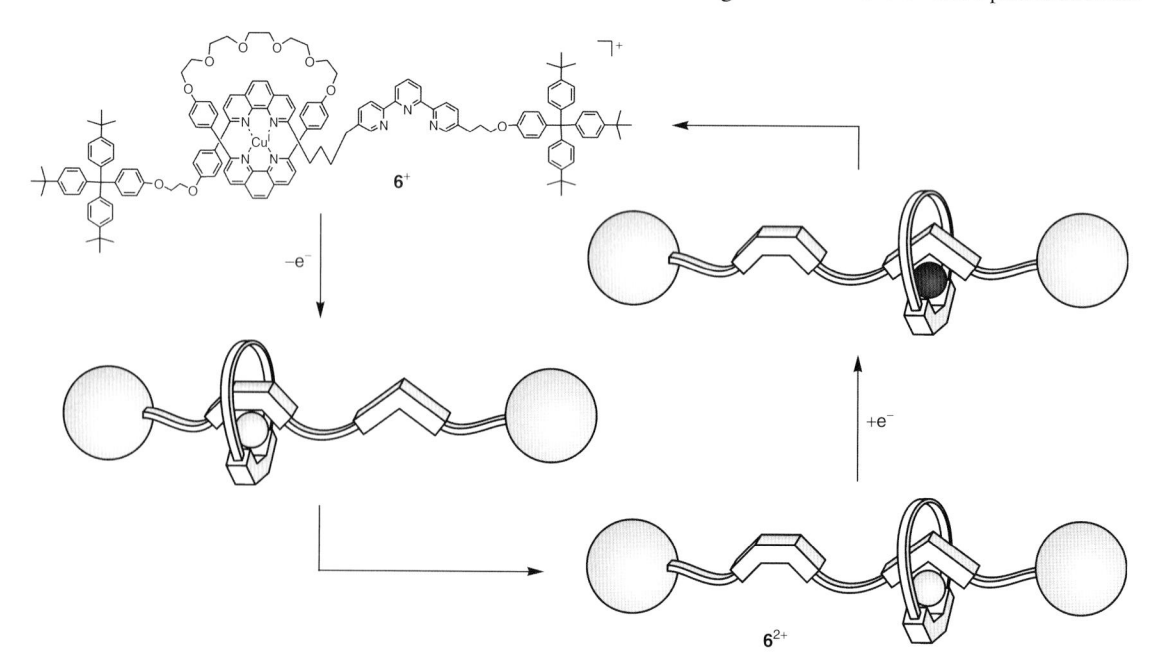

Figure 9 Shuttling of the macrocyclic component of 6^+ along its dumbbell-shaped component can be controlled electrochemically by oxidizing-reducing the metal center [89]. Dark and light circles represent Cu(I) and Cu(II), respectively. Reproduced with permission from Balzani V, Credi A, and Venturi M (2008) *Molecular Devices and Machines – Concepts and Perspectives for the Nanoworld*, p. 428. Weinheim: Wilwy-VCH.

(phen) and a terpyridine (tpy) unit in its axle component [89]. It also incorporates a Cu(I) center coordinated tetrahedrally by the phen ligand of the axle together with the phen ligand of the macrocycle. Oxidation of the tetracoordinated Cu(I) center of **6⁺** to a tetracoordinated Cu(II) ion occurs on electrolysis (+1.0 V relative to the saturated calomel electrode, SCE) of an acetonitrile solution of the rotaxane. In response to the preference of Cu(II) for a pentacoordination geometry, the macrocycle shuttles away from the bidentate phen ligand of the axle and encircles the terdentate tpy ligand instead. In this conformation, the Cu(II) center adopts a pentacoordination geometry that is significantly more stable than the tetracoordination one associated with the original conformation. Consistently the cyclic voltammogram shows the disappearance of the reversible wave (+0.68 V) associated with the tetracoordinated Cu(II)/Cu(I) redox couple and the concomitant appearance of a reversible wave (−0.03 V) corresponding to the pentacoordinated Cu(II)/Cu(I) redox couple. A second electrolysis (−0.03 V) of the solution of the rotaxane reduces the pentacoordinated Cu(II) center back to a pentacoordinated Cu(I) ion. In response to the preference of Cu(I) for a tetracoordination geometry, the macrocycle moves away from the terdentate tpy ligand and encircles the bidentate phen ligand. The cyclic voltammogram recorded after the second electrolysis shows the original redox wave (+0.68 V) corresponding to the tetracoordinated Cu(II)/Cu(I) redox couple.

Very recently, this system has been improved by replacing the highly shielding and hindering phenathroline moiety contained in the ring with a nonhindering biisoquinoline unit [90]. In the new rotaxane, the electrochemically driven shuttling of the ring is, indeed, at least 4 orders of magnitude faster than in the previous phen-based system.

8.3.1.3 Shuttles powered by light energy

Initial attempts aimed at using light to power artificial molecular machines involved the assembly–disassembly of host–guest complexes triggered by photo-induced isomerization, dissociation, proton-transfer, and electron-transfer processes. On the basis of the experience gained with these model systems, several examples of photochemically driven molecular shuttles have been developed [4,5,7,41,42,46–51].

One remarkable example is the bistable rotaxane **7⁶⁺** (**Figure 10**), which was specifically designed [91] to achieve reversible ring shuttling in solution triggered by photo-induced electron-transfer processes [92]. This compound has a modular structure; its ring component R is a π-electron-donating bis-p-phenylene[34]crown-10, whereas its axle component is made of several covalently linked units. They are a Ru(II) polypyridine complex (P^{2+}), a p-terphenyl-type rigid spacer (S), a 3,3′-dimethyl-4,4′-bipyridinium (A$_2^{2+}$) and a 4,4′-bipyridinium (A$_1^{2+}$) π-electron-accepting stations, and a tetraarylmethane group as the terminal stopper (T). The Ru-based unit plays the dual role of a light-fueled power station and a stopper, whereas the mechanical switch consists of the two electron-accepting stations and the electron-donating macrocycle. Six PF$_6^-$ ions are present as the counteranions of the positively charged rotaxane. The stable translational isomer of rotaxane **7⁶⁺** is the one in which the R component encircles the A$_1^{2+}$ unit, in keeping with the fact that this station is a better electron acceptor than A$_2^{2+}$.

The strategy devised to obtain the photo-induced shuttling movement of R between the two stations A$_1^{2+}$ and A$_2^{2+}$ is based on a four-stroke synchronized sequence of electron-transfer and molecular-rearrangement processes, as illustrated in **Figure 10** [93]. Light excitation of the photoactive unit P^{2+} (**Figure 10(a)**, process 1) is followed by the transfer of an electron from this unit to A$_1^{2+}$ (process 2) which competes with the intrinsic decay of the P^{2+} excited state (process 3). After the reduction of A$_1^{2+}$, with the consequent deactivation of this station, the ring moves (**Figure 10(b)**, process 4) by 1.3 nm to encircle A$_2^{2+}$, a step that is in competition with the back electron-transfer from A$_1^+$ (still encircled by R) to the oxidized unit P^{3+} (process 5). Eventually, a back electron-transfer from the free reduced station A$_1^+$ to the oxidized unit P^{3+} (**Figure 10(c)**, process 6) restores the electron-acceptor power to this radical cationic station. As a consequence of the electronic reset, thermally activated back movement of the ring from A$_2^{2+}$ to A$_1^{2+}$ takes place (**Figure 10(d)**, process 7).

Steady-state and time-resolved spectroscopic experiments complemented by electrochemical measurements in acetonitrile solution showed [93] that the absorption of a visible photon by **7⁶⁺** causes the occurrence of the forward and backward ring movement, that is, a full mechanical cycle according to the mechanism illustrated in

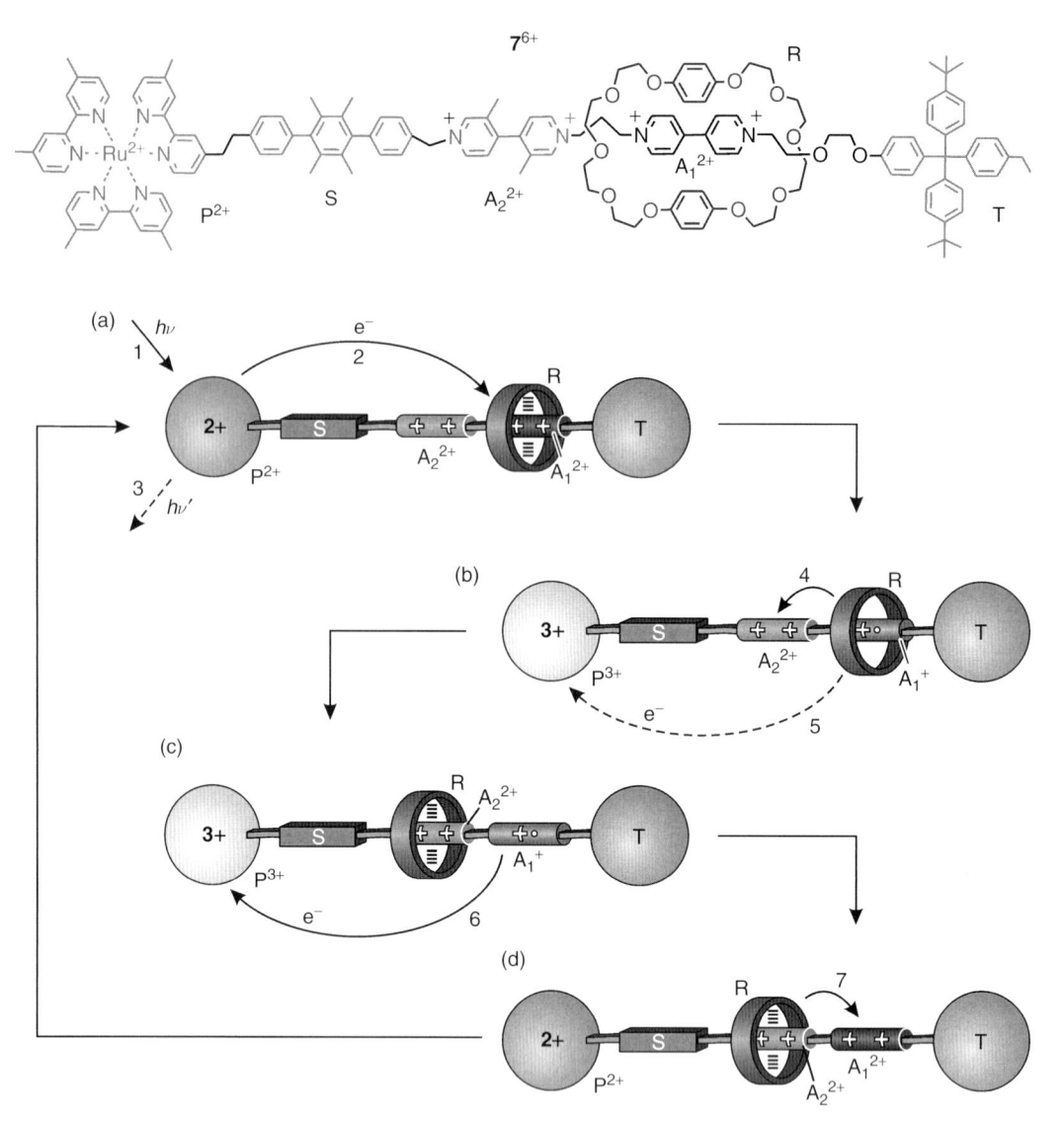

Figure 10 Structure formula of rotaxane 7^{6+} and schematic representation of its working mechanism as an autonomous four-stroke molecular shuttle powered by visible light [91, 93]. Reproduced with permission from Silvi S, Venturi M, and Credi A (2009) *Journal of Materials Chemistry* 19: 2285.

Figure 10. The somewhat disappointing quantum yield for shuttling (2%) is compensated by the fact that the investigated system gathers together the following features: (1) it is powered by visible light (in other words, sunlight); (2) it exhibits autonomous behavior, like motor proteins; (3) it does not generate waste products; (4) its operation can rely only on intramolecular processes, allowing, in principle, operation at the single-molecule level; (5) it can be driven at a frequency of about 1 kHz; (6) it works in mild environmental conditions (i.e., fluid solution at ambient temperature); and (7) it is stable for at least 10^3 cycles.

A thorough computational investigation on 7^{6+} revealed [94] that the rate-limiting step for the shuttling motion (process 4 in **Figure 10**) could be related to the detachment of the PF_6^- counteranions from the station that has to receive the ring (A_2^{2+}). If such a station were not hindered by anions, the shuttling motion would be almost barrierless and occur with a time constant as fast as 20 ns at 300 K. Hence, the shuttling quantum yield of 7^{6+} could be substantially improved by adopting weakly coordinating counteranions or by changing the solvent. The latter choice, however, would also affect the energetics and kinetics of the electron-transfer processes.

Unfortunately, experiments in these directions are not easy because of difficulties related to counter-anion exchange for 7^{6+}, and solubility issues. Systems related to 7^{6+} in which either the position of the A_1^{2+} and A_2^{2+} stations is exchanged with respect to the P^{2+} unit [95], a different Ru-based moiety is employed [96], or the photosensitizer is connected noncovalently to an electron-accepting station [97, 98] have also been investigated.

More recently, the second-generation molecular shuttle 8^{4+} (**Figure 11**) was designed and constructed [99]. The system is composed of two devices: a bistable

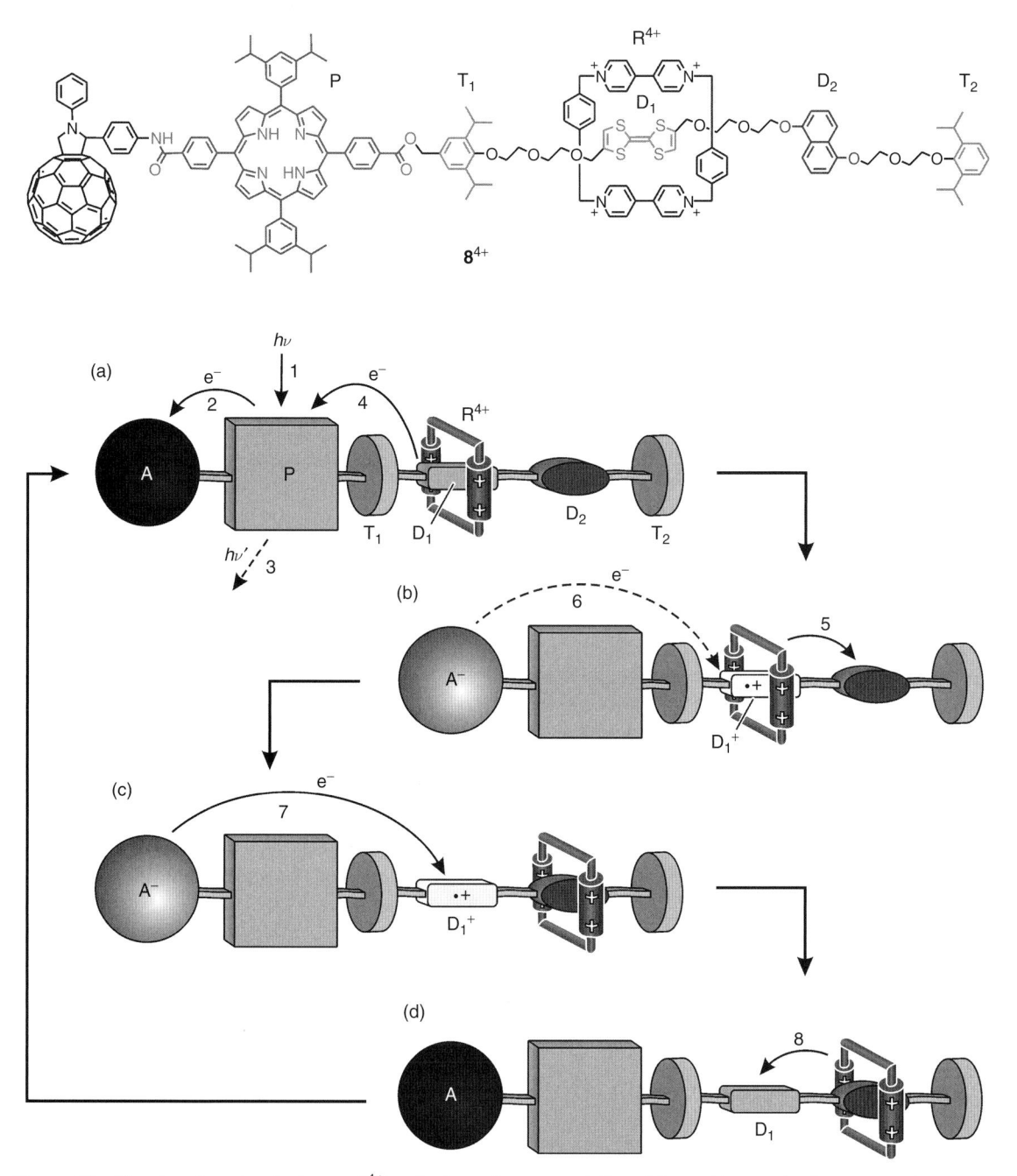

Figure 11 Structure formula of rotaxane 8^{4+} and schematic representation of its working mechanism as an autonomous four-stroke molecular shuttle powered by visible light [99]. Reproduced with permission from Balzani V, Credi A, and Venturi M (2009) *Chemical Society Reviews* 38: 1548.

redox-driven molecular shuttle, and a module for photo-induced charge separation. In the stable translational isomer the electron-accepting ring R^{4+}, which is confined in the region of the axle delimited by the two stoppers T_1 and T_2, encircles the better electron-donor tetrathiafulvalene (TTF) station. The operation scheme of 8^{4+} (**Figure 11**) is similar to that of 7^{6+} (**Figure 10**): in solution, excitation of the porphyrin unit with visible light should cause an electron transfer to C_{60}. Then, an electron shift from TTF to the oxidized porphyrin is expected to destabilize the original structure, causing the displacement of R^{4+} from the oxidized TTF unit to the 1,5-dioxynaphthalene (DNP) station. Subsequent back electron transfer (from C_{60}^- to TTF^+) and macrocycle replacement (from DNP to TTF) regenerates the starting isomer.

Rotaxane 8^{4+} is expected to exhibit a better performance as a light-driven autonomous molecular shuttle compared with the first-generation compound 7^{6+} for the following three reasons. First, by using a triad approach [100], a relatively long-lived charge-separated state should be obtained with a high efficiency. Second, the electrostatic repulsion between the photo-oxidized TTF^+ station and the R^{4+} ring is expected to speed up the displacement of the latter. Third, the hampering effect of the counteranions discussed for 7^{6+} is not expected to be dramatic in the case of 8^{4+} because the stations are originally uncharged.

Spectroscopic and voltammetric experiments revealed [99] remarkable electronic interactions between the various units of 7^{4+}, pointing to the existence of folded conformations in solution. Interestingly, the TTF unit can be electrochemically oxidized only in a limited fraction of the rotaxane molecules; in these species, TTF oxidation causes the shuttling of the R^{4+} ring away from this station. Most likely, 8^{4+} occurs as conformations in which the electroactive TTF unit is buried inside a complex molecular structure and is therefore protected against oxidation performed by an electric potential applied externally. Such a behavior limits the efficiency for the operation of 7^{4+} as a redox-driven molecular shuttle. The possibility of oxidizing TTF by an electric potential generated internally through intramolecular photo-induced electron transfer is currently under investigation.

In general terms, these results indicate that, as the structural complexity increases, the overall properties of the system cannot be easily rationalized solely on the basis of the type and sequence of the functional units incorporated in the molecular framework – that is, its primary structure. Higher-level conformational effects, which are reminiscent of those related to the secondary and tertiary structure of biomolecules [2], have to be taken into consideration. The comprehension of these effects constitutes a stimulating scientific problem, and a necessary step for the design of novel artificial molecular devices and machines.

8.3.2 Ring Switching in Catenanes

As for rotaxanes, the general strategy for preparing catenanes in high yields is based on the template effect, which relies on the presence of molecular recognition sites between the components to be assembled [76]. This kind of synthesis usually leads to ring components that carry two identical or different recognition sites (**Figure 12**). If the two recognition sites of each ring in a catenane are identical, the system can undergo degenerate conformational change when a macrocycle spontaneously circumrotates (**Figure 12(a)**). When at least one of the rings of a catenane carries two different recognition sites, the system can occur in two different conformations that can be interchanged (**Figure 12(b)**) by use of appropriate stimuli (chemical, electrochemical, and photochemical in nature). Such a bistable behavior is reminiscent of that of controllable molecular shuttles (Section 8.3.1). In a diagram of potential energy against angle of rotation of the dissymmetric macrocycle, the two conformations correspond to energy minima, provided by the

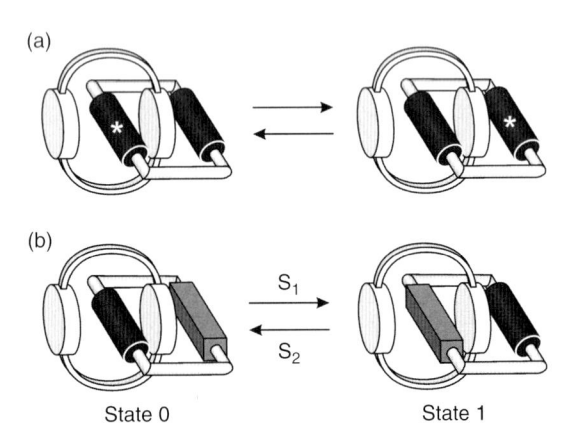

Figure 12 (a) Degenerate conformational changes in a catenane with identical recognition sites in the ring components. Asterisks are used to highlight the exchange of position of identical units; (b) conformational changes induced by appropriate stimuli (S_1 and S_2) in a catenane comprising a ring which carries two different recognition sites.

intercomponent noncovalent bonding interactions. It should be remarked that repeated switching between the two states does not need to occur through a full rotation. In fact, because of the intrinsic symmetry of the system, both movements from state 0 to state 1 and that from state 1 to state 0 can occur with equal probabilities along a clockwise or anticlockwise direction. This kind of switching motion is often called circumrotation [4,5,42]. A full (360°) rotation movement can occur only in ratchet-type systems, that is, in the presence of dissymmetry elements which can be structural or functional in nature (Section 8.3.3.3).

8.3.2.1 Chemically driven switching

Catenane 9^{4+} (**Figure 13**) incorporates a bipyridinium-based tetracationic cyclophane and a π-electron rich macrocyclic polyether comprising a TTF and a DNP units [101]. The ^{1}H NMR spectrum (acetonitrile, 298 K) of 9^{4+} indicates that the TTF unit resides preferentially inside the cavity of the tetracationic cyclophane, whereas the DNP unit is positioned alongside. The tendency of o-chloroanil **10** to stack against TTF can be exploited to lock this unit alongside the cavity of the tetracationic cyclophane (**Figure 13**). On addition of a mixture of Na$_2$S$_2$O$_5$ and NH$_4$PF$_6$ in H$_2$O, the adduct formed between the TTF unit and o-chloroanil is destroyed, and the original conformation with tetrathiafulvalene inside the cavity of the tetracationic cyclophane is then restored.

8.3.2.2 Electrochemically driven switching

Electrochemically driven switching processes have been observed for several metal-based catenanes [4,5,41,42,46–51]. Catenane 11^+ (**Figure 14**), for instance, incorporates two identical macrocyclic components comprising tpy and phen ligands. The Cu(I) ion is coordinated tetrahedrally by the two phen ligands, whereas the two tpy ligands are located well away from each other [102]. The cyclic voltammogram of 11^+ contains a reversible wave (+0.63 V relative to SCE) associated with the tetracoordinated Cu(II)/Cu(I) redox couple. The visible absorption spectrum of the catenane contains a metal-to-ligand charge-transfer band at 439 nm for the tetracoordinated Cu(I) chromophore. On electrochemical oxidation of 11^+, the tetracoordinated Cu(I) center is converted into a tetracoordinated Cu(II) ion which has an absorption band at 670 nm. The intensity of this band decreases with time, however. Indeed, in response to the preference of the Cu(II) ion for a coordination number higher than four, one of the two macrocycles circumrotates through the cavity of the other, affording a pentacoordinated Cu(II) ion. Subsequently, the other macrocycle undergoes a similar circumrotational process, yielding a

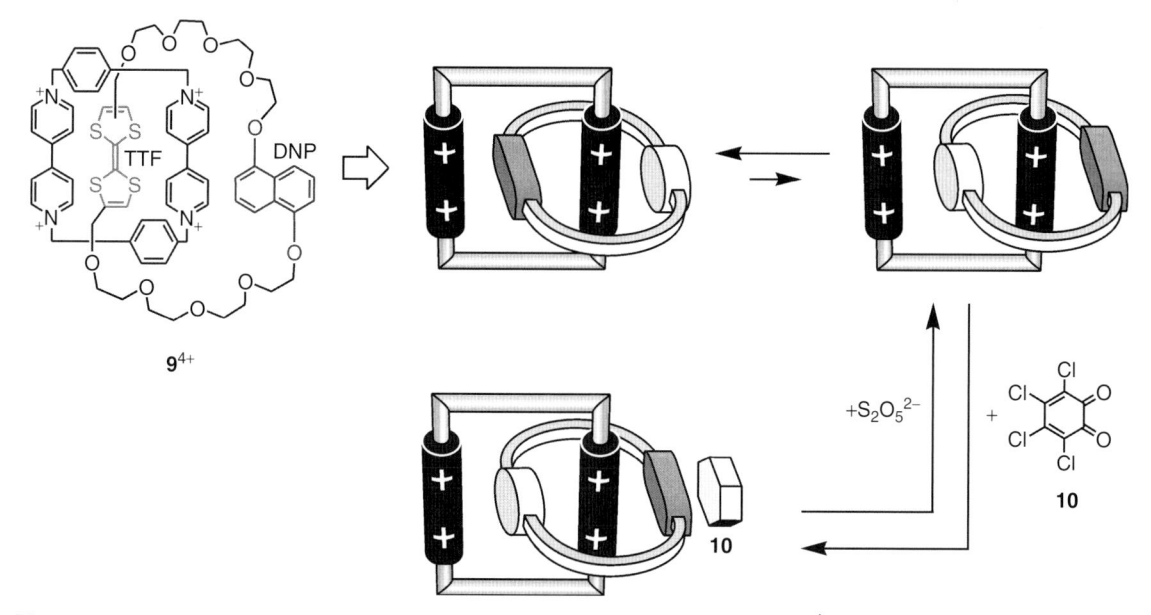

Figure 13 The circumrotation of the macrocyclic polyether component of catenane 9^{4+} can be controlled reversibly by adding or removing o-chloroanil (**10**) which forms a charge-transfer adduct with the tetrathiafulvalene unit of these catenanes [101]. The adduct can be disrupted by reducing o-chloroanil with Na$_2$S$_2$O$_5$.

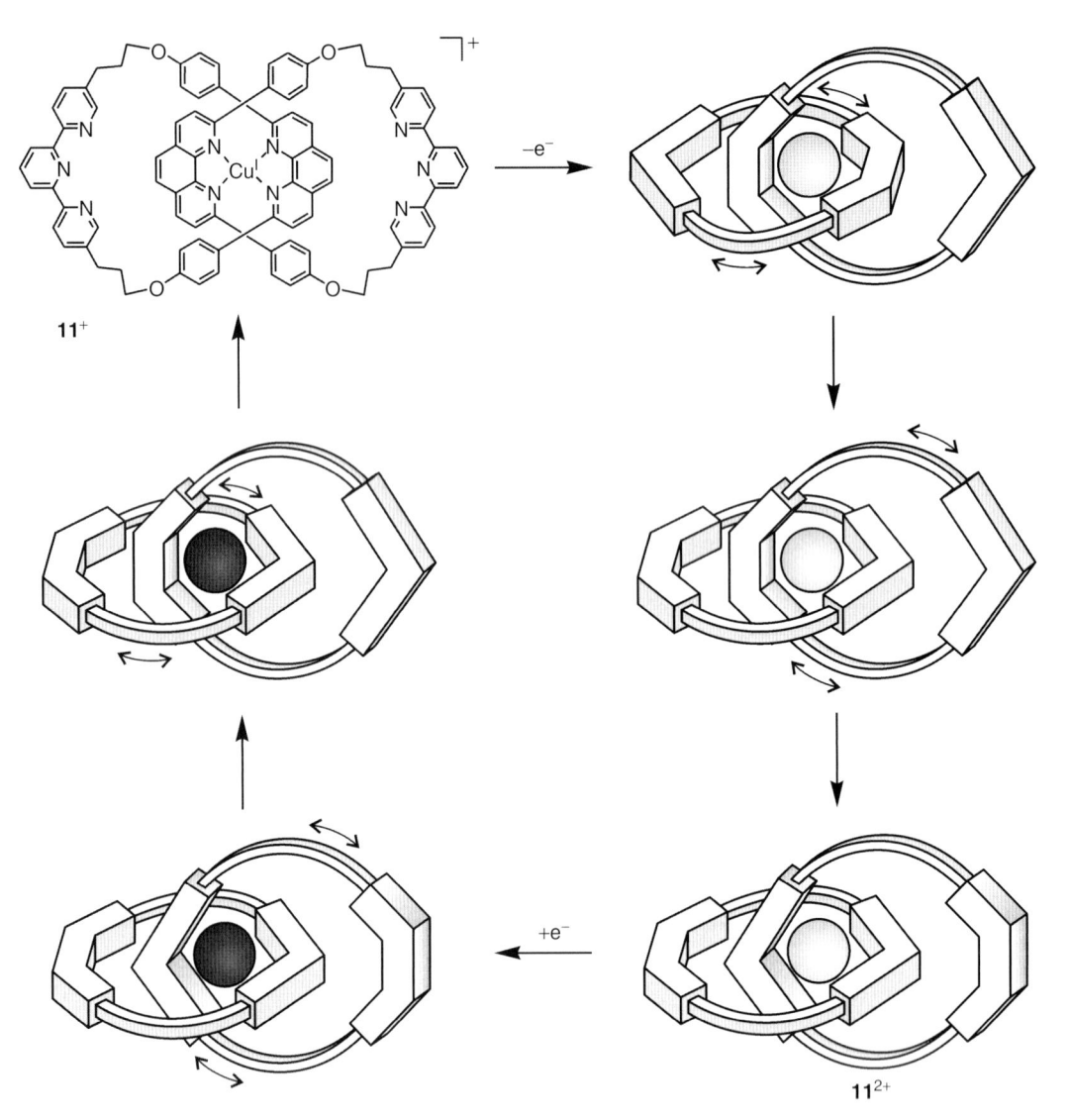

Figure 14 Circumrotation of the macrocyclic components of catenane **11**$^+$ can be controlled reversibly by oxidizing or reducing the metal center [102]. Dark and light circles represent Cu(I) and Cu(II), respectively. Reproduced with permission from Balzani V, Credi A, and Venturi M (2008) *Molecular Devices and Machines – Concepts and Perspectives for the Nanoworld*, p. 468. Weinheim: Wiley-VCH.

hexacoordinated Cu(II) ion which gives, instead, a weak absorption band at 687 nm. Electrolysis (−1.0 V) of the acetonitrile solution of the catenane reduces the hexacoordinated Cu(II) center back to a hexacoordinated Cu(I) ion. In response to the preference of Cu(I) for a tetracoordinated geometry the two macrocycles circumrotate through the cavity of each other in turn, affording the original conformation quantitatively (**Figure 14**).

The conformational motion associated with catenane **9**$^{4+}$ (**Figure 13**) can also be controlled electrochemically by reversible oxidation–reduction of the TTF unit [101]. The cyclic voltammogram of the free macrocyclic polyether contains a reversible wave (∼+0.3 V relative to SCE) for the monoelectronic oxidation of the TTF unit. In the catenanes, this unit is located inside the cavity of the tetracationic cyclophane and its monoelectronic oxidation occurs at more positive potentials. Large separation between the anodic and cathodic peaks associated with this process is observed. This separation varies as the scan rate is changed. On increasing the scan rate, the anodic peak moves to more positive potentials, whereas the cathodic one shifts to less positive values. These observations indicate that the oxidation–reduction of the TTF unit is accompanied by the circumrotation of the macrocyclic polyether through the cavity of the tetracationic cyclophane and that this conformational change is occurring on the timescale of the electrochemical experiment. Indeed, after oxidation the newly formed monocationic tetrathiafulvalene unit is expelled from the cavity of the tetracationic cyclophane and is replaced by the neutral DNP unit. After reduction, the original conformation is restored as the neutral TTF unit replaces the DNP unit inside the cavity of the tetracationic cyclophane.

8.3.2.3 Photochemically driven switching

Catenanes 12^{2+} and 13^{2+} (**Figure 15(a)**) were synthesized using an octahedral Ru(II) center as the template [103]. Compound 12^{2+} consists of a 50-membered ring which incorporates two phenanthroline units and a 42-membered ring which contains a 2,2′-bipyridine (bpy) unit. Compound 13^{2+} contains the same bpy-incorporating ring as 12^{2+}, but the other ring is a 63-membered ring. Clearly, 13^{2+} is more adapted to molecular motions than 12^{2+} which has a relatively tight structure. The light-induced motions and the thermal back reactions (both quantitative) exhibited by 12^{2+} and 13^{2+} are illustrated in **Figure 15(b)** [103]. The photoproducts **12′** and **13′** contain two disconnected rings because the photoexcitation of the Ru(II) moiety leads to decomplexation of the bpy unit contained in the 42-membered ring.

8.3.3 Rotary Motors

Artificial, molecular-level rotary motors are systems capable of undergoing unidirectional and repetitive rotation under the action of external energy inputs. The bottom–up construction of molecular-level rotary motors poses several challenges, particularly because it is difficult to satisfy the unidirectional rotation requirement. In the last few years, however, several different approaches have been followed to reach the goal of unidirectional rotation and in some cases the research has been fully successful.

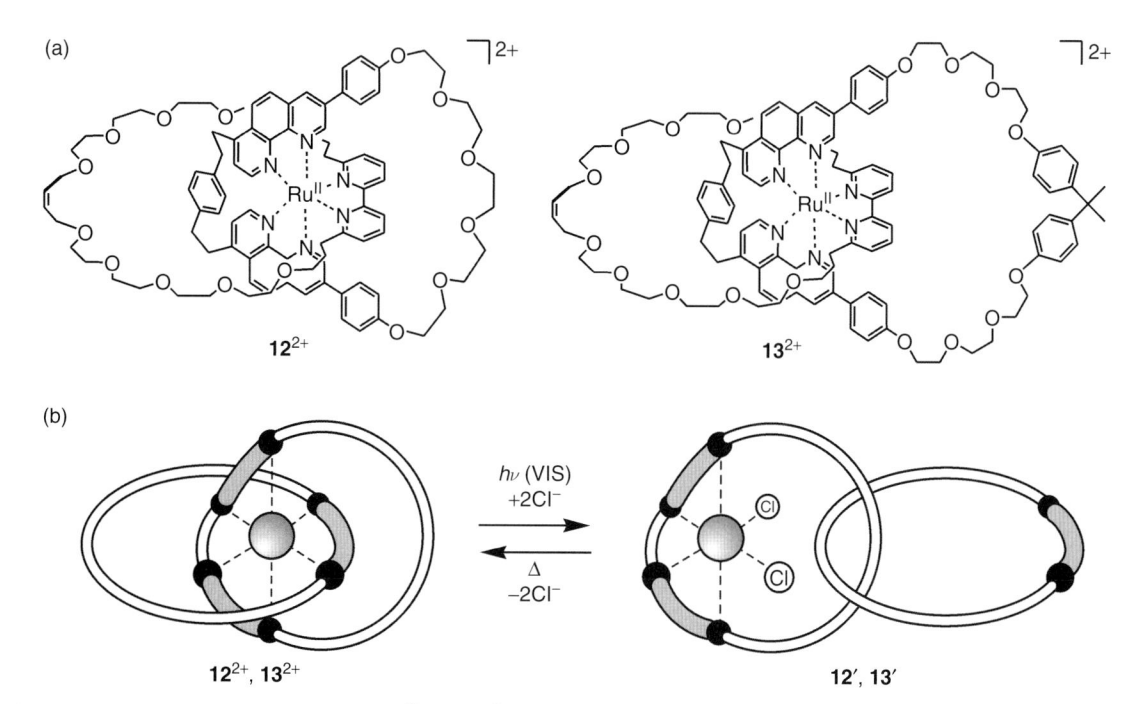

Figure 15 (a) Structures of catenanes 12^{2+} and 13^{2+}; (b) schematic representation of their reversible rearrangement on irradiation with visible light [103]. Reproduced with permission from Balzani V, Credi A, and Venturi M (2008) *Molecular Devices and Machines – Concepts and Perspectives for the Nanoworld*, p. 477. Weinheim: Wiley-VCH.

8.3.3.1 Chemically driven rotary motors

An interesting and very clever attempt to construct a molecular rotary motor was developed [104,105], starting from a suitably functionalized triptycene–helicene system [30]. The energy to produce the motion was obtained from a chemical fuel, phosgene. The strategy used to obtain unidirectional rotation is illustrated in **Figure 16**. The steric hindrance associated with the helicene group of **14** inhibits rotation about the single bond connecting this unit to the 9-triptycyl ring system. Treatment of **14** with Cl$_2$CO and Et$_3$N gives the isocyanate **15**, which is chemically armed to react with the hydroxypropyl tether attached to the helicene. At those instants when clockwise rotation of the triptycene brings the isocyanate and the hydroxyl group sufficiently close to react intramolecularly (**15a**), urethane **16** is formed in the shape of a highly strained conformation. The strain is released by clockwise rotation of the 9-triptycyl ring system about the single bond connecting it to the helicene, yielding **16a** (**Figure 16**). Cleavage of the urethane linkage regenerates compound **14** in a conformation, **14a**, different from original one and that has been obtained as a result of unidirectional 120° rotation induced by the formation of a covalent bond. The process can be monitored by ^{1}H NMR spectroscopy using the bridgehead proton of the 9-triptycyl ring system as a probe. The half-life for rotation of the system is approximately 3 h. To accelerate the speed of rotation a shorter tether attached to helicene was used (two instead of three methylene units) because calculations showed that in the modified compound the energy barrier to rotation is smaller. The speed of rotation for the compound with the 2-carbon tether was indeed much faster (half-life ∼5 min), and urethane formation became the rate-limiting step [105].

The unidirectional rotation achieved with this system is, indeed, a good result. It should be noted, however, that such a rotation is limited to 120°. To achieve full rotation, the system should be modified as follows [106]: (1) an amino group should be incorporated on each blade of the triptycene; (2) a means for selectively delivering phosgene to the amine in the firing position should be devised; (3) a phosgene-fueled 120° rotation of the triptycene should be brought about by formation of an intramolecular urethane; and (4) the remains of the phosgene should be removed by cleavage of the urethane to allow subsequent repetition of the three preceding steps.

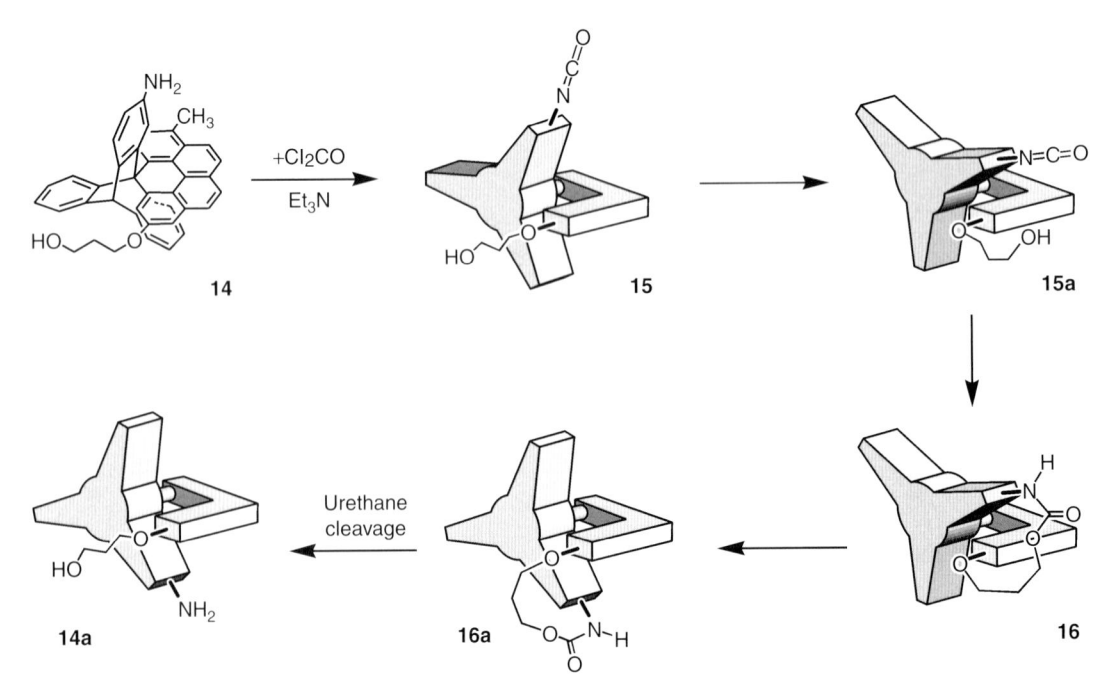

Figure 16 Sequence of events causing unidirectional rotation of 120° in a triptycene–helicene system powered by phosgene as a chemical fuel [104]. Reproduced with permission from Balzani V, Credi A, and Venturi M (2008) *Molecular Devices and Machines – Concepts and Perspectives for the Nanoworld*, p. 459. Weinheim: Wiley-VCH.

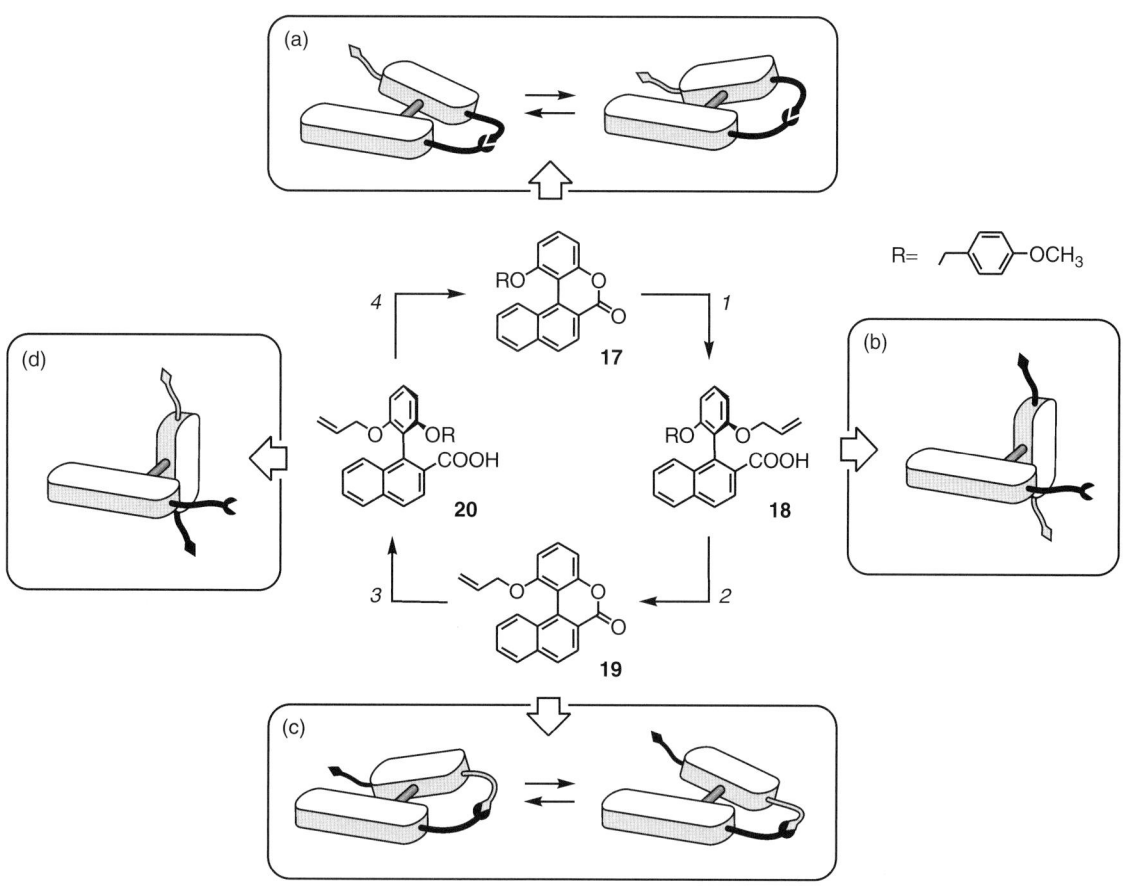

Figure 17 Structural formulas and schematic representation of the four states (a–d) involved in the chemically driven reversible unidirectional rotation around a single bond [107]. See text for details. Reproduced with permission from Balzani V, Credi A, and Venturi M (2008) *Molecular Devices and Machines – Concepts and Perspectives for the Nanoworld*, p. 461. Weinheim: Wiley-VCH.

With a second-generation compound, the first two objectives have been achieved, but new problems have appeared [106].

Unidirectional rotary motion has been achieved in the system illustrated in **Figure 17** [107]. The basis of the rotation is the movement of a phenyl rotor relative to a naphthyl stator about the single-bond axis. The sense of rotation is governed by the choice of chemical reagents that power the motor through four distinct intermediates **17–20** (**Figures 17(a)–17(d)**). In each of these intermediates, rotation around the biaryl bond is restricted by covalent attachment in (a) and (c), and because of steric hindrance in (b) and (d). Interchange between these intermediates requires a stereoselective bond-breaking reaction in steps 1 and 3, and a regioselective bond-formation reaction in steps 2 and 4.

8.3.3.2 Rotary motors powered by light energy

trans–cis Photoisomerization around a carbon–carbon double bond is one of the most extensively studied photochemical reactions. In suitably designed alkene-type compounds containing chiral centers (chiroptical switches), the relative direction of the movement leading to geometrical isomerization can be controlled by the wavelength of the light and depends on the chirality of the molecule [108]. The concerted action of two chiral elements in a single chemical (or physical) event, by virtue of their diastereomeric nature, can, furthermore, lead to unique handedness [109]. A light-driven molecular motor has been designed and constructed on the basis of these principles [31].

As each of the two helical subunits of the resulting compound **21** (**Figure 18**) can adopt a right-handed

268 Molecular Machines and Motors

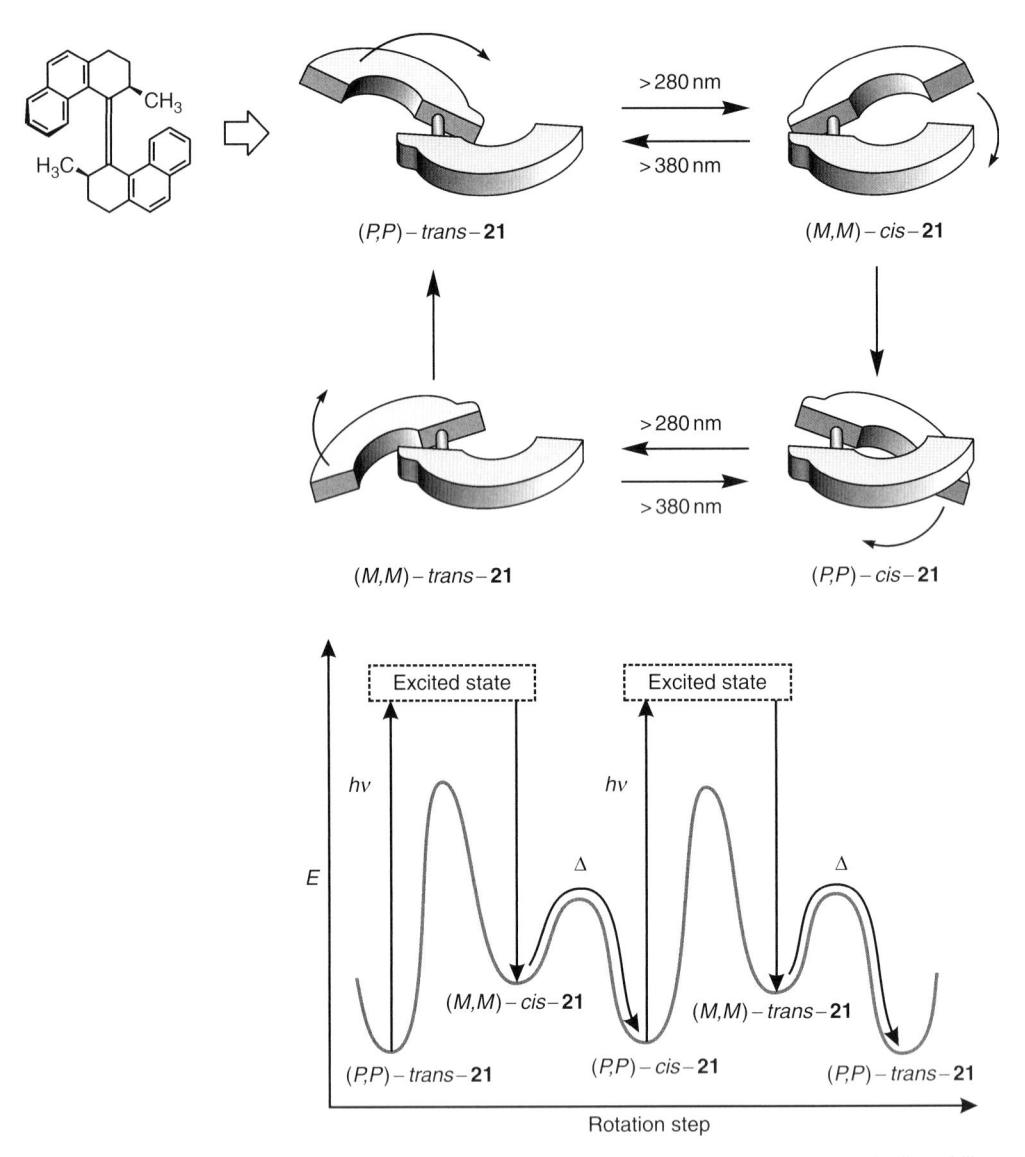

Figure 18 Compound **21** undergoes unidirectional rotation in four steps; each light-driven, energetically uphill process is followed by a thermal, energetically downhill process [31]. Reproduced with permission from Balzani V, Credi A, and Venturi M (2008) *Molecular Devices and Machines – Concepts and Perspectives for the Nanoworld*, p. 473. Weinheim: Wiley-VCH.

(*P*) or a left-handed (*M*) helicity, a total of four stereoisomers are possible for this compound. The *cis–trans* isomerizations are reversible and occur on irradiation at appropriate wavelengths. In contrast, the inversions of helicities, while maintaining a *cis* or a *trans* configuration, occur irreversibly under the influence of thermal energy. On irradiation (≥280 nm, 218 K) of a solution of (*P*, *P*)-*trans*-**21**, a mixture of (*P*, *P*)-*trans*-**21** and (*M*, *M*)-*cis*-**21** is obtained in a ratio of 5:95. When the solution is warmed to 293 K (*M*, *M*)-*cis*-**21** is converted irreversibly to (*P*, *P*)-*cis*-**21**. Subsequent irradiation (≥280 nm) of the solution produces a mixture of (*P*, *P*)-*cis*-**21** and (*M*, *M*)-*trans*-**21** in a ratio of 10:90. When the temperature is increased further (333 K), (*M*, *M*)-*trans*-**21** is converted irreversibly to the original isomer (*P*, *P*)-*trans*-**21**. Thus, a sequence of light- and temperature-induced isomerizations can be exploited to move this molecular rotor in one direction only (**Figure 18**). Indeed, when (*P*, *P*)-*trans*-**21** is irradiated (≥280 nm) at 293 K, a clockwise 360° rotation occurs spontaneously [31]. The overall process can be followed by monitoring the change in the intensity of the absorption band at 217-nm circular dichroism trace. The unidirectional motion in this system is dictated by the stereogenic centers associated with the two methyl substituents. As a result of *trans–cis* isomerization, the axial methyl substituents of (*P*, *P*)-*trans*-21 are forced to adopt a less favorable equatorial orientation in (*M*, *M*)-*cis*-21. The strain associated with the equatorial methyl substituents is, however, released on thermal conversion of (*M*, *M*)-*cis*-21 to the more stable isomer (*P*, *P*)-*cis*-21. The subsequent *cis–trans* isomerization forces the methyl groups to adopt, once again, equatorial orientations in the isomer (*M*, *M*)-*trans*-21. Finally, the thermal conversion of (*M*, *M*)-*trans*-21 to the original isomer (*P*, *P*)-*trans*-21 is accompanied by a change from equatorial to the more stable axial orientations for the methyl substituents.

As the photoisomerization process in such systems is extremely fast (picosecond timescale) [108], the rate-limiting step is the slowest of the thermally activated isomerization (relaxation) reactions. The effect of molecular structure on this rate has been investigated in a series of derivatives. An ethyl substituent at the two stereogenic centers results in a moderate increase in the rates of the thermal relaxation, probably because of greater steric hindrance when these groups occupy the unfavorable equatorial position [110]. When an isopropyl substituent was used, a further increase in the rate of the thermal relaxation step between the *cis* diastereoisomers was observed, but the equivalent step between the two *trans* diastereoisomers was significantly retarded. This result enabled the isolation of an intermediate which demonstrated that the thermal relaxation step comprises two consecutive helix inversions [110].

The rotary motor was then redesigned so that it had distinct upper and lower parts (**Figure 19**) [111]. In a perspective, the lower half of these molecular motors could be connected to other molecules or surfaces, thereby playing the role of a stator, while the upper half acts as a rotor. In such second-generation compounds (e.g., 22 and 23 in **Figure 19**), a single stereogenic center is present. The results obtained showed that the presence of a single stereogenic center is, indeed, a sufficient condition for unidirectional rotation. For all the second-generation motors the slowest thermal isomerization step has a lower kinetic barrier than in 21. Rates up to 44 rotations per second have been achieved. In general, the presence of smaller bridging groups at Y and, in particular, X positions decreases the activation barrier, but unexpected results have also been obtained. For example, the activation energy for motor 23 is higher than that for its phenanthrene analog 22 (X=CH$_2$; Y=CH=CH; R$_1$=R$_2$=H) [111c]. A derivative of this family bearing donor and acceptor substituents (X=Y=S; R$_1$=NMe$_2$; R$_2$=NO$_2$) can operate with visible light [111].

The third-generation compound 24 (**Figure 19**) is based on cyclopentane rings, which allow higher synthetic yields than the six-ring compounds. This compound operates again as a unidirectional rotary motor, but with somewhat lower photon efficiency. It has been found recently that structure 25 (**Figure 19**) is capable of achieving unidirectional rotation at 3 MHz at room temperature under suitable irradiation conditions [112].

An attempt to control the speed of rotation around a single bond by light excitation has been based on compound 26 (**Figure 20**) [113], structurally related to 22. It was envisaged that photoisomerization of *cis*-26 to *trans*-26 would result in a distinct decrease in steric hyndrance to biaryl rotation. The results obtained showed, however, that the *trans* isomer has a slightly higher barrier than the *cis* isomer.

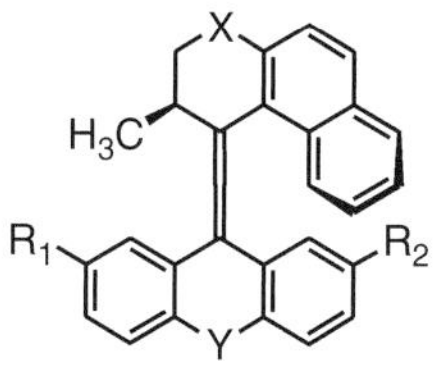

22

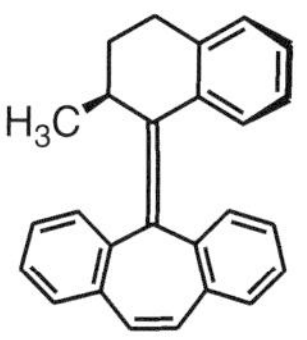

X = S, CH$_2$
X = S, O, C(CH$_3$)$_2$, CH=CH
R$_1$, R$_2$ = H, OCH$_3$, N(CH$_3$)$_2$, NO$_2$

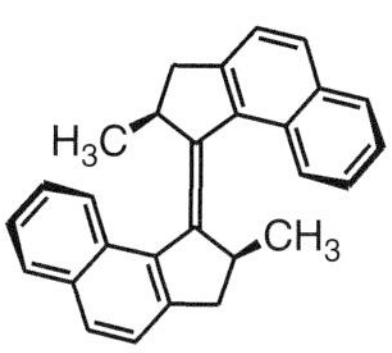

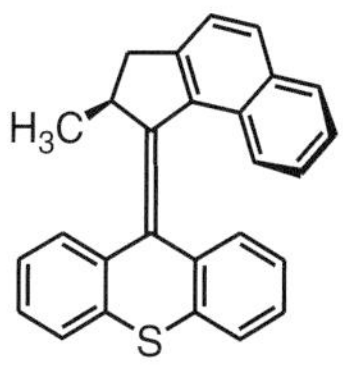

23

24

25

Figure 19 Structures of second- (**22, 23**) and third- (**24, 25**) generation photochemically driven molecular rotary motors [111,112].

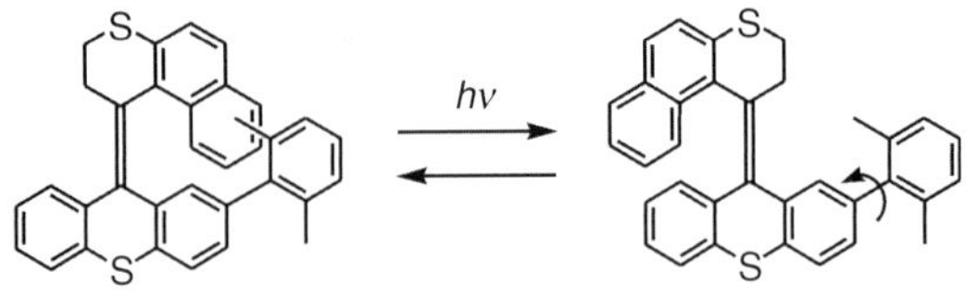

Figure 20 Structure of compound **26** which represents an attempt to control the speed of rotation around a single bond by light excitation [113]. Reproduced with permission from Balzani V, Credi A, and Venturi M (2008) *Molecular Devices and Machines – Concepts and Perspectives for the Nanoworld*, p. 475. Weinheim: Wiley-VCH.

Presumably, the methyl substituents of the rotor moiety interfere with the methylene groups in the upper part in *trans*-**26**, whereas in the *cis* isomer the naphthalene unit easily bends away to enable passage for the rotor.

These results are extremely interesting because they demonstrate the occurrence of unidirectional and repetitive rotary motion in a fully artificial molecular device. The motor is powered by excitation with light, even in the visible region, it works autonomously and it does not generate waste products. Moreover, the rate of rotation can be changed by suitable chemical modification. Because of its chemical structure, the system can be functionalized so it can be connected with other molecules (e.g., polymers), [114] nanoparticles, [115] or macroscopic surfaces [116,117].

An interesting development of this research line involves the operation of molecular motors in liquid crystalline environments. A host–guest system comprising a nematic liquid crystal film (4-pentyloxy-4′-biphenylcarbonitrile) doped with the chiral light-driven molecular rotor **23** was assembled [118]. Irradiation of the film results in unidirectional rotary motion of the molecular rotor, which leads to a molecular reorganization of the mesogenic molecules and, as a consequence, a change in the color of the film. This result, however, is likely related to the state of the system (i.e., the distribution of its various forms) modified by the photo-induced rotary motion, rather than to the unidirectional trajectory of the motor. In other words, the effect of the system is related to its switching properties.

On doping a liquid-crystal film with a light-driven unidirectional rotor of the third-generation series discussed above, the helical organization induced by the dopant results in a fingerprint-like structure to the surface of the film [119]. Irradiation of the film with light changes the distribution of the isomers and, as they have different helical-twisting power, the organization of the liquid crystal is changed. This results in a rotational reorganization of the surface structure, which can be followed by a glass rod sitting on the film. When a photostationary state is reached, the rod's rotatory motion ceases. Removing the light excitation enables the population of the unstable isomer to decay, returning the system to its starting state, accompanied by rotation of the rod in the opposite direction. Similar to the above-discussed color switching, the observed effect is not related to the unidirectional trajectory of the motor, but to the change in the distribution of dopant isomers on irradiation. The directionality of the rod's movement is due to the original chirality of the liquid-crystalline phase and dopant.

Chiral C=N compounds, such as imines, also fulfill the criteria of two-step light-driven unidirectional molecular motors [120].

8.3.3.3 Unidirectional ring rotation in catenanes

Unidirectional rotation in a catenane requires a careful design of the system [4,5,121,122]. A bistable catenane can be a starting point to make a rotary motor, but an additional control element has to be added, as illustrated in **Figure 21** [4,121]. The track ring of the catenane should contain, besides two different recognition sites A and B, a hindering group K and a blocking group X. In the starting conformation (I), the moving ring surrounds the most efficient site (A) on the track ring. On application of the stimulus S_1, site A is switched off (A′) and the ring moves from it. The system has to reach the new stable conformation (II) wherein the ring surrounds site B. The presence of a blocking group X makes anticlockwise rotation faster compared with clockwise rotation. At this stage, application of stimulus S_2 causes the cleavage of the blocking group, and a reset stimulus S_{-1} restores the recognition ability of site A. The system has now to reach the starting conformation wherein the moving ring surrounds site A. The presence of the hindering group K makes again anticlockwise rotation faster compared with clockwise rotation. The original catenane structure is then obtained with a reset stimulus S_{-2} by which the blocking-group X is put back in place. Unidirectional rotation in such a catenane occurs by a 'flashing ratchet' mechanism [82,122], which is based on a periodic change of the potential energy surface viewed by the moving part (**Figure 21**) by

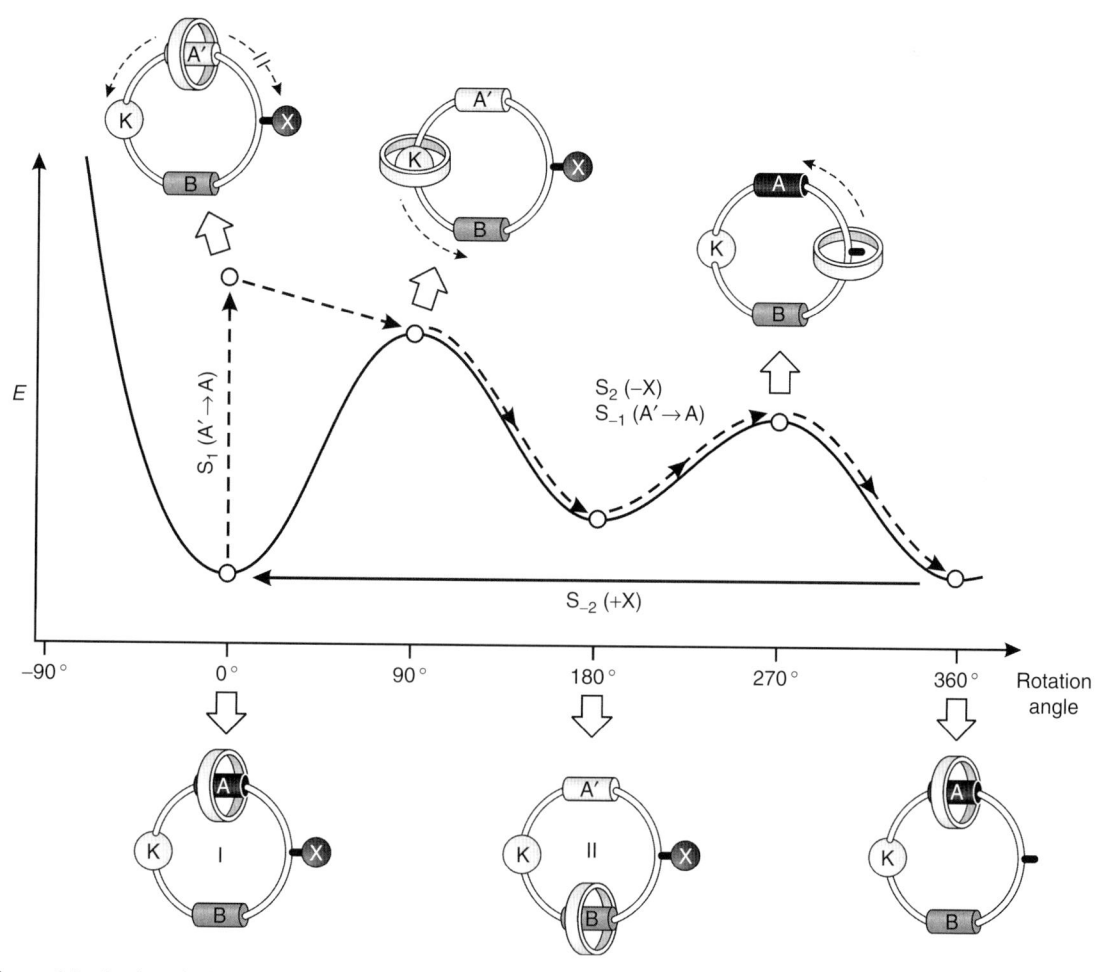

Figure 21 Design of a bistable catenane that performs as a molecular rotary motor controlled by two pairs of independent stimuli. The working scheme is based on the potential energy changes expected for the chemical reactions and conformational rearrangements brought about by stimulation with independent inputs. See text for details. Reproduced with permission from Balzani V, Credi A, and Venturi M (2008) *Molecular Devices and Machines – Concepts and Perspectives for the Nanoworld*, p. 478. Weinheim: Wiley-VCH.

orthogonal (i.e., independent) reactions. It is worth noting that the direction of rotation can be inverted by reversing the order of the two input stimuli.

This concept was cleverly realized using light as a stimulus with catenane **27** shown in **Figure 22(a)** [56]. Its larger ring contains two recognition sites for the smaller ring – namely, a photoisomerizable fumaramide unit (A) and a succinamide unit (B) – and two bulky substituents that can be selectively detached–reattached – namely, a silyl (X_1) and a triphenylmethyl (X_2) group. In the starting isomer (I in **Figure 22(b)**) the smaller ring surrounds the fumaramide site. On $E \rightarrow Z$ photoisomerization of such a unit with 254-nm light and subsequent desilylation (II), the smaller ring moves in the clockwise direction to surround the succinamide site. At this point, a silyl group is reattached at the original position (III). Piperidine-assisted back-isomerization of the maleamide unit to the fumaramide one, followed by removal of the triphenylmethyl group (IV), causes another half-turn of the smaller ring in the clockwise direction to surround the fumaramide unit. Reattachment of a triphenylmethyl substituent regenerates the starting isomer (I). The overall result is a net clockwise rotation of the smaller ring about the larger one. Exchanging the order in which the two blocking groups are manipulated produces an equivalent anticlockwise rotation of the smaller ring. The structures of the compounds obtained after each of the above reaction steps, and particularly the position of the smaller ring, were determined by ^{1}H NMR spectroscopy [56]. This system is more complex than

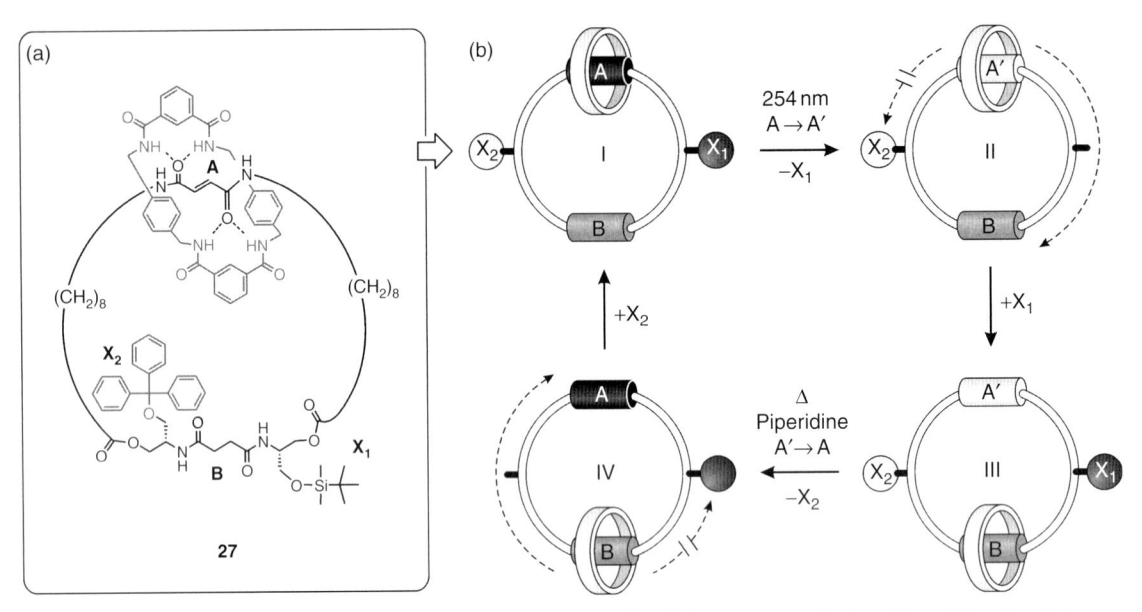

Figure 22 (a) Structure of catenane **27**; (b) schematic representation of the processes that enable unidirectional ring rotation [56]. Reproduced with permission from Balzani V, Credi A, and Venturi M (2008) *Molecular Devices and Machines – Concepts and Perspectives for the Nanoworld*, p. 479. Weinheim: Wiley-VCH.

that described in **Figure 21** because it contains two independently addressable blocking groups. Hence, unidirectional rotation is achieved with three pairs of different stimuli (one for driving the conformational rearrangement, and two for the ratcheting of the energy barriers).

Other strategies to obtain unidirectional ring rotation in catenanes have been explored. Catenane **28** (**Figure 23(a)**) consists of a benzylamide macrocycle threaded into a larger macrocycle which carries three different binding sites (stations) [123]. Stations A and B are two photoisomerizable fumaramide units with different macrocycle binding affinities. Station B, being a methylated fumaramide residue, has a lower affinity for the macrocycle than station A. The third station C (a succinic amide ester) is not photoactive and is intermediate in macrocycle binding affinity between the two fumaramide stations (*E*) and their maleamide (*Z*) counterparts. In the initial conformation (I in **Figure 23(b)**) the small macrocycle resides on the nonmethylated fumaramide station A. This station is located next to a benzophenone unit, which enables its selective, photosensitized isomerization by irradiation at 350 nm (A → A′). Photoisomerization of station A destabilizes the system and the macrocycle finds its new energy minimum on station B (form II). Subsequent direct photoisomerization of this station by irradiation with light at 254 nm (B → B′) moves the macrocycle onto the succinic amide ester station

C (III). Finally, heating the catenane (or treating it with photo-generated bromine radicals or piperidine) results in isomerization of both the *Z*-olefins back to their *E*-form so that the original order of binding affinities is restored and the macrocycle returns to its original position (I). The ^{1}H NMR spectra for each diastereoisomer show excellent positional integrity of the small macrocycle in this three-way switch at all stages of the process, but the rotation is not directional. Over a complete sequence of reactions, an equal number of macrocycles go from A, through B and C, back to A again in each direction.

As discussed for catenane **27**, to bias the direction the macrocycle takes from station to station in **28**, kinetic barriers are required to restrict Brownian motion in one direction at each stage. Such a situation is intrinsically present in catenane **29** (**Figure 24(a)**) [123], consisting of two benzylamide rings interlocked with the same large macrocycle of catenane **28**. In this case, however, and in contrast to what happens for **28**, the isolated amide group incorporated in the macrocycle (D in **Figure 24(a)**), which can make fewer hydrogen bonding contacts than A, B, and C, plays a significant role. In the initial conformation (I in **Figure 24(b)**) the two small macrocycles reside on the fumaramide stations, A and B. Irradiation at 350 nm switches off station A and causes anticlockwise (as drawn) rotation of the macrocycle M_1 to the succinic amide ester station C

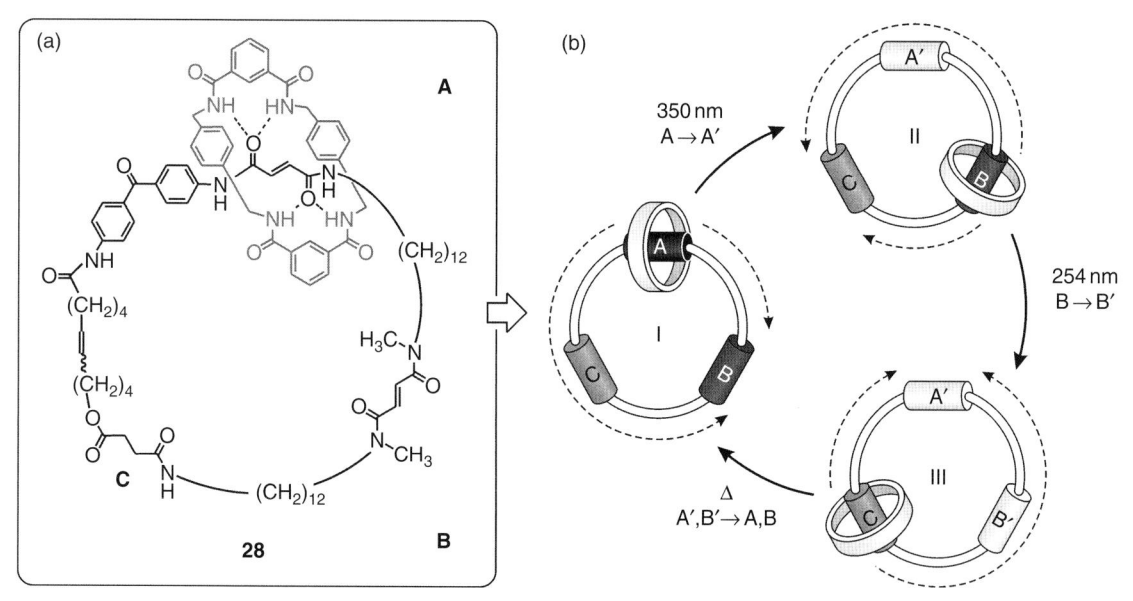

Figure 23 (a) Structure of catenane **28**; (b) stimuli-induced sequential movement of the small macrocycle between three different stations located on the large macrocycle [123]. Reproduced with permission from Balzani V, Credi A, and Venturi M (2008) *Molecular Devices and Machines – Concepts and Perspectives for the Nanoworld*, p. 480. Weinheim: Wiley-VCH.

(II). Photoisomerization with 254-nm light of the remaining fumaramide (B → B′) causes the other macrocycle M$_2$ to relocate to the single amide station D (III) and again this occurs anticlockwise because the clockwise direction is blocked by the other macrocycle. This follow-the-leader process, in which each macrocycle in turn moves and then blocks a direction of passage for the other macrocycle, is repeated throughout the sequence of transformations shown in **Figure 24(b)**. After three diastereoisomer conversions, the *E, E* form of **29** is again obtained, but 360° rotation of each of the small rings has not yet occurred, they have only swapped places (IV in **Figure 24(b)**). Complete unidirectional rotation of both small rings occurs only after the three-step sequence leading from structures I–IV has been completed twice (structures IV, V, VI, and back to I). In this system, the yield of direction fidelity is high (>99%), but the efficiency of the motor is poor (<1% based on absorbed photons).

The timescales and number of reactions involved for unidirectional ring rotation in catenanes **27** and **29** make their operation as rotary motors somewhat unpractical. Nevertheless, the analysis of the thermodynamic and kinetic aspects of the operation mechanisms provides a fundamental insight on how energy inputs can be used to harness thermal fluctuations and drive unidirectional motion.

8.3.4 Systems Based on DNA

Since the 1940s it has been known that the genetic heritage of living organisms is stored in DNA molecules. The perception of DNA as a physical object underwent an important change in the early 1980s, when it was proposed [124] that DNA molecules could play as building blocks for assembling nanoscale materials and devices by exploiting the highly specific interactions between the base pairs. A few years later it was shown that base pairing in DNA molecules could also be used to process complex information [125].

As the specific interactions between two nucleic acid strands can be programmed into their base sequence, made-to-order DNA with predetermined recognition properties can be synthesized. This possibility, combined with the mechanical features of single-stranded (ss)- and double-stranded (ds)-DNA, opens the way to the design and construction of artificial nanosystems with specific structural and/or functional properties [4,45,52]. For these reasons, and because nucleic acids are, in general, easier to be chemically synthesized and more convenient to handle than proteins, DNA (and to a minor extent RNA) has been preferentially used for the assembly of functional nanomaterials.

The simple hybridization of single-stranded oligonucleotides to form a linear duplex does not enable

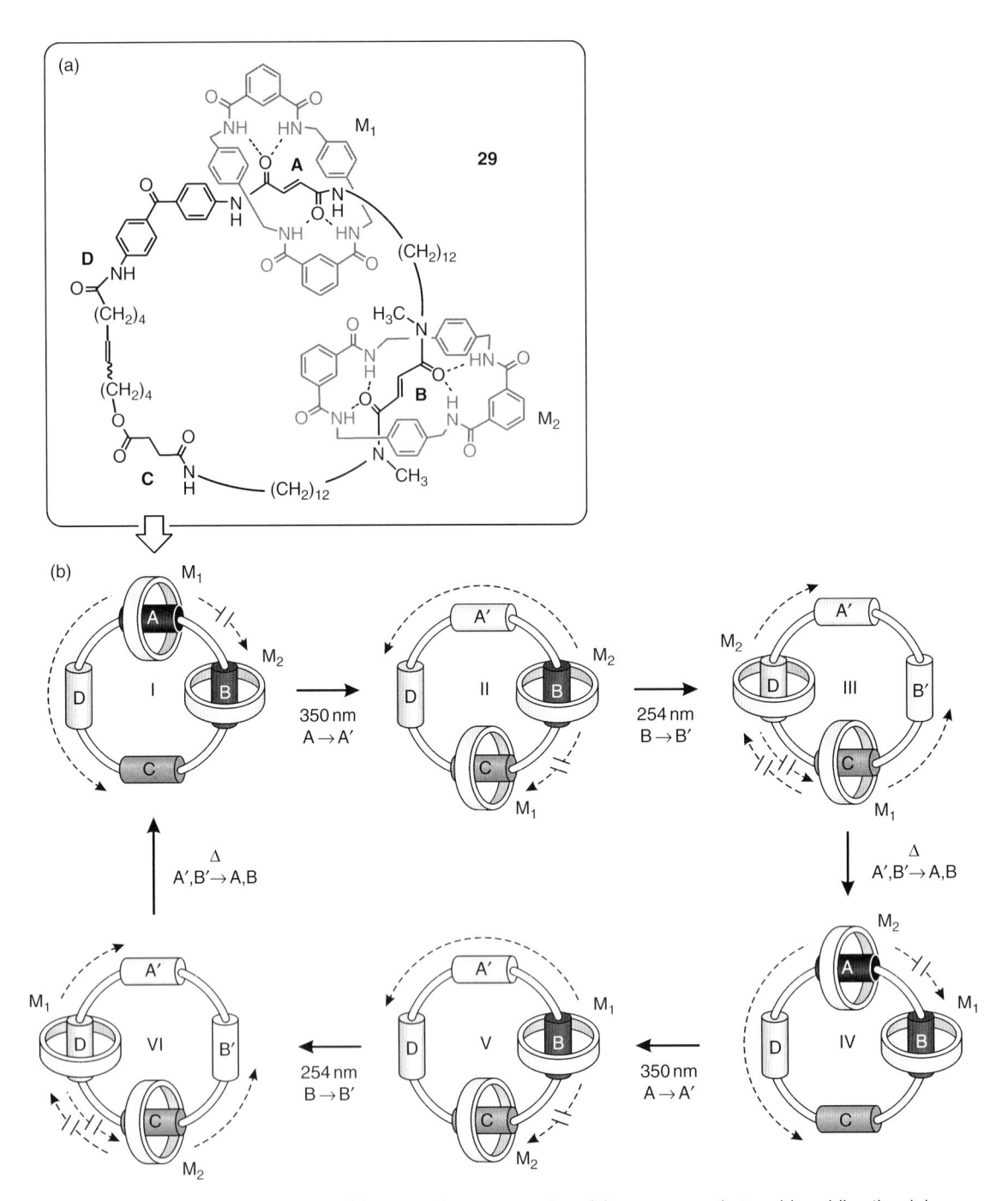

Figure 24 (a) Structure of catenane **29**; (b) schematic representation of the processes that enable unidirectional ring rotation [123]. Reproduced with permission from Balzani V, Credi A, and Venturi M (2008) *Molecular Devices and Machines – Concepts and Perspectives for the Nanoworld*, p. 481. Weinheim: Wiley-VCH.

to make two- and three-dimensional structures. However, DNA can be used to form rigid building blocks for the construction of complex nanostructures. For example, short single-stranded portions (sticky ends) can serve as toeholds for strand exchange (**Figure 25(a)**): a complementary ss DNA which binds first to the toehold can displace the shorter strand of the original duplex. This process, known as 'branch migration', is driven by the greater stability of the duplex formed by the incoming

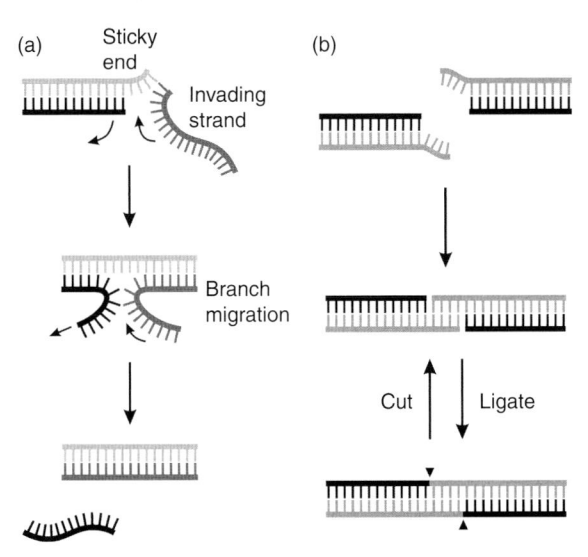

Figure 25 (a) Exchange of two DNA strands in a duplex. The invading DNA strand binds first to the sticky end of the duplex and then displaces the original strand by branch migration; (b) Two double-stranded DNA molecules can be joined together by hybridization of complementary sticky ends, and then covalently connected by the enzyme DNA ligase. Other enzymes, such as restriction endonucleases, can cut the DNA strands (the triangles indicate the cleavage sites).

strand, which results from additional base-pairing in the toehold region. Most DNA-based mechanical devices use strand exchange to drive conformational changes. Short, single-stranded sticky ends can also be used to join two segments of ds DNA (**Figure 25(b)**). Hybridization between complementary sticky ends leaves nicks in the backbone. These nicks can be repaired by the enzyme DNA ligase, which catalyzes the formation of a phosphodiester bond between adjacent 3′-hydroxyl and 5′-phosphate groups.

In the following sections, a few examples of DNA nanomachines are described. Apart from their interest for basic science, this kind of systems could find applications as biosensors and intelligent drug-delivery systems.

8.3.4.1 Tweezers
DNA-based devices in which two arms can approach each other and move away in a controlled manner can be viewed as nanometric tweezers. One example is represented in **Figure 26** [126]. The system results from the hybridization of three oligonucleotide strands, A, B, and C. Strand A is labeled at the 5′ and 3′ ends with two different dyes. When a suitable closing strand F is added, it hybridizes with the sticky ends of strands B and C, pulling the tweezer closed. Reopening of the tweezer is achieved by adding strand G, which is a complement to the entire length of the closing strand F. Therefore, G starts hybridization with the overhang section of F and eventually removes it, forming a double-stranded waste product, FG. The overall cycle, which can be repeated many times, is monitored by measuring the fluorescence intensity, which drops by a factor of 6 on going from the open to closed state because of fluorescence resonance energy transfer (FRET). This corresponds to a difference of 6 nm in the distance between the two ends, and to an angle of ~50° between the open tweezer arms. Not surprisingly, given the complex hybridization processes associated with switching, the operation time scale is rather slow (~13 s in the experimental conditions employed). Tweezer systems showing autonomous behavior have been developed using a catalytic approach [127,128].

8.3.4.2 Walking devices
The realization of artificial molecular machines that can move in a defined direction under control of external stimuli is a hard yet, fascinating challenge for nanoscience. Inspired by naturally occurring linear molecular motors such as kinesin and myosin (Section 8.2.1), several studies have addressed the design and construction of DNA-based machines capable of walking unidirectionally on a nanoscale track made from nucleic acid strands.

The first-reported example of a DNA walker is shown in **Figure 27** [37]. The device consists of two components: a track comprising three stations and a

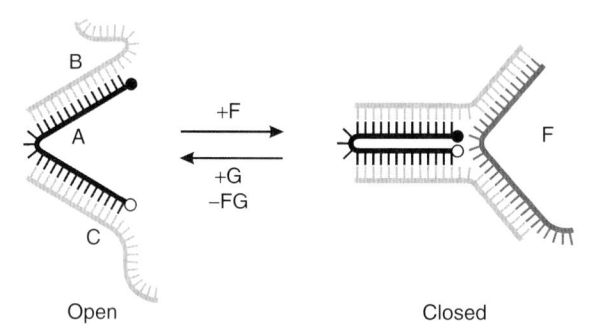

Figure 26 A DNA-based tweezer operated by hybridization of complementary DNA strands [126]. The open and filled circles attached to strand A represent dyes whose separation distance changes on switching between open and closed forms of the tweezer. Reproduced with permission from Balzani V, Credi A, and Venturi M (2008) *Molecular Devices and Machines – Concepts and Perspectives for the Nanoworld*, p. 388. Weinheim: Wiley-VCH.

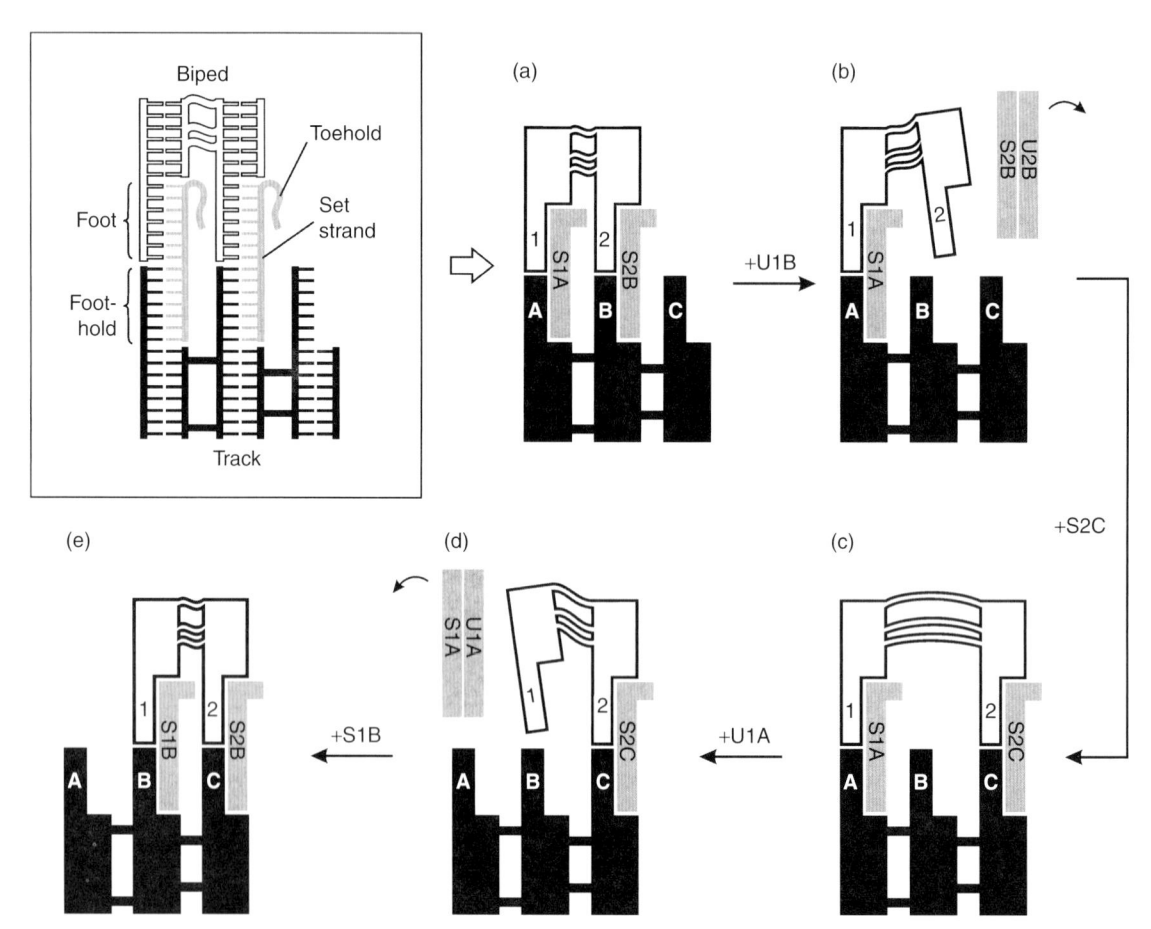

Figure 27 A biped walking device based on DNA and cartoon representation of its operation [37]. See text for details. Reproduced with permission from Balzani V, Credi A, and Venturi M (2008) *Molecular Devices and Machines – Concepts and Perspectives for the Nanoworld*, p. 391. Weinheim: Wiley-VCH.

biped machine consisting of two legs connected by flexible linkers. Each station of the track and each leg of the biped terminate with sticky ends, called 'footholds' (A—C) and 'feet' (1 and 2), respectively, that are available to base-pair with complementary strands of DNA. The sequences of the feet and footholds have been carefully selected to minimize complementarity between them. A foot attaches to a foothold when a set strand S complementary to both is added to the solution. Each of these linking strands has an 8-base toehold, which is not complementary to any of the feet or footholds. The toehold allows the set strand to be removed by pairing with a successively added unset strand U and the system is designed so that after the unset procedure all the unset strands and the set strands with which they are paired can be removed from the solution.

The operation of the biped is based on the following sequence of processes (**Figure 27**) [37]. At the beginning (a), strand set S1A links foothold A with foot 1, and set strand S2B links foothold B with foot 2. The first step consists in releasing foot 2 from foothold B, which is obtained by introducing in the solution unset strand U2B that binds to the toehold of S2B and leads to complete pairing of S2B with U2B (b). Once foot 2 has been freed from foothold B, it is free to be set to foothold C by an appropriate set strand S2C (c). At this stage, foot 1 can be freed from foothold A by an appropriate unset strand U1A (d) and set to foothold B by an appropriate set strand S1B (e). Hence, Brownian motion provides the movement, and the order of adding set and unset strands provides directionality.

The state of the system is observed by taking an aliquot of the solution and exposing it to UV light that causes a cross-link reaction between different strands. Subsequent analysis by denaturation (unpairing of base-paired strands) enables to establish which feet were attached to which footholds. Because of the relatively short lengths of the flexible linkers that connect the two legs, the biped moves by an inchworm-type mechanism, while natural walking proteins, such as kinesin and myosin V, move by a hand-over-hand mechanism that resemble human gait (Section 8.2.1, **Figure 2**). By modifying the design of the system, a DNA biped that walks by a hand-over-hand gait was realized and its processive motion monitored by multichannel fluorescence measurements [129].

Autonomously moving DNA walkers have been developed relying on enzymatic cleavage of their DNA [130] or DNA–RNA track [131]. The operation principle of these systems is shown schematically in **Figure 28**. The walker, or cargo (C), is a

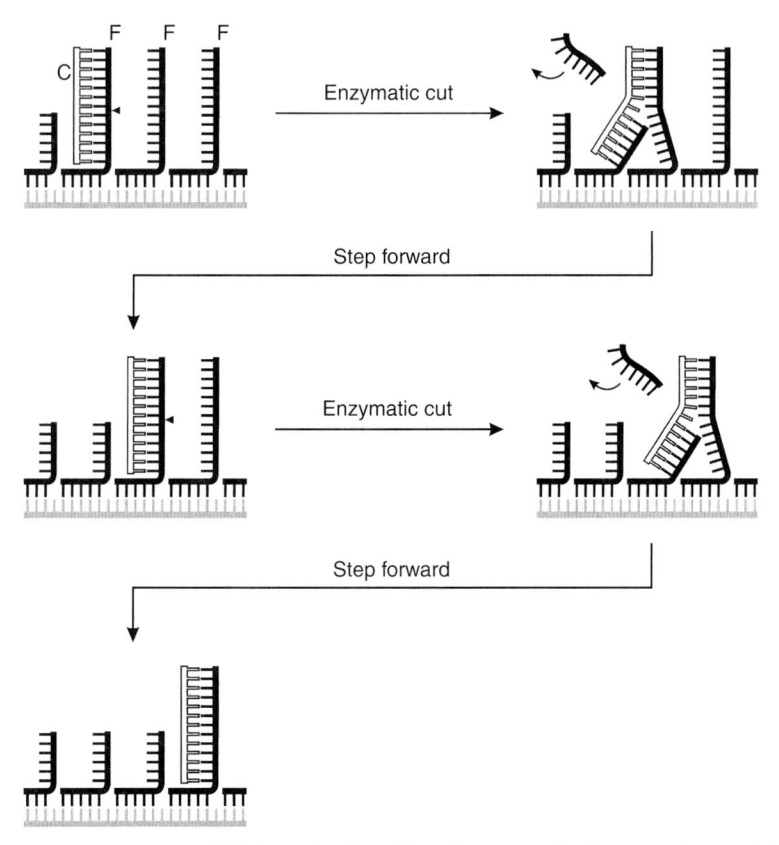

Figure 28 Autonomous movement in a DNA-based walker driven by enzymatic cleavage of the nucleic acid track [130]. The passage of the cargo C irreversibly destroys the footholds F. The triangles indicate the cleavage sites.

DNA enzyme capable to cleave specific sequences of DNA or RNA. The track consists of repeated identical single-stranded footholds attached to a double-stranded backbone. Binding of the walker (C) to a foothold (F) enables the enzyme to cleave the foothold. A short fragment of the latter is released, leaving the cargo with a single-stranded toehold that can bind to the intact foothold ahead of it. The cargo can then step forward by a branch migration reaction. The operation of these systems was monitored by fluorescence measurements [130] or gel electrophoresis [131].

Other examples of autonomous DNA walkers are based on repeated cycles of enzymatic ligation and cleavage reactions [132] or catalytic DNA hybridization processes [133,134]. Walking molecular devices that can move unidirectionally along a predefined nanoscale route are expected to find applications in nanorobotics, diagnostics, medicine – for example, as carriers for molecules or even cells – and computing. As a matter of fact, a molecular machine moving on and altering a DNA instruction tape, according to its internal state and transition rules, would correspond to a nanoscale version of a Turing machine.

8.3.5 Other Systems

8.3.5.1 Rotation around a metal ion in sandwich-type compounds

In contrast to metallocenes or double-decker porphyrin-based compounds, metallocarboranes tend to have rather high barriers to rotation around the metal–ligand axis. Dicarbollide ligand 30^{2-} (**Figure 29(a)**) [135] is characterized by two adjacent carbon atoms on its bonding face which convey a dipole moment perpendicular to the metal–ligand axis. Usually, the resulting carborane complexes adopt a transoid configuration to minimize electrostatic repulsion. In the case of Ni(IV), however, the structure of the [Ni(**30**)$_2$] complex is cisoid (**Figure 29(b)**). This complex can thus operate as a reversible rotary switch on reduction of the Ni(IV) and successive oxidation of the Ni(III) center [134]. The system is reported to switch also on photoexcitation.

An electroactive organometallic rotor consisting of a tris(indazolyl)borate and a heteropolytopic penta(4-ethynylphenyl)cyclopentadiene moieties coordinated to a Ru ion has been synthesized and designed to function on a surface [136].

8.4 Hybrid Systems

Recent scientific advances in molecular biology, supramolecular chemistry, and nanofabrication techniques have opened up the possibility of building functional hybrid devices based on natural motors. One long-term objective of this research is to utilize the finest attributes associated with the worlds of both biological and synthetic materials for the creation of nanomechanical systems that are powered by biological motors [137,138].

Molecular shuttles based on kinesin and microtubules have been constructed by using two different approaches: either the microtubules are fixed to a surface and kinesin is moving similar to cars driving

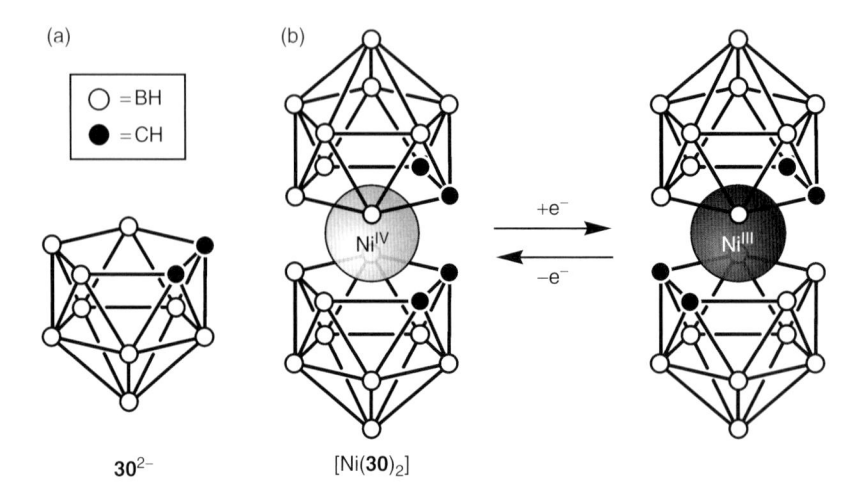

Figure 29 (a) Dicarbollide ligand 30^{2-}; (b) cisoid structure of the [Ni(**30**)$_2$] complex [134]. Reproduced with permission from Balzani V, Credi A, and Venturi M (2008) *Molecular Devices and Machines – Concepts and Perspectives for the Nanoworld*, p. 471. Weinheim: Wiley-VCH.

on a highway and transporting kinesin-coated objects [139], or the kinesin is bound to the surface and the microtubules are propelled by the kinesin analogous to a linear motor [140]. An interesting application of the second technique is a statistical approach to surface imaging using fluorescent microtubules moving as probe robots across a surface coated with kinesin. Motor proteins can find application in artificial microfluidic systems, for example, driving into either one of two arms of a Y junction, the direct transport of microtubules under control of an external electric field [141]. The energy of molecular motors can also be used to promote self-organization processes of nanoparticles [142].

8.4.1 A Hybrid Nanomechanical Device Powered by ATP Synthase

An interesting nanomechanical device was obtained by coupling nanofabrication techniques with biochemical engineering of a motor protein [143]. Such nanomechanical device consists of three elements, namely: nanofabricated substrates of nickel posts, each 50–120 nm in diameter and 200 nm high; F_1-ATP synthase molecules, specifically modified for selective interfacement with the nanofabricated structures; and nanofabricated Ni rods (150 nm in diameter and 750–1500 nm long). The F_1-ATP synthase molecules were attached to the Ni posts using histidine tags introduced into the β-subunits. Streptavidin was bound to the biotin residue on the γ-subunit tip, and the Ni nanorods, coated with biotinylated histidine-rich peptides, were then attached to the substrate-mounted F_1-ATP synthase motors through a biotin–streptavidin linkage. Rotation of the nanopropellers, which was observed in a flow cell with a charge-coupled device (CCD) camera, was initiated by addition of ATP, and inactivated by addition of NaN_3, a F_1-ATP synthase inhibitor. Although only 5 of the 400 total observed propellers were found to rotate, probably because of incorrect assembly of the components, these experiments demonstrate the possibility to integrate biomolecular motors with nanoengineered systems to produce nanomechanical or micro-mechanical machines. From a chemical point of view, however, it should be noted that this device, unlike that described in Section 8.4.3 below, does not contain any artificial active component.

8.4.2 Mechanically Driven Synthesis of ATP

In most cases, natural–artificial hybrid devices convert chemical or biological energy into mechanical work [136,137]. Particularly interesting is the reverse conversion of mechanical energy into chemical energy, obtained with an ATP-based system [144]. A magnetic bead was attached to the γ-subunit of isolated F_1-ATP synthase molecules on a glass surface, and the bead was induced to rotate using electrical magnets. Rotation in the appropriate direction resulted in the appearance of ATP in the medium, as detected by the luciferin–luciferase assay. This outstanding result shows that a vectorial force (torque) working at one particular point on a protein machine can drive a chemical reaction, occurring in physically remote catalytic sites, far from equilibrium.

8.4.3 An Artificial Photosynthetic Membrane for the Light-Driven Production of ATP

A multicomponent molecular system (tetrad) specifically designed to achieve photo-induced charge separation was vectorially incorporated in the bilayer membrane of a liposome. It was demonstrated [145] that, by visible light irradiation of this system under proper experimental conditions, protons were pumped into the liposome from the outside water solution. In principle, such a proton-motive force generated by photons can be utilized to perform mechanical work. This result was beautifully achieved [146] by incorporating F_0F_1-ATP synthase – with the ATP-synthesizing portion extending out into the external aqueous solution – in liposomes containing the components of the proton-pumping photocycle. Irradiation of the membrane with visible light leads to the charge-separation process that causes proton translocation, with generation of a proton-motive force. On accumulation of sufficient proton-motive force, protons flow through the F_0F_1-ATP synthase, with the formation of ATP from ADP and inorganic phosphate (Section 8.2.2).

The above-described system is the first complete biomimetic one which effectively couples electrical potential, derived from photo-induced electron transfer, to the chemical potential associated with the ADP–ATP conversion, thereby mimicking the entire process of bacterial photosynthesis. It constitutes a synthetic biological motor that, in principle, can be used to power anything which requires a proton gradient or ATP to work, or even future nanomachines. It might also be advantageous to use artificial, light-driven systems to produce ATP to carry out enzymatic reactions in the absence of interfering biological materials and without the need of living cells.

8.5 Artificial Nanomachines in Device-Like Settings

As for any device invented and constructed for a specific purpose, the function performed by a molecular machine represents its most important feature. Indeed, Nature uses molecular machines to perform a large number of important functions, including energy conversion, cell division, intracellular displacement, and muscle contraction [1–3]. Artificial nanomachines, in contrast, have found no technological applications until now, despite the remarkable progresses of the field in the past 15 years. The reason is that, while chemists have reached a good understanding on the design principles and can master efficient synthetic methodologies, they are still in the process of learning how these (admittedly complicated) chemical systems can be brought to a next level of structural complexity. In other words, investigations are necessary to unravel how molecular machines can be transferred from fluid solutions to more organized and sophisticated environments (e.g., solid supports, interfaces, membranes, and porous materials), to operate them with some degree of control in space and in time, and possibly coordinate their performance on larger scales.

Despite extensive efforts in the past decade [4–7, 43], the identification of viable strategies for depositing molecular machines to solid supports has proven to be challenging. Even more difficult is to find reliable methods that allow a full characterization of their state and the direct observation of the stimuli-induced motions on surfaces or other solid-state interfaces. The main problem, however, is to ensure that the mechanical switching properties observed in the solution phase are preserved in such environments [147,148]. The two main strategies employed to anchor molecular machines on solid surfaces involve the formation of Langmuir–Blodgett films and self-assembled monolayers. Examples of systems embedded in polymer matrices, liquid crystals, and gels have also been reported [6,7]. These studies have led to impressive achievements, as described in the next sections.

8.5.1 Nanovalves

A macroscopic valve is a machine constructed by combining a controllable component that regulates the flow of gases or liquids with a reservoir which can also serve as a supporting platform for the movable element. Construction of such a device at the molecular scale requires the integration of a stable and inert nanocontainer with an appropriate moving part that can act as a gatekeeper to regulate the molecular transport in to and out of the container. The construction of functioning nanovalves [149] can be useful for controlling drug delivery, signal transduction, nanofluidic systems, and sensors.

A recently investigated approach is based on the use of the mechanical motion of molecular shuttles to regulate the access to nanocontainers [35,150]. The redox switchable bistable rotaxane 31^{4+} (**Figure 30(a)**) was covalently attached to mesoporous silica nanoparticles. As shown schematically in **Figure 30(b)**, the silica nanopores can be closed and opened by moving the mechanically interlocked ring component of the rotaxane closer to and away from the pores' orifices, respectively [149]. When the ring sits on the preferred TTF station, access to the interior of the nanoparticles is unrestricted and diffusion of the solute (a luminescent iridium complex) from the surrounding solution can occur. Chemically induced oxidation of the TTF unit results in a shuttling of the ring closer to the solid surface to encircle the DNP unit, thereby blocking the access to the pores and trapping any solute molecule inside. Back reduction of the TTF unit reverses the mechanical motion, thus releasing the guest molecules and returning the system to its initial state. More recently, it has been shown that the properties of these rotaxane-based valves can be fine-tuned by changing the length of the linkers between the surface and the rotaxane molecules and the location of the movable components on the nanoparticles [151]. Nanovalves controlled by pH changes [152] or by enzymatic reactions [153] have also been constructed.

8.5.2 Microcantilever Bending by Molecular Muscles

As discussed in Section 8.3.1.1, an exciting development in the field of molecular machines has been the construction of molecular muscles based on rotaxane

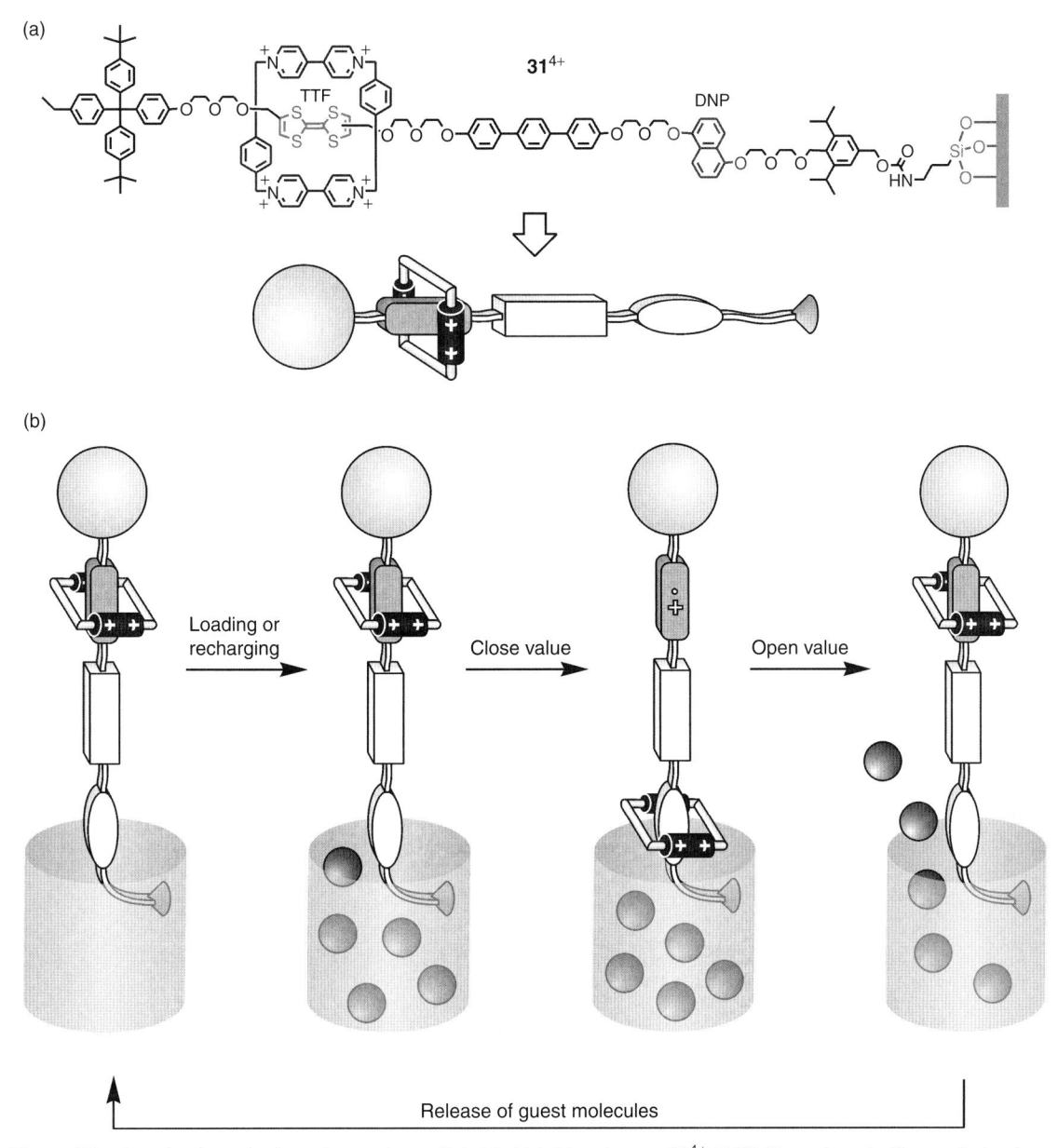

Figure 30 A molecular valve based on redox switchable bistable rotaxane **31**$^{4+}$ [149]. Reproduced with permission from Balzani V, Credi A, and Venturi M (2008) *Molecular Devices and Machines – Concepts and Perspectives for the Nanoworld*, p. 494. Weinheim: Wiley-VCH.

dimer topology which can undergo contraction and stretching movements in solution [34,85,86]. Cumulative nanoscale movements within surface-bound molecular muscles based on rotaxanes have been harnessed to perform large-scale mechanical work.

The investigated system [154,155] consists of the palindromic bistable rotaxane **32**$^{8+}$ with two rings and four stations, namely two TTF and two oxynaphthalene (ONP) moieties (**Figure 31**). Tethers attached to each ring anchor the whole rotaxane system as a self-assembled monolayer to a gold surface deposited on silicon microcantilever beams. Each microcantilever ($500 \times 100 \times 1$ μm) is covered with approximately 6 billion randomly oriented rotaxane molecules. The setup was then inserted into a fluid cell. Oxidation of the ring-preferred TTF station with iron(III) perchlorate results in shuttling of the two rings onto the ONP stations, significantly shortening the inter-ring separation

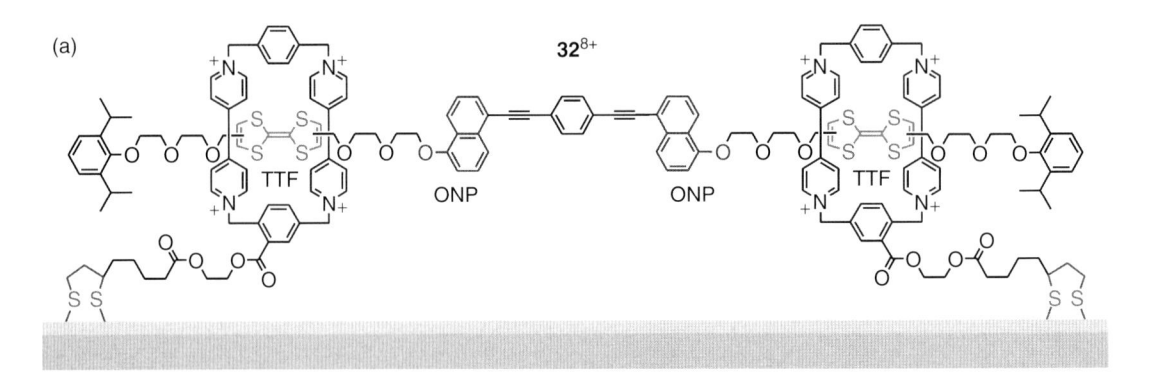

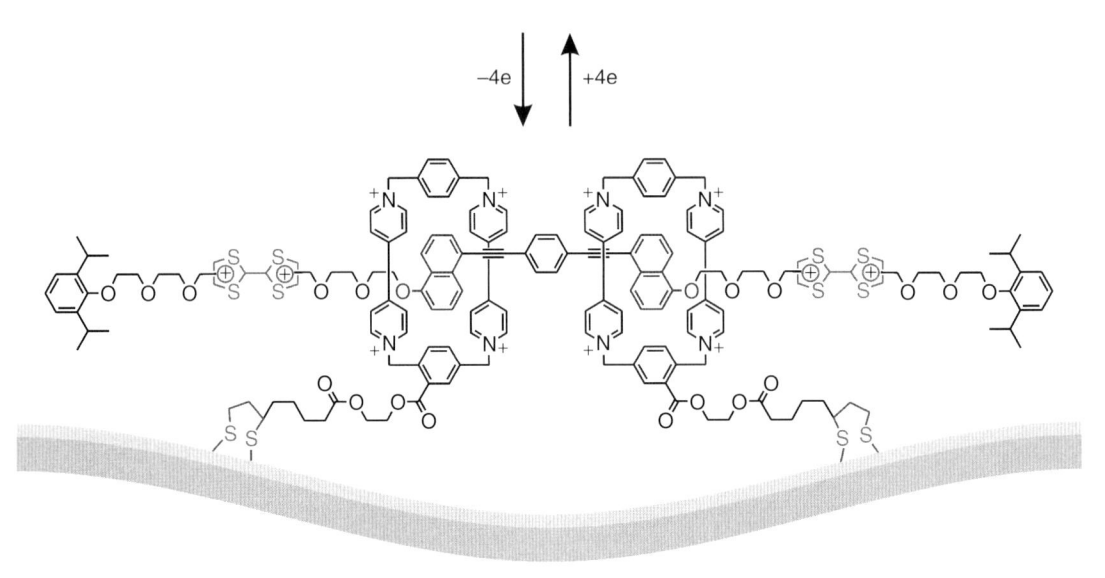

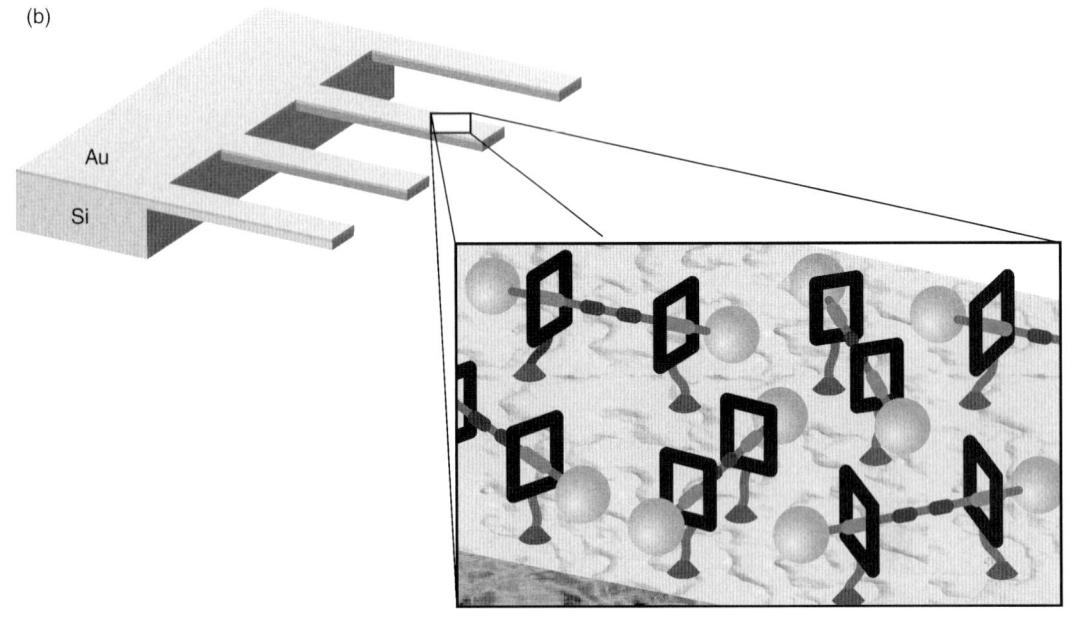

Figure 31 (a) A molecular muscle based on a self-assembled monolayer of the palindromic rotaxane 32^{8+}; (b) anchored on the gold surface of microcantilever beams[153, 154]. Reproduced with permission from Silvi S, Venturi M, and Credi A (2009) *Journal of Materials Chemistry* 19: 2290.

(from 4.2 to 1.4 nm). This change in the inter-ring distance induces a tensile stress on the gold surface of the microcantilever which causes an upward bending of the beam by about 35 nm. Under the same conditions, a monolayer of the axle component alone is inactive. Back reduction of the TTF station with ascorbic acid returns the cantilever to its original position. The process can be repeated over several cycles, albeit with gradually decreasing amplitude. Indeed, this important experiment demonstrates the possibility of harnessing cumulative nanoscale movements within surface-bound molecular shuttles to perform larger-scale mechanical work.

8.5.3 Solid-State Electronic Circuits

The redox switching behavior observed for solid-supported thin films of bistable catenanes and rotaxanes (see previous sections) [4–7] encouraged attempts to incorporate such molecules in electrically addressable solid-state devices [156].

A Langmuir monolayer of the TTF-DNP rotaxane 33^{4+} (Figure 32(a)) was transferred onto a photolithographically patterned polycrystalline silicon electrode [157]. The patterning was such that the film was deposited along several parallel lines of poly-Si on the electrode. A second set of orthogonally oriented wires was then deposited on top of the first set such that a crossbar architecture is obtained. This second set of electrodes consisted of a 5-nm thick layer of Ti, followed by a 100-nm thick layer of Al. By this approach, an array of junctions, each one addressable individually, was constructed (Figure 32(b)). In the first setup, the wire electrodes were a few micrometers wide, but the scalability of the fabrication method allowed the construction of wires less than 100 nm in width, yielding junctions with areas of 0.005–0.01 μm^2 and containing about 5000 rotaxane molecules [155].

The mechanism for conduction is by electron tunneling through the single-molecule thick layer between the junction electrodes. Thus, any change in the electronic characteristics of the interelectrode medium is expected to affect the tunneling efficiency and change the resistance of the junction. It should be noticed that such devices are conductors, not capacitors. Experiments were carried out by applying a series of voltage pulses (between +2.0 and −2.0 V) and reading, after each pulse, current through the device at a small voltage (between +0.2 and −0.2 V) that does not affect switching. The current (read)–voltage (write) curve displays a highly

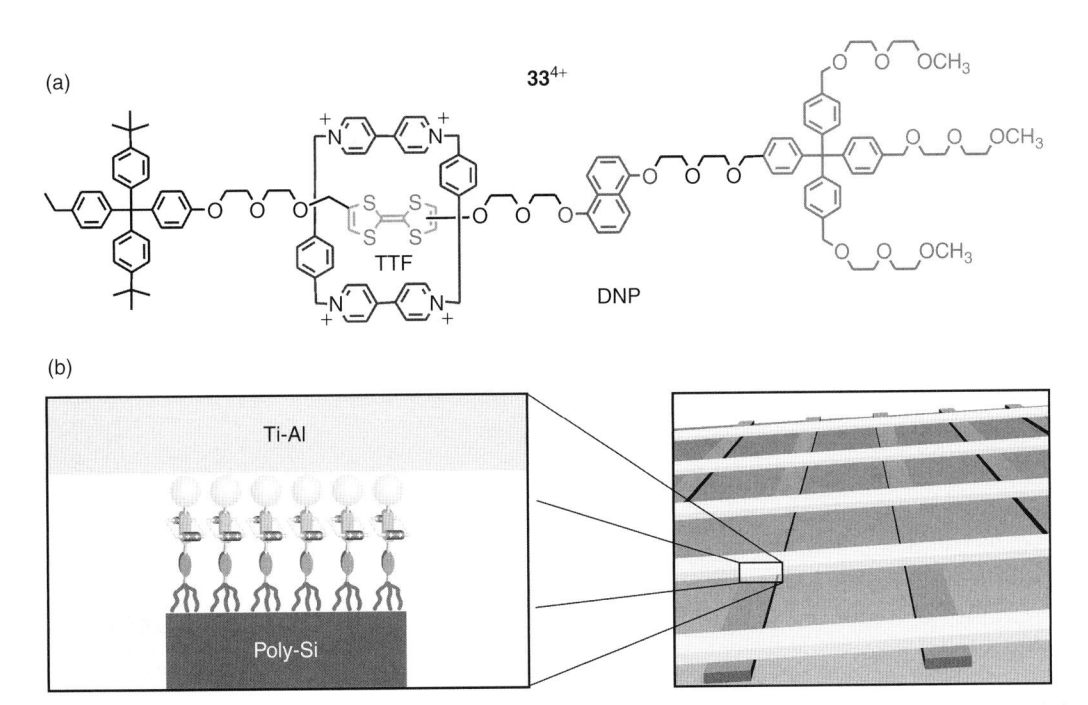

Figure 32 (a) Structure formula of bistable rotaxane 33^{4+}; (b) schematic representation of solid-state junctions consisting of a monolayer of 33^{4+} sandwiched between poly-Si and Ti–Al crossbar electrodes [155, 156]. Reproduced with permission from Silvi S, Venturi M, and Credi A (2009) *Journal of Materials Chemistry* 19: 2291.

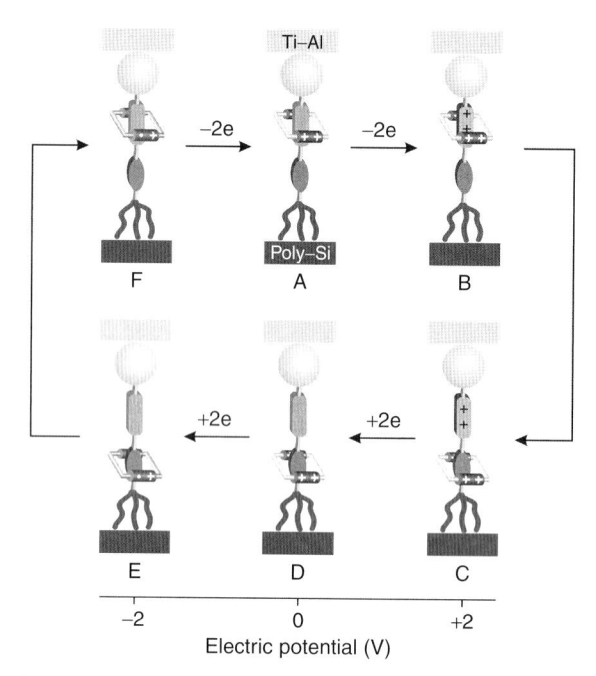

Figure 33 Operation mechanism of solid-state molecular electronic switching devices based on bistable molecular shuttles like **33**$^{4+}$. Reproduced with permission from Silvi S, Venturi M, and Credi A (2009) *Journal of Materials Chemistry* 19: 2291.

histeretic profile, making the rotaxane junction device interesting for potential use in random access memory (RAM) storage.

The current–voltage curve was interpreted on the basis of the mechanism illustrated in **Figure 33**, which is derived from the behavior of the same rotaxane in solution [158]. Co-conformation A is the switch-open state and co-conformation D the switch-closed state of the device. When the TTF unit of **33**$^{4+}$ is oxidized (+2 V, state B), a coulombic repulsion inside the tetracationic cyclophane component is generated. This results in the displacement of the latter and formation of state C in which the ring encircles the DNP unit (note that, in solution at +2 V relative to SCE, TTF undergoes two-electron oxidation and DNP is also oxidized) [101]. When the voltage is reduced to near-zero bias, a metastable state D is obtained which, however, does not return to state A (see also **Figure 7**). The initial co-conformation can, in fact, be restored only via states E and F in which the bipyridinium units of the cyclophane component are reduced (in solution, at the potential value used, −2 V, each bipyridinium unit undergoes two-electron reduction) [101]. Most likely, the reduction of the bipyridinium units weakens the charge-transfer interaction with the DNP unit, thereby decreasing the barrier that hinders the replacement of the cyclophane on the TTF site. An analogous mechanism was used to interpret the behavior of solid-state devices containing other TTF-based bistable interlocked molecules [155, 156]. The metastable state corresponding to co-conformation D (**Figure 33**) was, in fact, observed for a number of different bistable rotaxanes and catenanes in a variety of environments (solution, self-assembled monolayer, and solid-state polymer matrix) [147].

More recently, by use of the same paradigms described above, a molecular electronic memory with an amazingly high density of 10^{11} bits cm^{-2} was constructed by sandwiching a monolayer of a bistable rotaxane structurally related to **31**$^{4+}$ (**Figure 30(a)**) between arrays of nanoelectrodes in a crossbar arrangement [159]. The realization of this device relies on a method for producing ultra-dense, highly aligned arrays, and crossbars of metal or semiconductor nanowires with high aspect ratios [160]. It was estimated that each junction acting as a memory element consists of approximately 100 rotaxane molecules. For practical reasons, only 128 (16 × 8 contacts) of the 160 000 memory cells (400 × 400 nanowires) contained in the circuit were tested [158]. The measurements showed that 25% of the tested cells displayed good and reproducible switching, whereas 35% failed because of bad contacts or shorts, and the remaining 40% showed poor switching. This work is a compelling demonstration that the combination of top-down and bottom–up nanofabrication methods can lead to outstanding technological achievements. However, several aspects – such as stability, reliability, and ease of fabrication – need to be optimized before these systems can find real industrial applications [161].

8.6 Conclusion

The results described here show that, by taking advantage of careful incremental design strategies, of the tools of modern synthetic chemistry, of the paradigms of supramolecular chemistry, as well as of inspiration by natural systems, it is possible to produce compounds capable of performing nontrivial mechanical movements and exercising a variety of different functions upon external stimulation.

In the previously mentioned address to the American Physical Society [9], R. P. Feynman concluded his reflection on the idea of constructing molecular machines as follows:

> What would be the utility of such machines? Who knows? I cannot see exactly what would happen, but I can hardly doubt that when we have some control of the rearrangement of things on a molecular scale we will get an enormously greater range of possible properties that substances can have, and of different things we can do.

This text, pronounced in 1959, is still an appropriate comment to the work described in this article. The results achieved enable to devise future developments, which are under investigation in several laboratories: (1) the design and construction of more sophisticated artificial molecular motors and machines; (2) the use of such systems to perform tasks such as molecular-level transportation, catalysis, and mechanical gating of molecular channels; and (3) the possibility of exploiting their logic behavior for information processing at the molecular level and, in the long run, for the construction of chemical computers.

It should also be noted that the majority of the artificial molecular motors developed so far operates in solution, that is, in an incoherent fashion and without control of spatial positioning. The solution studies of complicated chemical systems such as molecular motors and machines are indeed of fundamental importance to understand their operation mechanisms; moreover, for some use (e.g., drug delivery) molecular machines will have to work in liquid solution. In this regard, it should be recalled that motor proteins operate in – or at least in contact with – an aqueous solution. It seems reasonable, however, that before such systems can find applications in many fields of technology, they have to be interfaced with the macroscopic world by ordering them in some way. The next generation of molecular machines and motors will need to be organized at interfaces, deposited on surfaces, incorporated into polymers, or immobilized into membranes or porous materials so that they can behave coherently and can be addressed in space.

Apart from foreseeable applications related to the development of nanotechnology [162], investigations on molecular machines and motors are important to increase the basic understanding of the processes that determine the behavior of nanoscale objects, as well as to develop reliable theoretical models. This research has also the important merit of stimulating the ingenuity of chemists, thereby conveying new incitements to chemistry as a scientific discipline.

Acknowledgments

Financial support from Ministero dell'Istruzione, dell'Università e della Ricerca, Fondazione Carisbo and Università di Bologna is gratefully acknowledged.

References

1. Jones RAL (2004) *Soft Machines: Nanotechnology and Life*. Oxford: Oxford University Press.
2. Goodsell DS (2004) *Bionanotechnology. Lessons from Nature*. Hoboken, NJ: Wiley.
3. Schliwa M (ed.) (2003) *Molecular Motors*. Weinheim: Wiley-VCH.
4. Balzani V, Credi A, and Venturi M (2008) *Molecular Devices and Machines – Concepts and Perspectives for the Nanoworld*, 2nd edn. Weinheim: Wiley-VCH.
5. Kay ER, Leigh DA, and Zerbetto F (2007) Synthetic molecular motors and mechanical machines. *Angewandte Chemie International Edition* 46: 72.
6. Balzani V, Credi A, and Venturi M (2008) Molecular machines working on surfaces and at interfaces. *ChemPhysChem* 9: 202.
7. Silvi S, Venturi M, and Credi A (2009) Molecular shuttles: from concepts to devices. *Journal of Materials Chemistry* 19: 2279.
8. International Technology Roadmap for Semiconductors (2008) Update. http://www.itrs.net (accessed July 2009).
9. Feynman RP (1960) There's plenty of room at the bottom. *Engineering and Science* 23: 22.
10. Drexler KE (1992) *Nanosystems. Molecular Machinery, Manufacturing, and Computation*. New York: Wiley.
11. Smalley RE (2001) Of chemistry, love and nanobots - How soon will we see the nanometer-scale robots envisaged by

K. Eric Drexler and other molecular nanotechologists? The simple answer is never. *Scientific American* 285: 76.
12. Joachim C and Launay J-P (1984) Sur la possibilité d'un traitement moleculaire du signal. *Nouveau Journal de Chimie* 8: 723.
13. Balzani V, Moggi L, and Scandola F (1987) Towards a supramolecular photochemistry. Assembly of molecular components to obtain photochemical molecular devices. In: Balzani V (ed.) *Supramolecular Photochemistry*, p. 1. Dordrecht: Reidel.
14. Lehn JM (1990) Perspectives in supramolecular chemistry - From molecular recognition towards molecular information-processing and self-organization. *Angewandte Chemie International Edition* 29: 1304.
15. Metzger RM (2003) Unimolecular rectifiers. *Chemical Reviews* 103: 3803.
16. Lehn JM (1995) *Supramolecular Chemistry: Concepts and Perspectives*. Weinheim: Wiley-VCH.
17. Atwood JL and Steed JW (eds.) (2004) *Encyclopedia of Supramolecular Chemistry*. New York: Dekker.
18. Samorì P (ed.) (2006) *Scanning Probe Microscopies beyond Imaging*. Weinheim: Wiley-VCH.
19. Shinkai S, Nakaji T, Ogawa T, Shigematsu K, and Manabe O (1981) Photoresponsive crown ethers. 2. Photocontrol of ion extraction and ion transport by a bis(crown ether) with a butterfly-like motion. *Journal of the American Chemical Society* 103: 111.
20. Iwamura H and Mislow K (1988) Stereochemical consequences of dynamic gearing. *Accounts of Chemical Research* 21: 175.
21. Shinkai S, Ikeda M, Sugasaki A, and Takeuchi M (2001) Positive allosteric systems designed on dynamic supramolecular scaffolds: toward switching and amplification of guest affinity and selectivity. *Accounts of Chemical Research* 34: 494.
22. Bedard TC and Moore JS (1995) Design and synthesis of molecular turnstiles. *Journal of the American Chemical Society* 117: 10662.
23. Khuong TAV, Nunez JE, Godinez CE, and Garcia-Garibay MA (2006) Crystalline molecular machines: a quest toward solid-state dynamics and function. *Accounts of Chemical Research* 39: 413.
24. Skopek K, Hershberger MC, and Gladysz JA (2007) Gyroscopes and the chemical literature: 1852–2002. *Coordination Chemistry Reviews* 251: 1723.
25. Koga N, Kawada Y, and Iwamura H (1983) Recognition of the phase relationship between remote substituents in 9, 10-bis(3-chloro-9-triptycyloxy)triptycene molecules undergoing rapid internal rotation cooperatively. *Journal of the American Chemical Society* 105: 5498.
26. Kelly TR, Bowyer MC, Bhaskar KV, et al. (1994) A molecular brake. *Journal of the American Chemical Society* 116: 3657.
27. Muraoka T, Kinbara K, and Aida T (2007) Reversible operation of chiral molecular scissors by redox and UV light. *Chemical Communications* 2007: 1441.
28. Muraoka T, Kinbara K, and Aida T (2006) Mechanical twisting of a guest by a photoresponsive host. *Nature* 440: 512.
29. Kai H, Nara S, Kinbara K, and Aida T (2008) Toward long-distance mechanical communication: studies on a ternary complex interconnected by a bridging rotary module. *Journal of the American Chemical Society* 130: 6725.
30. Kelly TR (2001) Progress toward a rationally designed molecular motor. *Accounts of Chemical Research* 34: 514.
31. Koumura N, Zijlstra RWJ, van Delden RA, Harada N, and Feringa BL (1999) Light-driven monodirectional molecular rotor. *Nature* 401: 152.
32. Bissell A, Córdova E, Kaifer AE, and Stoddart JF (1994) A chemically and electrochemically switchable molecular shuttle. *Nature* 369: 133.
33. Badjic JD, Balzani V, Credi A, Silvi S, and Stoddart JF (2004) A molecular elevator. *Science* 303: 1845.
34. Jiménez-Molero MC, Dietrich-Buchecker C, and Sauvage J-P (2000) Towards synthetic molecular muscles: contraction and stretching of a linear rotaxane dimer. *Angewandte Chemie International Edition* 39: 3284.
35. Saha S, Leug KC-F, Nguyen TD, Stoddart JF, and Zink JI (2007) Nanovalves. *Advanced Functional Materials* 17: 685.
36. Thordarson P, Bijsterveld EJA, Rowan AE, and Nolte RJM (2003) Epoxidation of polybutadiene by a topologically linked catalyst. *Nature* 424: 915.
37. Sherman WB and Seeman NC (2004) A precisely controlled DNA biped walking device. *Nano Letters* 4: 1203.
38. Shirai Y, Morin JF, Sasaki T, Guerrero JM, and Tour JM (2006) Recent progress on nanovehicles. *Chemical Society Reviews* 35: 1043.
39. Paxton WF, Sen A, and Mallouk TE (2005) Motility of catalytic nanoparticles through self-generated forces. *Chemistry – A European Journal* 11: 6462.
40. Pantarotto D, Browne WR, and Feringa BL (2008) Autonomous propulsion of carbon nanotubes powered by a multienzyme ensemble. *Chemical Communications* 2008: 1533.
41. Balzani V, Credi A, Raymo FM, and Stoddart JF (2000) Artificial molecular machines. *Angewandte Chemie International Edition* 39: 3348.
42. Sauvage J-P (2005) Transition metal-complexed catenanes and rotaxanes as molecular machine prototypes. *Chemical Communications* 2005: 1507.
43. Kottas GS, Clarke LI, Horinek D, and Michl J (2005) Artificial molecular rotors. *Chemical Reviews* 105: 1281.
44. Kinbara K and Aida T (2005) Toward intelligent molecular machines: directed motions of biological and artificial molecules and assemblies. *Chemical Reviews* 105: 1377.
45. Simmel FC and Dittmer WU (2005) DNA Nanodevices. *Small* 1: 284.
46. Credi A (2006) Artificial molecular motors powered by light. *Australian Journal of Chemistry* 59: 157.
47. Browne WR and Feringa BL (2006) Making molecular machines work. *Nature Nanotechnology* 1: 25.
48. Mateo-Alonso A, Guldi DM, Paolucci F, and Prato M (2007) Fullerenes: multitask components in molecular machinery. *Angewandte Chemie International Edition* 46: 8120.
49. Stoddart JF (ed.) (2001) *Special Issue: Molecular Machines. Accounts of Chemical Research* 34(6).
50. Sauvage J-P (ed.) (2001) *Special Volume: Molecular Machines and Motors. Structure and Bonding* 99.
51. Kelly TR (ed.) (2005) *Special Volume: Molecular Machines. Topics in Current Chemistry* 262.
52. Willner I (ed.) (2006) *Special Issue: DNA-Based Nanoarchitectures and Nanomachines. Organic and Biomolecular Chemistry* 4(18).
53. Credi A and Tian H (eds.) (2007) *Special Issue: Molecular Machines and Switches. Advanced Functional Materials* 17(5).
54. Coffey T and Krim J (2006) C-60 molecular bearings and the phenomenon of nanomapping. *Physical Review Letters* 96: 186104.
55. Feringa BL (ed.) (2001) *Molecular Switches*. Weinheim: Wiley-VCH.
56. Hernandez JV, Kay ER, and Leigh DA (2004) A reversible synthetic rotary molecular motor. *Science* 306: 1532.

57. Balzani V, Credi A, and Venturi M (2008) Processing energy and signals by molecular and supramolecular systems. *Chemistry – A European Journal* 14: 26.
58. Balzani V (2003) Photochemical molecular devices. *Photochemical and Photobiological Sciences* 2: 479.
59. Kaifer AE and Gomez-Kaifer M (1999) *Supramolecular Electrochemistry*. Weinheim: Wiley-VCH.
60. Lakowicz JR (2006) *Principles of Fluorescence Spectroscopy*, 3rd edn. New York: Springer.
61. Armaroli N and Balzani V (2007) The future of energy supply: challenges and opportunities. *Angewandte Chemie International Edition* 46: 52.
62. Vale RD and Milligan RA (2000) The way things move: Looking under the hood of molecular motor proteins. *Science* 288: 88.
63. Boyer PD (1998) Energy, life and ATP. *Angewandte Chemie International Edition* 37: 2296.
64. Walker JE (1998) ATP synthesis by rotary catalysis. *Angewandte Chemie International Edition* 37: 2308.
65. Yildiz A, Park H, Safer D, et al. (2004) Myosin VI steps via a hand-over-hand mechanism with its lever arm undergoing fluctuations when attached to actin. *Journal of Biological Chemistry* 279: 37223.
66. Shiroguchi K and Kinosita K (2007) Myosin V walks by lever action and Brownian motion. *Science* 316: 1208.
67. Sakamoto T, Webb MR, Forgacs E, White HD, and Sellers JR (2008) Direct observation of the mechanochemical coupling in myosin Va during processive movement. *Nature* 455: 128.
68. Yildiz A, Tomishige M, Vale RD, and Selvin PR (2004) Kinesin walks hand-over-hand. *Science* 303: 676.
69. Mori T, Vale RD, and Tomishige M (2007) How kinesin waits between steps. *Nature* 450: 750.
70. Yildiz A (2006) How molecular motors move. *Science* 311: 792.
71. Alonso MC, Drummond DR, Kain S, Hoeng J, Amos L, and Cross RA (2007) An ATP gate controls tubulin binding by the tethered head of kinesin-1. *Science* 316: 120.
72. Bustamante C, Keller D, and Oster G (2001) The physics of molecular motors. *Accounts of Chemical Research* 34: 412.
73. Rondelez Y, Tresset G, Nakashima T, et al. (2005) Highly coupled ATP synthesis by F-1-ATPase single molecules. *Nature* 433: 773.
74. Noji H, Yasuda R, Yoshida M, and Kinosita KJ, Jr. (1997) Direct observation of the rotation of F-1-ATPase. *Nature* 386: 299.
75. Itoh H, Takahashi A, Adachi K, et al. (2004) Mechanically driven ATP synthesis by F-1-ATPase. *Nature* 427: 465.
76. Sauvage J-P and Dietrich-Buchecker C (eds.) (1999) *Catenanes, Rotaxanes and Knots*. Weinheim: Wiley-VCH.
77. Fyfe MCT and Stoddart JF (1997) Synthetic supramolecular chemistry. *Accounts of Chemical Research* 30: 393.
78. Anelli PL, Spencer N, and Stoddart JF (1991) A molecular shuttle. *Journal of the American Chemical Society* 113: 5131.
79. Chatterjee MN, Kay ER, and Leigh DA (2006) Beyond switches: ratcheting a particle energetically uphill with a compartmentalized molecular machine. *Journal of the American Chemical Society* 128: 4058.
80. Chen N-C, Lai C-C, Liu Y-H, Peng S-M, and Chiu S-H (2008) Parking and restarting a molecular shuttle in situ. *Chemistry – A European Journal* 14: 2904.
81. Serreli V, Lee C-F, Kay ER, and Leigh DA (2007) A molecular information ratchet. *Nature* 445: 523.
82. Astumian RD and Hänggi P (2002) Brownian motors. *Physics Today* 55: 33.
83. Ashton PR, Ballardini R, Balzani V, et al. (1998) Acid-Base controllable molecular shuttles. *Journal of the American Chemical Society* 120: 11932.
84. Garaudée S, Silvi S, Venturi M, Credi A, Flood AH, and Stoddart JF (2005) Shuttling dynamics in an acid-base-switchable [2]rotaxane. *ChemPhysChem* 6: 2145.
85. Wu J, Leung KC-F, Benitez D, et al. (2008) An acid-base-controllable [c2]daisy chain. *Angewandte Chemie International Edition* 47: 7470.
86. Dawson RE, Lincoln SF, and Easton CJ (2008) The foundation of a light driven molecular muscle based on stilbene and alpha-cyclodextrin. *Chemical Communications* 2008: 3980.
87. Balzani V, Clemente-León M, Credi A, et al. (2003) Controlling multivalent interactions in triply-threaded two-component superbundles. *Chemistry – A European Journal* 9: 5348.
88. Badjic JD, Ronconi CM, Stoddart JF, Balzani V, Silvi S, and Credi A (2006) Operating molecular elevators. *Journal of the American Chemical Society* 128: 1489.
89. Armaroli N, Balzani V, Collin J-P, Gaviña P, Sauvage J-P, and Ventura B (1999) Rotaxanes incorporating two different coordinating units in their thread: synthesis and electrochemically and photochemically induced molecular motions. *Journal of the American Chemical Society* 121: 4397.
90. Durola F and Sauvage J-P (2007) Fast electrochemically induced translation of the ring in a copper-complexed [2]rotaxane: the biisoquinoline effect. *Angewandte Chemie International Edition* 46: 3537.
91. Ashton PR, Ballardini R, Balzani V, et al. (2000) A photochemically driven molecular-level abacus. *Chemistry – A European Journal* 6: 3558.
92. Brouwer AM, Frochot C, Gatti FG, et al. (2001) Photoinduction of fast, reversible translational motion in a hydrogen-bonded molecular shuttle. *Science* 291: 2124.
93. Balzani V, Clemente-León M, Credi A, et al. (2006) Autonomous artificial nanomotor powered by sunlight. *Proceedings of the National Academy of Sciences of the United States of America* 103: 1178.
94. Raiteri P, Bussi G, Cucinotta CS, Credi A, Stoddart JF, and Parrinello M (2008) Unravelling the shuttling mechanism in a photoswitchable multicomponent bistable rotaxane. *Angewandte Chemie International Edition* 47: 3536.
95. Balzani V, Clemente-León M, Credi A, et al. (2006) A comparison of shuttling mechanisms in two constitutionally isomeric bistable rotaxane-based sunlight-powered nanomotors. *Australian Journal of Chemistry* 59: 193.
96. Davidson GJE, Loeb SJ, Passaniti P, Silvi S, and Credi A (2006) Wire-type ruthenium(II) complexes with terpyridine-containing [2]rotaxanes as ligands: synthesis, characterization, and photophysical properties. *Chemistry – A European Journal* 12: 3233.
97. Rogez G, Ferrer Ribera B, Credi A, et al. (2007) A molecular plug-socket connector. *Journal of the American Chemical Society* 129: 4633.
98. Ferrer B, Rogez G, Credi A, et al. (2006) Photoinduced electron flow in a self-assembling supramolecular extension cable. *Proceedings of the National Academy of Sciences of the United States of America* 103: 18411.
99. Saha S, Flood AH, Stoddart JF, et al. (2007) A redox-driven multicomponent molecular shuttle. *Journal of the American Chemical Society* 129: 12159.
100. Gust D, Moore TA, and Moore AL (2001) Mimicking photosynthetic solar energy transduction. *Accounts of Chemical Research* 34: 40.
101. Asakawa M, Ashton PR, Balzani V, et al. (1998) A chemically and electrochemically switchable [2]catenane

incorporating a tetrathiafulvalene unit. *Angewandte Chemie International Edition* 37: 333.
102. Cárdenas DJ, Livoreil A, and Sauvage J-P (1996) Redox control of the ring-gliding motion in a cu-complexed catenane: a process involving three distinct geometries. *Journal of the American Chemical Society* 118: 11980.
103. Mobian P, Kern J-M, and Sauvage J-P (2004) Light-driven machine prototypes based on dissociative excited states: photoinduced decoordination and thermal recoordination of a ring in a ruthenium(II)-containing [2]catenane. *Angewandte Chemie International Edition* 43: 2392.
104. Kelly TR, De Silva H, and Silva RA (1999) Unidirectional rotary motion in a molecular system. *Nature* 401: 150.
105. Kelly TR, Silva RA, De Silva H, Jasmin S, and Zhao Y (2000) A rationally designed prototype of a molecular motor. *Journal of the American Chemical Society* 122: 6935.
106. Kelly TR, Cai X, Damkaci F, et al. (2007) Progress toward a rationally designed, chemically powered rotary molecular motor. *Journal of the American Chemical Society* 129: 376.
107. Fletcher SP, Dumur F, Pollard MP, and Feringa BL (2005) A reversible, unidirectional molecular rotary motor driven by chemical energy. *Science* 310: 80.
108. Feringa BL, van Delden RA, Koumura N, and Geertsema EM (2000) Chiroptical molecular switches. *Chemical Reviews* 100: 1789.
109. Feringa BL (2001) In control of motion: from molecular switches to molecular motors. *Accounts of Chemical Research* 34: 504.
110. ter Wiel MKJ, van Delden RA, Meetsma A, and Feringa BL (2005) Light-driven molecular motors: stepwise thermal helix inversion during unidirectional rotation of sterically overcrowded biphenanthrylidenes. *Journal of the American Chemical Society* 127: 14208.
111. Pollard MM, Klok M, Pijper D, and Feringa BL (2007) Rate acceleration of light-driven rotary molecular motors. *Advanced Functional Materials* 17: 718.
112. Klok M, Boyle N, Pryce MT, Meetsma A, Browne WR, and Feringa BL (2008) MHz Unidirectional rotation of molecular rotary motors. *Journal of the American Chemical Society* 130: 10484.
113. Schoevaars AM, Kruizinga W, Zijlstra RWJ, Veldman N, Spek AL, and Feringa BL (1997) Toward a switchable molecular rotor. Unexpected dynamic behavior of functionalized overcrowded alkenes. *Journal of Organic Chemistry* 62: 4943.
114. Pijper D and Feringa BL (2007) Molecular transmission: controlling the twist sense of a helical polymer with a single light-driven molecular motor. *Angewandte Chemie International Edition* 46: 3693.
115. van Delden RA, ter Wiel MKJ, Pollard MM, Vicario J, Koumura N, and Feringa BL (2005) Unidirectional molecular motor on a gold surface. *Nature* 437: 1337.
116. Pollard MM, Lubonska M, Rudolf P, and Feringa BL (2007) Controlled rotary motion in a monolayer of molecular motors. *Angewandte Chemie International Edition* 46: 1278.
117. London G, Carroll GT, Landaluce TF, Pollard MM, Rudolf P, and Feringa BL (2009) Light-driven altitudinal molecular motors on surfaces. *Chemical Communications* 1712.
118. van Delden RA, Koumura N, Harada N, and Feringa BL (2002) Unidirectional rotary motion in a liquid crystalline environment: color tuning by a molecular motor. *Proceedings of the National Academy of Sciences of the United States of America* 99: 4945.
119. Eelkema R, Pollard MM, Vicario J, et al. (2006) Nanomotor rotates microscale objects. *Nature* 440: 163.
120. Lehn J-M (2006) Conjecture: imines as unidirectional photodriven molecular motors-motional and constitutional dynamic devices. *Chemistry – A European Journal* 12: 5910.
121. Ballardini R, Balzani V, Credi A, Gandolfi MT, and Venturi M (2001) Artificial molecular-level machines: which energy to make them work?. *Accounts of Chemical Research* 34: 445.
122. Astumian RD (2005) Chemical peristalsis. *Proceedings of the National Academy of Sciences of the United States of America* 102: 1843.
123. Leigh DA, Wong JKY, Dehez F, and Zerbetto F (2003) Unidirectional rotation in a mechanically interlocked molecular rotor. *Nature* 424: 174.
124. Seeman NC (1982) Nucleic acid junctions and lattices. *Journal of Theoretical Biology* 99: 237.
125. Adleman L (1994) Molecular computation of solutions to combinatorial problems. *Science* 266: 1021.
126. Yurke B, Turberfield AJ, Mills AP, Jr., Simmel FC, and Neumann JL (2000) A DNA-fuelled molecular machine made of DNA. *Nature* 406: 605.
127. Chen Y, Wang M, and Mao C (2004) An autonomous DNA nanomotor powered by a DNA enzyme. *Angewandte Chemie International Edition* 43: 3554.
128. Chen Y and Mao C (2004) Putting a brake on an autonomous DNA nanomotor. *Journal of the American Chemical Society* 126: 8626.
129. Shin J-S and Pierce NA (2004) A synthetic DNA walker for molecular transport. *Journal of the American Chemical Society* 126: 10834.
130. Bath J, Green SJ, and Turberfield AJ (2005) A free-running DNA motor powered by a nicking enzyme. *Angewandte Chemie International Edition* 44: 4358.
131. Tian Y, He Y, Chen Y, Yin P, and Mao C (2005) Molecular devices - A DNAzyme that walks processively and autonomously along a one-dimensional track. *Angewandte Chemie International Edition* 44: 4355.
132. Yin P, Yan H, Daniell XG, Turberfield AJ, and Reif JH (2004) A unidirectional DNA walker that moves autonomously along a track. *Angewandte Chemie International Edition* 43: 4906.
133. Turberfield AJ, Mitchell JC, Yurke B, Mills AP, Blakey MI, and Simmel FC (2003) DNA fuel for free-running nanomachines. *Physical Review Letters* 90: 118102.
134. Omabegho T, Sha R, and Seeman NC (2009) A bipedal DNA Brownian motor with coordinated legs. *Science* 324: 67.
135. Hawthorne MF, Zink J, Skelton JM, et al. (2004) Electrical or photocontrol of the rotary motion of a metallacarborane. *Science* 303: 1849.
136. Carella A, Coudret C, Guirado G, Rapenne G, Vives G, and Launay J-P (2007) Electron-triggered motions in technomimetic molecules. *Dalton Transactions* 2007: 177.
137. Hess H, Bachand GD, and Vogel V (2004) Powering nanodevices with biomolecular motors. *Chemistry – A European Journal* 10: 2110.
138. Xi J, Schmidt JJ, and Montemagno CD (2005) Self-assembled microdevices driven by muscle. *Nature Materials* 4: 180.
139. Muthukrishnan G, Hutchins BM, Williams ME, and Hancock WO (2006) Transport of semiconductor nanocrystals by kinesin molecular motors. *Small* 2: 626.
140. Ramachandran S, Ernst K-H, Bachand GD, Vogel V, and Hess H (2006) Selective loading of kinesin-powered molecular shuttles with protein cargo and its application to biosensing. *Small* 2: 330.
141. van den Heuvel MGL, de Graaff MP, and Dekker (2006) Molecular sorting by electrical steering of microtubules in Kinesin-coated channels. *Science* 312: 910.

142. Hess H (2006) Self-assembly driven by molecular motors. *Soft Matter* 2: 669.
143. Soong RK, Bachand GD, Neves HP, Olkhovets AG, Craighead HG, and Montemagno CD (2000) Powering an inorganic nanodevice with a biomolecular motor. *Science* 290: 1555.
144. Itoh H, Takahashi A, Adachi K, *et al.* (2004) Mechanically driven ATP synthesis by F_1-ATPase. *Nature* 427: 465.
145. Steinberg-Yfrach G, Liddell PA, Hung SC, Moore AL, Gust D, and Moore TA (1997) Conversion of light energy to proton potential in liposomes by artificial photosynthetic reaction centres. *Nature* 385: 239.
146. Steinberg-Yfrach G, Rigaud J-L, Durantini EN, Moore AL, Gust D, and Moore TA (1998) Light-driven production of ATP catalysed by F_0F_1-ATP synthase in an artificial photosynthetic membrane. *Nature* 392: 479.
147. Raehm L, Kern J-M, Sauvage J-P, Hamann C, Palacin S, and Bourgoin J-P (2002) Disulfide- and thiol-incorporating copper catenanes: Synthesis deposition onto gold, and surface studies. *Chemistry – A European Journal* 8: 2153.
148. Choi JW, Flood AH, Steuerman DW, *et al.* (2006) Ground-state equilibrium thermodynamics and switching kinetics of bistable [2]rotaxanes switched in solution, polymer gels, and molecular electronic devices. *Chemistry – A European Journal* 12: 261.
149. Kocer A, Walko M, Meijberg W, and Feringa BL (2005) A light-actuated nanovalve derived from a channel protein. *Science* 309: 755.
150. Nguyen TD, Tseng H-R, Celestre PC, *et al.* (2005) A reversible molecular valve. *Proceedings of the National Academy of Sciences of the United States of America* 102: 10029.
151. Nguyen TD, Liu Y, Saha S, Leug KC-F, Stoddart JF, and Zink JI (2007) Design and optimization of molecular nanovalves based on redox-switchable bistable rotaxanes. *Journal of the American Chemical Society* 129: 626.
152. Nguyen TD, Leug KC-F, Liong M, Pentecost CD, Stoddart JF, and Zink JI (2006) Construction of a pH-driven supramolecular nanovalve. *Organic Letters* 8: 3363.
153. Patel K, Angelos S, Dichtel WR, *et al.* (2008) Enzyme-responsive snap-top covered silica nanocontainers. *Journal of the American Chemical Society* 130: 2382.
154. Huang TJ, Brough B, Ho C-M, *et al.* (2004) A nanomechanical device based on linear molecular motors. *Applied Physics Letters* 85: 5391.
155. Liu Y, Flood AH, Bonvallet PA, *et al.* (2005) Linear artificial molecular muscles. *Journal of the American Chemical Society* 127: 9745.
156. Collier CP, Mattersteig G, Wong EW, *et al.* (2000) A [2]catenane-based solid state electronically reconþgurable switch. *Science* 289: 1172.
157. Luo Y, Collier CP, Jeppesen JO, *et al.* (2002) Two-dimensional molecular electronics circuits. *ChemPhysChem* 3: 519.
158. Steuerman DW, Tseng H-R, Peters AJ, *et al.* (2004) Molecular-mechanical switch-based solid-state electrochromic devices. *Angewandte Chemie International Edition* 43: 6486.
159. Green JE, Choi JW, Boukai A, *et al.* (2007) A 160-kilobit molecular electronic memory patterned at 10^{11} bits per square centimetre. *Nature* 445: 414.
160. Melosh NA, Boukai A, Diana F, *et al.* (2003) Ultrahigh-density nanowire lattices and circuits. *Science* 300: 112.
161. Ball P (2007) A switch in time. *Nature* 445: 362.
162. van den Heuvel MGL and Dekker C (2007) Motor proteins at work for nanotechnology. *Science* 317: 333.

INDEX

NOTES:

Cross-reference terms in italics are general cross-references, or refer to subentry terms within the main entry (the main entry is not repeated to save space). Readers are also advised to refer to the end of each article for additional cross-references - not all of these cross-references have been included in the index cross-references.

The index is arranged in set-out style with a maximum of three levels of heading. Major discussion of a subject is indicated by bold page numbers. Page numbers suffixed by *t* and *f* refer to Tables and Figures respectively. *vs.* indicates a comparison.

This index is in letter-by-letter order, whereby hyphens and spaces within index headings are ignored in the alphabetization. Prefixes and terms in parentheses are excluded from the initial alphabetization.

A

acoustic DoD printing, 185*t*, 187
adenosine triphosphate (ATP) synthase
    ATP synthase-powered nanomechanical devices, 279
    characteristics, 252, 253*f*
    light-driven production/artificial photosynthetic membranes, 279
    mechanically driven synthesis, 279
alanine, 95, 95*f*, 98*f*, 105
alcohols, 95
amino acids, 95, 95*f*, 98*f*, 105
atomic force microscopy (AFM)
    multivalency and molecular printboards, 213*f*
    nanoparticle characterization analyses, 82, 83*f*
    single-molecule interaction analyses, 239, 240*f*, 240*f*, 240*f*
    top-down nanopatterning techniques, 155

B

bit-patterned media (BPM), 135, 136*f*

C

Cahn–Ingold–Prelong nomenclature convention, 91–92, 92*f*
carbon (C)
    carbon nanotubes (CNTs), 139*f*
    highly oriented pyrolytic graphite (HOPG), 3
carboxylic acids, 95
catenanes
    basic concepts, 252, 262
    chemically driven devices, 263, 263*f*
    electrochemically driven devices, 263, 264*f*
    photochemically driven devices, 265, 265*f*
    schematic diagram, 253*f*
    structural characteristics, 272*f*, 273*f*, 274*f*
    unidirectional ring rotation, 270, 271*f*, 272*f*, 273*f*, 274*f*
chirality, **91–119**
    chiral heterogeneous catalysis, 111, 113*f*, 114*f*
    definition, 91
    desorption kinetics
        achiral surfaces, 108
        chiral surfaces, 110, 110*f*, 111*f*
        templating/modification effects, 111
    general discussion, 91, 115
    nomenclature conventions, 91, 92*f*, 92*f*
    schematic diagram, 92*f*
    surface chirality
        achiral molecules–achiral surfaces, 93, 94*f*, 94*f*
        chiral amplification and recognition
            recognition processes, 102, 102*f*
            two-dimensional amplification, 101
        chiral molecules–achiral surfaces
            adsorption without substrate modification, 95, 95*f*, 96*f*, 97*f*, 98*f*, 99*f*
            general discussion, 94
            substrate modifications, 100, 101*f*, 101*f*
        chiral molecules–chiral surfaces
            adsorption processes, 105, 107*f*, 108*f*, 109*f*, 109*f*
            general discussion, 103
            substrate geometries, 103, 103*f*, 104*f*
condensation polymerization via dehydration, 43, 46*f*, 46*f*, 47*f*
copper (Cu)
    multivalency and molecular printboards, 225*f*
    structural characteristics, 3, 4*f*
    surface chirality, 100, 101*f*, 102*f*, 103, 109*f*
cyclodextrin (CD), *See* molecular printboards; multivalency
cysteine, 95, 95*f*, 99*f*, 105, 108*f*

D

diacetylenes, 31, 34*f*, 36*f*
digital lithography, 196
directed assembly, **1–56**
    basic concepts
        general discussion, 2
        molecule–surface interactions
            common surfaces, 3
            general discussion, 2
        noncovalent molecular interactions
            hydrogen bonding, 2, 3*t*
            metal–organic coordination, 2
    covalently bonded structures
        condensation polymerization via dehydration, 43, 46*f*, 46*f*, 47*f*
        dendronized polymers, 32*f*
        diacetylenes, 31, 34*f*
        general discussion, 31

292   Index

directed assembly (*continued*)
    polyphenylene lines, 38, 41*f*
    polythiophene formation, 35, 37*f*
    porphyrin networks, 41, 42*f*
    STM tip-induced polymerization, 31, 35*f*, 36*f*
    tetraazaperopyrene (TAPP) dimerization, 38, 39*f*
    two-dimensional polymers, 41
    Ulmann dehalogenation, 38, 41*f*
    ultraviolet (UV) light-induced polymerization, 31, 36*f*
  future outlook, 47
  general discussion, 1
  molecule–surface patterned bonding
    chemical chain reactions, 5, 5*f*, 7*f*, 8*f*
    general discussion, 5
    selectively patterned surfaces, 6, 8*f*, 9*f*
  self-assembled inclusion networks
    fullerenes (C$_{60}$), 21, 23*f*, 23*f*, 24*f*, 26*f*, 26*f*, 27*f*
    host–guest network design, 20, 21*f*, 26*f*, 27*f*
    non-fullerene guests, 28, 29*f*, 30*f*
  supramolecular assembly guidelines
    general discussion, 8
    hydrogen-bonded architectures, 9, 11*f*, 12*f*, 13*f*, 14*f*, 15*f*, 18*f*, 18*f*
    metal–organic coordination, 17, 19*f*, 20*f*, 23*f*, 23*f*
    solubility, 16*f*, 18*f*
  surface-confined polymerization
    condensation polymerization via dehydration, 43, 46*f*, 46*f*, 47*f*
    dendronized polymers, 32*f*
    diacetylenes, 31, 34*f*, 36*f*
    general discussion, 31
    polyphenylene lines, 38, 41*f*
    polythiophene formation, 35, 37*f*
    porphyrin networks, 41, 42*f*
    STM tip-induced polymerization, 31, 35*f*, 36*f*
    tetraazaperopyrene (TAPP) dimerization, 38, 39*f*
    two-dimensional polymers, 41
    Ulmann dehalogenation, 38, 41*f*
    ultraviolet (UV) light-induced polymerization, 31, 36*f*, 36*f*
  templated physisorption, 20
DNA strands
  nanomachines
    basic concepts, 273, 275*f*
    tweezers, 275, 276*f*
    walking devices, 275, 276*f*, 277*f*
  nanoparticle molecules, 65, 66*f*, 67*f*, 68*f*, 69*f*
  nanoparticle superlattices
    nonspecific DNA-based assembly, 74, 76*f*, 77*f*
    programmable DNA-based assembly, 70, 70*f*, 71*f*, 73*f*, 75*f*

E

effective molarity (EM), 210, 210*f*, 211*f*
electrohydrodynamic-jet printing, 185*t*, 187
electron beam lithography, **121–148**
  background information, 121
  e-beam tools
    advanced electron optics, 124
    basic concepts
      backscatter–forward scatter interactions, 122*f*
      electron optical column diagram, 123*f*, 124*f*
      Gaussian exposure diagram, 123*f*
      general discussion, 121
    current trends
      general discussion, 127
      laboratory setup, 129*f*
      point spread simulation data, 129*t*, 130*f*
      SEM image, 128*f*
    e-beam writer evolution, 125, 126*f*, 127*f*
    high-throughput issues, 130
    modern electron sources, 124
    resists, 131, 133*t*, 134*f*
  general discussion, 143
  nanostructure applications
    bit-patterned media (BPM), 135, 136*f*
    general discussion, 132
    high frequency electronics, 137, 138*f*
    molecular electronics, 134, 135*f*
    nanotubes/nanofibers/nanowires, 139, 139*f*, 139*f*, 140*f*, 140*f*, 141*f*
    photonic devices, 135, 136*f*, 136*f*
    spintronics, 132, 134*f*, 135*f*
    X-ray optics, 137, 137*f*, 138*f*
  proximity corrections
    basic concepts, 140
    corrected line ends, 145*f*
    dot array corner, 144*f*
    dot spacing, 142*f*
    slotted ring resonator, 145*f*, 146*f*
    underexposed dots, 143*f*
    underexposed line ends, 144*f*
A Tale of Two Cities (Dickens), 122*f*
energy-dispersive X-ray (EDX) spectroscopy, 78

F

fullerenes (C$_{60}$), 8*f*, 21, 23*f*, 23*f*, 24*f*, 26*f*, 26*f*, 27*f*

G

gel electrophoresis, 79, 80*f*, 81*f*
glutamic acid, 95*f*
glycine, 95*f*, 105
gold (Au)
  cysteine adsorption, 95, 99*f*
  multivalency and molecular printboards
    functionalized nanoparticle assembly, 233*f*, 235*f*
    interface chemistry, 210, 212*f*, 212*f*, 213*f*
  ordered nanoparticle superstructure assembly
    genetically engineered proteins, 77–78, 79*f*
    nonspecific DNA-based assembly, 76*f*, 77*f*
    peptide-based assembly, 77, 78*f*
    programmable DNA-based assembly, 75*f*
  structural characteristics, 3, 4*f*
graphite (C), 3

H

Hamaker potential, 63–64, 64*f*
high frequency electronics, 137, 138*f*
highly oriented pyrolytic graphite (HOPG), 3
hydrogen-bonded architectures
  general discussion, 9
  structural characteristics, 11*f*, 12*f*, 13*f*, 14*f*, 15*f*, 18*f*, 18*f*

I

ink-jet printing, **183–208**
  applications
    background information, 193
    digital lithography, 196
    direct ink-jet etching/polymeric substrate, 196
    display industry, 194, 194*f*
    organic thin film transistor fabrication, 195
  characteristics, 183
  drop on demand (DoD) ink-jet printing
    acoustic DoD printing, 185*t*, 187
    basic concepts, 184, 184*f*
    drop volume–drop diameter comparisons, 185*t*
    electrohydrodynamic-jet printing, 185*t*, 187
    piezoelectric DoD printing, 185, 185*t*, 186*f*
  future outlook, 206
  organic thin film transistors

Index

fabrication methods, 195
self-aligned printing
  general discussion, 196
  high-resolution ink-jet printing, 196, 197f
  lithography-free processes, 197, 198f, 199f
  metal nanoparticles inks, 199, 200f, 201f
piezoelectric DoD printing
  basic concepts, 185, 186f
  drop volume–drop diameter comparisons, 185t
  tools and materials
    ink-jet printers, 188, 189f
    piezoelectric printheads, 189, 190f
    printable metallic inks, 191, 191t
polymer thin film transistors
  downscaling requirements, 201, 202f
  self-aligned printing
    gate printing, 203, 204f, 205f, 206f
    thin gate dielectrics, 201, 202f, 203f, 203f
self-aligned printing
  organic thin film transistors
    high-resolution ink-jet printing, 196, 197f
    lithography-free processes, 197, 198f, 199f
    metal nanoparticles inks, 199, 200f, 201f
  polymer thin film transistors
    downscaling requirements, 201, 202f
    gate printing, 203, 204f, 205f, 206f
    thin gate dielectrics, 201, 202f, 203f, 203f

K

kinesin, 251, 252f

L

lithographic techniques
  digital lithography, 196
  electron beam lithography, **121–148**
    bit-patterned media (BPM), 135, 136f
    general discussion, 132
    high frequency electronics, 137, 138f
    molecular electronics, 134, 135f
    nanotubes/nanofibers/nanowires, 139, 139f, 139f, 140f, 140f, 141f
    photonic devices, 135, 136f, 136f
    spintronics, 132, 134f, 135f
    X-ray optics, 137, 137f, 138f
  nanoimprint lithography (NIL), 234, 235f, 236f, 238f
  roller imprint lithography, 157, 157f
  sheet imprint lithography, 157, 157f
  soft lithography, 155f, 155–156
  supramolecular dip-pen nanolithography (DPN), 217, 218f, 218f
  ultraviolet (UV) imprint lithography, **149–181**

M

methionine, 105
microscopy
  atomic force microscopy (AFM)
    multivalency and molecular printboards, 213f
    nanoparticle characterization analyses, 82, 83f
    single-molecule interaction analyses, 239, 240f, 240f, 240f
    top-down nanopatterning techniques, 155
  confocal microscopy, 222f
  fluorescence microscopy, 232f
  scanning electrochemical microscopy (SECM) –, 233f
  scanning emission microscopy (SEM), 78, 127, 128f
  transmission electron microscopy (TEM), 78
molecular electronics, 134, 135f
molecular machines and motos, **247–289**
  advantages, 285

artifical systems
  application challenges, 280
  basic concepts, 252, 253f
  DNA-based systems
    basic concepts, 273, 275f
    tweezers, 275, 276f
    walking devices, 275, 276f, 277f
  microcantilever bending, 280, 282f
  molecular shuttles
    basic concepts, 254
    chemically driven devices, 255, 257f
    electrochemically driven devices, 258, 258f
    light energy-powered devices, 259, 260f, 261f
    operation diagram, 254f, 256f
    structural characteristics, 255f
  nanovalves, 280, 281f
  organometallic complexes, 278, 278f
  ring switching (catenanes)
    basic concepts, 262
    chemically driven devices, 263, 263f
    conformational changes, 262f
    electrochemically driven devices, 263, 264f
    photochemically driven devices, 265, 265f
  rotary motors
    basic concepts, 265
    chemically driven devices, 266, 266f, 267f
    light energy-powered devices, 267, 268f, 269f, 270f
    unidirectional ring rotation, 267f, 268f, 270, 271f, 272f, 273f, 274f
  solid-state electronic circuits, 283, 283f, 284f
basic concepts
  definitions and terminology, 248, 249f
  energy supply and monitoring signals, 249
  motion characteristics, 250
bottom-up (supramolecular) approaches, 247
general discussion, 247
historical background, 247
hybrid systems
  adenosine triphosphate (ATP) synthase
    ATP synthase-powered nanomechanical devices, 279
    light-driven production/artificial photosynthetic membranes, 279
    mechanically driven synthesis, 279
  characteristics, 278
natural systems
  adenosine triphosphate (ATP) synthase, 252, 253f
  general discussion, 251
  kinesin, 251, 252f
  myosin, 251, 252f
molecular printboards, **209–245**
  basic concepts, 209
  interface chemistry, 209
  multivalency
    advantages, 241
    background information, 209
    basic concepts, 210, 210f, 211f
    guest immobilization, 214, 214f, 215f, 216f, 216f
    interface chemistry, 210, 212f, 212f, 213f, 213f
  nanoparticle assembly
    basic concepts, 233
    convective assembly diagrams, 237f
    functionalized nanoparticle assembly, 233f, 234f, 235f, 239f
    integrated nanofabrication scheme, 235f
    layer-by-layer (LBL) assembly scheme, 233, 235f, 235f, 236f
    nanoimprint lithography (NIL), 234, 235f, 236f, 238f
    patterned assembly, 236f
    SEM images, 238f
    substrate-related adsorption and assembly diagram, 237f

molecular printboards (continued)
    single-molecule interaction analyses, 239, 240f, 240f, 240f
    stepwise assembly/stimulus-dependent desorption
        assembly adsorption schemes, 225f, 226f, 227f
        basic concepts, 222
        bionanostructure/protein assembly, 228f, 228f, 229f, 229f, 230f, 231f
        capsule building blocks, 223f
        confocal microscopy images, 222f
        copper-based supramolecular assembly, 225f
        cytochrome c immobilization diagram, 232f
        fluorescence microscopy images, 232f
        scanning electrochemical microscopy (SECM) experiment, 232, 233f
        vancomysin/fluorescein group absorption processes, 224f
    two-dimensional/three-dimensional nanofabrication, 233
    writing patterns
        basic concepts, 217
        friction force images, 218f, 219f
        guest binding modes, 220f
        guest image, 221f
        guest line array, 220f
        guest molecular structure, 219f, 220f
        schematic diagram, 218f
        synthesis scheme, 221f
multivalency
    background information, 209
    basic concepts, 210, 210f, 211f
    interface chemistry, **209–245**
    molecular printboards
        advantages, 241
        guest immobilization, 214, 214f, 215f, 216f, 216f
        interface chemistry, 210, 212f, 212f, 213f, 213f
        nanoparticle assembly
            basic concepts, 233
            convective assembly diagrams, 237f
            functionalized nanoparticle assembly, 233f, 234f, 235f, 239f
            integrated nanofabrication scheme, 235f
            layer-by-layer (LBL) assembly scheme, 233, 235f, 235f, 236f
            nanoimprint lithography (NIL), 234, 235f, 236f, 238f
            patterned assembly, 236f
            SEM images, 238f
            substrate-related adsorption and assembly diagram, 237f
        single-molecule interaction analyses, 239, 240f, 240f, 240f
        stepwise assembly/stimulus-dependent desorption
            assembly adsorption schemes, 225f, 226f, 227f
            basic concepts, 222
            bionanostructure/protein assembly, 228f, 228f, 229f, 229f, 230f, 231f
            capsule building blocks, 223f
            confocal microscopy images, 222f
            copper-based supramolecular assembly, 225f
            cytochrome c immobilization diagram, 232f
            fluorescence microscopy images, 232f
            scanning electrochemical microscopy (SECM) experiment, 232, 233f
            vancomysin/fluorescein group absorption processes, 224f
        two-dimensional/three-dimensional nanofabrication, 233
        writing patterns
            basic concepts, 217
            friction force images, 218f, 219f
            guest binding modes, 220f
            guest image, 221f
            guest line array, 220f
            guest molecular structure, 219f, 220f
            schematic diagram, 218f
            synthesis scheme, 221f
myosin, 251, 252f

N

nanofibers, 139, 139f, 140f
nanomachines, **247–289**
    advantages, 285
    artifical systems
        application challenges, 280
        basic concepts, 252, 253f
        DNA-based systems
            basic concepts, 273, 275f
            tweezers, 275, 276f
            walking devices, 275, 276f, 277f
        microcantilever bending, 280, 282f
        molecular shuttles
            basic concepts, 254
            chemically driven devices, 255, 257f
            electrochemically driven devices, 258, 258f
            light energy-powered devices, 259, 260f, 261f
            operation diagram, 254f, 256f
            structural characteristics, 255f
        nanovalves, 280, 281f
        organometallic complexes, 278, 278f
        ring switching (catenanes)
            basic concepts, 262
            chemically driven devices, 263, 263f
            conformational changes, 262f
            electrochemically driven devices, 263, 264f
            photochemically driven devices, 265, 265f
        rotary motors
            basic concepts, 265
            chemically driven devices, 266, 266f, 267f
            light energy-powered devices, 267, 268f, 269f, 270f
            unidirectional ring rotation, 267f, 268f, 270, 271f, 272f, 273f, 274f
        solid-state electronic circuits, 283, 283f, 284f
    basic concepts
        definitions and terminology, 248, 249f
        energy supply and monitoring signals, 249
        motion characteristics, 250
    bottom-up (supramolecular) approaches, 247
    general discussion, 247
    historical background, 247
    hybrid systems
        adenosine triphosphate (ATP) synthase
            ATP synthase-powered nanomechanical devices, 279
            light-driven production/artificial photosynthetic membranes, 279
            mechanically driven synthesis, 279
        characteristics, 278
    natural systems
        adenosine triphosphate (ATP) synthase, 252, 253f
        general discussion, 251
        kinesin, 251, 252f
        myosin, 251, 252f
nanoparticles, **57–90**
    biofunctionalization processes
        1:1 functionalized nanoparticles, 61
        anisotropically functionalized nanoparticles, 61
        basic concepts, 60
        N:1 functionalized nanoparticles, 60
    bonding interactions
        general discussion, 62
        nonspecific physical bonding interactions, 63, 65f
        specific chemical bonding interactions, 62, 63f, 64f
    characterization methods
        atomic force microscopy (AFM), 82, 83f
        gel electrophoresis, 79, 80f, 81f

microscopy, 78, 79f
optical spectroscopy, 80, 81f
small-angle X-ray scattering (SAXS), 81, 82f
future outlook, 83
general discussion, 57
metallic inks, 199, 200f, 201f
multivalency and molecular printboards
  basic concepts, 233
  functionalized nanoparticle assembly, 233f, 234f, 235f, 239f
ordered nanoparticle superstructure assembly
  nanoparticle molecules, 65, 66f, 67f, 68f, 69f
  nanoparticle superlattices
    genetically engineered proteins, 77–78, 79f
    nonspecific DNA-based assembly, 74, 76f, 77f
    peptide-based assembly, 77, 78f
    programmable DNA-based assembly, 70, 70f, 71f, 73f, 75f
wet chemistry synthesization processes, 58, 59f
nanostructures
  directed assembly, **1–56**
    basic concepts, 2
    covalently bonded structures
      condensation polymerization via dehydration, 43, 46f, 46f, 47f
      dendronized polymers, 32f
      diacetylenes, 31, 34f
      general discussion, 31
      polyphenylene lines, 38, 41f
      polythiophene formation, 35, 37f
      porphyrin networks, 41, 42f
      STM tip-induced polymerization, 31, 35f, 36f
      tetraazaperopyrene (TAPP) dimerization, 38, 39f
      two-dimensional polymers, 41
      Ulmann dehalogenation, 38, 41f
      ultraviolet (UV) light-induced polymerization, 31, 36f
    future outlook, 47
    general discussion, 1
    molecule–surface patterned bonding
      chemical chain reactions, 5, 5f, 7f, 8f
      general discussion, 5
      selectively patterned surfaces, 6, 8f, 9f
    self-assembled inclusion networks
      fullerenes (C$_{60}$), 21, 23f, 23f, 24f, 26f, 26f, 27f
      host–guest network design, 20, 21f, 26f, 27f
      non-fullerene guests, 28, 29f, 30f
    supramolecular assembly guidelines
      general discussion, 8
      hydrogen-bonded architectures, 9, 11f, 12f, 13f, 14f, 15f, 18f, 18f
      metal–organic coordination, 17, 19f, 20f, 23f, 23f
      solubility, 16f, 18f
    surface-confined polymerization
      condensation polymerization via dehydration, 43, 46f, 46f, 47f
      dendronized polymers, 32f
      diacetylenes, 31, 34f, 36f
      general discussion, 31
      polyphenylene lines, 38, 41f
      polythiophene formation, 35, 37f
      porphyrin networks, 41, 42f
      STM tip-induced polymerization, 31, 35f, 36f
      tetraazaperopyrene (TAPP) dimerization, 38, 39f
      two-dimensional polymers, 41
      Ulmann dehalogenation, 38, 41f
      ultraviolet (UV) light-induced polymerization, 31, 36f, 36f
    templated physisorption, 20
  electron beam lithography, **121–148**
    bit-patterned media (BPM), 135, 136f
    general discussion, 132
    high frequency electronics, 137, 138f
    molecular electronics, 134, 135f
    nanotubes/nanofibers/nanowires, 139, 139f, 139f, 140f, 140f, 141f
    photonic devices, 135, 136f, 136f
    spintronics, 132, 134f, 135f
    X-ray optics, 137, 137f, 138f
  ink-jet printing, **183–208**
  nanoparticles, **57–90**
  ultraviolet (UV) imprint lithography, **149–181**
nanotubes, 139, 139f, 140f
nanovalves, 280, 281f
nanowires, 139, 141f

O

optical spectroscopy, 80, 81f
organic thin film transistors
  fabrication methods, 195
  self-aligned printing
    general discussion, 196
    high-resolution ink-jet printing, 196, 197f
    lithography-free processes, 197, 198f, 199f
    metal nanoparticles inks, 199, 200f, 201f

P

photolithographic techniques, 154
photonic devices, 135, 136f, 136f
piezoelectric DoD printing
  basic concepts, 185, 186f
  drop volume–drop diameter comparisons, 185t
  tools and materials
    ink-jet printers, 188, 189f
    piezoelectric printheads, 189, 190f
    printable metallic inks, 191, 191t
platinum (Pt), 103
polymers
  polymer thin film transistors
    downscaling requirements, 201, 202f
    self-aligned printing
      gate printing, 203, 204f, 205f, 206f
      thin gate dielectrics, 201, 202f, 203f, 203f
  two-dimensional polymers, 41
polymethylmethacrylate (PMMA)
  electron beam lithography, 131, 133t, 137
  nanoparticle assembly methods, 234, 236f
  polymer thin film transistors, 201, 202f, 203f, 203f
polyphenylene lines, 38, 41f
polythiophene formation, 35, 37f
porphyrin networks, 41, 42f

Q

quantum dots, 140f
quartz crystals, 103, 103f

R

roller imprint lithography, 157, 157f
rotaxanes
  basic concepts, 252
  electrochemically driven devices, 258, 258f
  light energy-powered devices, 259, 260f, 261f
  microcantilever bending, 282f
  nanovalves, 281f
  operation diagram, 254f, 256f
  schematic diagram, 253f
  solid-state electronic circuits, 283f
  structural characteristics, 255f
R,S nomenclature convention, 91–92, 92f

Index

S

scanning emission microscopy (SEM), 78, 127, 128f
self-assembled monolayers (SAMs), *See* molecular printboards; multivalency
serine, 95f, 105
sheet imprint lithography, 157, 157f
silicon (Si), 3
silver (Ag), 100, 101f
small-angle X-ray scattering (SAXS), 81, 82f
soft lithography, 155f, 155–156
spectroscopy
    energy-dispersive X-ray (EDX) spectroscopy, 78
    optical spectroscopy, 80, 81f
spintronics, 132, 134f, 135f
supramolecular dip-pen nanolithography (DPN), 217, 218f, 218f
supramolecular microcontact printing (μCP), 217, 218f, 218f, 221f

T

A Tale of Two Cities (Dickens), 122f
tartaric acid, 95, 95f, 96f, 97f
tetraazaperopyrene (TAPP) dimerization, 38, 39f
thermal nanoimprinting, 155f, 156
transmission electron microscopy (TEM), 78
two-dimensional polymers, 41

U

Ulmann dehalogenation, 38, 41f
ultraviolet (UV) imprint lithography, **149–181**
    building blocks
        general discussion, 159
        imprint materials
            adhesion layer, 169, 169f
            adhesion measurements, 167f
            characteristics, 162
            general discussion, 162
            imprint resists, 155f, 163, 165f, 165f, 166f, 168f, 176
            longevity improvements, 168f
            process diagram, 160f
            standard tone/reverse tone etch process, 162f, 163f
            thin film characterization, 169–170, 170f
        imprint tools
            air evacuation process, 171f
            alignment systems, 172
            drop local volume compensation, 170f
            general discussion, 170
            process diagram, 160f
            schematic diagram, 160f
            spin-on imprinting, 172f
            UV imprint masks, 160, 160f, 161f, 171f, 175
    future outlook, 178
    importance
        biomedical devices, 151f
        general discussion, 149
        hard-disk substrates, 150f
        magnetic media, 150f
        photonic crystals, 150f
        semiconductor devices, 150f
        throughput, 152f
    nanoscale manufacturing requirements
        general discussion, 149
        pattern density variations, 151f, 152f
        patterning challenges, 153t
    patterning challenges, 151, 153t
    top-down nanopatterning options
        general discussion, 152
        nanoimprint technique comparison studies
            general discussion, 158
            spin-on–DoD (drop on demand) resist deposition processes, 159t, 172f
            step and repeat (S and R)–whole wafer (WW) imprint processes, 159t
            thermal–UV imprint processes, 158t
        photon-based patterning techniques, 154
        proximity mechanical techniques, 155, 155f, 157f
    UV nanoimprint lithography process results
        alignment and overlay, 174, 175t, 176f
        CD control, 173, 175f
        cost considerations, 177
        defects, 175
        imprint resolution, 173, 174f, 174f
        line edge roughness, 173
        mask life, 175
        three-dimensional template and pattern, 178f
        throughput, 152f, 176, 177f, 177t
        top-down nanopatterning techniques, 155f, 156

X

X-ray methods
    energy-dispersive X-ray (EDX) spectroscopy, 78
    small-angle X-ray scattering (SAXS), 81, 82f
    X-ray optics, 137, 137f, 138f